"十四五"时期国家重点出版物
出版专项规划项目

磷科学前沿与技术丛书

# 手性膦配体合成及应用

Synthesis and Application of
Chiral Phosphine Ligands

徐利文 | 编著

·北京·

## 内容简介

本书为"磷科学前沿与技术丛书"分册之一。手性膦配体在金属不对称催化反应中起着至关重要的作用。本书主要介绍了手性膦配体及其不对称催化反应应用，从新的视角总结和描述手性膦配体的历史与发展，系统介绍膦配体的结构特点、分类、制备方法，包括手性单齿膦配体、手性双膦配体、手性膦-杂原子双齿配体、手性多官能化多齿型膦配体等内容，并探讨了若干手性膦配体的合成与应用实例。同时，还介绍了手性膦配体在材料合成、药物合成、农药合成中的最新应用。本书适合化学、化工、材料、医药、环境、农药、生命科学及相关专业大专院校师生、科研人员参考阅读。

## 图书在版编目（CIP）数据

手性膦配体合成及应用／徐利文编著．—北京：化学工业出版社，2022.12（2024.10重印）
（磷科学前沿与技术丛书）
ISBN 978-7-122-41911-8

Ⅰ.①手… Ⅱ.①徐… Ⅲ.①膦-不对称有机合成-络合物-研究 Ⅳ.①O613.62

中国版本图书馆CIP数据核字（2022）第150102号

责任编辑：曾照华
文字编辑：姚子丽　师明远
责任校对：宋　玮
装帧设计：王晓宇

出版发行：化学工业出版社
　　　　　（北京市东城区青年湖南街13号　邮政编码100011）
印　　装：北京建宏印刷有限公司
710mm×1000mm　1/16　印张27　彩插1　字数464千字
2024年10月北京第1版第2次印刷

购书咨询：010-64518888
售后服务：010-64518899
网　　址：http://www.cip.com.cn

凡购买本书，如有缺损质量问题，本社销售中心负责调换。

定　　价：198.00元　　版权所有　违者必究

## 磷科学前沿与技术丛书 编委会

**主　　任**　赵玉芬

**副 主 任**　周　翔　　张福锁　　常俊标　　夏海平　　李艳梅

**委　　员**（以姓氏笔画为序）

　　　　　　王佳宏　　石德清　　刘　艳　　李艳梅　　李海港
　　　　　　余广鳌　　应见喜　　张文雄　　张红雨　　张福锁
　　　　　　陈　力　　陈大发　　周　翔　　赵玉芬　　郝格非
　　　　　　贺红武　　贺峥杰　　袁　佳　　夏海平　　徐利文
　　　　　　徐英俊　　高　祥　　郭海明　　梅　毅　　常俊标
　　　　　　章　慧　　喻学锋　　蓝　宇　　魏东辉

# 丛书序

磷是构成生命体的基本元素,是地球上不可再生的战略资源。磷科学发展至今,早已超出了生命科学的范畴,成为一门涵盖化学、生物学、物理学、材料学、医学、药学和海洋学等学科的综合性科学研究门类,在发展国民经济、促进物质文明、提升国防安全等诸多方面都具有不可替代的作用。本丛书希望通过"磷科学"这一科学桥梁,促进化学、化工、生物、医学、环境、材料等多学科更高效地交叉融合,进一步全面推动"磷科学"自身的创新与发展。

国家对磷资源的可持续及高效利用高度重视,国土资源部于2016年发布《全国矿产资源规划(2016—2020年)》,明确将磷矿列为24种国家战略性矿产资源之一,并出台多项政策,严格限制磷矿石新增产能和磷矿石出口。本丛书重点介绍了磷化工节能与资源化利用。

针对与农业相关的磷化工突显的问题,如肥料、农药施用过量、结构失衡等,国家也已出台政策,推动肥料和农药减施增效,为实现化肥农药零增长"对症下药"。本丛书对有机磷农药合成与应用方面的进展及磷在农业中的应用与管理进行了系统总结。

相较于磷化工在能源及农业领域所获得的关注度及取得的成果，我们对精细有机磷化工的重视还远远不够。白磷活化、黑磷在催化新能源及生物医学方面的应用、新型无毒高效磷系阻燃剂、手性膦配体的设计与开发、磷手性药物的绿色经济合成新方法、从生命原始化学进化过程到现代生命体系中系统化的磷调控机制研究、生命起源之同手性起源与密码子起源等方面的研究都是今后值得关注的磷科学战略发展要点，亟需我国的科研工作者深入研究，取得突破。

本丛书以这些研究热点和难点为切入点，重点介绍了磷元素在生命起源过程和当今生命体系中发挥的重要催化与调控作用；有机磷化合物的合成、非手性膦配体及手性膦配体的合成与应用；计算磷化学领域的重要理论与新进展；磷元素在新材料领域应用的进展；含磷药物合成与应用。

本丛书可以作为国内从事磷科学基础研究与工程技术开发及相关交叉学科的科研工作者的常备参考书，也可作为研究生及高年级本科生等学习磷科学与技术的教材。书中列出大量原始文献，方便读者对感兴趣的内容进行深入研究。期望本丛书的出版更能吸引并培养一批青年科学家加入磷科学基础研究这一重要领域，为国家新世纪磷战略资源的循环与有效利用发挥促进作用。

最后，对参与本套丛书编写工作的所有作者表示由衷的感谢！丛书中内容的设置与选取未能面面俱到，不足与疏漏之处请读者批评指正。

2023 年 1 月

# 前言 PREFACE

  手性膦配体在金属不对称催化反应中起着至关重要的作用，不仅是手性膦配体产生手性诱导和选择性控制的源泉，同时也是调节手性催化剂的催化活性和稳定性的核心要素。近50年来，设计和发展高效的新型手性膦配体及其金属催化剂始终是不对称催化研究的核心内容之一，其进展是推动不对称催化反应快速发展的重要组成部分，对手性农医药、环境科学、香料以及功能材料等与国民经济相关的学科发展起着不可忽视的推动作用。至今，手性膦配体化学仍是一个具有强大生命力的研究方向。本书起因于2019年9月在云南昆明举行的中国化学会第十二届全国磷化学化工学术讨论会，得到了中国科学院院士赵玉芬教授的关心和帮助。为便于读者更多地了解手性膦配体的发展，本书主要介绍了手性膦配体及其不对称催化反应应用，简明扼要地总结了国内外近30年来关于手性膦配体的研究进展，按照膦配体从结构较为简单的手性单膦配体到结构较为复杂的多功能化手性膦配体的顺序进行编写，并为了读者更好地了解和合成其中一些膦配体，第6章单独提供了30余类手性膦配体的合成过程和具体应用反应实例，期望能够在手性膦配体化学研究方面对大家的学习和研究工作有所帮助。

本书第 1 章概括地介绍了手性膦配体的发展简史、基本知识以及手性膦配体的结构与分类，并简单地介绍了不对称催化反应。第 2 章讨论了各种结构的手性单齿磷配体。第 3 章介绍了一些重要的手性双膦配体，这部分的内容相对较为详细。第 4 章介绍了手性膦-杂原子双齿配体及在不对称金属催化反应中的应用。第 5 章探讨了手性多官能化多齿型膦配体。在第 2～5 章的基础上，第 6 章进一步探讨了若干手性膦配体的合成与应用实例。本书的编写目的不仅是提供手性膦配体的研究现状，而且期望本书能够作为一本手性膦配体领域的小型工具书，便于读者查阅其非常具体的合成方法和应用场景。

因编写工作量大，时间较为匆忙，实难以一人之力来完成，故编著者动员了本课题组的所有师生参与本书的编写工作。徐利文教授负责大纲的制定以及全书的统稿、补充和修改工作，并负责文献的查阅与搜集工作。郑战江、曹建、崔玉明等为本书的编写提供了很大的帮助，在此向他们表示衷心感谢。

由于手性膦配体数量众多，而且其应用领域数不胜数，手性膦科学发展迅速，加之我们知识水平有限，书中难免有疏漏和不妥之处，敬请读者批评指正，不胜感激！

徐利文
2023 年 6 月

# 目录 CONTENTS

## 1 绪论      001

   1.1   手性膦配体的发展简史    002

   1.2   手性膦配体的基本知识    004

   1.3   手性膦配体的结构与分类    005

   1.4   膦配体与过渡金属的配位反应    006

   1.5   不对称催化反应简介    006

   参考文献    009

## 2 手性单齿膦配体      013

   2.1   单齿叔膦配体    014

       2.1.1   环磷联萘骨架叔膦配体（BINEPINE）    016

       2.1.2   环磷螺环骨架叔膦配体（SITCP）    020

       2.1.3   非环联萘骨架叔膦配体（MOP）    022

       2.1.4   磷杂二氢苯并呋喃骨架叔膦配体    024

   2.2   亚磷酸衍生物单磷配体    029

       2.2.1   联萘二酚骨架    030

       2.2.2   螺环二酚骨架    034

       2.2.3   酒石酸衍生的二醇骨架    038

   2.3   单膦手性配体的小结及延伸阅读    046

   参考文献    046

# 3 手性双膦配体 053

## 3.1 轴手性双膦配体 054
### 3.1.1 联萘及八氢联萘骨架 054
### 3.1.2 联苯及联芳杂骨架 057
### 3.1.3 轴手性双膦配体的应用 062

## 3.2 手性螺环双膦配体 137

## 3.3 碳手性双膦配体 141

## 3.4 二茂铁双膦配体 144

## 3.5 磷手性双膦配体 146

## 3.6 膦-亚磷酸酯及膦-亚磷酰胺酯 149

## 参考文献 152

# 4 手性膦-杂原子双齿配体 167

## 4.1 手性膦-杂原子双齿配体的合成 168
### 4.1.1 膦/氮配体 168
### 4.1.2 膦/硫配体 199
### 4.1.3 膦/氧配体 203

## 4.2 手性膦-杂原子双齿配体的应用 207
### 4.2.1 烯丙基化反应 207
### 4.2.2 氢化反应 210
### 4.2.3 共轭加成反应 212
### 4.2.4 环加成反应 214
### 4.2.5 芳基化反应 216
### 4.2.6 氢硅化反应 217

  4.2.7 A3 偶联反应 218

 参考文献 219

# 5  手性多官能化多齿型膦配体 231

 5.1 多官能化单膦配体 232

  5.1.1 多官能 P,N 配体 233

  5.1.2 多官能 P,O 配体 238

  5.1.3 多官能含 S 多齿单膦配体 239

  5.1.4 多官能化 P,N,N 型手性配体 246

  5.1.5 多官能化 N,P,N 型手性配体 250

  5.1.6 多官能化 P,O,O 型手性配体 250

  5.1.7 多官能化 P,N,O 型手性配体 253

  5.1.8 多官能化 P,N,S 型手性配体 257

  5.1.9 多官能化 N,P,O 型手性配体 257

 5.2 多官能化双膦配体 260

  5.2.1 P,P,N 和 P,N,P 型手性配体 260

  5.2.2 P,P,N,N 型多齿配体 264

  5.2.3 P,N,N,P 型配体 266

  5.2.4 N,P,P,N 型多齿配体 270

 5.3 P,P,P 型三齿配体 272

 参考文献 274

# 6  手性膦配体的合成及应用实例 287

 6.1 手性单膦配体的代表性合成及应用实例 288

  6.1.1 Ph-BINEPINE 配体的合成及应用实例 288

|  |  |  |
|---|---|---|
| | 6.1.2 Monophosphine-olefin 配体的合成及应用实例 | 292 |
| | 6.1.3 Ding's secondary phosphine oxide（SPO）配体的合成及应用实例 | 295 |
| | 6.1.4 Taddol 类亚磷酰胺配体的合成及应用实例 | 301 |
| | 6.1.5 Me-BI-DIME 配体的合成及应用实例 | 304 |
| | 6.1.6 N- 芳基亚磷酰胺配体的合成及应用实例 | 307 |
| | 6.1.7 Feringaphos 配体的合成及应用实例 | 309 |
| 6.2 | 手性双膦配体的代表性合成及应用实例 | 313 |
| | 6.2.1 BINAP 配体的合成及应用实例 | 313 |
| | 6.2.2 SegPhos 配体的合成及应用实例 | 316 |
| | 6.2.3 Spiro Diphosphines（SDP）配体的合成及应用实例 | 319 |
| | 6.2.4 Spiroketal-based diphosphine（SKP）配体的合成及应用实例 | 323 |
| | 6.2.5 $C_3$-TunePhos 配体的合成及应用实例 | 327 |
| | 6.2.6 DuanPhos 配体的合成及应用实例 | 331 |
| | 6.2.7 QuinoxP 配体的合成及应用实例 | 335 |
| | 6.2.8 DuPhos 配体的合成及应用实例 | 338 |
| | 6.2.9 DIOP 配体的合成及应用实例 | 341 |
| | 6.2.10 （S,S）-f-Binaphane 配体的合成及应用实例 | 344 |
| | 6.2.11 Josiphos 配体的合成及应用实例 | 347 |
| | 6.2.12 DIPAMP 配体的合成及应用实例 | 350 |
| | 6.2.13 （R,S）-Binaphos 配体的合成及应用实例 | 353 |
| 6.3 | 手性膦 - 杂原子双齿配体的代表性合成及应用实例 | 358 |
| | 6.3.1 iPr-BiphPHOX 配体的合成及应用实例 | 358 |
| | 6.3.2 MeO-MOP 配体的合成及应用实例 | 362 |
| | 6.3.3 PHOX 配体的合成及应用实例 | 366 |
| | 6.3.4 SIPHOX 配体的合成及应用实例 | 368 |
| | 6.3.5 TF-BiphamPhos 配体的合成及应用实例 | 374 |

  6.3.6 Spiro Phosphino-oxazine 配体的合成及应用实例 378
 6.4 手性多官能化多齿型膦配体的代表性合成及应用实例 383
  6.4.1 Xing-Phos 配体的合成及应用实例 383
  6.4.2 Tao-Phos 配体的合成及应用实例 387
  6.4.3 HZNU-Phos 配体的合成及应用实例 392
  6.4.4 Ming-Phos 配体的合成及应用实例 396
  6.4.5 由金鸡纳碱衍生的手性膦酰胺配体的合成及应用实例 400
  6.4.6 Fei-Phos 配体的合成及应用实例 402
  6.4.7 SIOCPhox 配体的合成及应用实例 406
  6.4.8 Trost 配体的合成及应用实例 410

**索引** 412

# 1 绪论

1.1 手性膦配体的发展简史
1.2 手性膦配体的基本知识
1.3 手性膦配体的结构与分类
1.4 膦配体与过渡金属的配位反应
1.5 不对称催化反应简介

Synthesis and Application of Chiral Phosphine Ligands

## 1.1 手性膦配体的发展简史

经过近半个世纪的发展,过渡金属催化的不对称反应已成为合成手性化合物最有效的方法之一。在不对称催化中,手性配体起着至关重要的作用,一方面它控制着反应的立体化学,另一方面它调节着手性催化剂的催化活性和稳定性,因此高效、高选择性手性配体的设计与发展便成为不对称催化的关键要素。在诸多的手性配体中,手性膦配体是研究得最早,应用最广泛的一类配体,并已有成功应用于工业化生产的范例[1]。

1968 年,Knowles 和 Horner 等人[2]首次将手性单膦配体应用于不对称催化氢化反应中,虽然该反应产物的对映体过量(ee)值只有不到 15%,对映选择性较差,但这个研究结果引起了化学家对手性膦配体的重视并推动了此后数十年的迅速发展[3]。1972 年,Knowles 等发展出了手性单膦配体 CAMP,该配体可在铑催化下实现脱氢氨基酸的不对称氢化反应,并获得了中等到良好的对映选择性(ee 值高达 90%)[4]。与此同时,化学家们发现单膦配体与金属配位后形成的催化剂通常存在配体与金属中心作用不够强,其构型易发生改变等不足。因此在 20 世纪 90 年代以前,单膦配体的成功例子不多,直到 Hayashi 等[5]发展出了萘骨架的手性单膦配体,并将其成功应用于钯催化的烯烃不对称硅氢加成反应中,获得了高光学纯的产物。

CAMP    DIOP    DIPAMP    BINAP

在手性膦配体的发展史中,双膦配体更为人们所关注。一个非常重要的进展是 Kagan 等人[6]在 1971 年报道的具有 $C_2$ 对称性的手性膦配体 DIOP,该配体在铑催化下可以良好的对映选择性(ee 值为 80%)催化脱氢氨基酸类化合物的不对称氢化反应,这是具有里程碑式的研究结果。Knowles 等人[7]设计并发展出了手性双膦配体 DIPAMP,该配体在铑催化的脱氢氨基酸的不对称氢化反应中取得了高达96%的ee值。1980年，Noyori等人[8]发展出了具有轴手性的联萘骨架双膦配体BINAP，此配体由于连接两个萘环的C—C单键旋转受阻而使整个分子具有了光学活性。深入研究表明，BINAP配体可在钌或铑等催化下，实现脱氢氨基酸衍生物、简单酮、酮酸酯等多种底物的不对称氢化反应，取得了良好的对映选择性，并成功应用于手性药物和香料的工业化生产。Knowles和Noyori也因为在不对称催化氢化方面的开创性突出贡献与Sharpless一起获得了2001年的诺贝尔化学奖。

在Kagan、Knowles、Noyori等化学家的成就激励下，许多高效的双膦配体随后逐渐被开发出来，很长一段时间里都是过渡金属催化合成反应的首选配体，早期比较有代表性的工作如20世纪90年代初，Trost[9]开发出了环己二胺衍生的酰胺膦配体，是钯催化烯丙基化反应的优势配体；Achiwa等人[10]开发出的阻转异构体BIMOP、FUPMOP、MOC-BIMOP等，可用于钌和铑催化的不对称氢化反应；Togni等人[11]发展的二茂铁类手性膦配体，可用于不对称氢化、烯丙基化及氢硼化反应。如1995年，Burk等人[12]发展了手性中心在碳原子上的DuPhos和BPE类手性双膦配体[(R,R)-iPr-BPE]，该配体在钌催化下可控制β-羰基酯的不对称氢化反应，以很高的对映选择性得到相应的手性仲醇产物（ee值高达99.4%）。

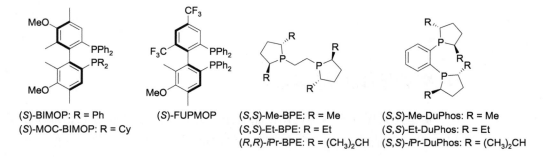

(S)-BIMOP: R = Ph
(S)-MOC-BIMOP: R = Cy

(S)-FUPMOP

(S,S)-Me-BPE: R = Me
(S,S)-Et-BPE: R = Et
(R,R)-iPr-BPE: R = (CH$_3$)$_2$CH

(S,S)-Me-DuPhos: R = Me
(S,S)-Et-DuPhos: R = Et
(S,S)-iPr-DuPhos: R = (CH$_3$)$_2$CH

进入20世纪90年代以来，受药物化学及不对称催化等学科迅速发展的影响，越来越多的研究小组，如Pfaltz、Feringa、Fu、Carreira等著名科学家致力于发展更为高效的手性膦配体，不仅对已知的配体进行合理的修饰和改进，而且还设计发展出了大量新骨架的高效膦配体，形成的海量研究数据极大地有助于人们理解配体的结构、位阻和电子效应对不对称催化反应的影响[13-23]。代表性的几类高效配体包括：①具阻转异构的联芳环类手性膦配体；②含碳手性双膦配体；③含 $C_2$ 对称轴的手性膦配体；④手性二茂铁基膦配体；⑤手性螺环配体；⑥氮-膦、氧-膦类等膦/杂原子双齿型手性膦配体等。

在国内，一大批卓越的有机化学家为手性膦配体的发展做出了巨大贡献[24-41]，如中国科学院上海有机化学研究所的戴立信教授[26]、丁奎岭教授[29]（现上海交通大学）、侯雪龙教授[26]、游书力教授[30]和汤文军教授[31]，香港理工大学的陈新滋教授[27]，南开大学的周其林教授[28]，中国科学院成都有机化学研究所的蒋耀忠教授[32]，南方科技大学的张绪穆教授[33]，中国科学院化学研究所范青华教授[34]，上海交通大学张万斌教授[35]，厦门大学高景星教授[36]，复旦大学的张俊良教授[37]，中国科学院大连化学物理研究所周永贵教授[38]、胡向平教授[39]，武汉大学王春江教授[40]等一大批科学家[41]。本书的编著者徐利文多年来也致力于新骨架膦配体的设计与合成，取得了一些重要进展[42]。总之，我国科学家在利用手性膦配体发展不对称催化氢化新反应方面取得了一系列突破性进展，提高了反应的效率与选择性，并围绕新类型的烯烃、羰基化合物、亚胺化合物和芳香杂环化合物等挑战性底物建立了具有潜在应用价值的不对称催化氢化反应新过程，同时还发展了不对称催化氢化反应的新策略和新方法，实现了 $4.55\times10^6$ 的转化数，达到世界领先水平。

# 1.2
## 手性膦配体的基本知识

磷是元素周期表中第三周期第五主族元素，原子序数为 15，原子量为 30.97，磷的核外电子排布式为 $1s^22s^22p^63s^23p^3$。跟氮类似，它的价电子层含有 5 个电子，常见的化合价为-3、+3 及 +5 价。手性的磷化合物在许多科学领域中都起着重要作用，包括生物活性药物、农用化学品以及作为手性配体用于过渡金属催化反应等。因此，手性磷中心化合物的构型稳定性受到了广泛的关注，与具有不对称氮中心化合物相比，四面体型磷中心手性化合物一般较稳定而不易消旋，如手性磷烷一般在室温下稳定，而无环手性胺则容易消旋。三价磷化合物以金字塔形连接三个不同的取代基，同时含有一对未共用电子对时可能会发生构型翻转，三价磷化合物的消旋程度取决于它们的结构，含有吸电子取代基将降低磷手性的构型稳定性，非环手性磷化合物的翻转能垒大约为 150 kJ/mol，并受与磷相连的原子的电负性的影响，而非环手性胺的翻转能垒大约为 30 kJ/mol。三价磷形成的化合物通常为三角锥形，显然它的翻转能垒比氮高，因此在室温下能够以某种特定的构型存在，而三级胺在室温条件下则可以快速翻转。然而，有些情况下手性磷连接吸电子取代基时也会发生消旋。

因此，手性有机磷化合物的出现使得化学领域取得了巨大成就，尤其是大量手性膦配体的发现，极大地促进了不对称金属催化领域的发展。膦配体是指 $PH_3$ 有机衍生物配体，配位原子磷通过 σ-π 配键与过渡金属形成配位化合物。手性膦配体是在膦配体的基础上发展起来的，并在不同类型过渡金属催化的不对称合成中都起着重要的作用，原则上，手性中心可以是磷原子，也可以是和磷原子相连的侧链上的碳原子，还可以是不含手性原子，但整个分子有手性轴或手性面的分子等。

# 1.3
## 手性膦配体的结构与分类[43]

手性膦配体根据配位的 P 原子种类不同分为单膦配体、双膦配体和膦-杂原子配体。代表性的单膦配体包括膦配体、亚膦酸酯配体及其他类型的单膦配体。

膦　　　亚膦酸酯

手性单膦配体分为手性中心在 P 原子上和手性中心在碳原子上的单膦配体。手性双膦配体是指含有两个配位点的双齿型膦配体，手性中心可以在 P 原子上或者在 P 原子附近的碳原子上。手性膦-杂原子配体主要是指配位点在 P 原子以及其他杂原子(主要包括 O、N 或 S 等)上的配体。在手性配体中，除具有手性原子的配体外，还有一类手性配体，本身不具有任何手性原子，但由于分子具有 $C_2$ 对称性而具有手性。

一般而言，可根据配体骨架的不同，分为螺环骨架膦配体、联芳环类膦配体(如联苯类、联萘类膦配体等)、含二茂铁基膦配体等。

## 1.4 膦配体与过渡金属的配位反应

膦配体是在均相催化甚至整个金属有机化学中最重要的一类配体，所有的过渡金属，尤其是后过渡金属都可以和三价磷配位，P 原子上较强的给电子能力可以和软的低价金属很好地匹配，P 上的取代基可以显著影响中心金属的性质和反应活性，目前已有大量的报道涉及含膦配体的金属配合物。通常而言，含有二个、三个或四个配位原子的配体分别称为二齿、三齿或四齿配体。这些配体和金属以螯合的方式配位，因此也被统称为螯合配体。

三烷基膦配体和金属的配位作用主要是通过路易斯酸-碱作用实现的，膦原子软的延展性好的孤对电子云可以作为强的路易斯碱和软的过渡金属路易斯酸起作用，一般来说，三烷基膦是膦配体中给电子能力最强的，芳基膦配体的给电子能力稍弱，这种趋势是因为芳基取代基的 $sp^2$ 杂化轨道中较大的 s 成分使它的给电子能力要弱于烷基取代基。类似的，亚膦酸酯配体中的三个烷氧基的给电子能力弱于烷基或芳基，导致亚膦酸酯配体的给电子能力弱于常见的膦配体。

叔膦或亚膦酸酯也可以作为 π 受体，过去认为，P 的空的 d 轨道可以接受金属的 d 轨道电子形成 dπ-dπ 反馈 π 键。近期的研究表明，P-R 键的 σ* 轨道起着 π 受体的作用。

## 1.5 不对称催化反应简介

手性是自然界和生命的基本属性之一，作为自然界生命基础的生物大分子都具有手性，参与生命活动的许多物质，如蛋白质、核酸、氨基酸和酶等都具有手性特征。人们使用的药物也大多具有手性，在美国药典所列出的药物中，大约有一半药物分子中含有一个以上的手性中心。尽管对映体间物理化学性质几乎完全相同，但它们的生化和药理作用却往往不同。因此，合成单一手性的化合物对于生命科学和药物化学研究以及人类的健康具有十分重要的意义。

手性分子的两个对映体中，每个对映体都能把平面偏振光旋转到一定的角度，两个角度数值相同但方向相反，这种性质称为光学活性，测量旋光性的仪器为旋光仪。在不对称催化中，衡量反应产物的光学活性的一个重要标准就是利用 ee 值来判断，ee 值可通过比旋光来计算，但最常用的计算方法即利用手性高效液相色谱进行拆分后计算两个对映体的峰的比例。

不对称催化反应产物的对映体组成可用对映体过量(enantiomeric excess, ee) 值来描述。它表示一个对映体（如 R，以 [R] 表示其浓度）对另一个对映体（如 S，以 [S] 表示其浓度）的过量百分数，如下式所示：

$$\text{ee} = \frac{[R]-[S]}{[R]+[S]} \times 100\%$$

获得手性化合物的常见方法有三种：外消旋体拆分、化学计量不对称合成和不对称催化反应。外消旋体的拆分是传统的经典方法，即利用自然界存在的光学纯的手性化合物，通过它和待拆分的手性外消旋化合物进行反应，生成一对非对映异构体，通过对非对映异构体的分离最终得到起始化合物的两个对映异构体，这一过程被称作外消旋体拆分法。通过这种方法来拆分外消旋体，应满足的主要条件包括：①拆分试剂应该廉价、容易回收而且回收后光学纯度不降低，拆分试剂应该和外消旋体容易反应完全等；②拆分试剂和外消旋体形成的非对映异构体应该能通过色谱法、结晶法或蒸馏法容易地进行分离。化学计量不对称合成即一个单元的手性因子只能产生一个单元的手性产物，这个过程中手性单元不能增值，它主要包括三种方法：①手性源法，通过手性底物中已有手性单元的分子内诱导获得产物；②手性辅基法，是指在非手性底物上引入手性辅助基团，从而进行不对称诱导，最后再除去手性辅基即可；③手性试剂法，是指将手性试剂与非手性底物直接反应得到手性化合物。不对称催化合成，是通过使用手性催化剂来实现不对称合成的方法，即在非手性底物进行不对称反应时加入少量手性催化剂，使它与反应底物形成高反应活性中间体，催化剂作为手性模板控制反应的对映面，从而可以将大量前手性底物选择性地转化成特定构型的产物，实现手性增值和手性放大的效果。因此，不对称催化是生产大量手性化合物的最经济、最实用的技术，从而为学术界和工业界所高度重视，不少手性药物的生产都采用了不对称催化方法。

不对称催化发展至今，研究得比较多的反应包括不对称氢化反应和不对称环氧化反应，2001年的诺贝尔化学奖就授予了在不对称催化氢化和环氧化反应方面做出突出贡献的化学家 W. S. Knowles、R. Noyori 和 K. B. Sharpless。不对称氢化反应是指在手性配体及金属催化剂的存在下，对官能团化的烯烃和酮等进行不对称氢化还原的反应，该过程可将 $sp^2$-C 转化为 $sp^3$-C，不对称氢化反应的底物通常包括烯烃、羰基化合物、亚胺和烯胺等。反应的氢源主要是氢气，也包括其他的氢给体（如甲酸、异丙醇等）。首例不对称氢化的工业化生产是孟山都开发的用于治疗帕金森病的 L-多巴胺（L-DOPA）的生产过程[44]。

当前，新型手性膦配体仍在不断被设计和发展出来[45]，加之成书时间匆忙和篇幅限制，本书难以将数千篇研究论文一一列举（见图 1-1），将主要展示一些性能优异的具有一定代表性的手性膦配体，旨在抛砖引玉，期望有助于读者进一步查阅原始文献和启发同行开发性能更好的新配体。

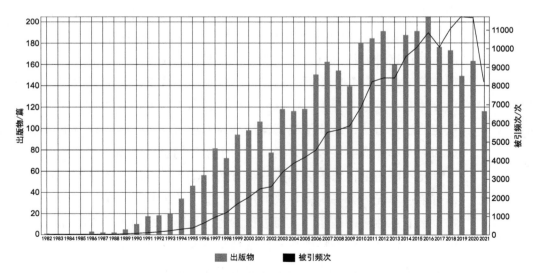

图 1-1 近 40 年里发表的手性膦配体数据一览表（经 web of science 检索关键词——chiral phosphine）

## 参考文献

[1] (a) Enthaler S, Erre G, Junge K, et al. Development of practical rhodium phosphine catalysts for the hydrogenation of $\beta$-dehydroamino acid derivatives. Organic Process Research & Development, 2007, 11(3): 568 – 577. (b) Blaser H U, Malan C, Pugin B;et al. Selective hydrogenation for fine chemicals: recent trends and new developments. Advanced Synthesis & Catalysis, 2003, 345(1/2): 103-151.

[2] (a) Knowles W S, Sabacky M. Catalytic asymmetric hydrogenation employing a soluble, optically active, rhodium complex. Chemical Communications (London), 1968 (24): 1445-1446. (b) Horner L, Siegel H, Büthe H. Asymmetric catalytic hydrogenation with an optically active phosphinerhodium complex in homogeneous solution. Angewandte Chemie International Edition, 1968, 7 (12): 942-942.

[3] (a) Hayashi T, Kumada M. Asymmetric synthesis catalyzed by transition-metal complexes with functionalized chiral ferrocenylphosphine ligands.Accounts of Chemical Research, 1982, 15: 395-401. (b) Tang W, Zhang X. New chiral phosphorus ligands for enantioselective hydrogenation. Chemical Reviews, 2003, 103(8): 3029-3070.

[4] Knowles W S, Sabacky M J, Vineyard B D. Catalytic asymmetric hydrogenation. Journal of the Chemical Society, Chemical Communications, 1972 (1): 10-11.

[5] (a) Uozumi Y, Hayashi T. Catalytic asymmetric synthesis of optically active 2-alkanols via hydrosilylation of 1-alkenes with a chiral monophosphine-palladium catalyst. Journal of the American Chemical Society, 1991, 113, 9887-9888. (b) Hayashi T. Chiral Monodentate Phosphine Ligand MOP for Transition-Metal-Catalyzed Asymmetric Reactions. Accounts of Chemical Research, 2000, 33(6): 354-362.

[6] (a) Dang T. P, Kagan H G. The asymmetric synthesis of hydratropic acid and aminODacids by homogeneous catalytic hydrogenation. Journal of the Chemical Society D: Chemical Communications, 1971, 10:481.(b) Dang T P, Kagan H G.Asymmetric catalytic reduction with transition metal complexes. I. catalytic system of rhodium( I ) with (-) -2,3-(0)-isopropyli dene-2,3-dihydroxy-1,4-bis(diphenylphosphino) butane, a new chiral diphosphine. Joirnal of the American Chemical Society, 1972,94(18):6429-6433.

[7] (a) Knowles W S, Sabacky M J, Vineyard B D, et al. Asymmetric hydrogenation with a complex of rhodium and a chiral bisphosphine. Journal of the American Chemical Society, 1975, 97(9): 2567-2568. (b) Knowles W S. Asymmetric Hydrogenation. Accounts of Chemical Research, 1983, 16(3): 106-112.

[8] Miyashita A, Yasuda A, Takaya H,et al. Synthesis of 2,2′-bis(diphenylphosphino)-1,1′-binaphthyl (BINAP), an atropisomeric chiral bis(triaryl)phosphine, and its use in the rhodium( I )-catalyzed asymmetric hydrogenation of $\alpha$-(acylamino)acrylic acids. Journal of the American Chemical Society, 1980, 102(27): 7932-7934.

[9] Trost B M. Designing a receptor for molecular recognition in a catalytic synthetic reaction: allylic alkylation. Accounts of Chemical Research, 1996, 29(8): 355-364.

[10] (a) Yamamoto N, Murata M, Morimoto T, et al. Synthesis of axially dissymmetric biphenyl-Bisphophine ligands, bimops and asymmetric hydrogenations of $\beta$-keto ester and $\alpha,\beta$-unsaturated carboxylic acid catalyzed by their ruthrnium ( II ) complexes. Chemical and Pharmaceutical Bulletin, 1991, 39(4): 1085-1087. (b) Yoshikawa K, Yamamoto N, Murata M, et al. A new type of atropisomeric biphenylbisphosphine ligand, (R)-MOCBIMOP and its use in efficient asymmetric hydrogenation of o-aminoketone and itaconic acid.Tetrahedron: Asymmetry, 1992, 3 (1): 13-16.

[11] Togni A, Breutel C, Schnyder A, et al. A novel easily accessible chiral ferrocenyldiphosphine for highly enantioselective hydrogenation, allylic alkylation, and hydroboration reactions. Journal of the American Chemical Society, 1994, 116 (9): 4062-4066.

[12] Burk M J, Tharper G P, Christopher S. Highly enantioselective hydrogenation of $\beta$-keto esters under mild conditions. Journal of the American Chemical Society, 1995, 117(15): 4423-4424.

[13] Helmchen G, Pfaltz A. Phosphinooxazolines-a new class of versatile, Modular P,N-ligands for asymmetric catalysis. Accounts of Chemical Research, 2000, 33(6): 336-345.

[14] Feringa A. Phosphoramidites: marvellous ligands in catalytic asymmetric conjugate addition. Accounts

of Chemical Research, 2000, 33(6): 346-353.
[15] Breit B. Synthetic aspects of stereoselective hydroformylation. Accounts of Chemical Research, 2003, 36(4): 264-275.
[16] Genet J P. Asymmetric catalytic hydrogenation. Design of new Ru catalysts and chiral ligands: from laboratory to industrial applications. Accounts of Chemical Research, 2003, 36(12): 908-918.
[17] Leung P H. Asymmetric synthesis and organometallic chemistry of functionalized phosphines containing stereogenic phosphorus centers. Accounts of Chemical Research, 2004, 37, 169-177.
[18] Fu G C. Applications of planar-chiral heterocycles as ligands in asymmetric catalysis. Accounts of Chemical Research, 2006, 39(11): 853-860.
[19] Klosin J, Landis C R. Ligands for practical rhodium-catalyzed asymmetric hydroformylation. Accounts of Chemical Research, 2007, 40(12): 1251–1259.
[20] Minnaard A J, Feringa B L, Lefort L, et al. Asymmetric hydrogenation using monodentate phosphoramidite ligands. Accounts of Chemical Research, 2007, 40(12): 1267–1277.
[21] Dieguez M, Pamies O. Biaryl phosphites new efficient adaptative ligands for Pd-catalyzed asymmetric allylic substitution reactions. Accounts of Chemical Research, 2010, 43(2): 312-322.
[22] Rössler S L, Petrone D A, Carreira E M. Iridium-catalyzed asymmetric synthesis of functionally rich molecules enabled by (phosphoramidite,olefin) ligands. Accounts of Chemical Research, 2019, 52: 2657 – 2672.
[23] Cabré A, Riera A, Verdaguer X. P-stereogenic amino-phosphines as chiral ligands: from privileged intermediates to asymmetric catalysis. Accounts of Chemical Research, 2020, 53: 676 – 689.
[24] Liu Y, Li W, Zhang J. Chiral ligands designed in China. National Science Review, 2017, 4(3): 326-358.
[25] Zhou Q L. Privileged Chiral Ligands and Catalysts.Weinheim: Wiley-VCH Verlag GmbH & Co. KGaA, 2011.
[26] Dai L X, Tu T, You S L, et al. Asymmetric catalysis with chiral ferrocene ligands. Accounts of Chemical Research, 2003,36(9): 659-667.
[27] Wu J, Chan A S C. P-Phos: a family of versatile and effective atropisomeric dipyridylphosphine ligands in asymmetric catalysis. Accounts of Chemical Research, 2006, 39(10): 711-720.
[28] Xie J H, Zhou Q L. Chiral diphosphine and monodentate phosphorus ligands on a spiro scaffold for transition-metal-catalyzed asymmetric reactions. Accounts of Chemical Research, 2008, 41(5): 581-593.
[29] (a) Ding K L. Synergistic effect of binary component ligands in chiral catalyst library engineering for enantioselective reactions. Chemical communications, 2008 (8): 909-921. (b) Wang X M, Han Z B, Wang Z, et al. A type of structurally adaptable aromatic spiroketal based chiral diphosphine ligands in asymmetric catalysis. Accounts of Chemical Research, 2021, 54: 668 – 684.
[30] (a) Zhuo C X. Zheng C, You S L. Transition-metal-catalyzed asymmetric allylic dearomatization reactions. Accounts of Chemical Research, 2014, 47: 2558-2573. (b) Gao D W, Gu Q, Zheng C, et al. Synthesis of planar chiral ferrocenes via transition-metal-catalyzed direct C-H bond functionalization. Accounts of Chemical Research, 2017, 50(2): 351-365.
[31] Xu G Q, Senanayake C H, Tang W J. P-chiral phosphorus ligands based on a 2,3-dihydrobenzo[d][1,3] oxaphosphole motif for asymmetric catalysis. Accounts of Chemical Research, 2019, 52(4): 1101-1112.
[32] Hu W H, Yan M, Lau C P, et al. A highly effective rhodium spirocyclic phosphinite catalyst for the asymmetric hydrogenation of enamides. Tetrahedron Letters, 1999, 40(5): 973-976.
[33] (a) Zhang W C, Chi Y X, Zhang X M. Developing chiral ligands for asymmetric hydrogenation. Accounts of Chemical Research, 2007, 40(12): 1278–1290. (b) Zhao Q Y, Chen C Y, Wen J L, et al. Noncovalent interaction-assisted ferrocenyl phosphine ligands in asymmetric catalysis. Accounts of Chemical Research, 2020, 53: 905 – 1921. (c) Wan F, Tang W J. Phosphorus ligands from the Zhang Lab: design, asymmetric hydrogenation, and industrial applications. Chinese Journal of Chemistry, 2021, 39(4): 954-968.
[34] He Y M, Feng Y, Fan Q H. Asymmetric hydrogenation in the core of dendrimers. Accounts of Chemical

Research, 2014, 47(10): 2894-2906.
- [35] Butt N A, Zhang W B. Transition metal-catalyzed allylic substitution reactions with unactivated allylic substrates. Chemical Society Reviews, 2015, 44 (22): 7929-7967.
- [36] Li Y Y, Yu S L, Shen W Y, et al. Iron-, cobalt-, and nickel-catalyzed asymmetric transfer hydrogenation and asymmetric hydrogenation of ketones. Accounts of Chemical Researc, 2015, 48(9): 2587-2598.
- [37] Li W B, Zhang J L. Recent developments in the synthesis and utilization of chiral $\beta$-aminophosphine derivatives as catalysts or ligands. Chemical Society Reviews, 2016, 45: 1657-1677.
- [38] Zhou Y G. Asymmetric hydrogenation of heteroaromatic compounds. Accounts of Chemical Research, 2007, 40(12): 1357-1366.
- [39] 侯传金, 刘小宁, 夏英, 等. 非对称杂化的手性膦-亚磷酰胺酯配体在不对称催化反应中的应用进展. 有机化学, 2012, 32: 2239-2247.
- [40] Wei L, Chang X, Wang C J. Catalytic asymmetric reactions with *N*-metallated azomethine ylides. Accounts of Chemical Research, 2020, 53: 1084-1100.
- [41] (a) 徐利文, 夏春谷, 孙伟, 等. 手性膦配体的合成及其在不对称催化反应中的应用研究进展. 有机化学, 2003, 23(9): 919-932. (b) Li Y M, Kwong F Y, Yu, W Y, et al. Recent advances in developing new axially chiral phosphine ligands for asymmetric catalysis. Coordination Chemistry Reviews, 2007, 251: 2119–2144. (c) 赵文献, 杨代月, 张玉华. 高效手性膦配体的研究进展. 有机化学, 2016, 36: 2301-2316. (d) 张树辛, 冯宇, 范青华. 过渡金属催化的不对称氢化反应的国内研究进展. 高等学校化学学报, 2020, 41(10): 2107-2136.
- [42] Ye F, Xu Z, Xu L W. The discovery of multifunctional chiral P ligands for the catalytic construction of quaternary carbon/silicon and multiple stereogenic centers. Accounts of Chemical Research, 2021, 54(2): 452 – 470.
- [43] Muephy P J. Organophosphorus Reagents. A Practical Approach in chemistry. New York: Oxford University Press, 2004.
- [44] Knowles W S. Asymmetric Hydrogenations (Nobel Lecture). Angewandte Chemie International Edition, 2002,41 (12): 1998-2007
- [45] Margalef J, Biosca M, de la Cruz Sánchez P, et al. Evolution in heterodonor P-N, P-S and P-O chiral ligands for preparing efficient catalysts for asymmetric catalysis. From design to applications. Coordination Chemistry Reviews, 2021, 446: 214120.

PH♦SPHORUS 磷科学前沿与技术丛书　　　　　　　　手性膦配体合成及应用

# 2

# 手性单齿膦配体

2.1 单齿叔膦配体
2.2 亚磷酸衍生物单磷配体
2.3 单膦手性配体的小结及延伸阅读

Synthesis and Application of Chiral Phosphine Ligands

不对称金属催化的研究最早是从单齿膦配体开始的，然而之后效果更好的双齿螯合膦配体的出现使得单齿膦配体很长一段时间处于被忽视的境地。尽管双齿螯合膦配体在广泛的不对称金属催化反应中展示了强大的催化能力，但是仍有许多双齿螯合膦配体不能催化的反应，需要开发高效的单齿膦配体。近二十年来，许多高效的亚膦酸衍生物单膦配体的出现使得单齿膦配体在手性膦配体中重新占有一席之地。一些前人们认为只有双膦配体才能催化的反应现在用单齿膦配体也能取得很好的效果。根据与膦原子连接的原子种类不同，单齿膦配体可分为单齿叔膦配体 (**A**)、单齿亚膦酸酯配体 (**B**) 和单齿亚膦酰胺酯配体 (**C**) 等（图 2-1）。本章按照此分类方法对单齿膦配体进行介绍。

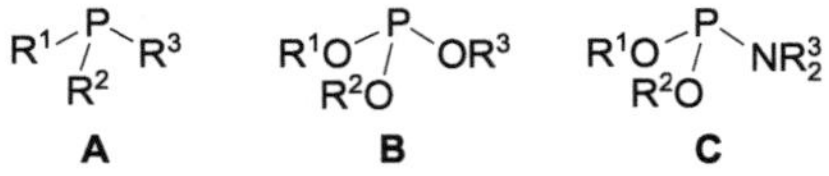

图 2-1　单齿膦配体分类

# 2.1 单齿叔膦配体

最早的不对称金属催化研究是从单齿叔膦配体开始的，经过 50 多年的发展已经积累了多种结构类型。在手性膦配体研究的早期，人们认为手性因素距离金属中心原子越近不对称催化效果越好，因而含有三个不同的碳取代基的磷手性中心叔膦最早被作为手性膦配体进行研究。例如 Horner、Knowles 等课题组制备了一系列磷手性中心叔膦配体（图 2-2），其中不对称催化效果较好的如叔膦配体 2-甲氧基苯基(环己基)甲基膦 (CAMP) **L1**[1]。

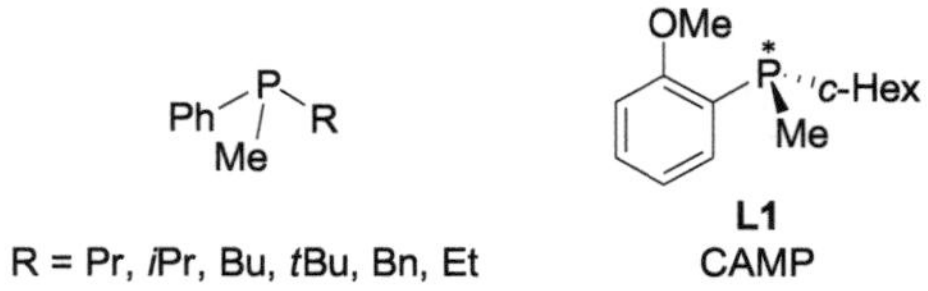

图 2-2　磷手性中心叔膦配体

此类磷手性中心叔膦配体合成较烦琐，多数需要拆分，而且不易得到光学纯的产物。若将含有碳手性中心的取代基引入磷原子上可以得到碳手性中心叔膦配体(图 2-3)，其中效果较好的如从新薄荷醇出发合成的叔膦——新蓋基二苯基膦 (NMDPP) **L7**[2]。

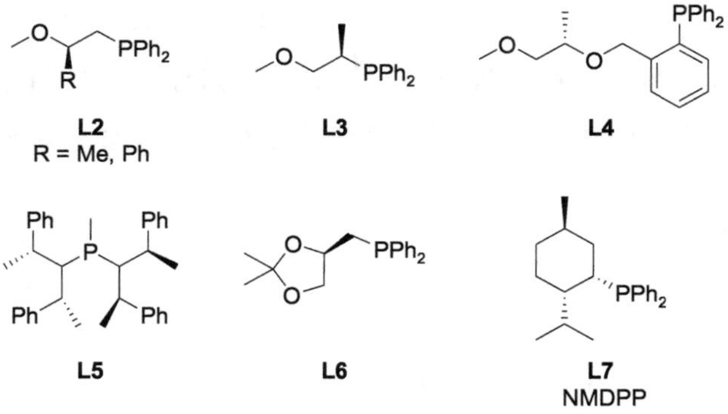

**图 2-3　碳手性中心叔膦配体**

若将磷原子设计在环上可以得到磷杂环叔膦配体，根据环的大小和取代基的位置和种类不同，磷杂环叔膦配体可以有多种结构类型(图 2-4)，其中使用较多的如 **L13**[3]。

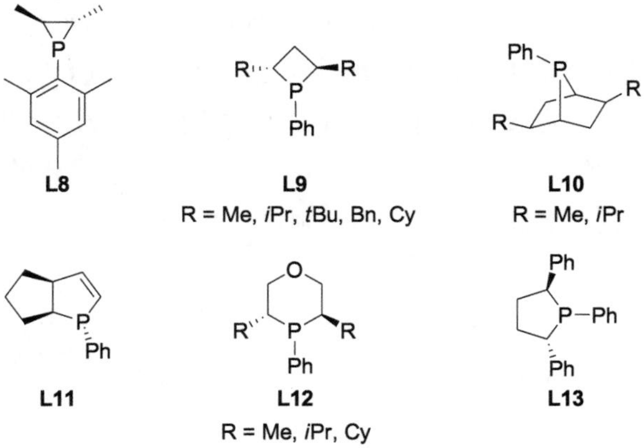

**图 2-4　磷杂环叔膦配体**

很多单齿叔膦配体虽然已被设计出来，然而其中很多在不对称催化中的应用尚待研究。目前已经显示出较为广泛的催化能力的是基于具有 $C_2$ 轴手性的联萘结构以及螺环结构衍生的叔膦配体等。下面对这些应用效果较好的单齿叔膦配体作着重介绍。

## 2.1.1　环磷联萘骨架叔膦配体（BINEPINE）

具有阻转异构结构的环磷联萘骨架叔膦配体(BINEPINE) **L14** 及 **L15** 是一类常见的轴手性单膦配体（图 2-5）。BINEPINE 在广泛的不对称碳-氢键、碳-碳键、碳-杂键构建反应中展示了显著的立体选择性。与双齿膦配体相比，BINEPINE 具有一些优势：可以从相对便宜的原料制备；制备路线适合引入结构多样性；可以用来设计含有单齿配体的金属配合物催化剂。

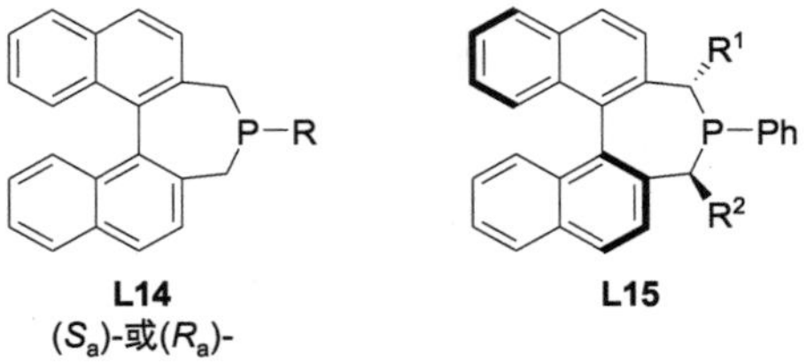

**L14**
($S_a$)-或($R_a$)-

**L15**

图 2-5　环磷联萘骨架叔膦配体

1994 年，Gladiali 等首先设计合成了具有环磷联萘骨架的叔膦配体 **L14**[4]。1-溴-2-甲基萘和由其自身制备的格氏试剂经过镍催化的 Kumada 偶联反应制得 2,2′-二甲基联萘，之后经过选择性的双锂化接着与苯基二氯化膦反应得到消旋的 **L14**。消旋配体与手性钯配合物形成非对映异构体，再通过结晶过程分离得到单一的非对映异构体，最后用双齿膦配体 dppe 解离得到光学纯的 BINEPINE（图 2-6）。此路线需要昂贵的拆分试剂，步骤烦琐，而且总产率低，不适合大量制备。

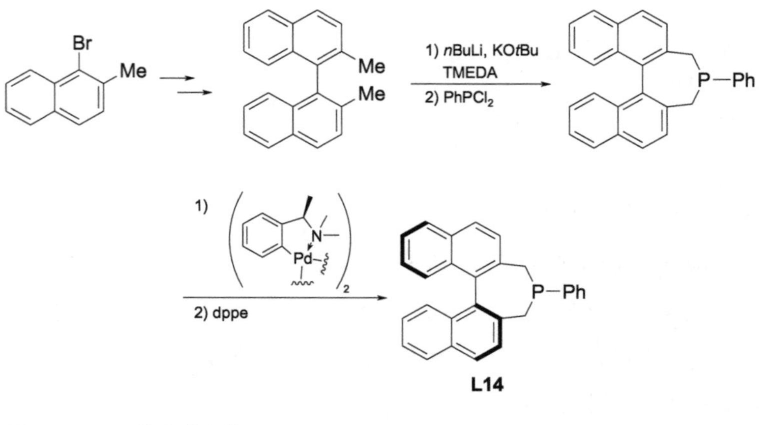

图 2-6　**L14** 的合成方法

之后 Beller 等[5,6]和张绪穆等[7]分别开发了不同的方法从光学纯的 ($S_a$)-BINOL（联萘二酚）出发制备了较为复杂的叔膦配体 **L14**。($S_a$)-BINOL 用三氟甲磺酸酐酯化后利用镍催化的 Kumada 偶联反应制得光学纯的 2,2'-二甲基联萘。2,2'-二甲基联萘可以经过双锂化和与烃基二氯化膦反应直接得到光学纯的 **L14**，但是商品化的烃基二氯化膦种类有限，因而限制了此路线的适用范围。若 2,2'-二甲基联萘经过双锂化之后与二乙氨基二氯化膦反应得到二乙氨基 **L14**，之后与氯化氢反应得到氯化 **L14**，最后用格氏试剂或锂试剂制得光学纯的 **L14**，此路线虽然步骤多，但是商品化的格氏试剂和锂试剂种类繁多，因而可以用以制备结构多样的 **L14**（图 2-7）。

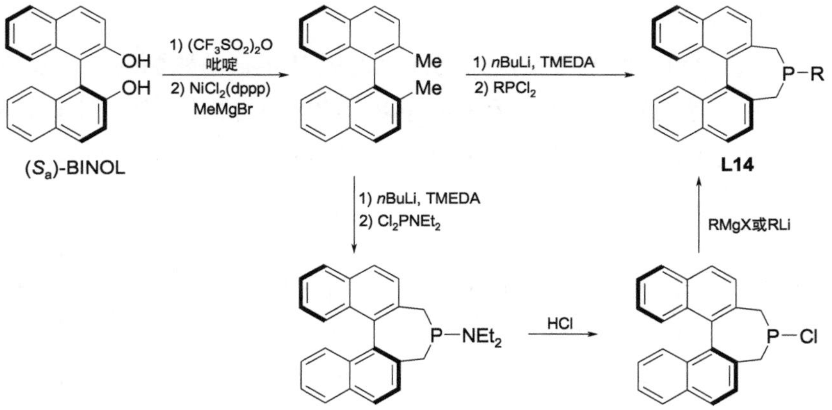

图 2-7　**L14** 改进的合成方法

在 **L14** 的基础上，Widhalm 等[8]和 Beller 等[9]分别用不同的方法引入 α-取代基，得到具有轴手性和手性碳原子的叔膦 **L15**。在磷原子边上引入新的手性中心必然将提高手性转化的效果。Widhalm 开发的方法将 **L14** 用单质硫氧化，然后经过两次锂化-烷基化得到 **L15**（图 2-8）。

叔膦 BINEPINE **L14** 和 **L15** 在许多金属催化的不对称转化中展示了优异的催化性能，研究较多的包括烯烃的氢化、转移氢化、氢甲酰化等。下面对典型反应进行举例介绍。

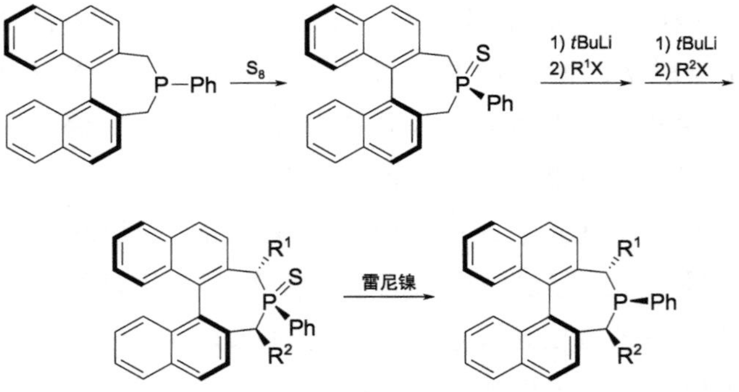

图 2-8　含有 α-取代基的 L15 的合成方法

**(1) 烯烃氢化反应**

Beller 课题组发现在铑催化的脱氢 α-氨基酸酯的氢化反应中，环磷联萘骨架叔膦配体展示了中等以上的对映选择性[5,6]。例如在下例中，简单的 **L14** 可以获得 90% 的 ee 值（图 2-9）。

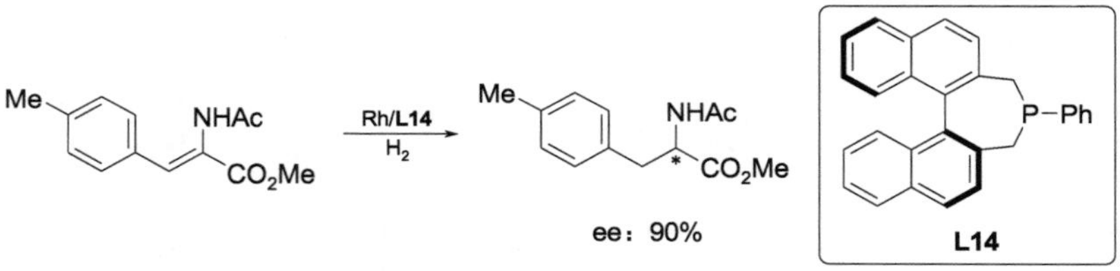

图 2-9　BINEPINE 在烯烃氢化反应中的应用

**(2) 酮的氢化反应**

Beller 课题组报道在钌催化 β-酮酸酯的氢化反应中，环磷联萘骨架叔膦配体展示了较好的催化活性和对映选择性[10,11]。例如在下例中，简单的 **L16** 可以获得 95% 的 ee 值（图 2-10）。

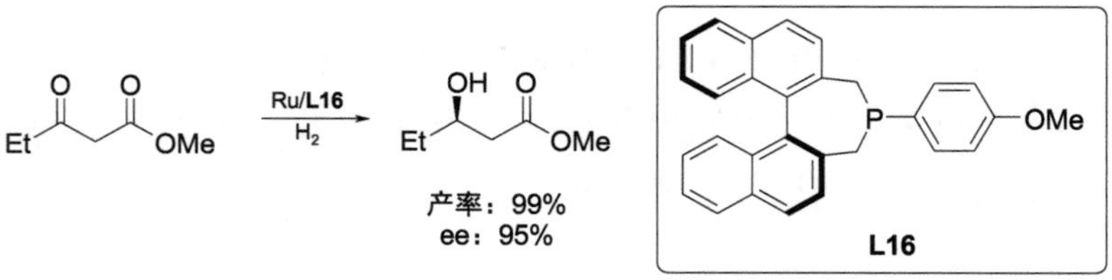

图 2-10　BINEPINE 在钌催化酮的氢化反应中的应用

2011 年，Beller 课题组报道了铜催化苯乙酮氢化为 1-苯乙醇[12]，底物扩展发现，对于一系列芳基、烷基、环状、杂环、脂肪酮类化合物，均有良好至优秀的产率和中等到良好的对映选择性，ee 值最高可达 89%（图 2-11）。

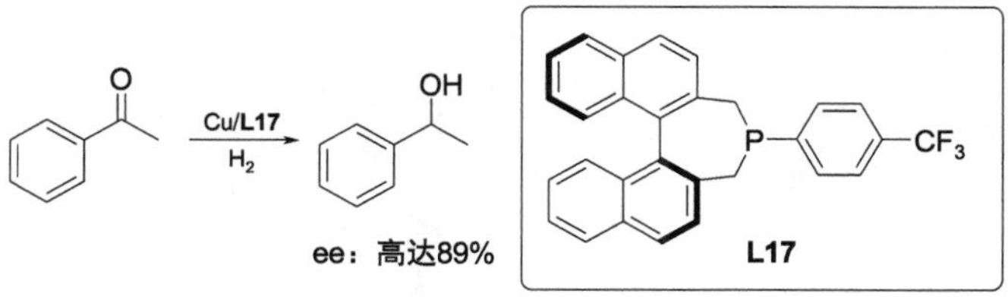

图 2-11 BINEPINE 在铜催化苯乙酮的氢化反应中的应用

（3）转移氢化反应

Gladiali 等探索了环磷联萘骨架叔膦配体在不饱和酸、脱氢氨基酸和酯的转移氢化反应中的应用效果，发现对衣康酸和酯的效果较好，而对脱氢氨基酸和酯的效果中等（图 2-12）。反应以铑为催化剂，甲酸为氢源[13]。

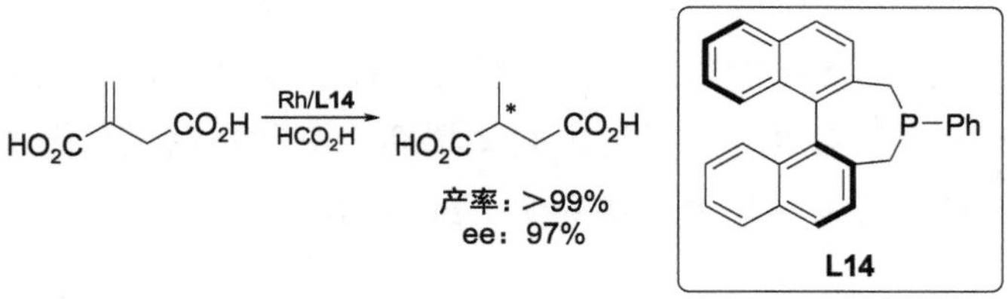

图 2-12 BINEPINE 在转移氢化反应中的应用

（4）碳-氢键活化

在碳-氢键活化反应中，Baudoin 等发现二茂铁取代的环磷联萘骨架叔膦配体 L18 展示了优异的催化活性、非对映选择性和对映选择性（图 2-13）[14,15]。

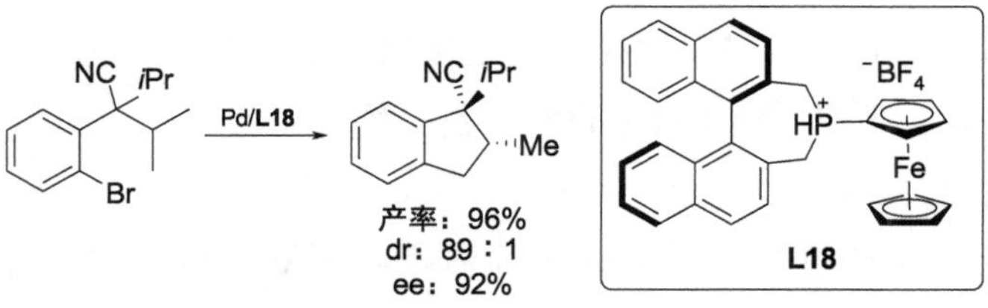

图 2-13 BINEPINE 在碳-氢键活化反应中的应用

(5) 醛的 α-芳基化

Mazet 课题组探索醛的 α-芳基化反应时发现环磷联萘骨架衍生的配体 **L19** 具有较好的结果，反应的产率较高，但是对映选择性不够理想（图 2-14）[16]。

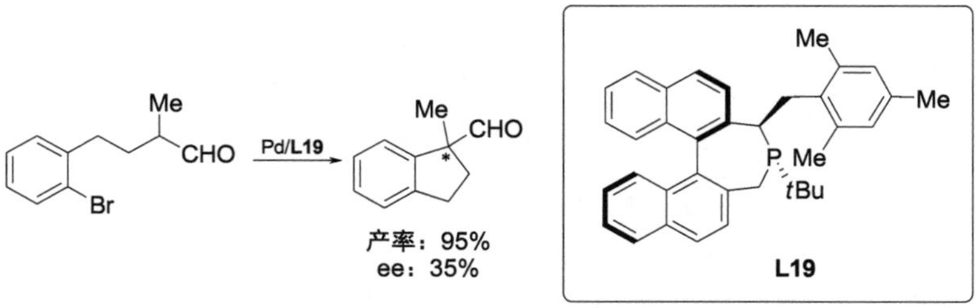

图 2-14　BINEPINE 在醛的 α-芳基化反应中的应用

## 2.1.2　环磷螺环骨架叔膦配体（SITCP）

周其林课题组开发了一系列基于螺环骨架的膦配体。除了著名的螺环双膦配体之外，单齿的环磷螺环骨架叔膦配体 **L20** 在一些反应中也已经显示了优异的催化能力[17,18]。从螺环二酚出发，经过三氟甲磺酸酐的酯化、钯催化甲氧羰基化得到中间体螺环二酯，然后经过还原、氯代和碳-磷键形成反应得到环磷叔膦配体 **L20**（图 2-15）。

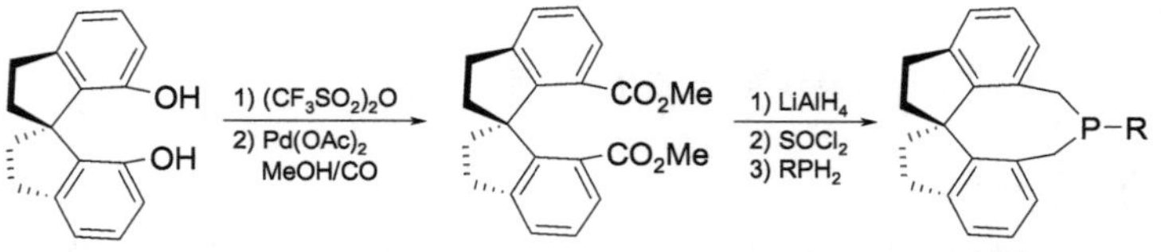

图 2-15　环磷螺环骨架叔膦配体 **L20** 的合成方法

环磷螺环骨架叔膦配体在一些金属催化的不对称转化中展示了优异的催化性能，下面对典型反应进行举例介绍。

(1) 烯丙基化反应

周其林课题组发现在钯催化烯丙醇与醛的烯丙基化反应中，配体 **L20** 能够使得反应顺利进行，并且得到良好的产率、非对映选择性以及较好的对映选择性。反应中三乙基硼作为极性反转试剂，推动反应进行

(图 2-16)[18]。

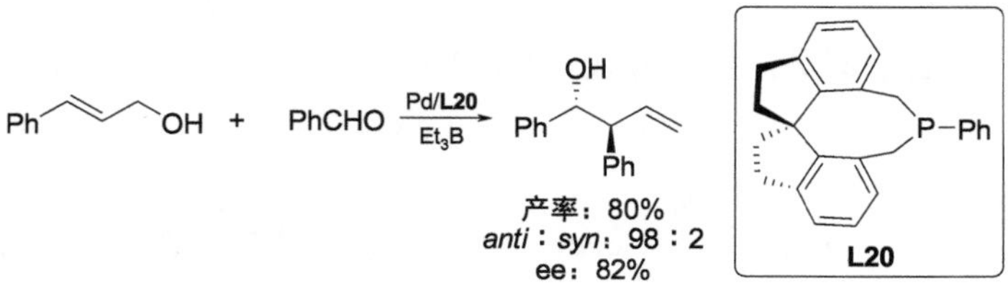

图 2-16 SITCP 在烯丙基化反应中的应用

**(2) 氧环双烯的开环反应**

在氧环双烯的开环反应中，环磷叔膦配体 **L21** 显示出较好的催化活性和立体选择性，反应以铜为催化剂，格氏试剂为亲核试剂(图 2-17)[17]。

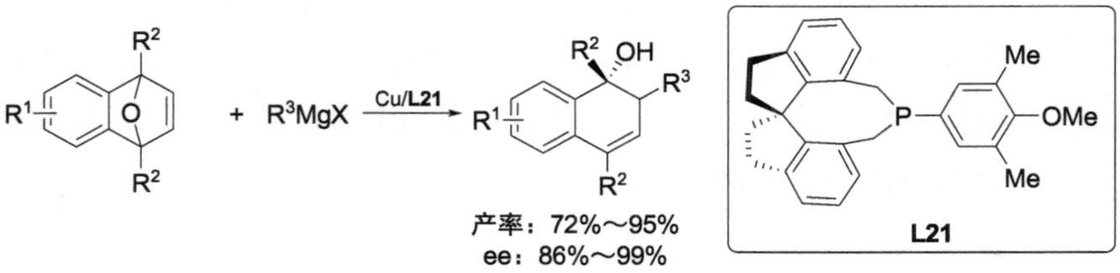

图 2-17 SITCP 在氧环双烯的开环反应中的应用

**(3) C—N 键偶联反应**

2016 年，Fu 课题组报道了光诱导铜催化的外消旋叔烷基氯化物亲电试剂与咔唑或吲哚的 C—N 键偶联反应。该反应在蓝色发光二极管的激发下于 $-40\ ℃$ 进行，铜与手性膦配体 **L20** 的配合物既充当光催化剂又充当不对称诱导源。反应可以从外消旋的底物得到单一的具有高对映选择性的产物(图 2-18)[19]。

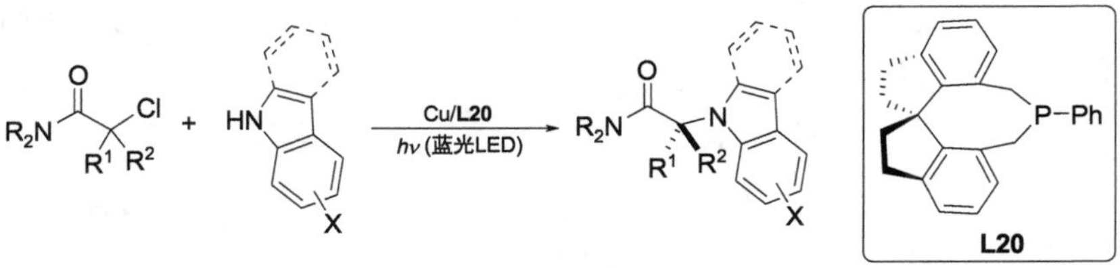

图 2-18 SITCP 在光催化 C—N 键偶联反应中的应用

之后 Fu 课题组报道了外消旋烷基碘代物与吲哚的 C—N 键偶联反应。与氯代烃不同，碘代烃的反应不要光照条件。铜与手性膦配体 **L20** 的配合物在室温下即可催化反应发生（图 2-19）[20]。

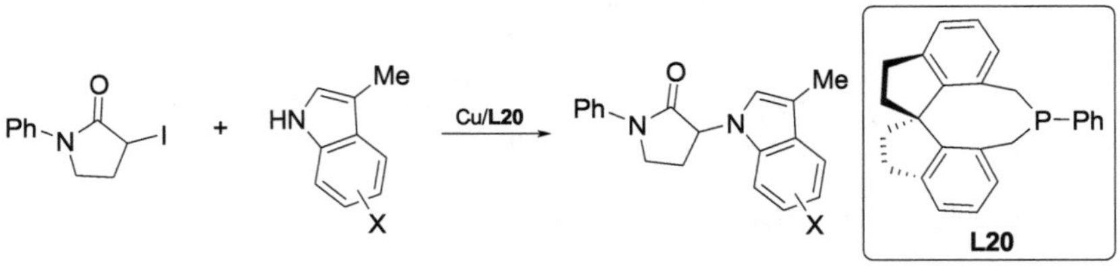

图 2-19　SITCP 在铜催化 C—N 键偶联反应中的应用

### 2.1.3　非环联萘骨架叔膦配体（MOP）

Hayashi 课题组开发了一系列非环联萘骨架叔膦配体（MOP），包括单齿膦配体 **L22**[21-24]，其中 2′-位含有烷氧基的衍生物 **L23** 是一类优秀的单齿膦配体[25-27]。MOP 中的氧在形成金属络合物以及催化过程中有重要的作用（图 2-20）。

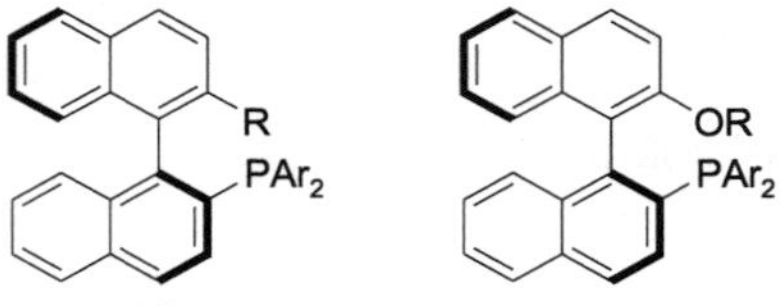

图 2-20　非环联萘骨架叔膦配体（MOP）

从 BINOL 与三氟甲磺酸酐的反应开始，经过钯催化与 $Ar_2POH$ 偶联，水解，2′-位烷氧基化，还原后得到 **L23**（图 2-21）。

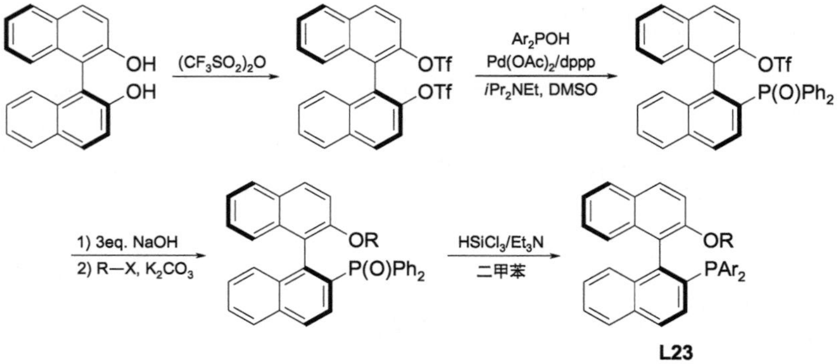

图 2-21　**L23** 的合成方法

早期 MOP 被发现在烯烃的氢硅化反应中效果优异，后来在其他反应中也得到了较好的应用。举例如下：

(1) 烯烃氢硅化反应

末端烯烃与三氯硅烷的钯催化氢硅化反应中，非环联萘骨架叔膦配体效果优异，区域选择性和立体选择性都很好。产物三氯硅烷用双氧水可以方便地转化成醇并测定 ee 值 (图 2-22)[26]。

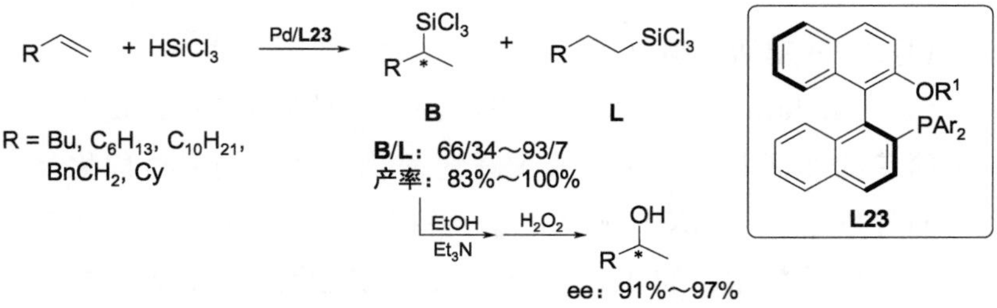

图 2-22 MOP 在烯烃氢硅化反应中的应用

(2) 酮的 α-芳基化反应

2016 年，Zhou、Hartwig 等人报道了 α-氟代茚酮的 α-芳基化反应。利用 **L24** 作为手性配体，消旋的 α-氟代茚酮与芳基溴或芳基三氟甲磺酸酯在钯催化的条件下得到含氟手性茚酮，产率和对映选择性良好 (图 2-23)[28]。

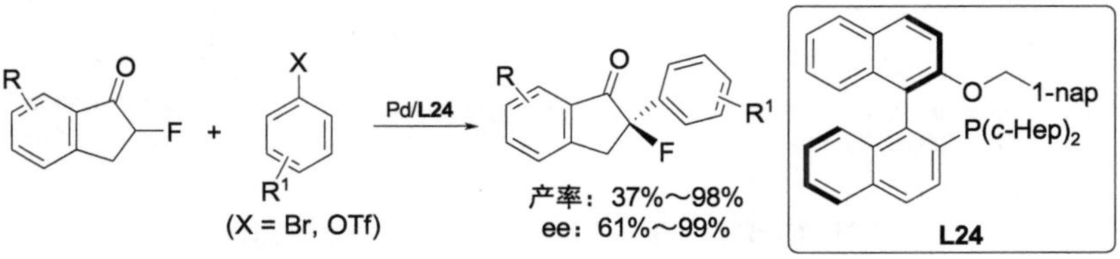

图 2-23 MOP 在酮的 α-芳基化反应中的应用

(3) 酮的加成反应

Hayashi 课题组报道了铑催化的靛红中的酮羰基与芳基或烯基硼酸的加成反应。在这个反应中，**L25** 表现出良好的催化能力和选择性，反应的适用性较广 (图 2-24)[29]。

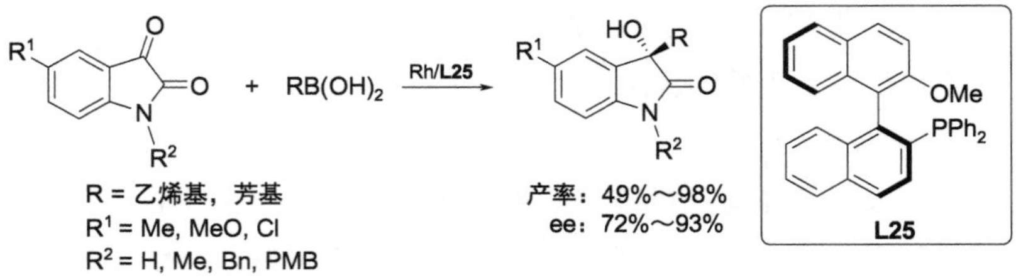

图 2-24 MOP 在酮的加成反应中的应用

**(4) C—H 键活化反应**

Baudoin 课题组在探索钯催化的去对称化 C—H 键活化反应时发现简单的非环联萘骨架叔膦配体对卤代芳香胺的去对称化反应的对映选择性很差。在 MOP 配体的结构基础上，作者设计了含有一个羧基的非环联萘骨架叔膦配体 **L26**。此衍生物作为一个双功能配体能够显著地提高反应对映选择性。羧基的作用可能是诱导反应的立体选择性的同时以协同金属化脱质子机理的方式实现 C—H 键的断裂。由于羧基和磷原子在一个分子中，反应中间体的构象能够被有效控制，从而实现高对映选择性的反应（图 2-25）[30]。

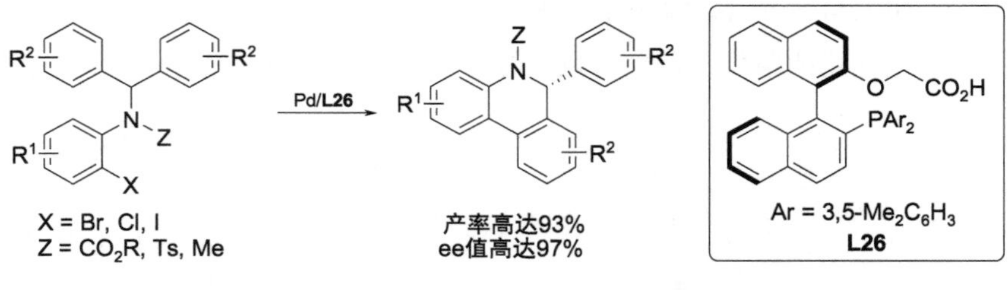

图 2-25 MOP 类含羧酸膦配体在 C—H 键活化反应中的应用

## 2.1.4 磷杂二氢苯并呋喃骨架叔膦配体

汤文军课题组近年来开发了一系列磷杂二氢苯并呋喃骨架叔膦配体 **L27**[31]。此类配体具有一些优势：大部分配体是在空气中稳定的固体，高温下也不发生消旋；此类配体富电子的特点使其能够活化惰性化学键；配体的空间位阻和电子效应可以通过改变取代基进行有效调控（图 2-26）。

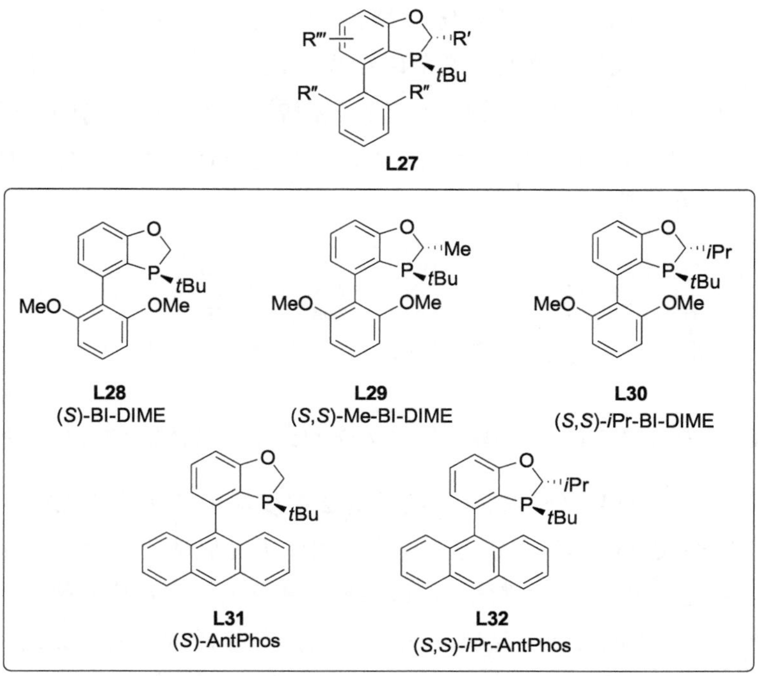

图 2-26 磷杂二氢苯并呋喃骨架叔膦配体

磷杂二氢苯并呋喃骨架叔膦配体的合成路线中的重要中间体是手性氧磷化合物。此消旋中间体可以从甲基二氯化磷或者叔丁基二氯化磷出发经过多步反应得到。之后用拆分的方法可以得到 ee > 99.5% 的中间体手性氧磷化合物。整个反应路线可以放大到千克级（图 2-27）。

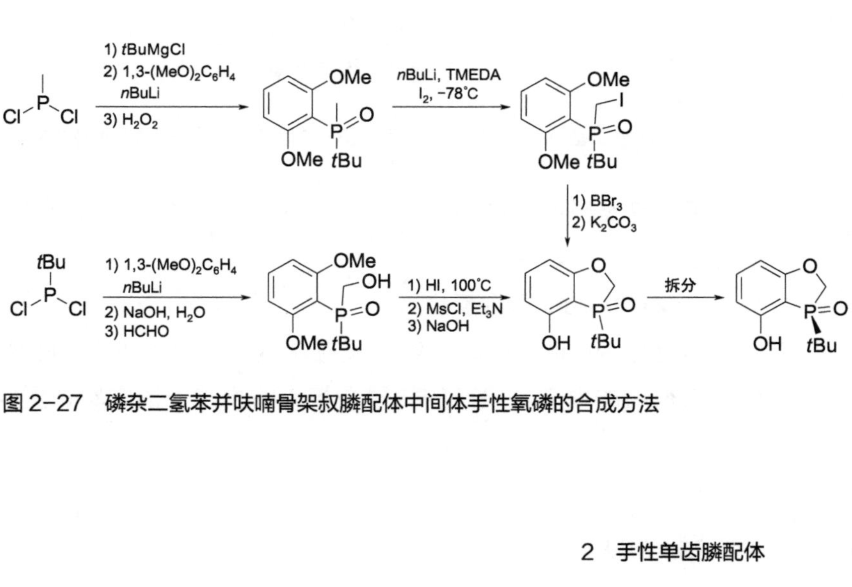

图 2-27 磷杂二氢苯并呋喃骨架叔膦配体中间体手性氧磷的合成方法

从中间体手性氧膦化合物出发，可以方便地得到 3-叔丁基-4-芳基-2,3-二氢苯并[d][1,3]氧杂磷杂戊环配体。合成方法：将中间体手性氧膦化合物的三氟甲磺酸酯与芳基硼酸进行 Suzuki-Miyaura 交叉偶联反应，再将其与聚甲基氢硅氧烷（PMHS）和 Ti(O$i$Pr)$_4$ 进行还原反应生成（图 2-28）[32]。

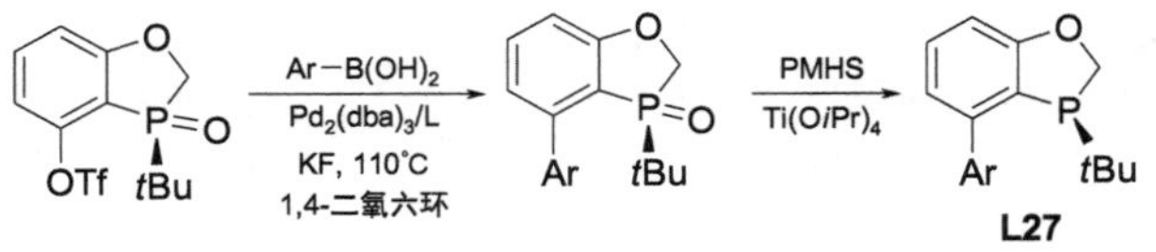

图 2-28　3-叔丁基-4-芳基-2,3-二氢苯并[d][1,3]氧杂磷杂戊环配体 L27 的合成方法

也可以很方便地制备 3-叔丁基-2-R′基-4-R 基-2,3-二氢苯并[d][1,3]氧杂磷杂戊环配体。合成方法：将中间体手性氧膦化合物中的 4-羟基进行官能化反应（羟基的醚化反应或将羟基转化为三氟甲磺酸酯与芳基硼酸进行 Suzuki-Miyaura 交叉偶联反应），然后与格氏试剂或 MeI 进行反应，再将其与聚甲基氢硅氧烷和 Ti(O$i$Pr)$_4$ 进行还原反应生成（图 2-29）[33]。

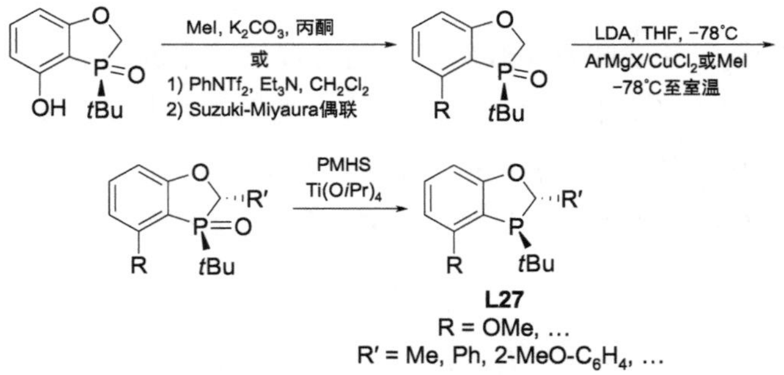

图 2-29　3-叔丁基-2-R′基-4-R 基-2,3-二氢苯并[d][1,3]氧杂磷杂戊环配体的合成方法

目前这类磷手性中心配体已经在多种金属催化反应中显示出较好的催化效果。举例如下：

(1) Suzuki-Miyaura 偶联反应

不对称 Suzuki-Miyaura 偶联反应是构建轴手性联萘的有效方法，然而位阻很大的轴手性联萘的构建是极大的挑战。汤文军课题组发现他们开发的 NitinPhos **L33** 对挑战性的四邻位取代的轴手性联萘的构建具有优异的立体选择性(图 2-30)[34]。

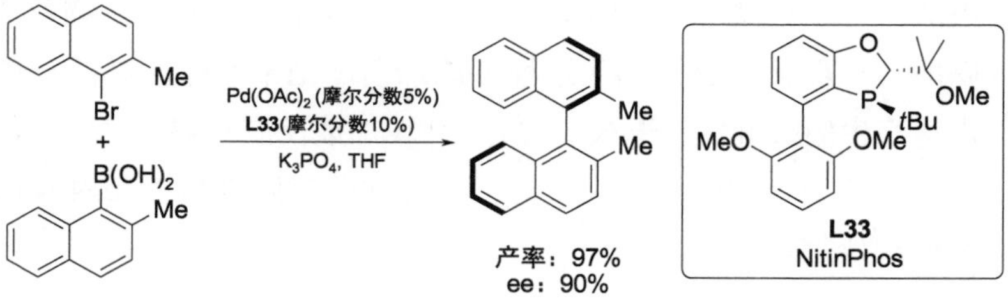

图 2-30 NitinPhos 在不对称 Suzuki–Miyaura 偶联反应中的应用

(2) 醛和硼酸的亲核加成反应

在钌催化的芳香醛和芳基硼酸的亲核加成反应中，磷杂二氢苯并呋喃骨架叔膦配体 **L34** 显示了良好的选择性(图 2-31)。例如 1-萘醛与苯基硼酸反应得到手性仲醇，产率为 96%，ee 值为 87%[35]。

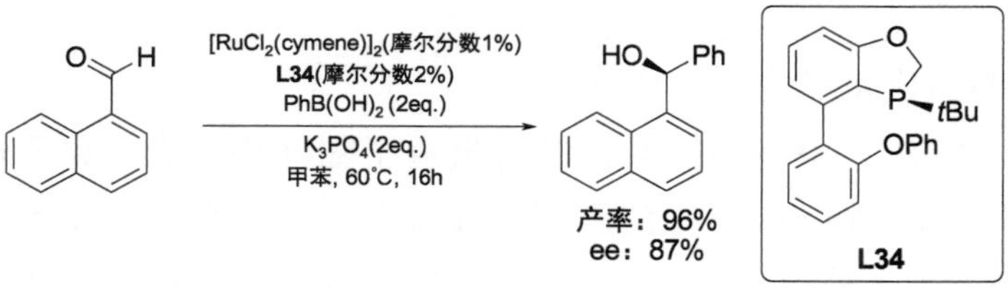

图 2-31 磷杂二氢苯并呋喃骨架叔膦配体在芳香醛和芳基硼酸的亲核加成反应中的应用

(3) 酮的 α-芳基化反应

磷杂二氢苯并呋喃骨架叔膦配体在酮的 α-芳基化反应中也显示了优异的性能。例如汤文军课题组在 (−)-Corynoline 的合成中的 α-芳基化反应步骤里应用 (R)-BI-DIME **L28** 为手性配体，产物 ee 值达到 92%(图 2-32)[36]。

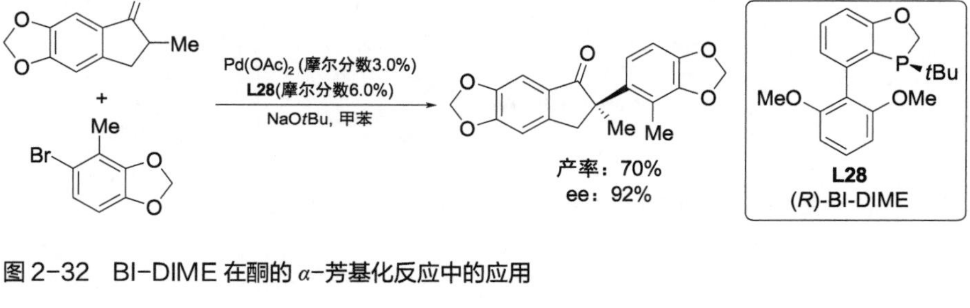

图 2-32 BI-DIME 在酮的 α-芳基化反应中的应用

### (4) 钯催化 [4+2] 环加成反应

邓卫平和汤文军等报道了钯催化的脱羧 [4+2] 环加成反应合成二氢喹啉酮。在手性膦配体 BI-DIME **L28** 的诱导下，消旋的底物乙烯基苯并噁嗪酮与钯催化剂进行脱羧过程形成烯丙基钯中间体，进而与羧酸和酰氯现场产生的酸酐发生形式上的 [4+2] 环加成反应，得到产物二氢喹啉酮并一步构建两个相邻的手性中心，反应的非对映选择性和对映选择性均非常优秀 (图 2-33)[37]。

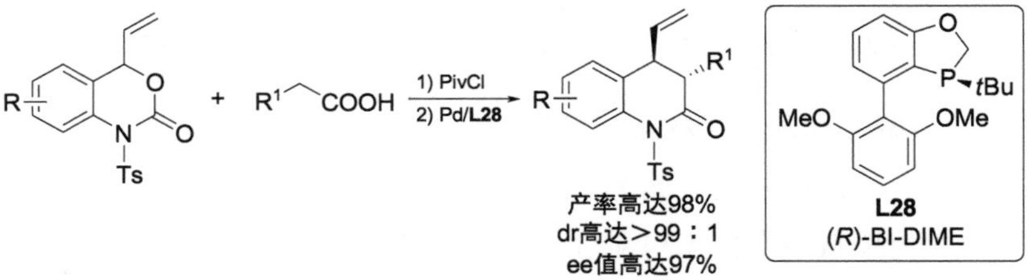

图 2-33 BI-DIME 在钯催化 [4+2] 环加成反应中的应用

### (5) 串联碳钯化-羰基化反应

2021 年，朱强和罗爽课题组报道了钯催化烯烃的串联碳钯化-羰基化反应。反应从简单的含有碘苯和碳-碳双键的联苯衍生物出发，经过 7-*exo* （外型）-trig 模式的分子内碳钯化得到含七元环的烷基钯中间体。在这个过程中 (*S*)-AntPhos 配体 **L31** 能够同时控制联苯部分的轴手性和碳中心手性。随后，烷基钯中间体与一氧化碳和醇或苯胺反应得到产物。理论计算得知轴手性转变造成的非对映异构体之间的能量差达到 2.8 kcal/mol (1cal=4.18J)，与实验观察到的非对映选择性相符合 (图 2-34)[38]。

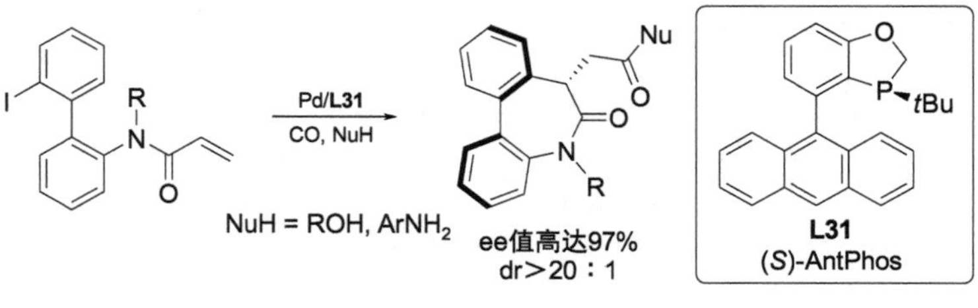

图 2-34 AntPhos 在串联碳钯化-羰基化反应中的应用

## 2.2 亚磷酸衍生物单磷配体

在不对称金属催化研究早期的二三十年里，人们的注意力都集中在开发含有三个磷-碳键的手性叔膦配体上，而含有磷-氧键、磷-氮键的化合物长期被忽略。直到近二十几年，一些手性亚磷酸酯和亚磷酰胺酯被发现具有巨大的催化潜力后，人们才认识到这一大类亚磷酸衍生物是值得开发的一类磷配体。相比叔膦配体，亚磷酸衍生物配体具有一些明显优势：①磷-杂键比磷-碳键的形成要容易得多，因此亚磷酸衍生物配体非常容易制备，亚磷酸衍生物大多数可以从手性醇、酚、胺等简单易得的手性物质经过简单的步骤合成，并且无需拆分；②由于含有磷-氧键或磷-氮键，此类磷配体不易氧化，因而具有较好的稳定性；③氧或氮原子的诱导效应使得这类配体具有较强的π-受体能力，因而能够稳定低氧化态金属。

含有磷-氧、磷-氮键的衍生物可以根据与磷相连的原子种类不同分为亚磷酸衍生物、亚膦酸衍生物和亚次膦酸衍生物三大类（八小类）（图2-35）。

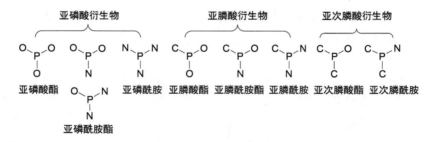

图 2-35　亚磷酸衍生物、亚膦酸衍生物、亚次膦酸衍生物分类

经过近二十多年的研究，人们发现亚磷酸衍生物，特别是亚磷酰胺酯类衍生物具有极为广泛的催化能力。亚磷酸衍生物的常用合成方法之一是用三氯化磷或六甲基亚磷酰三胺与手性二醇或二酚、二胺或氨基醇反应，之后再与醇或胺反应制备。另一个常用方法是将反应次序颠倒，先与醇或胺反应，再与手性二醇或二酚、二胺或氨基醇反应（图2-36）。

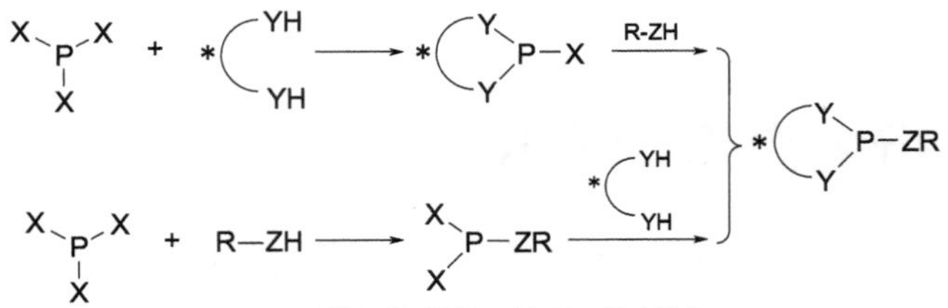


图 2-36　亚磷酸衍生物单磷配体的合成方法

按照此类配体的骨架类型，可分为联萘二酚、氢化联萘二酚、联苯二酚、螺环二酚、酒石酸衍生的二醇等。下面按照骨架类型对这类配体，尤其是对应用广泛的亚磷酰胺酯类配体进行介绍。

## 2.2.1　联萘二酚骨架

1994 年，Feringa 课题组制备了亚磷酰胺酯 MonoPhos **L35**[39]，之后又在 MonoPhos 的基础上设计了一系列衍生物 **L36**[40-52]。这类配体可以从手性联萘二酚出发，经过与三氯化磷和胺的几步简单反应方便地制备。若将胺换作醇则可以制备亚磷酸酯 **L37**[53,54]。从部分氢化的联萘出发可以合成 $H_8$-联萘二酚骨架的亚磷酰胺酯 **L38**[55,56]。这几类配体已经在多种不对称金属催化反应中表现出很好的活性(图 2-37)。

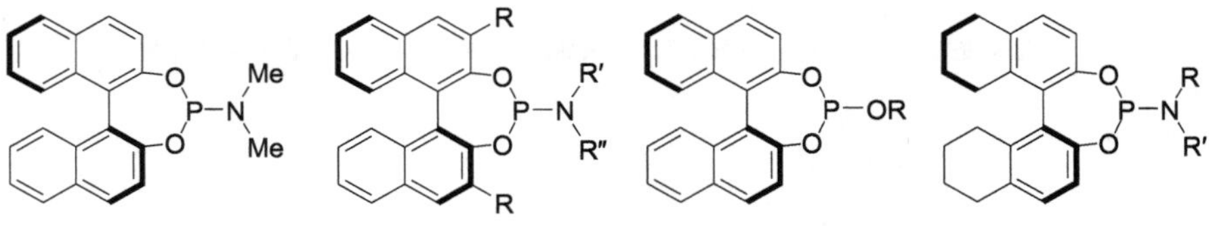

L35 MonoPhos　　L36　　L37　　L38

图 2-37　联萘二酚骨架的亚磷酸衍生物单磷配体

(1) 不饱和羰基化合物的共轭加成反应

最早的联萘二酚骨架的亚磷酰胺酯配体的应用研究是金属试剂与不饱和羰基化合物的共轭加成反应[48]。例如在 Feringa 等报道的不饱和酮与二乙基锌的不对称共轭加成反应中，亚磷酰胺酯配体 **L39** 具有优良的催化能力和对映选择性(图 2-38)[42]。

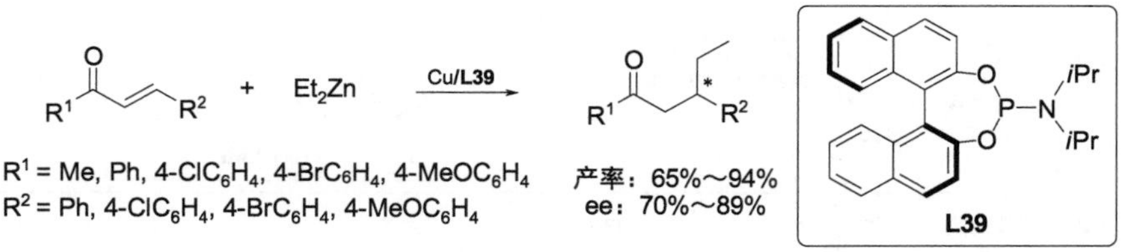

+ Et$_2$Zn
Cu/L39
R$^1$ = Me, Ph, 4-ClC$_6$H$_4$, 4-BrC$_6$H$_4$, 4-MeOC$_6$H$_4$
R$^2$ = Ph, 4-ClC$_6$H$_4$, 4-BrC$_6$H$_4$, 4-MeOC$_6$H$_4$
产率：65%～94%
ee：70%～89%
iPr
iPr
L39


图 2-38　联萘二酚骨架的亚磷酰胺酯配体在不饱和羰基化合物的共轭加成反应中的应用

(2) 烯烃氢化反应

长期以来，人们普遍认为双磷配体是对烯烃不对称氢化反应最有效的配体，然而近二十年亚磷酸衍生物配体也显示出相当好的活性。例如 Feringa 等报道，在脱氢 $\alpha$-氨基酸酯的氢化反应中 MonoPhos（**L35**）表现出优异的催化活性和对映选择性(图 2-39)[56]。

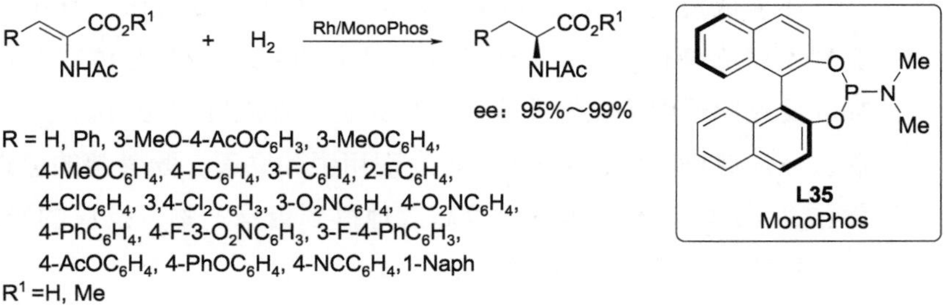

CO$_2$R$^1$
NHAc
+ H$_2$
Rh/MonoPhos
ee：95%～99%
R = H, Ph, 3-MeO-4-AcOC$_6$H$_3$, 3-MeOC$_6$H$_4$, 4-MeOC$_6$H$_4$, 4-FC$_6$H$_4$, 3-FC$_6$H$_4$, 2-FC$_6$H$_4$, 4-ClC$_6$H$_4$, 3,4-Cl$_2$C$_6$H$_3$, 3-O$_2$NC$_6$H$_4$, 4-O$_2$NC$_6$H$_4$, 4-PhC$_6$H$_4$, 4-F-3-O$_2$NC$_6$H$_3$, 3-F-4-PhC$_6$H$_3$, 4-AcOC$_6$H$_4$, 4-PhOC$_6$H$_4$, 4-NCC$_6$H$_4$, 1-Naph
R$^1$ = H, Me
Me
Me
L35
MonoPhos


图 2-39　MonoPhos 在烯烃氢化反应中的应用

(3) 烯丙基化反应

联萘二酚骨架的亚磷酰胺酯配体在铱催化的不对称烯丙基化反应中表现优异(图 2-40)。例如 Carreira 课题组发展的含有一个可配位双键的联萘二酚骨架的亚磷酰胺酯配体 **L40**，在一系列铱催化的不对称烯丙基化反应中都可以高产率、高区域选择性、高对映选择性地得到手性支链烯丙基化产物。亲核试剂可以是氮、氧、硫等杂原子亲核试剂，也可以是较硬的碳金属试剂，还可以是较软的碳亲核试剂[57]。

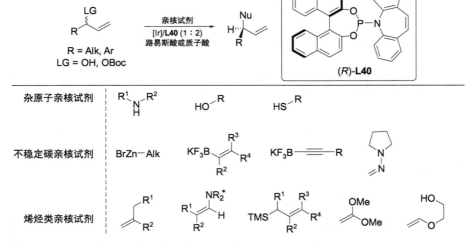

图 2-40 联萘二酚骨架的亚磷酰胺酯配体在铱催化的不对称烯丙基化反应中的应用

### (4) 环加成反应

环加成反应是构建碳环或杂环的一大类重要方法，其中大部分已经实现不对称催化合成。联萘二酚骨架的亚磷酰胺酯配体在多种不对称环加成反应中得到了较好的应用。例如在甲亚胺叶立德与缺电子烯烃的1,3-偶极环加成反应中，Sansano 等发现用银和联萘二酚骨架的亚磷酰胺酯配体 ($S_a,R,R$)-**L41** 催化体系可以高效地得到目标产物，反应的选择性优异，一步构建三个手性中心 (图 2-41)[58]。

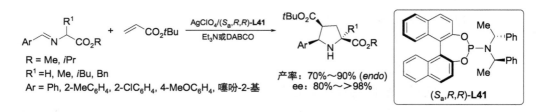

图 2-41 联萘二酚骨架的亚磷酰胺酯配体在不对称环加成反应中的应用

### (5) 烯烃氢硅化反应

烯烃氢硅化是一类非常有用的制备手性醇的反应，有很多高效的催化体系。Johannsent 等发现钯催化苯乙烯与三氯硅烷的氢硅化反应中，联萘二酚骨架的亚磷酰胺酯配体 ($R_a,S,S$)-**L42** 表现了很好的催化活性和选择性(图 2-42)[59]。

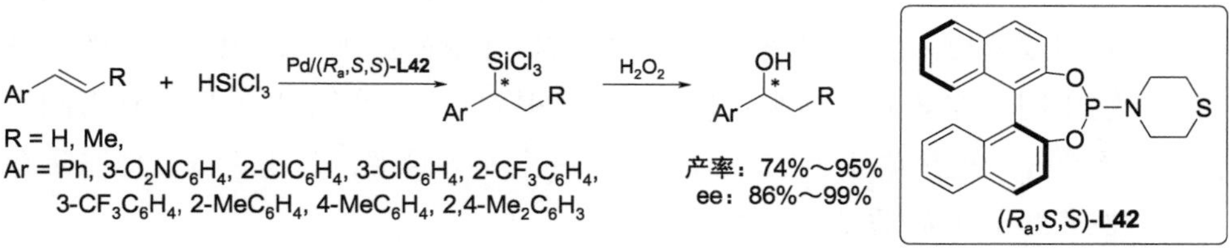

图 2-42 联萘二酚骨架的亚磷酰胺酯配体在不对称烯烃氢硅化反应中的应用

**(6) 酮的 α-芳基化反应**

MonoPhos 的衍生物在酮的 α-芳基化反应中也显示了良好的催化性能。例如在陆平课题组 2018 年报道的环丁酮的分子内 α-芳基化反应中，钯和 MonoPhos 的衍生物 L43 以及吡咯烷组成的催化体系高效地实现了转化，产率和立体选择性良好到优异(图 2-43)[60]。

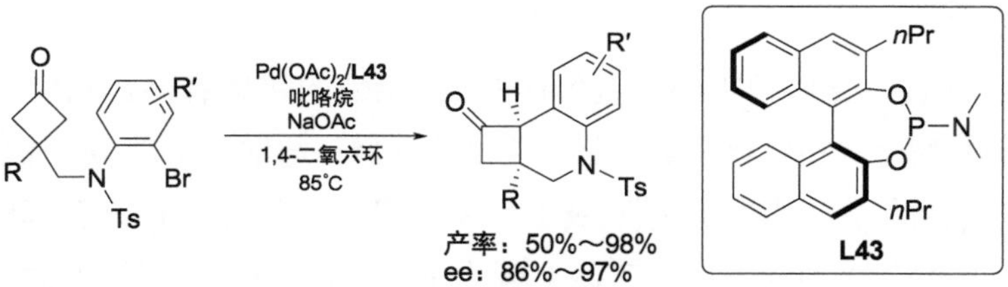

图 2-43 联萘二酚骨架的亚磷酰胺酯配体在酮的 α-芳基化反应中的应用

**(7) 钯催化 [4+2] 环加成反应**

赵洪武课题组报道了钯催化的 [4+2] 环加成反应。从消旋的底物乙烯基苯并噁嗪酮和亚烷基吡唑酮可以简便高对映选择性地合成手性螺吡唑酮。联萘二酚骨架的亚磷酰胺酯配体($R_a,R,R$)-L41 表现了很好的催化活性和选择性，反应的产率、非对映选择性和对映选择性都非常好(图 2-44)[61]。

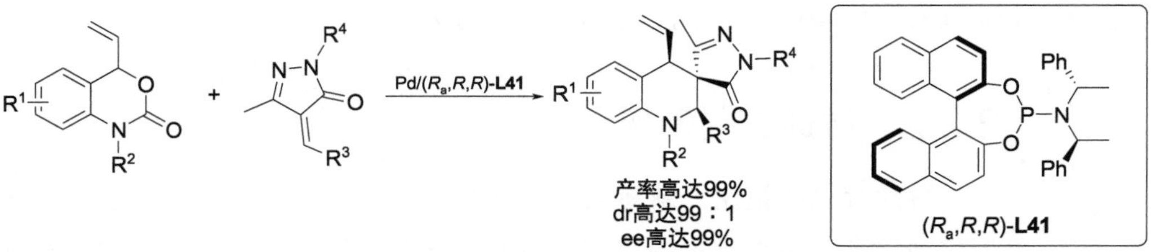

图 2-44 联萘二酚骨架的亚磷酰胺酯配体在钯催化 [4+2] 环加成反应中的应用

(8) 铜催化串联加成反应

林国强课题组报道了铜催化联硼酸酯与分子内联烯环己二烯酮的串联加成反应。反应过程首先是联烯与联硼酸酯的加成过程，之后烯丙基铜中间体与环己二烯酮发生 1,4-加成得到产物。该串联反应顺利得到了具有 3 个连续碳手性中心的六元环并五元环骨架，产率良好，对映选择性良好至优秀（图 2-45）[62]。

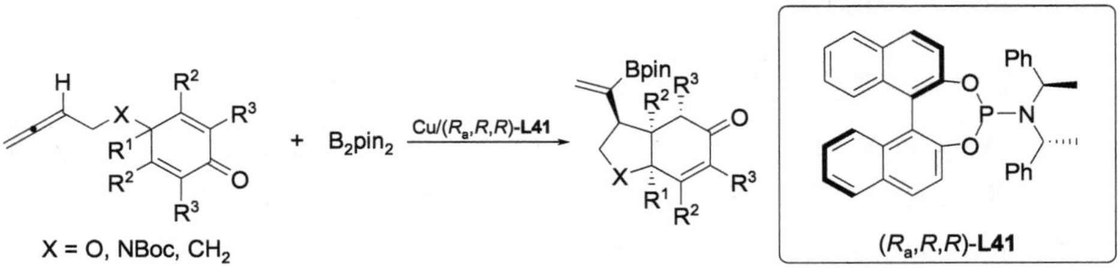

图 2-45　联萘二酚骨架的亚磷酰胺酯配体在铜催化串联加成反应中的应用

(9) 环丙烷化反应

2014 年，张俊良课题组报道了金催化的 1,6-烯炔的炔烃氧化/环丙烷化反应。一价金配合物在氧化剂喹啉氮氧化物的作用下与缺电子碳-碳三键形成金卡宾，之后与分子内的碳-碳双键发生环丙烷化得到五元环并三元环结构。联萘二酚骨架的亚磷酰胺酯配体 **L44** 表现了很好的催化活性和选择性，反应的产率、非对映选择性和对映选择性都非常好（图 2-46）[63]。

图 2-46　联萘二酚骨架的亚磷酰胺酯配体在环丙烷化反应中的应用

## 2.2.2　螺环二酚骨架

与联萘二酚相比，螺环二酚的骨架具有高度刚性的特点，在不对称催化中受到了越来越多的重视。周其林课题组基于螺环二酚的骨架开发了一系列亚磷酰胺酯配体 **L45** 和亚磷酸酯配体 **L46**（图 2-47）[64-66]。这类配体的合成方法与联萘二酚骨架配体类似，从光学纯的螺环二氢茚-1,7-二酚出发与亚磷酰三胺，或与三氯化磷和胺的锂盐反应制备。若将胺换作醇，可以制备亚磷酸酯配体。螺环二酚骨架的亚磷酰胺酯配体和亚磷酸酯配体与联萘二酚骨架的相应配体的应用范围类似，举例如下：

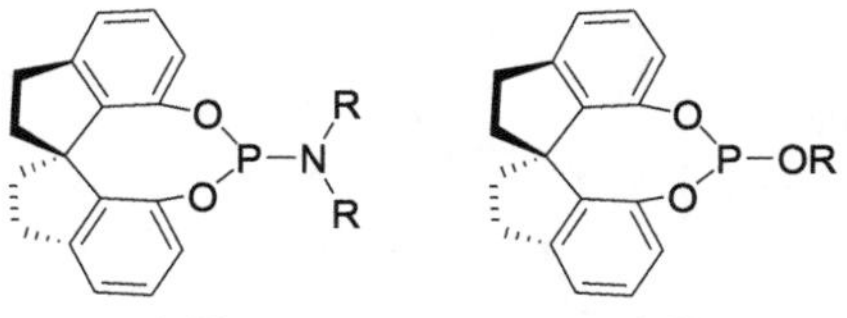

图 2-47　螺环二酚骨架的亚磷酰胺酯配体和亚磷酸酯配体

(1) 共轭加成反应

在铜催化的不饱和羰基化合物与二乙基锌的共轭加成反应中，螺环亚磷酰胺酯配体 (R)-**L47** 表现出较好的催化活性（图 2-48），如下例[65]：

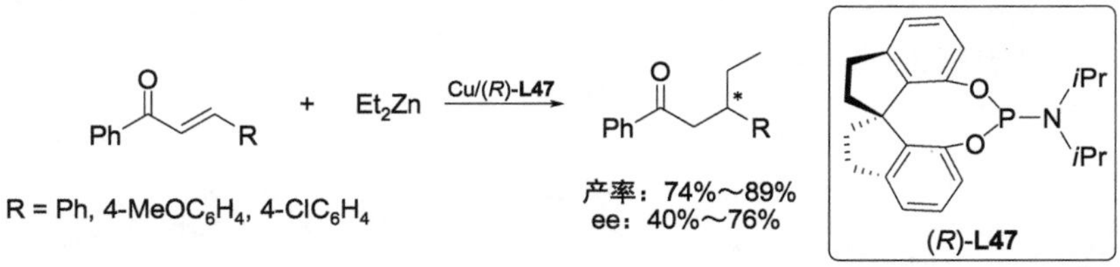

图 2-48　螺环二酚骨架的亚磷酰胺酯配体在共轭加成反应中的应用

(2) 烯丙基化反应

在烯丙基卤代物与二乙基锌试剂的烯丙基化反应中，铜催化剂与螺环亚磷酰胺酯配体 (R,R,R)-**L48** 催化体系具有中等以上的对映选择性（图 2-49）[64]。

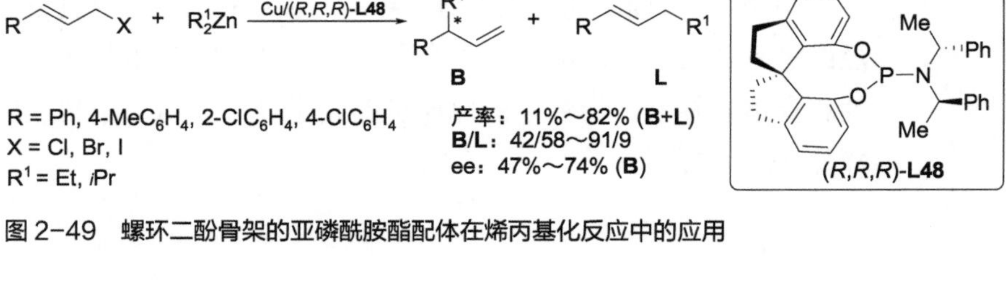

图 2-49　螺环二酚骨架的亚磷酰胺酯配体在烯丙基化反应中的应用

### (3) 烯烃氢化反应

在脱氢 α-氨基酸酯的铑催化氢化反应中，非常多的配体类型都表现出良好效果。其中，螺环亚磷酰胺酯配体 **L49** 也表现优异（图 2-50）[66,67]。

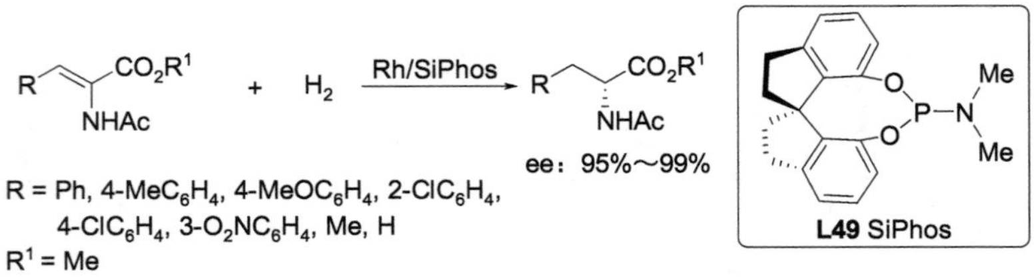

图 2-50　螺环二酚骨架的亚磷酰胺酯配体在烯烃氢化反应中的应用

### (4) 开环反应

氧杂双环烯的去对称开环反应是一类高效的制备多手性中心环状化合物的方法。周其林课题组在此反应中应用螺环亚磷酰胺酯配体 (*S,S,S*)-**L48**，与铜催化剂共同催化，得到很好的反式选择性和较好的对映选择性（图 2-51）[68]。

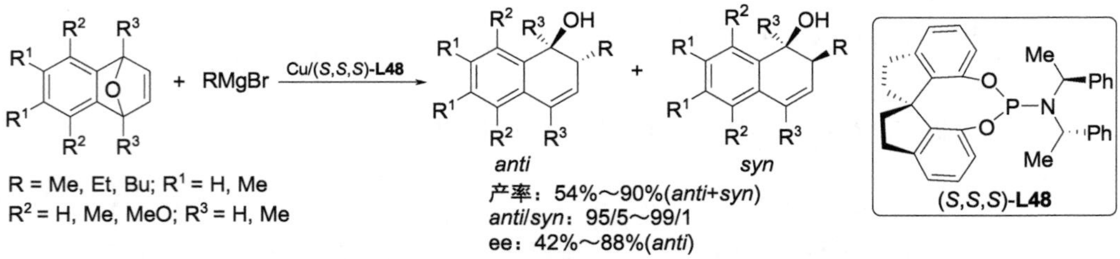

图 2-51　螺环二酚骨架的亚磷酰胺酯配体在开环反应中的应用

### (5) 烯烃的氢酰基化反应

2010 年，Dong 课题组报道了导向的区域和立体选择性的水杨醛和烯烃的氢酰基化反应。螺环亚磷酰胺酯配体 (*R,R,R*)-**L48** 展示了最佳催化活性和立体选择性。烯烃中的硫原子起到了导向作用，控制反应的区域选择性（图 2-52）[69]。

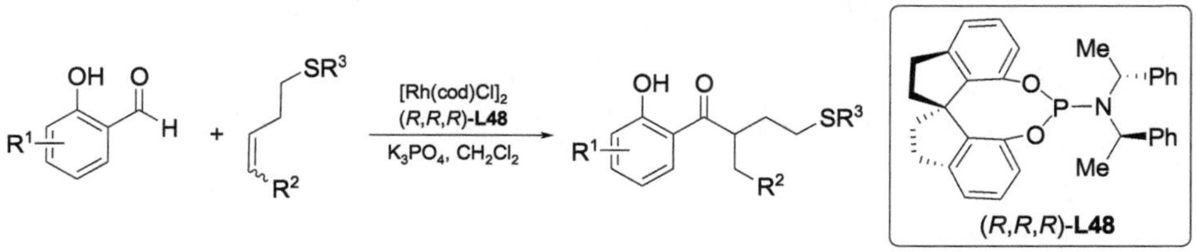

图 2-52　螺环二酚骨架的亚磷酰胺酯配体在烯烃的氢酰基化反应中的应用

**(6) 1,6-二烯环化**

2021 年，周其林课题组报道了镍催化的 1,6-二烯环化反应，形成含有两个手性中心的五元螺环。反应的催化活性物种是含镍氢键的配合物，其中的螺环手性单膦配体 **L50** 能够控制反应的非对映选择性和对映选择性（图 2-53）[70]。

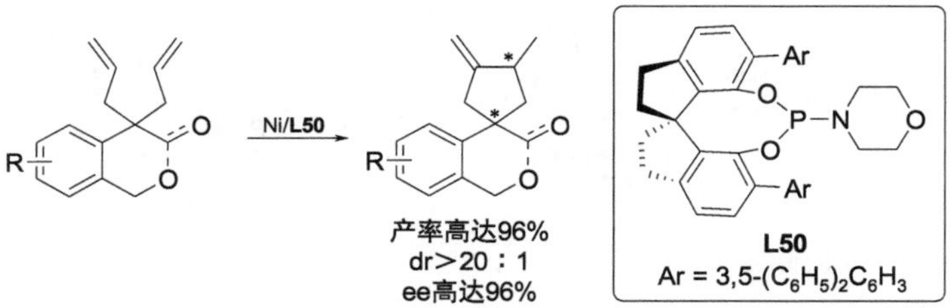

图 2-53　螺环二酚骨架的亚磷酰胺酯配体在 1,6-二烯环化反应中的应用

**(7) Aza-Heck 反应**

2019 年，Bower 课题组报道了钯催化的 Aza-Heck 反应。底物中的氮-氧键与零价钯氧化加成之后与碳-碳双键进行 Heck 反应，在双键插入过程中螺环配体 (*S*,*R*,*R*)-**L48** 有效控制手性中心的构型，得到四氢吡咯衍生物。对六元环底物，反应利用二异丙基取代的 SiPhos 配体 (*S*)-**L47** 也能够得到理想的结果（图 2-54）[71]。

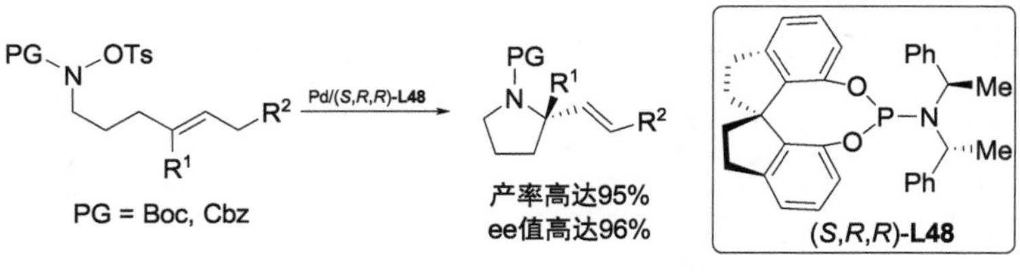

图 2-54

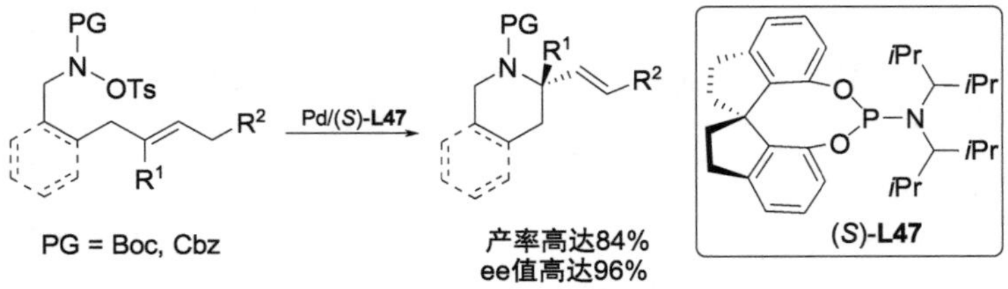

图 2-54 螺环二酚骨架的亚磷酰胺酯配体在 Aza-Heck 反应中的应用

(8) 1,3-二烯的氢磺酰化

2021 年，周其林课题组报道了钯催化 1,3-二烯的氢磺酰化反应。反应从 1,3-二烯和磺酰肼出发，首先产生烯丙基磺酰肼，螺环配体在这个过程中控制手性中心的构型。之后烯丙基磺酰肼在钯催化剂作用下脱去氢气和氮气，最后得到产物手性烯丙基砜，反应的产率和对映选择性均较高（图 2-55）[72]。

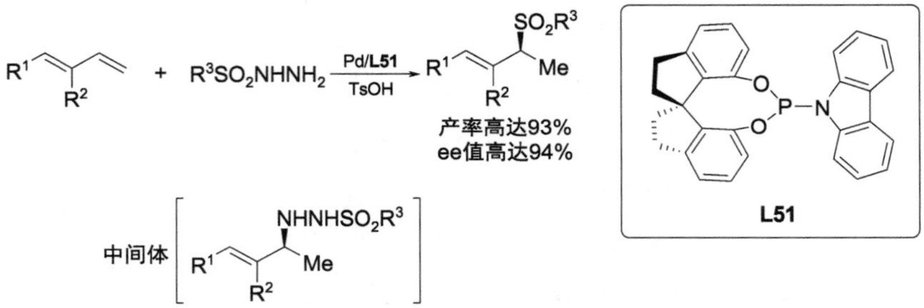

图 2-55 螺环二酚骨架的亚磷酰胺酯配体在 1,3-二烯的氢磺酰化中的应用

## 2.2.3 酒石酸衍生的二醇骨架

从廉价的酒石酸衍生得到的 TADDOL[（2, 2-二甲基-1, 3-二氧戊环-4, 5-二基）双（二苯基甲醇）]类二醇是一类重要的手性骨架。基于 TADDOL 骨架可以制备亚磷酰胺酯 **L52**[73-75] 和亚磷酸酯配体 **L53**（图 2-56）[76,77]。制备方法与联萘二酚骨架或螺环二酚骨架类配体类似。此类配体对空气和水的稳定性较好，目前在不对称金属催化中已得到广泛应用。

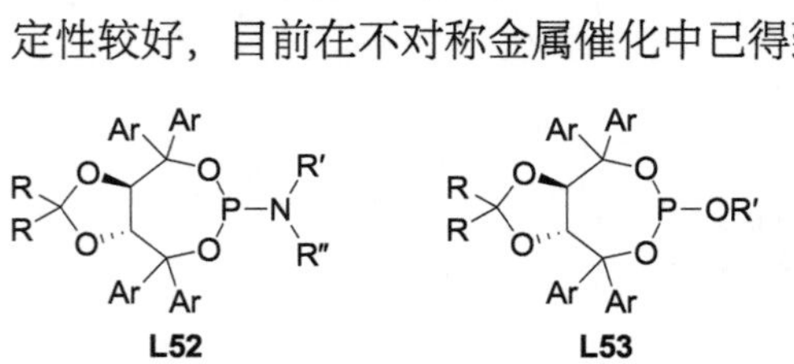

图 2-56 基于 TADDOL 骨架的亚磷酰胺酯和亚磷酸酯配体

(1)烯丙基化反应

在环状烯丙基化试剂与烷基锌试剂的钯催化烯丙基化反应中，TADDOL 骨架的亚磷酰胺酯配体 **L54** 效果良好，以优异的非对映选择性和对映选择性得到目标产物（图 2-57）[78]。

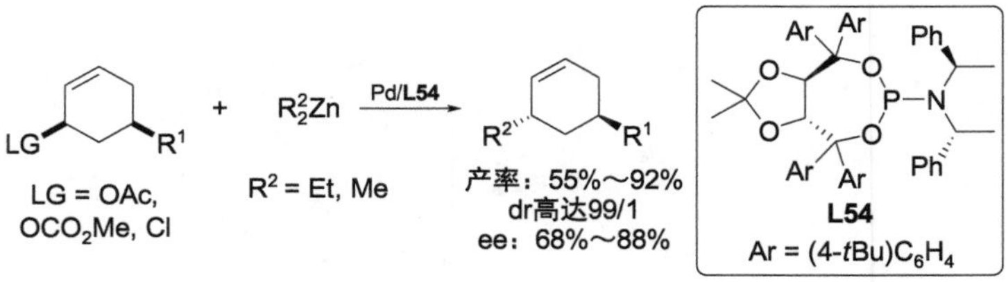

图 2-57　TADDOL 骨架的亚磷酰胺酯配体在钯催化烯丙基化反应中的应用

2014 年，Scheidt 课题组报道了烯丙基醋酸酯分子内的烯丙基化反应。同样，TADDOL 骨架的亚磷酰胺酯配体 **L55** 效果良好，以优异的产率和对映选择性得到目标产物手性 2-芳基苯并吡喃（图 2-58）[79]。

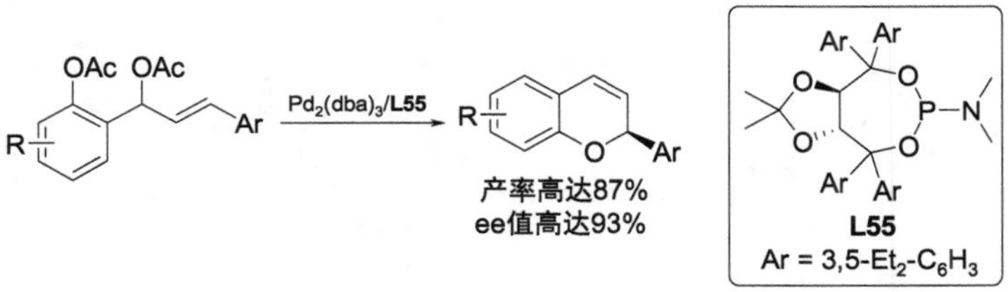

图 2-58　TADDOL 骨架的亚磷酰胺酯配体在分子内烯丙基化反应中的应用

2014 年，Morken 课题组报道了分子内烯丙基硼酸酯和氯代苯的烯丙基化反应。TADDOL 骨架的亚磷酰胺酯配体 **L56** 有较好效果，反应的产率和对映选择性中等到良好（图 2-59）[80]。

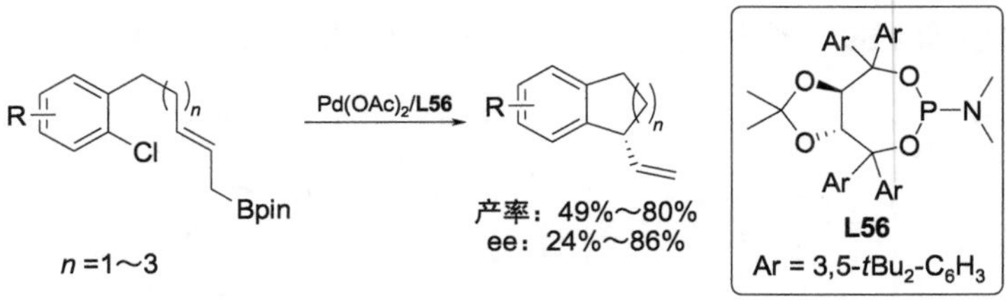

图 2-59　TADDOL 骨架的亚磷酰胺酯配体在钯催化分子内烯丙基化反应中的应用

### (2) 交叉偶联反应

Morken 报道了偕二硼试剂与芳基卤代物的不对称交叉偶联反应。在此反应中，TADDOL 骨架的亚磷酰胺酯配体 **L57** 效果最好，底物适用范围广，产率良好，对映选择性优异（图 2-60）[81]。

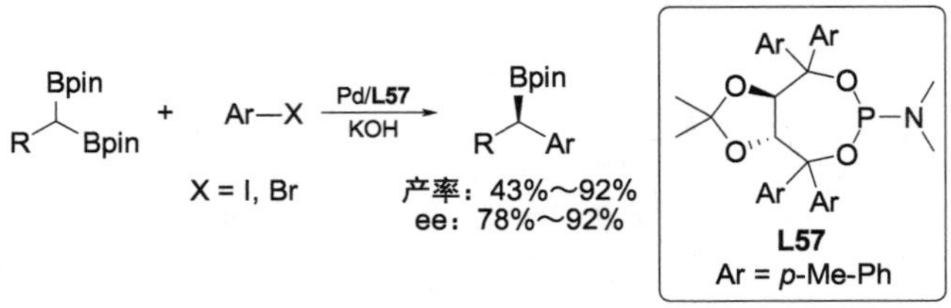

图 2-60　TADDOL 骨架的亚磷酰胺酯配体在交叉偶联反应中的应用

2016 年，顾振华课题组利用溴代萘和磺酰腙的钯催化交叉偶联反应合成了轴手性化合物。TADDOL 骨架的亚磷酰胺酯配体 **L58** 效果优异，反应的产率和对映选择性均优秀（图 2-61）[82]。

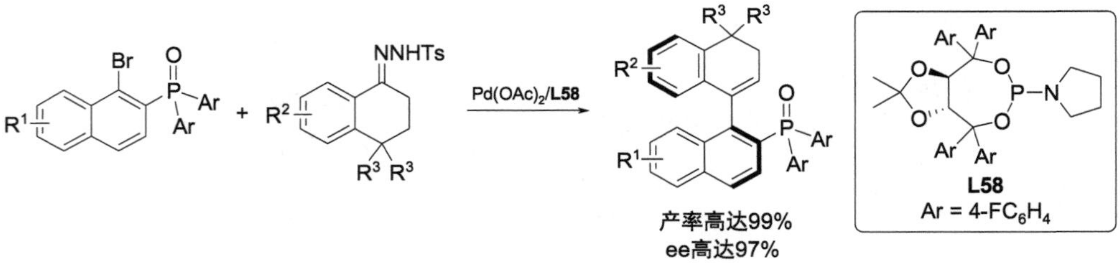

图 2-61　TADDOL 骨架的亚磷酰胺酯配体在溴代萘和磺酰腙的交叉偶联反应中的应用

### (3) 环加成反应

环加成反应类型众多，在有机合成中是一类重要的反应。在一些不对称催化的环加成反应中 TADDOL 骨架的亚磷酰胺酯显示了良好的催化活性。例如在 Fürstner 报道的分子内联烯-烯的 [2+2] 环加成反应中，亚磷酰胺酯 **L59** 与金形成的配合物具有很好的催化效果（图 2-62）[83]。

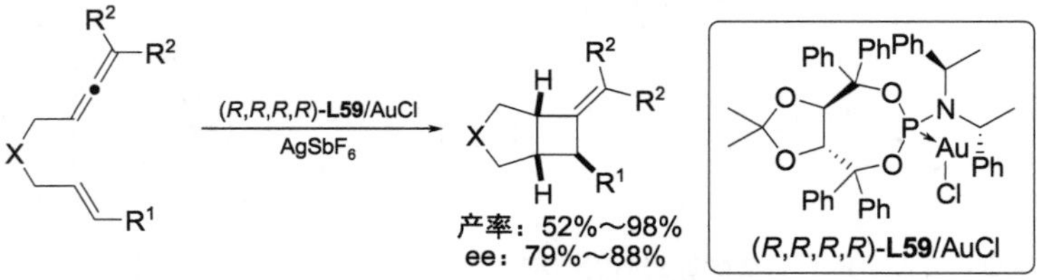

图 2-62　TADDOL 骨架的亚磷酰胺酯配体在 [2+2] 环加成反应中的应用

[4+2]环加成反应中，TADDOL 骨架的亚磷酰胺酯也展现出良好的催化活性。例如 Rovis 在 2011 年报道的烯基亚胺与异氰酸酯的铑催化 [4+2] 环加成反应中，TADDOL 骨架的亚磷酰胺酯 **L60** 被发现是最佳配体，一系列嘧啶酮被高产率、高对映选择性地制备出来（图 2-63）[84]。

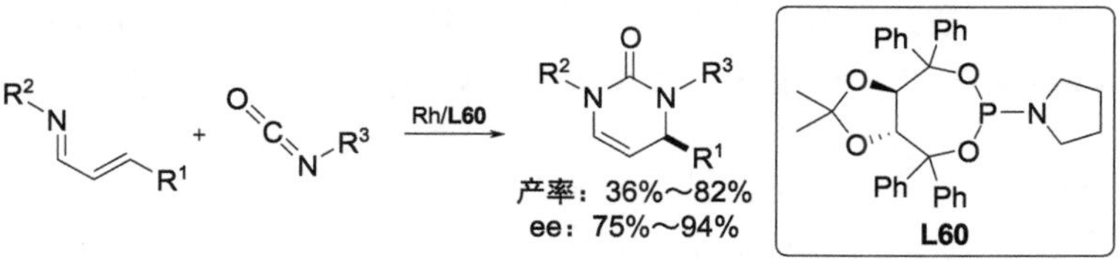

图 2-63　TADDOL 骨架的亚磷酰胺酯配体在 [4+2] 环加成反应中的应用

2013 年，Rovis 课题组报道了铑催化的含有碳-碳双键的异氰酸酯与端炔的 [2+2+2] 反应。异氰酸酯和端炔首先与铑配合物发生环金属化，之后羰基迁移和碳-碳双键插入得到产物。TADDOL 骨架的亚磷酰胺酯 **L61** 对该反应有效。X 射线单晶衍射、核磁和 DFT 计算研究显示五氟苯基与铑之间存在配位作用，使得催化效果有大幅度提高（图 2-64）[85]。

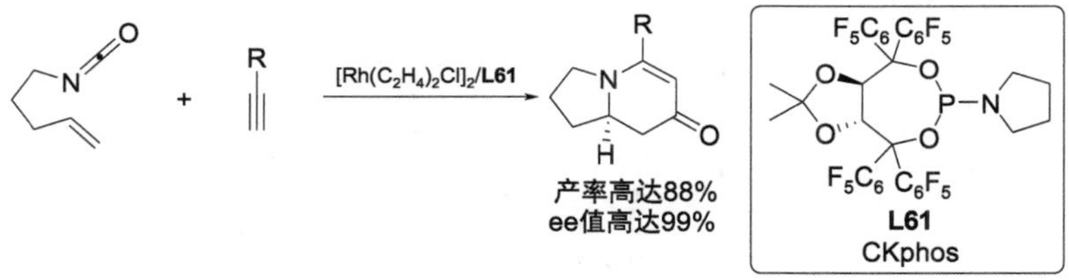

图 2-64　TADDOL 骨架的亚磷酰胺酯配体在 [2+2+2] 反应中的应用

（4）Heck 反应

分子内 Heck 反应是构建环状手性分子的有效手段。例如 Feringa 设计的分子内 Heck 反应构建多环化合物。TADDOL 骨架的亚磷酰胺酯 **L62** 的催化效果很好，产物的 ee 值高达 96%。钯与配体的比例为 2∶1 时效果最佳（图 2-65）[86]。

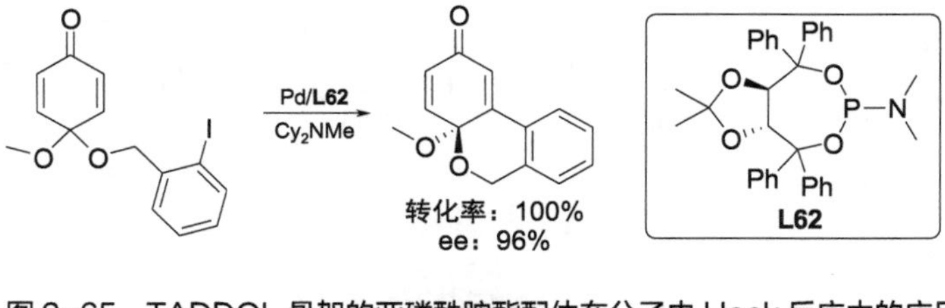

图 2-65　TADDOL 骨架的亚磷酰胺酯配体在分子内 Heck 反应中的应用

### (5) 碳-氢键活化反应

近二十年，碳-氢键活化迎来了研究高潮，甚至不对称碳-氢键活化也已经取得了丰硕的成果。在碳-氢键活化反应中，TADDOL 骨架的亚磷酰胺酯配体 **L63** 得到了很好的应用。例如，Cramer 等在 2009 年报道的烯基三氟甲磺酸酯与分子内的芳基碳-氢键的偶联中发现 TADDOL 骨架的亚磷酰胺酯配体是最佳配体，反应的产率和对映选择性良好到优秀（图 2-66）[87]。

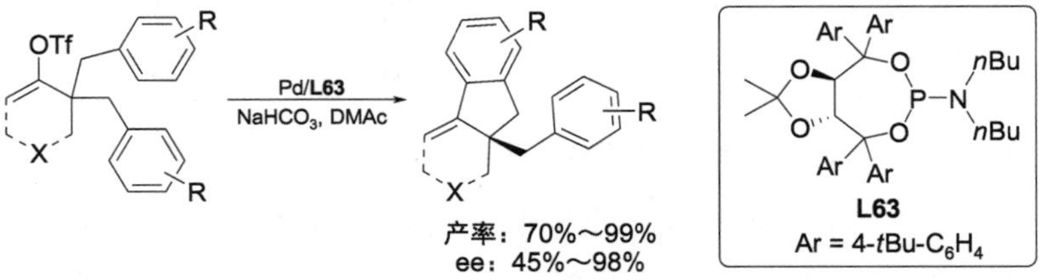

图 2-66　TADDOL 骨架的亚磷酰胺酯配体在碳-氢键活化反应中的应用

2015 年，Cramer 等报道了环丙烷的碳-氢键活化反应。在该反应中，TADDOL 骨架的亚膦酸酯配体 **L64** 是最佳配体，反应的产率和对映选择性良好到优秀（图 2-67）[88]。

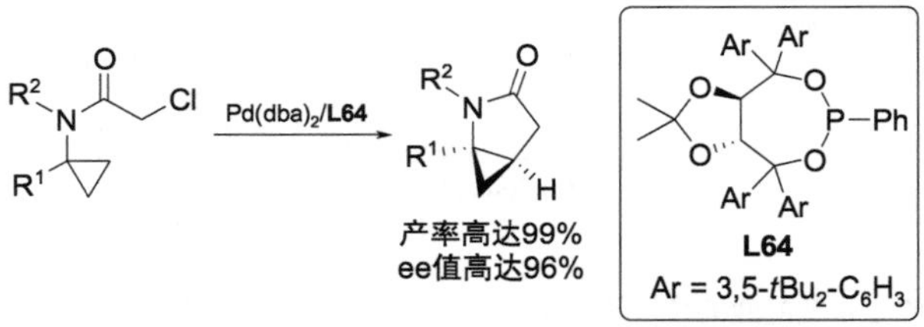

图 2-67　TADDOL 骨架的亚膦酸酯配体在环丙烷的碳-氢键活化反应中的应用

2017 年，顾振华课题组报道了吲哚的分子内碳-氢键活化反应，由于吲哚 3-位的大位阻萘基的阻碍，产物存在轴手性。TADDOL 骨架的亚磷酰胺酯配体 **L65** 能够很好地控制轴手性，反应的产率和对映选择性良好到优秀（图 2-68）[89]。

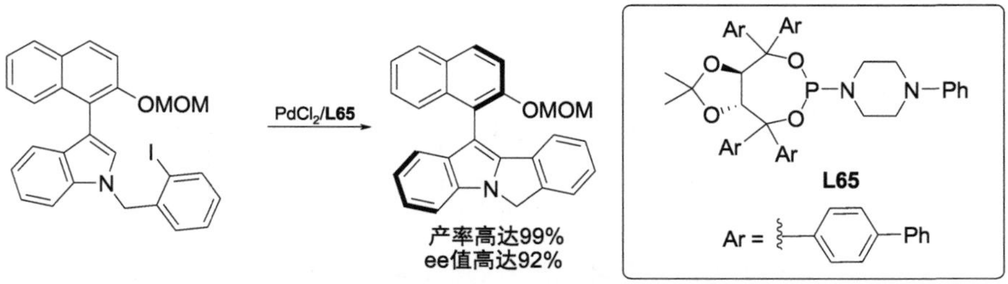

图 2-68　TADDOL 骨架的亚磷酰胺酯配体在轴手性构建反应中的应用

（6）碳-碳键活化反应

碳-碳键活化一直是有机化学的一个困难的挑战。近几年高张力小环分子的碳-碳键活化取得了令人瞩目的进展。在这类碳-碳键活化反应中 TADDOL 骨架的亚磷酰胺酯配体得到了很好的应用。例如 2019 年，徐利文和曹建等实现了一系列钯催化环丁酮的不对称开环-碳-碳/碳-杂键形成反应[90-93]。使用过渡金属钯作为催化剂和 TADDOL 衍生的亚磷酰胺酯类配体实现了碘迁移反应、环丙烷化反应、与苯硼酸和末端炔烃的交叉偶联以及 Cacchi 反应等，成功构建了茚酮类化合物。TADDOL 衍生的亚磷酰胺酯类配体 **L66** ～ **L69** 在这类反应中表现了优异的催化能力和选择性（图 2-69）。

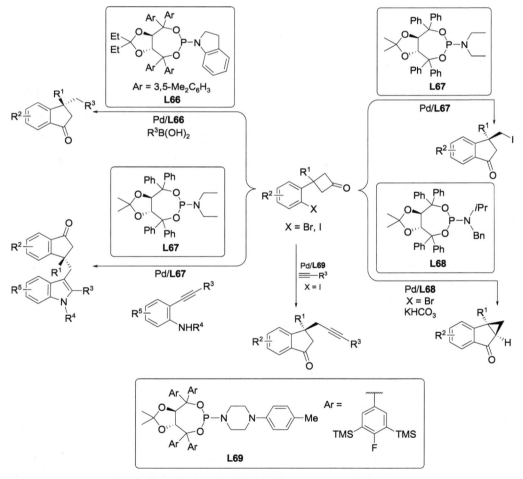

图 2-69　TADDOL 骨架的亚磷酰胺酯配体在环丁酮的碳-碳键活化反应中的应用

2019 年，顾振华课题组报道了钯催化二苯并环戊醇的碳-碳键断裂反应。TADDOL 衍生的亚磷酰胺酯类配体 **L70** 有效地实现了选择性的碳-碳键断裂，经过环戊醇的开环得到轴手性联苯衍生物(图 2-70)[94]。

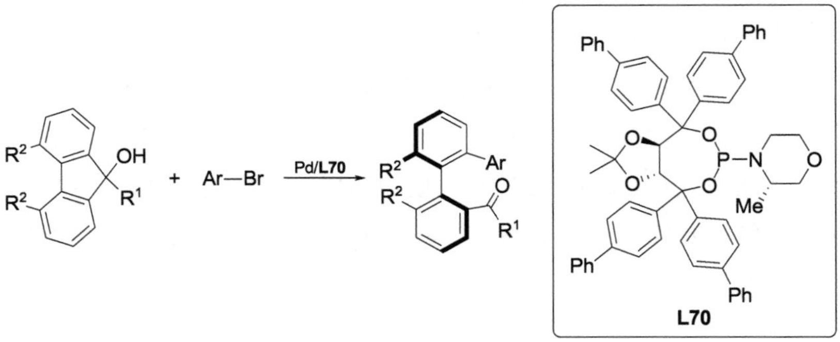

图 2-70　TADDOL 骨架的亚磷酰胺酯配体在二苯并环戊醇的碳-碳键活化反应中的应用

(7) 烯烃的氢硅化反应

2018 年，Morken 课题组报道了烯基硼酸酯的铂催化氢硅化反应。硅基区域选择性地加成在靠近硼原子的位置得到同位硅基硼酸酯。烯烃的取代基可以是含有多种官能团(例如：卤素、硅基、酯基、酮基、烯基、醚基、吡啶基等)的烷基，反应的产率良好，对映选择性优秀(图 2-71)[95]。

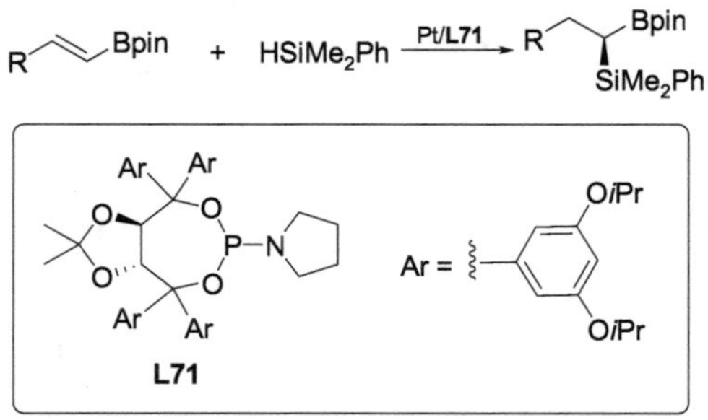

图 2-71　TADDOL 骨架的亚磷酰胺酯配体在烯烃的氢硅化反应中的应用

(8) 还原 Heck 反应

还原 Heck 反应是构建手性环状分子的常用策略。2017 年，de Vries 和 Minnaard 课题组报道了利用 TADDOL 衍生的亚磷酰胺酯类配体诱导的钯催化还原 Heck 反应。在该反应中，二环己基甲基胺（Cy$_2$NMe）作为还原剂提供氢，产率和对映选择性中等到良好（图 2-72）[96]。

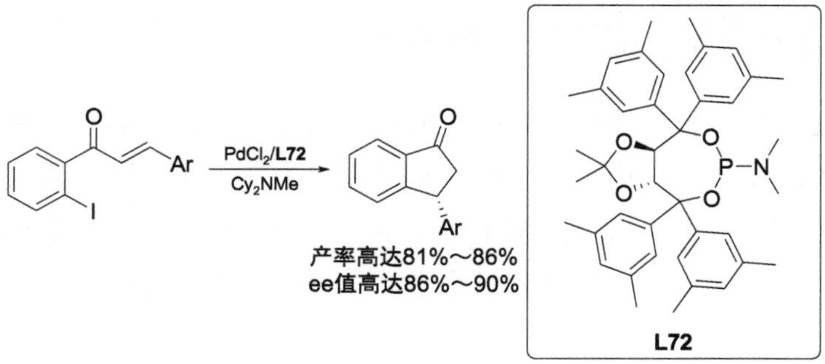

图 2-72　TADDOL 骨架的亚磷酰胺酯配体在还原 Heck 反应中的应用

(9) Si—C 键偶联反应

近年来手性硅化合物的合成受到越来越多的重视，硅-碳键偶联反应是构建手性硅的方法之一。例如 2012 年，Yamanoi 和 Nishihara 等报道了二氢硅烷与碘代芳烃的钯催化偶联反应，反应得到手性叔硅烷，产率和对映选择性中等 [图 2-73(a)][97]。2016 年，徐利文课题组对苯基二氢硅烷与碘代芳烃的钯催化硅-碳键偶联反应进行了研究。TADDOL 衍生的亚磷酰胺酯类配体 **L73** 对该反应有效，产率和对映选择性仅为中等，显示了硅手性构建的难度 [图 2-73(b)][98]。

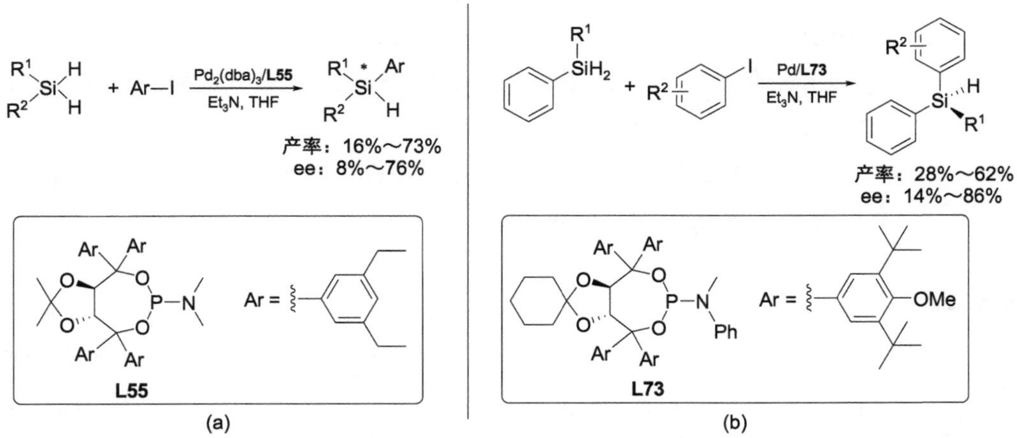

图 2-73　TADDOL 骨架的亚磷酰胺酯配体在 Si—C 键偶联反应中的应用

## 2.3 单膦手性配体的小结及延伸阅读

单膦手性配体是所有手性膦配体中发展最早也是当今仍在不断发展的一类骨架，尤其是磷手性中心的配体，至今方兴未艾，关于这方面的中文综述性文献不断涌现，限于篇幅本章将不再赘述。读者若有兴趣深入学习，可进一步学习学术期刊《有机化学》上先后发表的以下 8 篇综述性文章：

① 徐利文，夏春谷，孙伟，等．手性膦配体的合成及其在不对称催化反应中的应用研究进展．有机化学，2003, 23 (9): 919-932.

② 宋庆宝，东宇．二茂铁手性膦配体研究的一些新进展．有机化学，2007, 27 (01): 66-71.

③ 贾肖飞，王正，夏春谷，等．铑催化的烯烃不对称氢甲酰化反应研究进展．有机化学，2013, 33 (07): 1369-1381.

④ 徐广庆，赵庆，汤文军．发展高效的不对称 Suzuki-Miyaura 偶联反应及其合成应用．有机化学，2014, 34 (10): 1919-1940.

⑤ 赵文献，杨代月，张玉华．高效手性膦配体的研究进展．有机化学，2016, 36 (10): 2301-2316.

⑥ 张凤，刘祥华，刘玮，等．钯-单膦催化剂在烯烃不对称硅氢化反应中的应用．有机化学，2017, 37 (10): 2555-2568.

⑦ 李帅龙，李庄星，由才，等．不对称氢甲酰化反应研究进展．有机化学，2019, 39 (6): 1568-1582.

⑧ 许容华，杨贺，汤文军．P-手性膦配体促进的手性药物高效合成．有机化学，2020, 40 (6): 1409-1422.

**参考文献**

[1] Knowles W S, Sabacky M J, Vineyard B D. Catalytic asymmetric hydrogenation. Journal of the Chemical Society, Chemical Communications, 1972, (1): 10-11.

[2] Morrison J D, Burnett R E, Aguiar A M, et al. Asymmetric homogeneous hydrogenation with Rhodium(Ⅰ) complexes of chiral phosphines. Journal of the American Chemical Society, 1971, 93(5): 1301-1303.

[3] Guillen F, Fiaud J C. Enantiomerically pure 1,2,5-triphenylphospholane through the synthesis and resolution of the chiral trans-(2,5)-diphenylphospholanic acid. Tetrahedron Lett, 1999, 40(15): 2939-

2942.

[4] Gladiali S, Dore A, Fabbri D, et al. Novel atropisomeric phosphorus ligands: 4,5-dihydro-3H-dinaphtho[2,1-c;1',2'-e] phosphepine derivatives. Tetrahedron: Asymmetry, 1994, 5(4): 511-514.

[5] Junge K, Oehme G, Monsees A, et al. Synthesis of new chiral monodentate phosphines and their use in asymmetric hydrogenation. Tetrahedron Lett, 2002, 43(28): 4977-7980.

[6] Junge K, Hagemann B, Enthaler S, et al, Synthesis of chiral monodentate binaphthophosphepine ligands and their application in asymmetric hydrogenations. Tetrahedron: Asymmetry, 2004, 15(17): 2621-2631.

[7] Chi Y X, Zhang X M. Synthesis of novel chiral binaphthyl phosphorus ligands and their applications in Rh-catalyzed asymmetric hydrogenation. Tetrahedron Lett, 2002, 43(27): 4849-4852.

[8] Kasak P, Mereiter K, Widhalm M. Chiral α-branched mono phosphine auxiliaries, reversal of sense of asymmetric induction upon substitution. Tetrahedron: Asymmetry, 2005: 16(20), 3416-3426.

[9] Enthaler S, Erre G, Junge K, et al. Enantioselective rhodium-catalyzed hydrogenation of enol carbamates in the presence of monodentate phosphines. Tetrahedron: Asymmetry, 2007, 18(11): 1288-1298.

[10] Junge K, Hagemann B, Enthaler S, et al. Enantioselective hydrogenation of β-ketoesters with monodentate ligands. Angewandte Chemie International Edition, 2004, 43(38): 5066-5069.

[11] Hagemann B, Junge K, Enthaler S, et al. A general method for the enantioselective hydrogenation of β-ketoesters using monodentate binaphthophosphepine ligands. Advanced Synthesis & Catalysis, 2005, 347(15): 1978-1986.

[12] Junge K, Wendt B, Addis D, et al. Coppercatalyzed enantioselective hydrogenation of ketones. Chem. Eur. J., 2011, 17(1), 101- 105.

[13] Alberico E, Nieddu I, Taras R, et al. Expanding the scope of atropisomeric monodentate P-Donor ligands in asymmetric catalysis: hydrogen-transfer reduction of α,β-unsaturated acid derivatives by Rhodium/Ph-binepine catalysts. Helvetica Chimica Acta, 2006, 89(8): 1716-1729.

[14] Holstein P M, Vogler M, Larini P, et al. Efficient $Pd^0$-catalyzed asymmetric activation of primary and secondary C–H bonds enabled by modular binepine ligands and carbonate bases. ACS Catalysis, 2015, 5(7): 4300-4308.

[15] Martin N, Pierre C, Davi M, et al. Diastereo- and enantioselective intramolecular $C(sp^3)$-H arylation for the synthesis of fused cyclopentanes. Chem. Eur. J. 2012, 18(15): 4480-4484.

[16] Franzoni I, Guénée L, Mazet C. Chiral monodentate phosphine ligands for the enantioselective α and γ-arylation of aldehydes. Tetrahedron, 2014, 70(27-28): 4181-4190.

[17] Zhang W, Zhu S F, Qiao X C, et al. Highly enantioselective copper-catalyzed ring opening of oxabicyclic alkenes with Grignard reagents. Chemistry-An Asian Journal, 2008, 3(12): 2105-2111.

[18] Zhu S F, Yang Y, Wang L X, et al. Synthesis and application of chiral spiro pholane ligand in Pd-catalyzed asymmetric allylation of aldehydes with allylic alcohols. Organic Letters, 2005, 7(12): 2333-2335.

[19] Kainz Q M, Matier C D, Bartoszewicz A, et al. Asymmetric copper-catalyzed C-N cross-couplings induced by visible light. Science, 2016, 351(6274): 681-684.

[20] Bartoszewicz A, Matier C D, Fu G C. Enantioconvergent alkylations of amines by alkyl electrophiles: copper-catalyzed nucleophilic substitutions of racemic α-halolactams by indoles. J. Am. Chem. Soc. 2019, 141(37): 14864-14869.

[21] Uozumi Y, Tanahashi A, Lee S Y, et al. Synthesis of optically active 2-(diarylphosphino)-1,1'-binaphthyls, efficient chiral monodentate phosphine ligands. The Journal of Organic Chemistry, 1993, 58(7): 1945-1948.

[22] Hayashi T, Han J W, Takeda A, et al. Modification of chiral monodentate phosphine ligands (MOP) for palladium-catalyzed asymmetric hydrosilylation of cyclic 1,3-dienes. Advanced Synthesis & Catalysis, 2001, 343(3): 279-283.

[23] Uozumi Y, Suzuki N, Ogiwara A, et al. Preparation of optically active binaphthylmonophosphines (MOP's) containing various functional groups. Tetrahedron, 1994, 50(15): 4293-4302.

[24] Ohmura T, Taniguchi H, Suginome M. Palladium-catalyzed asymmetric silaboration of allenes. Journal of the American Chemical Society, 2006, 128(42): 13682-13683.

[25] Nandi M, Jin J, RajanBabu T V. Synergistic effects of hemilabile coordination and counterions in homogeneous catalysis: new tunable monophosphine ligands for hydrovinylation reactions. Journal of the American Chemical Society, 1999, 121(42): 9899-9900.

[26] Unozumi Y, Hayashi T. Catalytic asymmetric synthesis of optically active 2-alkanols via hydrosilylation of 1-alkenes with a chiral monophosphine-palladium catalyst. Journal of the American Chemical Society, 1991, 113(26): 9887-9888.

[27] Rajanbhabu T V, Nomura N, Jin J, et al. Heterodimerization of Olefins. 1. hydrovinylation reactions of olefins that are amenable to asymmetric catalysis. Organic Chemistry Frontiers, 2003, 68(22): 8431-8446.

[28] Jiao Z, Beiger J J, Jin Y, et al. Palladium-catalyzed enantioselective α-arylation of α-fluoroketones. Journal of the American Chemical Society, 2016, 138(49): 15980-15986.

[29] Shintani R, Inoue M, Hayashi T. Rhodium-catalyzed asymmetric addition of aryl-and alkenylboronic acids to Isatins. Angewandte Chemie International Edition, 2006, 45(20): 3353-3356.

[30] Yang L, Neuburger M, Baudoin O. Chiral bifunctional phosphine-carboxylate ligands for palladium(0)-catalyzed enantioselective C-H arylation. Angew andte Chemie International Edition, 2018, 57(5): 1394-1398.

[31] Fu W Z, Tang W J. Chiral monophosphorus ligands for asymmetric catalytic reactions. ACS Catalysis, 2016, 6(8): 4814-4858.

[32] Tang W, Capacci A G, Wei X, et al. A general and special catalyst for Suzuki-Miyaura coupling processes. Angewandte Chemie International Edition, 2010, 49(34): 5879-5883.

[33] Rodriguez S, Qu B, Haddad N, et al. Oxaphosphole-based monophosphorus ligands for palladium catalyzed amination reactions. Advanced Synthesis & Catalysis, 2011, 353(4): 533-537.

[34] Patel N D, Sieber J D, Tcyrulnikov S,et al. Computationally assisted mechanistic investigation and development of Pd-catalyzed asymmetric Suzuki-Miyaura and Negishi cross-coupling reactions for tetra-ortho-substituted biaryl synthesis. ACS Catalysis, 2018, 8(11), 10190-10209.

[35] Li K, Hu N F, Luo R S, et al. A chiral Ruthenium-monophosphine catalyst for asymmetric addition of arylboronic acids to aryl aldehydes. The Journal of Organic Chemistry, 2013, 78(12): 6350-6355.

[36] Rao X F, Li NK, Bai H, et al. Efficient synthesis of ( − )-corynoline by enantioselective palladium-catalyzed α-arylation with sterically hindered substrates. Angewandte Chemie International Edition, 2018, 57(38): 12328-12332.

[37] Jin J H, Wang H, Yang Z T, et al. Asymmetric synthesis of 3,4-dihydroquinolin-2-ones via a stereoselective palladium-catalyzed decarboxylative [4 + 2]-cycloaddition. Organic Letters, 2018, 20(1): 104-107.

[38] Hu H, Peng Y, Yu T, et al. Palladium-catalyzed enantioselective 7-exo-trig carbopalladation/carbonylation: cascade reactions to achieve atropisomeric dibenzo[b,d]azepin-6-ones. Org. Lett. 2021, 23(9): 3636-3640.

[39] Hulst R, de Vries N K, Feringa B L. α-Phenylethylamine based chiral phospholidines; new agents for the determination of the enantiomeric excess of chiral alcohols, amines and thiols by means of $^{31}$P NMR. Tetrahedron: Asymmetry 1994, 5(4): 699-708.

[40] de Vries A H M, Meetsma A, Feringa B L. Enantioselective conjugate addition of dialkylzinc reagents to cyclic and acyclic enones catalyzed by chiral copper complexes of new phosphorus amidites. Angewandte Chemie International Edition, 1996, 35(20): 2374-2376.

[41] Feringa B L, Pineschi M, Arnold L A, et al. Highly enantioselective catalytic conjugate addition and tandem conjugate addition–aldol reactions of organozinc reagents. Angewandte Chemie International Edition, 1997, 36(23): 2620-2623.

[42] Arnold L A, Imbos R, Mandoli A, et al. Enantioselective catalytic conjugate addition of dialkylzinc reagents using copper-phosphoramidite complexes; ligand variation and non-linear effects. Tetrahedron, 2000, 56(18): 2865-2878.

[43] Malda H, von Zijl A W, Arnold L A, et al. Enantioselective copper-catalyzed allylic alkylation with dialkylzincs using phosphoramidite ligands. Organic Letters, 2001, 3(8): 1169-1171.

[44] Pena D, Minnaard A J, de Vries J G, et al. Highly enantioselective rhodium-Catalyzed hydrogenation of β-fehydroamino acid derivatives using monodentate phosphoramidites. Journal of the American Chemical Society, 2002, 124(49): 14552-14553.

[45] Boiteau J G, Imbos R, Minnaard A J, et al. Rhodium-catalyzed asymmetric conjugate additions of boronic acids using monodentate phosphoramidite ligands. Organic Letters, 2003, 5(5): 681-684.

[46] Panella L, Aleixandre A M, Kruidhof G J, et al. Enantioselective Rh-catalyzed hydrogenation of N-formyl dehydroamino esters with monodentate phosphoramidite ligands. The Journal of Organic Chemistry, 2006, 71(5): 2026-2036.

[47] Greene N, Kee T P. Asymmetric silylphosphite esters: synthesis and reactivity of (rac-O,O-binaphtholato)POSiR$_3$ (R$_3$ = Ph$_3$, $^t$BuMe$_2$, Et$_3$). Synthetic Communications, 1993, 23(12): 1651-1657.

[48] Feringa B L. Phosphoramidites: marvellous ligands in catalytic asymmetric conjugate addition. Accounts of Chemical Research, 2000, 33(6): 346-353.

[49] Naasz R, Arnold L A, Pineschi M, et al. Catalytic enantioselective annulations via 1,4-addition-aldol cyclization of functionalized organozinc reagents. Journal of the American Chemical Society, 1999, 121(5): 1104-1105.

[50] Imbos R, Brilman M H G, Pinschi M, et al. Highly enantioselective catalytic conjugate additions to cyclohexadienones. Organic Letters, 1999, 1(4): 623-626.

[51] Pena D, Lopez F, Harutyunyan S R, et al. Highly enantioselective Cu-catalysed asymmetric 1,4-addition of diphenylzinc to cyclohexanone. Chemical Communications, 2004, 36(2): 1836-1837.

[52] Sebesta, Pizzuti M G, Minnaard A J, et al. Copper-atalyzed enantioselective conjugate addition of organometallic reagents to acyclic dienones. Advanced Synthesis & Catalysis, 2007, 349(11-12): 1931-1937.

[53] Reetz M T, Mehler G. Highly enantioselective Rh-catalyzed hydrogenation reactions based on chiral monophosphite ligands. Angewandte Chemie International Edition, 2000, 39(21): 3889-3890.

[54] Reetz M T, Mehler G, Meiswinkel A, et al. Enantioselective hydrogenation of enamides catalyzed by chiral rhodium–monodentate phosphite complexes. Tetrahedron Letters, 2002, 43(44): 7941-7943.

[55] Duursma A, Boiteau J G, Lefor L, et al. Highly enantioselective conjugate additions of potassium organotrifluoroborates to enones by use of monodentate phosphoramidite ligands. Organic Chemistry Frontiers, 2004, 69(23): 8045-8052.

[56] van den Berg M, Minnaard A J, Haak R M, et al. Monodentate phosphoramidites: A breakthrough in Rhodium-catalysed asymmetric hydrogenation of olefins. Advanced Synthesis & Catalysis, 2003, 345(1-2): 308-323.

[57] Rössler S L, Petrone D A, Carreira E M. Iridium-catalyzed asymmetric synthesis of functionally rich molecules enabled by (phosphoramidite,olefin) ligands. Accounts of Chemical Research, 2019, 52(9): 2657-2672.

[58] Najera C, Retamosa M G, Sansano J M. Catalytic enantioselective 1,3-dipolar cycloaddition reactions of azomethine ylides and alkenes by using phosphoramidite–Silver( I ) complexes. Angewandte Chemie International Edition, 2008, 47(32): 6055-6058.

[59] Jensen J F, Svendsen B Y, la Cour T V, et al. Highly enantioselective hydrosilylation of sromatic alkenes. Journal of the American Chemical Society, 2002, 124(17): 4558-4559.

[60] Wang M, Chen J, Chen Z J, et al. Enantioselective desymmetrization of cyclobutanones enabled by synergistic palladium/enamine catalysis. Angewandte Chemie International Edition, 2018, 57(10): 2707－2711.

[61] Guo J M, Fan X Z, Wu H H, et al. Asymmetric synthesis of spiropyrazolones via chiral Pd(0)/ligand complexcatalyzed formal [4+2] cycloaddition of vinyl benzoxazinanones with alkylidene pyrazolones. J. Org. Chem., 2021, 86(2): 1712-1720.

[62] Feng K R, Tan Y X, Ye W, et al. One-pot preparation of tricyclo[5.2.2.04,9]undecanes via Cu-catalyzed asymmetric carboboration of cyclohexadienone-tethered allenes. Org.Lett., 2021, 23(2): 607-611.

[63] Qian D Y, Hu H X, Liu F, et al. Gold( I )-catalyzed highly diastereo- and enantioselective alkyne oxidation/cyclopropanation of 1,6-enynes. Angew. Chem. Int. Ed., 2014, 53(50): 13751-13755.

[64] Shi W J, Wang L X, Fu Y, et al. Highly regioselective asymmetric copper-catalyzed allylic alkylation with dialkylzincs using monodentate chiral spiro phosphoramidite and phosphite ligands. Tetrahedron: Asymmetry, 2003, 14(24): 3867-3872.

[65] Zhou H, Wang W H, Fu Y, et al. Highly enantioselective copper-catalyzed conjugate addition of diethylzinc to enones using chiral spiro phosphoramidites as ligands. The Journal of Organic Chemistry, 2003, 68(4): 1582-1584.

[66] Fu Y, Xie J H, Hu A G, et al, Zhou Q L. Novel monodentate spiro phosphorus ligands for Rhodium-catalyzed hydrogenation reactions. Chemical Communications, 2002, (2): 480-481.

[67] Hu A G, Fu Y, Xie J H, et al. Monodentate chiral spiro phosphoramidites: efficient ligands for rhodium-catalyzed enantioselective hydrogenation of enamides. Angewandte Chemie International Edition, 2002, 41(13): 2348-2350.

[68] Shi W J, Zhang Q, Xie J H, et al. Highly enantioselective hydrovinylation of $\alpha$-alkyl vinylarenes. an approach to the construction of all-carbon quaternary stereocenters. Journal of the American Chemical Society, 2006, 128(9): 2780-2781.

[69] Coulter M M, Kou K G M, Galligan B, et al. Regio and enantioselective intermolecular hydroacylation: substrate-directed addition of salicylaldehydes to homoallylic sulfides. Journal of the American Chemical Society, 2010, 132(46): 16330-16333.

[70] Zhao T Y, Li K, Yang L L, et al. Nickel-catalyzed desymmetrizing cyclization of 1,6-dienes to construct quaternary stereocenters. Org. Lett.,2021, 23(10): 3814-3817.

[71] Ma X, Hazelden I R, Langer T, et al. Enantioselective aza-Heck cyclizations of $N$-(tosyloxy) carbamates: synthesis of pyrrolidines and piperidines. J. Am. Chem. Soc.,2019, 141(8): 3356-3360.

[72] Li M M, Cheng L, Xiao L J, et al. Palladium-catalyzed asymmetric hydrosulfonylation of 1,3-dienes with sulfonyl hydrazides. Angew. Chem. Int. Ed., 2021, 60(6): 2948-2951.

[73] Alexakis A, Burton J, Vastra J, et al. European Journal of Organic Chemistry, 2000, (24): 4011-4027.

[74] Alexakis A, Burton J V J, Benhaim C, et al. Asymmetric conjugate addition of diethyl zinc to enones with chiral phosphorus ligands derived from TADDOL. Tetrahedron Letters, 1998, 39(43): 7869-7872.

[75] Beck A K, Gysi P, Vecchia L L, et al, (4$R$,5$R$)-2,2-Dimethyl-$\alpha,\alpha,\alpha',\alpha'$-tetra(naphth-2-yl)-1,3-dioxolane-4,5-dimethanol from dimethyl tartrate and 2-naphthyl-magnesium bromide. Organic Syntheses, 1999, 76, 12.

[76] Alexakis A, Benhaim C. Enantioselective copper-catalyzed conjugate addition of dialkyl zinc to nitro-olefins. Organic Letters, 2000, 2(17): 2579-2581.

[77] Alexakis A, Benhaim C. Asymmetric conjugate addition to alkylidene malonates. Tetrahedron: Asymmetry, 2001, 12(8): 1151-1157.

[78] Misale A, Niyomchon S, Luparia M, et al. Asymmetric palladium-catalyzed allylic alkylation using dialkylzinc reagents: a remarkable ligand effect. Angewandte Chemie International Edition, 2014, 53(27): 7068-7073.

[79] Zeng B S, Yu X, Siu P W, et al. Catalytic enantioselective synthesis of 2-aryl-chromenes. Chem. Sci., 2014, 5(6): 2277-2281.

[80] Schuster C H, Coombs J R, Kasun Z A, et al. Enantioselective carbocycle formation through intramolecular Pd-catalyzed allyl－aryl cross-coupling. Org. Lett., 2014, 16(17): 4420-4423.

[81] Sun C, Potter B, Morken J P. A catalytic enantiotopic-group-selective Suzuki reaction for the construction of chiral organoboronates. Journal of the American Chemical Society, 2014, 136(18): 6534-6537.

[82] Feng J, Li B, He Y, et al. Enantioselective synthesis of atropisomeric vinyl arene compounds by palladium catalysis: a carbene strategy. Angew. Chem. Int. Ed., 2016, 55(6): 2186-2190.

[83] Teller H, Flügge S, Goddard R, et al. Enantioselective gold catalysis: opportunities provided by monodentate phosphoramidite ligands with an acyclic TADDOL backbone. Angewandte Chemie International Edition, 2010, 49(11): 1949-1953.

[84] Oberg K M, Rovis T. Enantioselective rhodium-catalyzed [4 + 2] cycloaddition of $\alpha,\beta$-unsaturated imines and isocyanates. Journal of the American Chemical Society, 2011, 133(13): 4785-4787.

[85] Dalton D M, Rappé A K, Rovis T. Perfluorinated Taddol phosphoramidite as an L,Z-ligand on Rh( I ) and Co(- I ): evidence for bidentate coordination via metal−$C_6F_5$ interaction. Chem. Sci., 2013, 4(5): 2062-2070.

[86] Imbos R, Minnaard A J, Feringa B L. A highly enantioselective intramolecular Heck reaction with a monodentate ligand. Journal of the American Chemical Society, 2002, 124(2): 184-185.

[87] Albicker M R, Cramer N. Enantioselective palladium-catalyzed direct arylations at ambient temperature: access to indanes with quaternary stereocenters. Angewandte Chemie International Edition, 2009, 48(48): 9139-9142.

[88] Pedroni J, Cramer N. Chiral γ-lactams by enantioselective palladium(0)-catalyzed cyclopropane functionalizations. Angew. Chem. Int. Ed. 2015, 54(40): 11826-11829.

[89] He C F, Hou M Q, Zhu Z X, et al. Enantioselective synthesis of indole-based biaryl atropisomers via palladium-catalyzed dynamic kinetic intramolecular C − H cyclization. ACS Catal., 2017, 7(8): 5316-5320.

[90] Sun Y L, Wang X B, Sun F N, et al. Enantioselective cross-exchange between C-I and C-C σ bonds. Angewandte Chemie International Edition, 2019, 58(20): 6747-6751.

[91] Cao J, Chen L, Sun F N, et al. Pd-catalyzed enantioselective ring opening/cross-coupling and cyclopropanation of cyclobutanones. Angewandte Chemie International Edition, 2019, 58(3): 897-901.

[92] Sun F N, Yang W C, Chen X B, et al. Enantioselective palladium/copper-catalyzed C–C σ-bond activation synergized with sonogashira-type $C(sp^3)$-C(sp) cross-coupling alkynylation. Chemical Science, 2019, 10(32): 7579-7583.

[93] Yang W C, Chen X B, Song K L, et al. Pd-catalyzed enantioselective tandem C-C bond activation/Cacchi reaction between cyclobutanones and o-ethynylanilines. Organic Letters 2021, 23(4): 1309-1314.

[94] Deng R X, Xi J W, Li Q G, et al. Enantioselective carbon-carbon bond cleavage for biaryl atropisomers synthesis. Chem, 2019, 5(7): 1834-1846.

[95] Szymaniak A A, Zhang C, Coombs J R, et al. Enantioselective synthesis of nonracemic geminal silylboronates by Pt-catalyzed hydrosilylation. ACS Catalysis, 2018, 8(4): 2897-2901.

[96] Mannathan S, Raoufmoghaddam S, Reek J N H, et al. Enantioselective intramolecular reductive Heck reaction with a palladium/monodentate phosphoramidite catalyst. ChemCatChem, 2017, 9(4): 551-554.

[97] Kurihara Y, Nishikawa M, Yamanoi Y, et al. Synthesis of optically active tertiary silanes via Pd-catalyzed enantioselective arylation of secondary silanes. Chemical Communication, 2012, 48(94): 11564-11566.

[98] Chen L, Huang J B, Xu Z, et al. Palladium-catalyzed Si–C bond-forming silylation of aryl iodides with hydrosilanes: an enhanced enantioselective synthesis of silicon-stereogenic silanes by desymmetrization. RSC Advances, 2016, 6(71): 67113-67117.

# 3

# 手性双膦配体

3.1 轴手性双膦配体
3.2 手性螺环双膦配体
3.3 碳手性双膦配体
3.4 二茂铁双膦配体
3.5 磷手性双膦配体
3.6 膦–亚磷酸酯及膦–亚磷酰胺酯

Synthesis and Application of Chiral Phosphine Ligands

手性双膦配体是膦配体中应用最为广泛，种类最为丰富的一类手性膦配体，本章将较为详细地介绍这方面的研究进展。

# 3.1 轴手性双膦配体

在手性双膦配体中，联芳基骨架是非常重要的骨架之一。其结构存在的阻转异构现象能够有效诱导手性的形成。由该类骨架所衍生的联萘、八氢联萘、联苯及其他联芳基等配体通常都有 $C_2$ 对称性，具有轴手性的一对对映体的构型可分别以 $R_a$ 和 $S_a$ 表示，联萘骨架双膦配体是这类配体中报道最早也是最重要的一类。

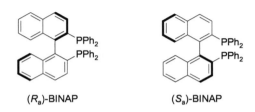

($R_a$)-BINAP      ($S_a$)-BINAP

## 3.1.1 联萘及八氢联萘骨架

1980 年 Noyori 等首次合成了著名的 BINAP 双膦配体并将其应用于不对称氢化反应中[1]，该手性 BINAP 依赖连接两个萘环的 C—C 单键旋转受阻而使整个分子具有旋光性。经过近几年的科学发展，该类配体的合成方法得到了不断的完善，目前公认的最实用的是以光学活性联萘二酚（BINOL）为起始原料的方法。消旋的 BINOL 可以轻松地进行千克级拆分，对映异构纯的 BINOL 首先转化为双三氟甲磺酸酯，然后通过 Ni/dppe 催化直接将二苯基膦基引入分子即可便利地得到 BINAP(**1a**)[2,3]。

由于 BINAP(**1a**) 配体在许多不对称反应中取得了极大的成功，通过修饰结构改变 BINAP 配体的立体和电子性质，从而实现对催化活性的调节以满足不同反应的需求并成为研究热点，因此许多 BINAP 的类似物被合成并在不对称反应中获得成功应用。其中最方便和直接的修饰 BINAP

的方法是利用其他取代膦基代替 BINAP 中的 2,2′-位的二苯基膦基，得到如下一系列结构的配体 **1b~1k**。

**1b**: R = 4-MeC$_6$H$_4$
**1c**: R = 4-$t$BuC$_6$H$_4$
**1d**: R = 4-MeOC$_6$H$_4$
**1e**: R = 3,5-Me$_2$C$_6$H$_3$
**1f**: R = 3,5-$t$Bu$_2$C$_6$H$_3$
**1g**: R = 4-MeO-3,5-Me$_2$C$_6$H$_2$
**1h**: R = Cy
**1i**: R = 呋喃-2-基
**1j**: R = 4-FC$_6$H$_4$
**1k**: R = 4-ClC$_6$H$_4$

另外一种方法是在 BINAP 配体的联萘骨架的不同位置上引入各种取代基得到空间及电子性质不尽相同的 BINAP 配体衍生物。例如 3,3′-双取代 BINAP 配体 **2a～2c**，4,4′-双取代配体 **3a～3l**、5,5′-位双取代 BINAP 衍生物 **4a**、**4b**、6,6′-位双取代化合物 **5**，7,7′-位双取代 **6a～6d**、**7a～7b**、**8a～8b** 以及衍生自甾体的 BINAP 配体类似物 **9**。

**3a**: R = Cl
**3b**: R = Br
**3c**: R = I
**3d**: R = Ph
**3e**: R = Me
**3f**: R = P(O)(OEt)$_2$
**3g**: R = P(O)(OH)$_2$
**3h**: R = SiMe$_3$
**3i**: R = Si($i$Pr)$_3$
**3j**: R = CPh$_2$(OH)
**3k**: R = 环戊醇-1-基
**3l**: R = CH$_2$NH$_2$

($S_a$)-或($R_a$)-**2**
**2a**: R = $i$Pr
**2b**: R = Me
**2c**: R = CO$t$Bu

($S_a$)-或($R_a$)-**3**

($S_a$)-或($R_a$)-**4**
**4a**: R = CH$_2$NH$_2$
**4b**: R = NH$_2$

3 手性双膦配体

此外，为了解决昂贵的催化剂回收再利用的问题，还发展了将 BINAP 配体与各种高分子材料键合的固载化 BINAP，如固载于聚苯乙烯高分子材料的配体 **10a**、**10b** 以及聚乙二醇 (PEG) 固载的配体 **11**。

1991 年，Takaya 将 BINAP 类配体进行部分氢化得到了含有八氢联萘骨架的 BINAP 配体衍生物 **12**($H_8$-BINAP)，该类配体在二面角方面与 BINAP 配体有明显差异，在不对称反应中同样获得了成功的应用，在一些不对称反应中取得了与 BINAP 配体相当甚至更高的对映选择结果。

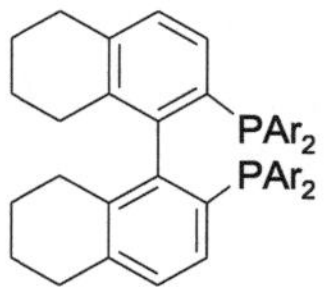

($S_a$)-或($R_a$)-**12a**, $H_8$-BINAP, Ar = Ph
**12b**, Xyl-$H_8$-BINAP, Ar = 3,5-$Me_2C_6H_3$

除了上述 BINAP 的类似物之外，还可在磷原子上引入 S、O 等原子，得到各种不同的 BINAP 配体衍生物，如 BINAP(O)、BINAP(S) 等。

BINAP(O):

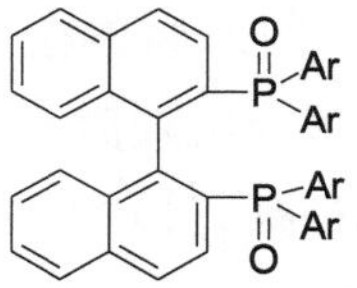

($S_a$)-或($R_a$)-**12**　a：Ar = Ph
　　　　　　　　　 b：Ar = 4-$MeC_6H_4$
　　　　　　　　　 c：Ar = 3,4-$Me_2C_6H_3$

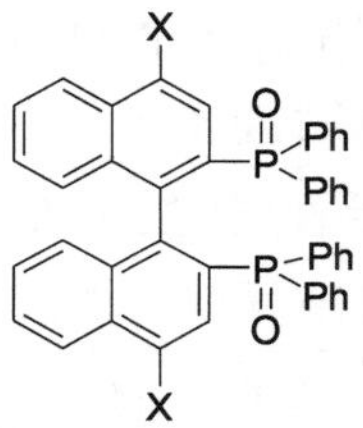

($S_a$)-或($R_a$)-**13**　a：X = Br
　　　　　　　　　 b：X = TMS
　　　　　　　　　 c：X = TES
　　　　　　　　　 d：X = TIPS

BINAP(S):

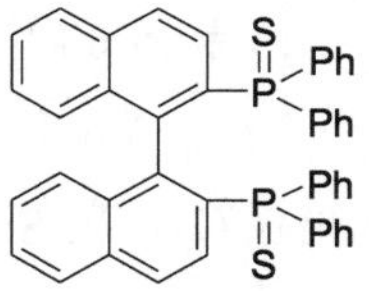

($S_a$)-或($R_a$)-**14**

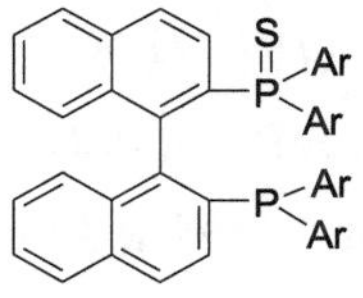

($S_a$)-或($R_a$)-**15**　a：Ar = Ph
　　　　　　　　　 b：Ar = 3,5-$(CH_3)_2C_6H_3$
　　　　　　　　　 c：Ar = 4-$CH_3C_6H_4$

在过去三十年中，BINAP 配体以及上述各种修饰的 BINAP 衍生物在许多金属催化不对称反应中获得了成功应用，取得了令人瞩目的研究成果。

## 3.1.2　联苯及联芳杂骨架

在联萘骨架的 BINAP 配体报道之后，与其相似的具有轴手性的联苯型双膦配体也随之出现。在联苯的 2,2′-位引入二苯基膦基即可得到最简单的联苯型双膦配体 Biphep(**16a**)。尽管 Biphep 及其类似物的一对对映体由于能垒较低，在室温条件下可以发生快速转化，但是近期研究发现，其形成的金属配合物在其他手性因素如手性二胺、二烯或手性阴离子配体的诱导下可以形成热力学稳定的单一非对映异构体，从而在不对称催化反应中得到了应用。

**16a**, Biphep, Ar = Ph
**16b**, Ar = 3,5-Me$_2$C$_6$H$_3$

如果在 Biphep 的 6,6′-位引入取代基，从而阻碍了两个苯环之间 σ-键的自由旋转，与 BINAP 一样形成一对对映的阻旋异构体，得到具有 C$_2$ 轴手性（构型用 $R_a$ 和 $S_a$ 表示）的联苯骨架双膦配体，通过拆分方法可得到对映异构纯的一对对映体。这类配体可以方便地通过改变苯环上取代基的类型、取代基的数目以及磷原子上取代基类型得到众多的在不对称金属催化中得以成功应用的双膦配体。例如 6,6′-位引入甲基得到的 Biphemp(**17a**) 以及其类似物 **17b**～**17i**，6,6′-位含有甲氧基的 MeO-Biphep(**18a**) 及其类似物 **18b**～**18q**，5,6-和 5,6′-并 1,3-二氧杂环戊烷得到的 SegPhos(**19a**) 及其类似物 **19b**～**19h**，5,6-和 5,6′-并 1,4-二氧杂环己烷或 1,4-氮氧杂环己烷得到的 SynPhos(**20**) 和 SolPhos(**21a**) 及其类似物 **21b**、**21c**，6,6′-位通过不同长度亚烷氧基桥联可以调节二面角大小的 C$_n$-TunaPhos(或 C$_n$-TunePhos，**22**)，苯环上引入拉电子基团的双膦配体 **23**、**24**，引入聚乙二醇结构单元可以方便地实现回收再用的配体 **25**，同时具有轴手性和中心手性的配体 **26**、**27** 以及具有双轴手性的配体 **28**、**29**。

| ($R_a$)-或($S_a$)-17 | R | R$^1$ | R$^2$ | R$^3$ | 备注 |
|---|---|---|---|---|---|
| a | H | H | H | Ph | Biphemp |
| b | H | H | H | Cy | |
| c | MeO | Me | H | Ph | |
| d | MeO | Me | H | 4-MeOC$_6$H$_4$ | |
| e | MeO | Me | H | Cy | Cy-Bimop |
| f | H | Me | Me | 3,5-Me$_2$C$_6$H$_3$ | |
| g | Me | Me | H | 3,5-Me$_2$C$_6$H$_3$ | |
| h | Me | Me | Me | Ph | |
| i | Me | Me | Me | Et | |

| ($R_a$)-或($S_a$)-18 | R | $R^1$ | $R^2$ | Ar | 备注 |
|---|---|---|---|---|---|
| a | H | H | H | Ph | MeO-Biphep |
| b | H | H | H | 3-$CF_3C_6H_4$ | |
| c | H | H | H | 4-$CF_3C_6H_4$ | |
| d | H | H | H | 4-$tBuC_6H_4$ | |
| e | H | H | H | 4-$MeOC_6H_4$ | |
| f | H | H | H | 4-$N(Me_2)C_6H_4$ | |
| g | H | H | H | 3,5-$Me_2C_6H_3$ | |
| h | H | H | H | 3,5-$tBu_2$-4-$MeOC_6H_2$ | |
| i | H | H | H | 3,5-$iPr_2$-4-$iPrOC_6H_2$ | |
| j | H | H | H | 3,4,5-$F_3C_6H_2$ | |
| k | H | H | H | 2,3,5,6-$F_4$-4-$CF_3C_6$ | |
| l | Cl | H | H | Ph | |
| m | H | H | Ph | Ph | |
| n | H | H | tBuO | Ph | |
| o | H | H | iPr | Ph | |
| p | MeO | MeO | MeO | Ph | |
| q | MeO | MeO | Ph | Ph | |

| ($R_a$)-或($S_a$)-19 | R | $R^1$ | 备注 |
|---|---|---|---|
| a | H | Ph | SegPhos |
| b | H | 3,5-$Me_2C_6H_3$ | |
| c | H | Cy | |
| d | H | 3,5-$tBu_2$-4-$MeOC_6H_2$ | |
| e | F | Ph | |
| f | Me | Ph | |
| g | Me | 4-$MeC_6H_4$ | |
| h | R,R = $(CH_2)_5$ | Ph | |

($S_a$)-**20a**, SynPhos, Ar = Ph
($S_a$)-**20b**, Ar = 3,5-$Me_2C_6H_3$

($S_a$)-**21a**, SolPhos, Ar = Ph
($S_a$)-**21b**, Ar = 3,5-$Me_2C_6H_3$
($S_a$)-**21c**, Ar = 4-MeO-3,5-$Me_2C_6H_2$

**22a**, n = 1, $C_1$-TunaPhos
**22b**, n = 2, $C_2$-TunaPhos
**22c**, n = 3, $C_3$-TunaPhos
**22d**, n = 4, $C_4$-TunaPhos
**22e**, n = 5, $C_5$-TunaPhos
**22f**, n = 6, $C_6$-TunaPhos

**23**    **24**    **25**

3 手性双膦配体

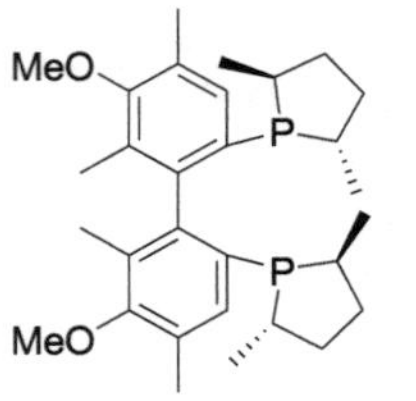

($R_a$,$S$,$S$)-**26a**
($S_a$,$S$,$S$)-**26b**

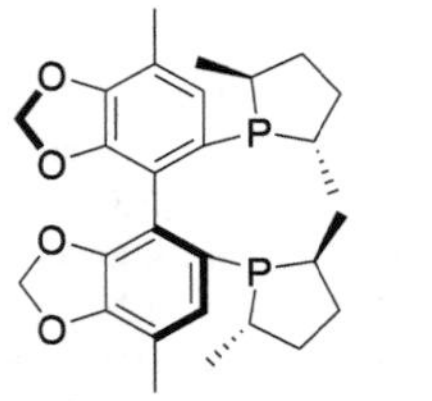

($S_a$,$S$,$S$)-**27c**

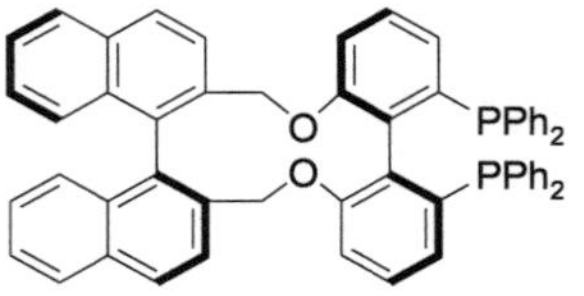

($S_a$,$S_a$)-**28**

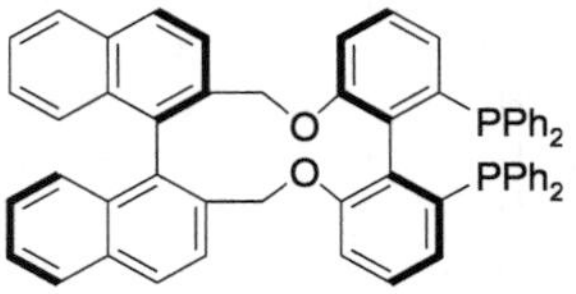

($R_a$,$S_a$)-**29**

以上具有旋光活性的联苯双膦配体的合成大体包括以下步骤：在苯环上引入含膦基团、将两个芳环进行偶联、拆分。如果引入的含膦基团是次膦酰基，最后还需要硅烷还原。当引入的基团是二苯氧基磷酰基时，也可以用 2,3-$O$-二苯甲酰基酒石酸（DBTA）拆分，然后与格氏试剂反应，还原得到目标产物。这样可以方便地在磷原子上引入各种取代基，合成不同含膦基团的双膦配体。

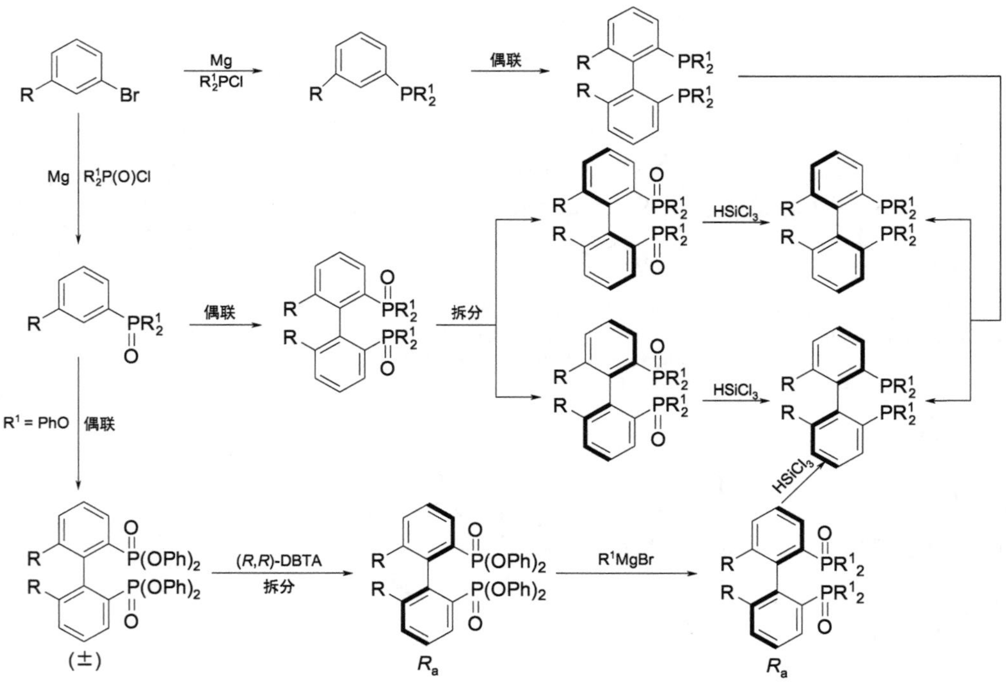

用非对映选择的方法可以实现同时含有中心手性和轴手性的联苯型配体 30、31 的合成。反应过程中巧妙地利用中心手性骨架诱导的分子内芳基偶联实现轴手性单元的构筑，反应具有非常优异的非对应选择性，几乎得到单一的立体异构体，从而避免了繁杂的拆分过程。

| 31 | $n$ | R | $R^1$ |
|---|---|---|---|
| a | 1 | H | Ph |
| b | 2 | H | Ph |
| c | 1 | H | 4-MeC$_6$H$_4$ |
| d | 1 | H | 3,5-Me$_2$C$_6$H$_3$ |
| e | 1 | H | 3,5-$t$Bu$_2$C$_6$H$_3$ |
| f | 1 | H | 4-MeO-3,5-$t$Bu$_2$C$_6$H$_2$ |
| g | 1 | H | Cy |
| h | 1 | $t$Bu | Ph |

($R_a$,$S$,$S$)-30    ($S_a$,$S$,$R$)-31

此外，从手性二醇出发，利用分离非对映异构体的方法也得到了同时具有中心手性和轴手性的双膦配体 32、33。

($S_a$,$R$,$R$)-32    ($R_a$,$R$,$R$)-32    ($S_a$,$S$,$S$)-33    ($R_a$,$S$,$S$)-33

Marinetti 小组同样以光学活性戊二醇为手性辅助剂通过非对映选择芳基偶联的方法合成了 C1 对称的配体 ($S_a$)-34 和 ($R_a$)-35，该类配体可以看成是由萘环和取代苯环偶联得到的联异芳基结构骨架双膦配体，是一类比较特殊的联苯型配体，其二面角正好介于联萘型配体 BINAP 和普通联苯型配体 MeO-Biphep 之间。

($S_a$)-34    ($R_a$)-35

### 3.1.3 轴手性双膦配体的应用

#### 3.1.3.1 氢化反应

过渡金属催化的不对称氢化是研究得最早、最广泛也是最有效的不对称反应,主要包括碳-氧双键、碳-碳双键及碳-氮双键的氢化。

(1) 碳-氧双键的氢化反应

1987 年 Noyori 等人使用 $RuX_2/(S_a)$-**1a**(X=Cl、Br)催化体系实现了 $\beta$-酮酸酯的不对称氢化,反应获得了很好的立体选择性,其中 $\beta$-芳基酮酸酯的立体选择性明显低于 $\beta$-烷基酮酸酯[4]。

R = Me, Et, Bu, Ph, iPr
R' = Me, Et, iPr, tBu
X = Cl, Br

产率:93%~>99.5%
ee:85%~100%

在此基础上,Noyori 等人使用 $Ru(OAc)_2$(**1a**)、$RuX_2$(**1a**)(X=Cl、Br)以及 $Ru_2Cl_4$(**1a**)·$2Et_3N$ 等手性钌配合物催化剂,进一步对各官能化的酮如 $\alpha$-氨基酮、$\alpha$-及 $\beta$-酮酸衍生物等的不对称氢化进行了研究。他们认为反应过程中对映识别的关键是羰基氧和分子中的另一个杂原子同时与中心金属铑进行配位形成五元或六元螯合环。

R = Me, iPr, $iPr_3SiOCH_2$, $BnOCH_2CH_2$, Ph
$R^1$ = $CH_2NMe_2$, $CH_2OH$, $CO_2CH_3$, $CH_2CH_2OH$,
$CH_2CO_2Et$, $CH_2CONMe_2$, $CH_2COSEt$,
$CH_2CO_2Me$, 2-$HOCOC_6H_4$, 2-$BrC_6H_4$,
$CH_2COCH_3$, $COCH_3$, $CH(CH_3)COCH_3$

产率高达100%
ee值高达100%

通过上述均相不对称氢化的方法,Noyori 小组首次实现了具有广泛生理和生物活性分子左旋肉碱(维生素 $B_T$)的高效化学合成[5]。

左旋肉碱
产率:50%;ee:99.9%

同样使用 RuCl$_2$/**1a** 为催化剂,Liu 等通过 N-甲基苯甲酰胺的不对称氢化反应实现了抗抑郁药物氟西汀(fluoxetin)一对对映的立体选择合成[6]。

$$\text{Ph-CO-CH}_2\text{-CONHMe} \xrightarrow[\text{H}_2]{\text{RuCl}_2/(R_a)\text{-或}(S_a)\text{-1a}} \text{Ph-CH(OH)-CH}_2\text{-CONHMe} \longrightarrow \text{氟西汀}$$

($R_a$)-**1a**:产率 50%;ee(S)>99.9%
($S_a$)-**1a**:产率 50%;ee(R)>99.9%

Noyori 应用钌催化 2-苯甲酰胺甲基乙酰乙酸甲酯的不对称氢化反应合成碳青霉烯抗生素(carbapenems),并实现了工业化[7]。

dr:94:6
ee(2S,3R):99%

碳青霉烯抗生素

Genêt 等报道了 Ru/**1a** 催化体系催化的 β-酮酸酯的不对称氢化,无论是脂肪族还是芳香族 β-酮酸酯均能得到很好的立体选择结果。此外,由于配体的独特性质使其在对含氟 β-酮酸酯或 1,3-二酮等富有挑战性的底物不对称氢化反应中取得了明显优于 BINAP 和 SegPhos 所获得的立体控制结果[8]。

R = Me, ClCH$_2$, Ph, 4-FC$_6$H$_4$, CF$_3$, C$_2$F$_5$
R$^1$ = OMe, OEt, CF$_3$

转化率:100%
ee:70%～99%

Noyori 小组报道了 RuCl$_2$[($S_a$)-**1e**][(S,S)-DPEN] 与 NaBH$_4$ 作用得到的手性钌配合物 trans-RuH($\eta^1$-BH$_4$)[($S_a$)-**1e**][(S,S)-DPEN] 同样是简单酮不对称氢化的高效催化剂,与以前双膦双胺催化体系中往往需要加入 KOH、iPrOK 或 tBuOK 等强碱不同,该催化剂可以在无碱条件下进行催化氢化,从而实现对碱敏感底物的氢化还原[9]。

$$\underset{R}{\overset{O}{\|}}\underset{R^1}{\overset{}{\|}} \xrightarrow[\text{H}_2]{\textit{trans}\text{-RuH}(\eta^1\text{-BH}_4)[(S_a)\text{-}\mathbf{1e}][(S,S)\text{-DPEN}]} \underset{R}{\overset{OH}{\|}}\underset{R^1}{\overset{}{\|}}$$

R = Ph, 4-EtO$_2$CC$_6$H$_4$, (E)-C$_5$H$_{11}$CH=CH

产率：95%～100%
ee：82%～99%

R$^1$ = Me, Me$_2$NCH$_2$CH$_2$

Saito 小组报道，Ru/($R_a$)-SegPhos(**19a**) 催化体系能高效地催化 $\beta$-羟基酮、$\alpha$-酮酸酯、$\beta$-酮酸酯、$\gamma$-酮酸酯的不对称氢化反应，大多数情况下可以获得比 Ru/BINAP 催化体系更好的结果[10]。

$$\underset{R}{\overset{O}{\|}}\underset{R^1}{\overset{}{\|}} \xrightarrow[\text{H}_2]{[\text{NH}_2\text{Me}_2][\{\text{RuCl}[(R_a)\text{-}\mathbf{19a}]\}_2(\mu\text{-Cl})_3]} \underset{R}{\overset{OH}{\|}}\underset{R^1}{\overset{}{\|}}$$

R = Me, BnCH$_2$, tBu, Ph, ClCH$_2$, BnOCH$_2$
R$^1$ = CH$_2$OH, CO$_2$Et, CH$_2$CO$_2$Et, (CH$_2$)$_2$CO$_2$Et

转化率：99%～100%
ee：94%～99%

Jessop 小组报道了一种不对称催化体系，通过酮的不对称合成 $\beta$-羟基羧酸，首先使用二氧化碳羧基化得到 $\beta$-酮羧酸，然后通过后续的 $\beta$-酮羧酸的不对称氢化得到 $\beta$-羟基羧酸。虽然 $\beta$-酮酸很容易发生脱羧反应，但是在 MeOH 或 CH$_2$Cl$_2$ 介质中，不对称氢化的速度明显快于脱羧，可以得到相应的手性 $\beta$-羟基酸。但是反应底物适用范围窄，对映选择性具有很强的底物依赖性。如苯甲酰基乙酸在 MeOH 或 CH$_2$Cl$_2$ 介质中反应得到 ee 值高达 99% 的氢化产物、4-氯取代的苯甲酰基乙酸以及 2-萘甲酰基乙酸也得到了很好的立体选择结果，而 2-甲基取代苯甲酰基乙酸则只给出几乎外消旋的氢化产物[11]。

Vinogradov 等人使用 RuCl$_3$/($R_a$)-**1a**/HCl 催化体系催化的 $\gamma$-羰基酸酯的不对称氢化反应实现了重要的天然香料以及合成中间体 $\gamma$-丁内酯的对映选择性合成。在催化体系中，酸性促进剂 HCl 对反应的产率影响很大，没有 HCl 的存在时反应的转化率急剧下降[12]。

$$\underset{\underset{O}{\|}}{R^1}\diagdown\diagup\diagdown\text{COOR}^2 \xrightarrow[\text{H}_2]{\text{RuCl}_3/(R_a)\text{-}\mathbf{1a},\text{ HCl}} \underset{\underset{OH}{\|}}{R^1}\diagdown\diagup\diagdown\text{COOR}^2 \longrightarrow \underset{R^1}{\overset{}{\bigcirc}}\!\!=\!\!O$$

R$^1$ = Me, Et, Pr, iPr, Ph, 4-ClC$_6$H$_4$
R$^2$ = Me, Et

产率：57%～88%
ee：91%～99%

进一步把底物扩展为酰基丁二酸酯可以实现 2-取代仲康酸(paraconic acid)衍生物的合成，由于酰基丁二酸酯同时可以看作是 $\gamma$-酮酸酯和 $\beta$-酮酸酯，反应过程中不需要酸性促进剂 HCl 的加入即能顺利进行氢化还原[13]。

R = Me, Et, Pr, iPr

产率：40%～75%
trans/cis：55:45～95:5
ee：98%～99.5%

Genêt 等报道，Rh/($R_a$)-MeO-Biphep(**18a**) 催化体系在对称 1,3-二酮的不对称氢化反应中表现出很好的催化效果，以极好的非对映及对映选择性得到了相应的光学活性 1,3-二醇，为该类手性化合物的合成提供了便利的途径[14]。

R = Me, Et, iPr, tBu, Cy, $C_5H_{11}$, Bn

de：>95%～>99%
ee：>95%～>99%

Zhang 等报道了 Ru/($S_a$)-**19f** 催化体系的 $\beta,\gamma$-不饱和酮酸酯和 $\alpha$-酮酸酯的不对称氢化反应，反应获得了很好的对映选择性，为合成 ACE(血管紧张素酶)抑制剂的重要合成中间体 (R)-2-羟基-4-苯基丁酸酯及其衍生物的合成提供了新途径[15]。

Ar = Ph, 4-$MeC_6H_4$, 4-$FC_6H_4$, 4-$ClC_6H_4$,
4-$BrC_6H_4$, 4-$MeOC_6H_4$, 3-$ClC_6H_4$,
2,4-$Cl_2C_6H_3$, 3,4-$(OCH_2O)C_6H_3$

100%转化率
ee：94%～96%

Zhang 小组使用 Ru/($S_a$)-**19h** 催化体系实现了对 $\beta$-羟基砜的不对称氢化反应。添加剂碘的存在可以显著提高反应的对映选择性，除了邻位有可配位原子取代的芳香族底物之外，其余无论是芳香族还是脂肪族底物均获得了很好的立体控制结果。研究表明反应过程中原位产生的无水 HI 才是真正发挥作用的添加剂，决定了反应的对映选择性[16]。

$$\underset{R}{\overset{O}{\|}}\underset{}{\overset{}{C}}\mathrm{SO_2Ph} + \mathrm{H_2} \xrightarrow{\mathrm{Ru}/(S_\mathrm{a})\text{-}\mathbf{19h}} \underset{R}{\overset{OH}{\underset{*}{C}}}\mathrm{SO_2Ph}$$

R = Ph, 4-FC$_6$H$_4$, 4-ClC$_6$H$_4$, 4-BrC$_6$H$_4$, 4-MeC$_6$H$_4$,
4-MeOC$_6$H$_4$, 3-ClC$_6$H$_4$, 2-MeC$_6$H$_4$, 2-ClC$_6$H$_4$,
Me, iPr, Cy

转化率：100%
ee：89.8%～99.1%

Clarke 等报道了 Ru/($S_\mathrm{a}$)-SynPhos(**20**) 催化体系在 1,3-二苯基-2,2-二氟-1,3-丙二酮的不对称氢化反应中表现出了明显优于 BINAP 类配体的不对称诱导能力，以中等程度非对映选择性和较好对映选择性得到相应的光学活性 1,3-二醇，进一步简单重结晶即可得到光学纯的产物[17]。

$$\mathrm{Ph}\underset{F\ F}{\overset{O\quad O}{\|\ \ \|}}\mathrm{Ph} + \mathrm{H_2} \xrightarrow[\text{转化率}100\%]{\mathrm{Ru}/(S_\mathrm{a})\text{-}\mathbf{20}} \mathrm{Ph}\underset{F\ F}{\overset{OH\ OH}{|\ \ |}}\mathrm{Ph} + \mathrm{Ph}\underset{F\ F}{\overset{OH\ OH}{|\ \ |}}\mathrm{Ph}$$

syn/anti：32/68
ee：72%

Ru/($R_\mathrm{a}$)-SynPhos(**20**) 催化体系在 β-羟基酮、α-酮酸酯、β-酮酸酯、β-羟基膦酸酯和 1,3-二酮的不对称氢化反应中观察到与 Ru/BINAP 催化体系相当或更好的立体选择性。

$$\underset{R}{\overset{O\quad O}{\|\ \ \|}}\mathrm{OR^1} + \mathrm{H_2} \xrightarrow{\mathrm{Ru}/(R_\mathrm{a})\text{-}\mathbf{20}} \underset{R}{\overset{OH\ \ O}{|\ \ \ \|}}\mathrm{OR^1}$$

R = Me, ClCH$_2$
R$^1$ = Me, Et, Bn

转化率：100%
ee：96.9%～99.5%

$$\underset{R}{\overset{O}{\|}}\mathrm{R^1} + \mathrm{H_2} \xrightarrow{\mathrm{Ru}/(R_\mathrm{a})\text{-}\mathbf{20}} \underset{R}{\overset{OH}{|}}\mathrm{R^1}$$

R = CF$_3$, C$_2$F$_5$, Ph, Me, Et, iPr
R$^1$ = CH$_2$OH, CH$_2$CO$_2$Et, CH$_2$CO$_2$Me,
CO$_2$Et, CH$_2$Ac, CH$_2$P(O)(OEt)$_2$

转化率：100%
ee：49%～99%

(2) 碳-碳双键的氢化反应

Donate 等报道了手性钌配合物 RuCl$_2$(p-cymene)[($R_\mathrm{a}$)-**1a**] 和手性铑配合物 Rh(cod)[($R_\mathrm{a}$)-**1a**]ClO$_4$ 能催化 2,3-二取代-γ-丁烯酸内酯的不对称氢化，但是反应的转化率不是很高。铑催化剂无论是在转化率还是立体选择性（非对映和对映选择性）方面的表现均优于钌催化剂[18]。

| 催化剂 | cis/trans | ee/% |
|---|---|---|
| $RuCl_2(p\text{-cymene})[(R_a)\text{-}\mathbf{1a}]$ | 80.4/19.6 | 87 |
| $Rh(cod)[(R_a)\text{-}\mathbf{1a}]ClO_4$ | 91.5/8.5 | 98 |

Maligres 等人使用手性钌催化的 $\alpha$-芳氧基-$\alpha,\beta$-不饱和羧酸的不对称氢化实现了具有广泛生物活性的 $\alpha$-芳氧羧酸衍生物的不对称合成。在 20%(摩尔分数) $RuCl_2[(S_a)\text{-}\mathbf{1a}]$ 的存在下，不同取代的 $\alpha$-芳氧基-$\alpha,\beta$-不饱和羧酸以中等至很好的对映选择性得到相应的氢化产物，将催化剂用量减低至 1%(摩尔分数)时，部分底物仍然获得了令人满意的结果，但其中 $\beta,\beta$-双取代底物会导致反应立体选择性的显著下降[19]。

R = H, 2-Me, 4-Me, 2-OMe, 2-F, 4-F,
　　2-Cl, 3-Br, 2-I, 3-I, 4-$NO_2$, 4-MeO
$R^1$ = Me, $nC_5H_{11}$
$R^2$ = H, Me

产率：65%~99%
ee：32%~95%

Takaya 等报道了 $H_8$-BINAP(12) 形成的手性钌配合物 $Ru(OAc)[(S_a)\text{-}H_8\text{-BINAP}]$ 在 $\alpha,\beta$-不饱和羧酸的催化氢化反应中同样获得了成功，取得了很好的对映选择性结果[20]。

R = H, Me, Et, 4-$tBuC_6H_4$
$R^1$ = H, Et, Pr, Ph, $CF_3$
$R^2$ = H, Me

产率：80%~99%
ee：93%~96%

2016 年，Mashima 课题组将铑(Ⅲ)催化剂和 (S)-DTBM-SegPhos 应用于简单烯烃的不对称氢化反应中。他们从铑(Ⅰ)前体 $[Rh(cod)_2Cl]_2$、手性二膦配体和盐酸合成了一系列新的氯化物桥连的双核铑(Ⅲ)配合物。与广泛使用的铑(Ⅰ)催化体系形成鲜明对比的是，该系列的配合物可有效地催化(E)-丙-1-烯-1,2-二基二苯及其衍生物进行不对称加氢反应，而无需任何导向基团。该催化体系不仅适用于烯丙基醇，还适用于烯基硼烷和不饱和环状砜的不对称氢化。根据对照实验，证明了双核铑(Ⅲ)配合物在该反应上优于典型的铑(Ⅰ)催化系统[21]。

$$\text{Ph} \overset{\text{Ph}}{=} \xrightarrow[\text{甲苯, 80°C, 20h}]{\underset{n\text{Bu}_4\text{NCl, H}_2}{(S)\text{-DTBM-SegPhos/[Rh(cod)}_2\text{Cl]}_2}} \text{Ph} \overset{*}{\underset{}{\frown}} \text{Ph}$$

产率：96%
ee：95%

2020年，Chang等使用[Rh(cod)$_2$Cl]$_2$和(R)-SegPhos形成的配合物、氢气作为还原剂,催化前手性烯烃的不对称氢化和醛的还原胺化反应。通过这种策略，合成了相应的手性胺化合物的效率大大提高了。该反应的产率和 ee 值均高达 98%，并在研究 α,β-不饱和醛和苯胺的官能团耐受性时，均得到良好的产率和 ee 值，因此该催化体系在该反应中有较高的普适性[22]。

$$\text{OHC} \overset{\text{Ph}}{=} + \text{PhNH}_2 \xrightarrow[\underset{\text{H}_2, 60°C}{\text{NaSbF}_6, \text{MeOAc/DMF},}]{\underset{\text{MS, 4'-ClC}_6\text{H}_4\text{SO}_3\text{H},}{\underset{(R)\text{-SegPhos}}{[\text{Rh(cod)}_2\text{Cl]}_2}}} \text{Ph} \overset{*}{\underset{\text{Ph}}{\frown}} \text{NHPh}$$

产率：98%
ee：98%

Scrivanti 小组以 Ru/($R_a$)-MeO-Biphep(18a) 催化的碳-碳双键的不对称氢化反应为关键步骤实现了柠檬香型芳香剂 3-甲基-5-苯基戊腈的不对称合成[23]。

$$\text{Ph}\frown\overset{}{=}\text{CO}_2\text{H} \xrightarrow{\text{Ru/}(S_a)\text{-MeO-Biphep}} \text{Ph}\frown\overset{*}{\frown}\text{CO}_2\text{H} \Longrightarrow \text{Ph}\frown\overset{*}{\frown}\text{CN}$$

ee：98%

Zhou 使用 Ir/($S_a$)-MeO-Biphep(18a) 催化体系实现了具有环外双键的四氢异喹啉类烯胺底物的不对称氢化反应。反应取得了很好的立体选择结果，在此催化体系中添加碘是反应成功的关键因素，如果没有碘的存在反应根本不能进行[24]。

R = H, F, Me
R¹ = Ph, Me, Pr, 4-CF$_3$C$_6$H$_4$, 4-iPrC$_6$H$_4$, 2-MeC$_6$H$_4$,
1-Naph, BnCH$_2$, OMe, NEt$_2$, 3,4-(MeO)$_2$C$_6$H$_3$

产率：91%～99%
ee：78%～96%

在铑催化不对称氢化反应中，双膦配体骨架的取代基效应起着关键作用，例如，在 ($S_a$)-MeO-Biphep(**18a**) 的 3,3′-位引入苯基后得到的 ($S_a$)-*o*-Ph-MeO-Biphep(**18m**) 在铑催化脱氢 α-氨基酸(酯)的不对称氢化反应中表现出比母体 ($S_a$)-MeO-Biphep 更好的手性诱导能力[25]。

L = **18a**, ee：21%
L = **18m**, ee：97%

Zhang 等报道了在用苯基替换 ($S_a$)-hexaMeO-Biphep(**18p**) 配体中 3,3′-位的甲氧基后得到的 ($S_a$)-*o*-Ph-hexaMeO-Biphep(**18q**) 在铑催化乙酰基保护烯胺的不对称氢化反应中表现出了比母体 ($S_a$)-hexaMeO-Biphep(**18p**)、MeO-Biphep(**18a**) 和 BINAP 更好的不对称诱导能力，但是反应具有一定的底物结构依赖性，部分底物仅获得中等程度的立体控制结果[26]。

($R_a$)-BINAP, ee：55%
($S_a$)-**18q**, ee：98%

(3) 碳-氮双键的氢化反应

2001 年，Uneyama 等人使用配体 **1a** 在 α-含氟烷基取代的亚胺酯的不对称氢化中获得了很好的结果，高产率地得到一系列光学活性含氟 α-氨基酸衍生物[27]。

```
         NPMP                              HN—PMP
          ‖      Pd(OCOCF₃)₂/1a              |
    Rf—C—CO₂R  ───────────────→     Rf—*C—CO₂R
                 H₂ (10MPa)                  |
                 CF₃CH₂OH
                 室温, 24h           产率：75%～>99%
PMP = 4-甲氧基苯基                    ee：30%～88%
Rf = CF₃, CClF₂, C₇F₁₅, CHF₂
R = Et, tBu, Bn
```

Zhou 等使用 Pd(OCOCF₃)₂ 与双膦配体 ($S_a$)-SegPhos(**19a**) 原位形成的手性钯配合物实现了 N-(二苯基次膦酰基) 芳香亚胺的不对称氢化反应。反应取得了很好的对映选择性，但部分底物的产率不太理想。由于二苯基次膦酰基很容易通过酸性水解或醇解脱除，可用于快速制备光学活性胺类化合物[28]。

```
         O                                        O
         ‖                                        ‖
         P—PPh₂                                   P—PPh₂
         |                                        |
         N              Pd(OCOCF₃)₂/(Sₐ)-19a    HN
         ‖     +   H₂  ──────────────────→        |
    Ar—C—R                                   Ar—*C—R  H
                                                  |
Ar = Ph, 4-MeC₆H₄, 4-FC₆H₄, 4-ClC₆H₄,    产率：29%～98%
     4-MeOC₆H₄, 3-MeOC₆H₄,                ee：87%～98.6%
     2-MeOC₆H₄, 2-Naph, 呋喃-2-基
```

2003 年，Zhou 等人利用 Ir/($R_a$)-MeO-Biphep(**18a**)/I₂ 催化体系首次实现了对 2-取代喹啉的不对称氢化，其中碘添加剂是获得高催化活性和对映选择性的关键所在。由于四氢喹啉环结构单元广泛存在于生物碱以及生物活性物质中，利用该催化体系可以方便地实现该类化合物的不对称合成，因而该方法在有机合成中具有重要意义[29]。

```
   ____                              ____
  /    \                            /    \
 |      |          Ir/(Rₐ)-18a/I₂  |      |
  \____/N   +  H₂ ────────────→   \____/N—H
        \                               \
         Me                              *Me

                                    产率：99%
                                    ee：94%
```

该小组利用上述催化体系催化的 2-取代喹啉的不对称氢化为关键步骤，成功实现了具有四氢喹啉环的天然生物碱 (−)-Galipeine 的不对称合成，并最终确定了其绝对构型[30]。该方法还可以应用于吡啶衍生物的不对称还原反应[31]。

```
                                                                     OH
   喹啉—CH₂CH₂—Ar(OBn)₂    Ir/(Sₐ)-18a    四氢喹啉—CH₂CH₂—Ar(OBn)₂ → 四氢喹啉—CH₂CH₂—Ar(OMe)
                           H₂, I₂
                                                           (−)-Galipeine(ee：96%)
```

2010 年，周永贵等以 (R)-Cl-MeO-Biphep (18l)/Pd(OCOCF$_3$)$_2$ 为催化剂、CF$_3$CH$_2$OH 为溶剂，在 4.1MPa 氢气压力下，成功实现了简单含氟亚胺的不对称氢化，高对映选择性地得到了各种手性含氟胺类化合物 (ee 值为 84%～94%)[32]。

$$\text{Ph}\underset{\text{CF}_3}{\overset{\text{N}^{\text{PG}}}{\diagdown}} \xrightarrow[\text{H}_2(4.1\text{MPa}), \text{TFE}, 4\text{Å MS}]{(R)\text{-Cl-MeO-Biphep (18l)} \atop \text{Pd}(\text{OCOCF}_3)_2} \text{Ph}\underset{\text{CF}_3}{\overset{\text{HN}^{\text{PG}}}{*}}$$

产率：99%
ee：93%

2011 年，周永贵等采用 Brønsted 酸 DTTA (二对甲基苯甲酰酒石酸) 活化底物的策略，将 (R)-C$_4$-TunePhos(22d)/Pd(OCOCF$_3$)$_2$ 用于 N-芳基亚胺的不对称氢化。虽然对于简单的 N-芳基亚胺底物，仅得到了中等的对映选择性 (ee 值为 60%～75%)。但是，对于含环外双键的 N-芳基亚胺底物有非常好的对映选择性 (ee 值为 86%～95%)[33]。

ee：60%~95%

Bunlaksananusorn 使用 Ru/($R_a$)-18l 催化体系在醋酸铵的存在下催化 β-酮酸酯的不对称还原氨化实现了一锅法制备 β-氨基酸酯。无论 β-位是烷基还是芳基取代的底物都不仅表现出很好的化学选择性 (β-氨基酸酯/β-羟基酸酯：94/6~99/1)，而且获得了很好的对映选择性 [34]。

R = Me, Ph, 3-MeOC$_6$H$_4$, 4-MeOC$_6$H$_4$,
3-ClC$_6$H$_4$, 4-FC$_6$H$_4$, 4-ClC$_6$H$_4$

产率：79%～88%
ee：96%～99%

手性苯并二氮杂酮是一类重要的七元杂环化物，具有广泛的生理和药理活性。然而，该类手性杂环化合物的不对称催化合成却鲜有报道。2012 年，周永贵等以 TunePhos(31a) 为配体，[Ir(cod)Cl]$_2$ 为金属前体，实现了 1,4-苯并二氮杂酮的不对称氢化，得了 ee 值为 77%～96% 的对映选择性 [35]。

2012 年，Ratovelomanana-Vidal 等将 (R)-3,5-Me$_2$-SynPhos 用于铱催化 1-芳基-3,4-二氢异喹啉的不对称氢化，获得了 ee 值为 81%～94% 的对映选择性，通过一次重结晶可将产物的 ee 值提高到 99%[36]。

1996 年，Charette 等以 (R)-BINAP/Ru(OAc)$_2$ 为催化剂，实现了活化的 N-磺酰基亚胺的氢化。相应产物的收率在 48%～82%，ee 值为 17%～84%[37]。

2015 年，Zhou 等通过利用位阻大、负电性强配体 (S)-Binapine 组成的 [NiCl$_2$(DME)]/(S)-Binapine 催化剂，在 HCO$_2$H/Et$_3$N 提供氢源的条件下，成功实现了苯甲酰基、乙酰基和苯基活化的肼类亚胺的高对映选择性转移氢化反应，ee 值高达 98%。根据氘代实验的结果，作者提出甲酸脱羧提供甲酰基的 H 作为负氢，生成镍-氢活性物种对活化亚胺共轭加成，随后质子解和互变异构得到氢化产物的机理。通过 DFT 计算，作者发现在过渡态的结构中，过渡态 TS 形成的凹面比过渡态 TS-S 的凹面与底物更能互补，同时配体 (S)-Binapine 叔丁基上的 C—H 键与肼类底物的氧原子存在弱的相互作用，正是催化剂与底物形状的互补和存在的弱相互作用导致了该不对称转移氢化的高立体选择性[38]。

## 3.1.3.2 硅氢还原反应

与吸电子基团共轭的 C=C 键的选择性还原是合成有机化学中重要的转化步骤。特别是选择性还原与氰基共轭的 C=C 长期以来一直是合成难题，Xu 等人报道了一种利用 $Cu(OAc)_2 \cdot H_2O/(S_a)$-1a 和 $PhSiH_3$ 原位形成的手性铜氢化物的催化剂，实现了对 β,β 二取代-α,β-不饱和二腈的不对称还原，当其中至少一个取代基为芳基时可以得到很高 ee 值的还原产物，当两个取代基同时为烷基时，则会导致反应选择性的急剧下降。此外苯环上取代基的电子性质对反应的对映选择性也有一定程度的影响，给电子基团的引入有利于选择性的提高，而吸电子基团的引入则不利于对反应的立体控制[39]。

Buchwald 等人报道了由 CuCl、NaO$t$Bu、$(S_a)$-1b 以及含氢硅油组成的催化体系能还原 α,β-不饱和酸酯的双键。反应不仅可以高产率地得到相应的还原产物，而且表现出很好的对映选择性。底物中的双键构型对选择性几乎没有影响。$Z$ 构型和 $E$ 构型烯烃得到相似的结果，但是产物的比旋光符号正好相反，因此可以方便地通过改变底物双键构型得到还原产物的一对对映体[40]。

[反应式: Ph-C(Me)=CH-C(O)OEt + CuCl/NaOtBu/(S$_a$)-1b, PHMS → Ph-*CH(Me)-CH$_2$-C(O)OEt]
产率：84%
ee：90%

Buchwald 小组进一步把该催化剂体系应用于 α,β-不饱和环酮双键的不对称还原，发现环戊烯酮是最适宜的反应底物，可得到很高的立体选择性，对于空间体积较大的异丙基取代底物以及环庚烯酮底物而言，反应速率相对较慢，但并不影响反应的立体控制结果。与前面的 α,β-不饱和酸酯的还原不同，还原试剂 PHMS 的用量应控制在 1eq. 以防止过度还原得到饱和环醇[41]。

[反应式: 环烯酮 + CuCl/NaOtBu/(S$_a$)-1b, PHMS → 饱和环酮]
R = Me, Bu, iPr, Bn, CH$_2$=CH(CH$_2$)$_4$,
    BnCH$_2$, BnO(CH$_2$)$_4$, MeO$_2$C(CH$_2$)$_2$
n = 1, 3
产率：78%～91%
ee：92%～98%

具有 β-手性中心的内酯或内酰胺是许多生物活性物质的重要结构单元，Buchwald 小组报道了铜催化的 α,β-不饱和内酯的不对称还原反应可以方便地实现该类化合物的合成，研究发现，使用在空气中稳定的 CuCl$_2$·H$_2$O 代替 CuCl 作为催化前体同样有效，而且质子性添加剂有助于提高底物的转化率。此外，该小组还成功地将该策略应用于抗抑郁药物 (−)-帕罗西汀的对映选择合成[42]。

[反应式: α,β-不饱和内酯/内酰胺 + CuCl$_2$·2H$_2$O/NaOtBu/(S$_a$)-1b, PHMS → 产物]
R = Ph, Bu, iPr, Et, Bn, 3,4-(MeO)$_2$C$_6$H$_3$CH$_2$
X = O, N(PMP)
n = 1, 2
产率：70%～94%
ee：71%～94%

[反应式: PMPN-内酰胺(4-氟苯基) + CuCl$_2$/NaOtBu/(S$_a$)-1b, PHMS, tAmOH → 手性内酰胺 → (−)-帕罗西汀]
产率：90%
ee：90%
(−)-帕罗西汀

Carretero 等人使用 CuCl/*t*BuONa 催化体系 ( 手性膦配体为 BINAP 或 SegPhos) 催化 $\beta,\beta$- 双取代-$\alpha,\beta$-不饱和吡啶-2- 基砜的还原，所用的双取代烯基砜均可高产率地得到相应的还原产物，但是当为萘 -2- 基取代底物时，产物的产率(89%)与 ee 值(70%)为最低，其他的烷基取代、芳基取代或环状取代均取得了比较满意的对映选择性。当用苯基代替吡啶 -2- 基后，不会发生还原反应，说明吡啶 -2- 基的存在是进行还原反应的必要条件 [43]。

L: *R*-BINAP　产率：98%；ee：94%
L: *R*-SegPhos　产率：98%；ee：94%

Zheng 等人利用 Cu(OAc)$_2$·H$_2$O 为催化剂前体，($S_a$)-**1a** 为配体，PMHS 为还原剂，叔丁醇作为质子性添加剂实现了 $\gamma$-邻苯二甲酰亚胺基取代-$\alpha,\beta$-不饱和羧酸酯的不对称还原反应，得到高 ee 值的 $\gamma$-氨基酸衍生物，并将该方法成功用于手性药物 (*R*)-Baclofen 的不对称合成 [44]。

Lipshutz 等人报道了 (Ph$_3$P)CuH/($R_a$)-**19d** 催化体系实现了对 $\beta$-取代 -$\alpha,\beta$-不饱和环酮的不对称还原。

产率：95%；ee：99%
产率：88%；ee：98.5%

当使用 CuCl/*t*BuONa 或 MeONa 代替 (Ph$_3$P)CuH 为催化剂前体，同样可以获得很好的结果，而且当反应底物与配体物质的量之比为 275000/1 时，虽然反应时间有所延长，但仍然可以实现完全转化并获得令人满意的立体控制结果 [45]。

在过渡金属催化的不对称氢硅烷还原反应中，除了上述活化烯烃的不对称碳-碳双键还原之外，其他含极化的碳-杂原子双键化合物如酮的还原也有报道。

Lipshutz 等人利用 CuCl 或 Cu(oAc)$_2$ 与手性膦配体 ($R_a$)-**19d** 形成的配合物，可在 *t*BuONa 和 PMHS 存在下构筑高效的 CuH 催化体系，实现了一系列官能团取代酮的对映选择还原，这些还原产物是合成一些具有重要生物活性物质或药物的关键中间体[46]。

$$\text{F}_3\text{C-C}_6\text{H}_3(\text{CF}_3)\text{-COCH}_3 \xrightarrow[\text{PMHS}]{\text{CuCl或Cu(OAc)}_2/(R_\text{a})\text{-19d}/t\text{BuONa}} \text{F}_3\text{C-C}_6\text{H}_3(\text{CF}_3)\text{-CH(OH)CH}_3$$

CuCl 产率：98%；ee：91.7%
Cu(OAc)$_2$ 产率：90%；ee：95%

Lipshutz 等人报道了 CuH 催化体系在催化氢硅烷对简单酮的硅烷还原中，如果选择适当的硅烷可以高产率地给出相应的还原产物醇的三烷基硅醚，为硅基保护醇的合成提供了便利的方法，当使用 Cu/($S_\text{a}$)-19d 催化体系时则可以实现对酮的不对称氢硅烷化得到光学活性醇的三烷基硅醚[47]。

$$\text{(2-OMe,4-OMe)C}_6\text{H}_3\text{COCH}_3 + \text{TBS-H} \xrightarrow[t\text{BuONa}]{\text{Cu/(}S_\text{a}\text{)-19d}} \text{(2-OMe,4-OMe)C}_6\text{H}_3\text{CH(OTBS)CH}_3$$

产率：90%；ee：80%

Riant 等人报道了一种 CuF$_2$/($S_\text{a}$)-1a/PhSiH$_3$ 催化体系，能有效地还原酮得到相应的光学活性仲醇。更有趣的是，该催化体系在空气中比在氩气氛围下表现出更好的催化活性以及对映选择性，芳香酮底物的反应速率和对映选择性均优于脂肪酮。此外，$\alpha$-酮酸酯作为底物时虽得到高产率的还原产物，但是 ee 值仅有 11%[48]。

$$\text{PhCOnPr} + \text{PhSiH}_3 \xrightarrow[\text{2. HCl aq.}]{\text{1. CuF}_2/(S_\text{a})\text{-1a/PhSiH}_3,\ \text{甲苯,室温}} \text{PhCH(OH)nPr}$$

产率：80%
ee：92%

Kantam 等人使用纳米氧化铜和 ($S_\text{a}$)-1a 为催化剂，PhSiH$_3$ 为还原剂，实现了芳香酮的高对映选择性还原，如果使用普通 CuO 替代纳米级 CuO 不仅会导致反应速率的大幅下降，而且反应的选择性也明显降低，此外，该催化体系的最大优势在于催化剂可回收再使用而不影响催化活性[49]。

$$\text{Ph-CO-CH}_3 + \text{PhSiH}_3 \xrightarrow[\text{甲苯, -20°C}]{\text{纳米CuO, }(S_a)\text{-1a}} \xrightarrow{\text{NaOH} \atop \text{MeOH}} \text{Ph-CH(OH)-CH}_3$$

产率：85%
ee：99%

Dagorne 等人报道了一种使用 CuCl/NaO$t$Bu/($R_a$)-1a 催化体系来还原芳香酮，发现还原剂氢硅烷种类对反应的对映选择性有明显的影响，其中 PhMeSiH$_2$ 的效果最好，均得到理想的立体选择性结果。他们从动力学角度分析发现反应速率不仅与催化剂、酮有关，而且与还原剂有关，表明还原剂参与了反应过渡态的形成，从而解释了选择性受还原试剂结构影响的原因[50]。

$$\text{Ph-CO-CH}_3 + \text{PhMeSiH}_2 \xrightarrow[\text{甲苯, -78°C, 18h}]{\text{CuCl/NaO}t\text{Bu}/(R_a)\text{-1a}} \xrightarrow{\text{NaOH} \atop \text{MeOH}} \text{Ph-CH(OH)-CH}_3$$

产率：99%
ee：93%

### 3.1.3.3 氢杂原子化反应

(1) 氢胺化

烃的不对称氢胺化反应虽然是制备光学活性胺类化合物便利的方法，但是该反应目前仍然是具有挑战性的研究课题之一。Togni 等人于 1997 年首次报道了铱催化的苯胺对降冰片烯的对映选择性加成，发现反应体系中存在裸露氟离子对反应活性及对映选择性有显著作用，当使用 [IrCl($S_a$)-1a]$_2$ 为催化剂没有氟离子存在时，只得到 12% 产率的氢胺化产物，ee 值为 57%，而在反应体系中 [F$^-$]/[Ir] 为 4 时，反应产率可以提高到 22%，产物 ee 值为 95%[51]。

| [F$^-$]/[Ir] | 产率/% | ee/% |
|---|---|---|
| 0 | 12 | 57 |
| 2 | 45 | 78 |
| 4 | 22 | 95 |

Hartwig 等人报道了 Pd(Ph$_3$P)$_4$/TfOH、Pd(O$_2$CCF$_3$)$_2$/DPPF/TfOH 以及 (DPPF)Pd(OTf)$_2$ 等催化体系能有效催化芳胺对苯乙烯衍生物的加成，高产率地得到相应的氢胺化产物。在此基础上，采用手性钯配合物 [($R_a$)-1a

Pd(OTf)$_2$ 为催化剂，可以实现对映选择性的氢胺化反应得到具有较高 ee 值的手性胺[52]。

PhNH$_2$ + [4-CF$_3$C$_6$H$_4$-CH=CH$_2$] $\xrightarrow{[(R_a)\text{-1a}]Pd(OTf)_2}$ (R)-4-CF$_3$C$_6$H$_4$-CH(NHPh)CH$_3$
产率：80%
ee：81%

PhNH$_2$ + [2-naphthyl-CH=CH$_2$] $\xrightarrow{[(R_a)\text{-1a}]Pd(OTf)_2}$ (R)-2-naphthyl-CH(NHPh)CH$_3$
产率：99%
ee：64%

Lin 等人对钯催化的苯胺对苯乙烯衍生物的不对称氢胺化反应进行了深入研究，考察了手性双膦配体 BINAP(1a) 中 4,4′-取代基对反应选择性的影响[53]。

Ar-CH=CH$_2$ + PhNH$_2$ $\xrightarrow{[Pd((S_a)\text{-1a})(NCMe)_2](OTf)_2}$ Ar-CH(NHPh)CH$_3$

Ar = Ph, 4-FC$_6$H$_4$, 4-ClC$_6$H$_4$, 4-MeOC$_6$H$_4$,
4-CF$_3$C$_6$H$_4$, 萘-1-基，萘-2-基，6-MeO-萘-2-基

产率：75%～97%
ee：27%～73%

Widenhoefer 等使用 Au/($S_a$)-18h 催化剂体系，该小组通过联烯胺的分子内氢胺化反应实现了光学活性四氢吡咯环的构筑，但是个别底物的对映选择性并不理想[54]。

$\xrightarrow{Au/(S_a)\text{-18h}}_{AgBF_4}$

R$^2$ = H, Ph
R$^1$ = H, Me, Et
PG = Cbz, Troc, Fmoc, CO$_2$Me, COMe

产率：61%～91%
ee：34%～91%

(2) 氢烷氧化反应

Widenhoefer 等进一步将 Au/($S_a$)-18h 催化剂体系应用于联烯醇的不对称分子内氢烷氧化反应同样获得了成功，为光学活性四氢呋喃及四氢呋喃衍生物的合成提供了新方法[55]。

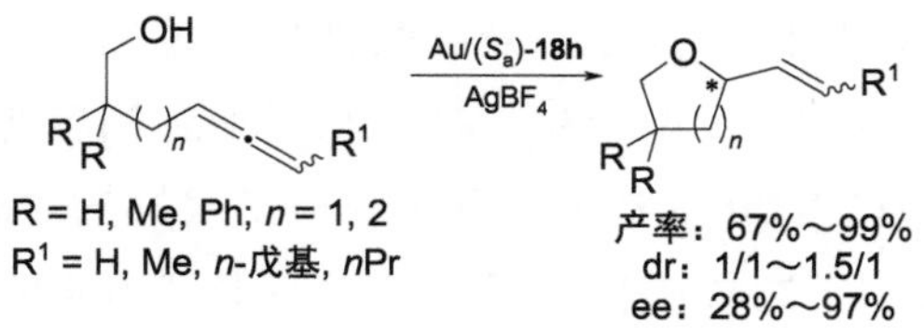
OH
Au/($S_a$)-18h
AgBF4
O
R1
R
R
n
R = H, Me, Ph; n = 1, 2
R1 = H, Me, n-戊基, nPr
产率：67%～99%
dr：1/1～1.5/1
ee：28%～97%

(3) 氢硫醇化反应

Dong 等人报道了一种 Rh 催化体系催化的 1,3- 环己二烯的氢硫醇化反应，获得了具有高化学、区域和对映体控制能力的环己烯 -3- 基硫化物[56]。

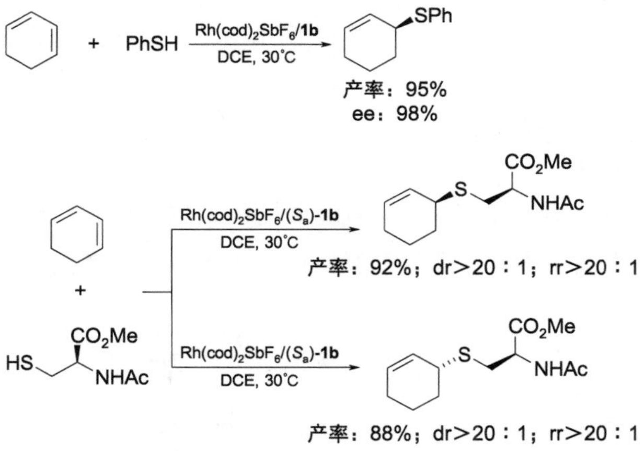
PhSH
Rh(cod)2SbF6/1b
DCE, 30°C
SPh
产率：95%
ee：98%
CO2Me
S
NHAc
Rh(cod)2SbF6/($S_a$)-1b
DCE, 30°C
产率：92%；dr>20：1；rr>20：1
HS
CO2Me
NHAc
Rh(cod)2SbF6/($S_a$)-1b
DCE, 30°C
产率：88%；dr>20：1；rr>20：1

(4) 硼氢化反应

Gevorgyan 等人报道了在配体 **1a** 或 **1b** 形成的手性铑催化剂的作用下，频哪醇硼烷对环丙烯衍生物进行氢硼化可以得到相应的手性环丙烷硼酸酯，其中环丙烯上的甲氧羰基具有很好的定位作用，不仅立体专一地得到 *cis*-异构体，并且具有很好的对映选择性。

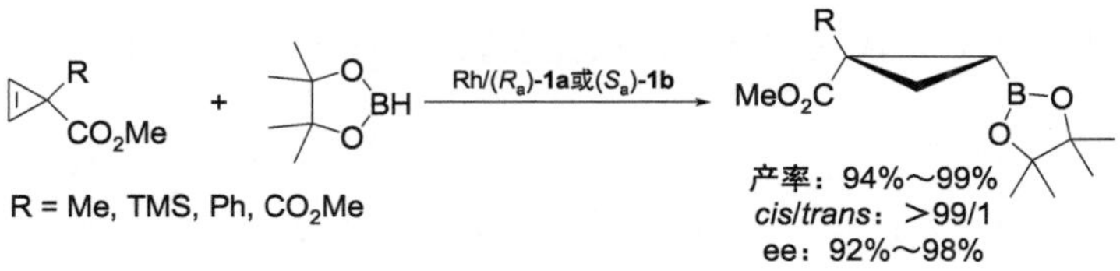
R
CO2Me
O
BH
O
Rh/($R_a$)-1a或($S_a$)-1b
R
MeO2C
B
O
O
R = Me, TMS, Ph, CO2Me
产率：94%～99%
cis/trans：>99/1
ee：92%～98%

此外，产物水解得到的硼酸可以与芳基及烯基碘代烃发生 Suzuki 偶联进一步转化为一系列手性环丙烷衍生物[57]。

### 3.1.3.4 还原偶联反应

氢化条件下的不对称还原偶联反应是一个氢气介入的不对称碳-碳键

偶联反应，Krische 等人报道了 1,3-丁二炔在铑催化的氢化过程中产生的烯基铑中间体可以被亲电的乙二醛衍生物捕获从而得到相应的还原偶联产物。在手性双膦配体 ($R_a$)-18l 存在下，反应获得了很好的对映选择性。此外，对于不对称的 1,3-丁二炔底物，反应同样获得了不错的区域选择性，得到主要在苯基取代一端进行还原偶联的产物 **A**[58]。

R = Ph, CH$_3$, Pr, tBu
R$^1$ = Ph, 萘-2-基, tBu, N-甲基吡咯烷-2-基
呋喃-2-基, 噻吩-2-基

产率：57%～80%
A/B：4/1～＞99/1
ee：83%～98%

Krische 小组实现了在 [Rh]/($S_a$)-18l 催化体系的催化下，磺酰亚胺与乙炔在氢化条件下发生不对称还原偶联可以得到光学活性烯丙胺衍生物[59]。

R = Ph, 4-MeC$_6$H$_4$, 4-ClC$_6$H$_4$, 4-NO$_2$C$_6$H$_4$, 3-MeOC$_6$H$_4$,
4-MeO$_2$CC$_6$H$_4$, 3-Br-4-FC$_6$H$_3$, 3-NO$_2$-4-BrC$_6$H$_3$,
5-NO$_2$-噻吩-2-基, Cy, iPr, cPr
ArSO$_2$：Ns, Ts
BArF = 四[3,5-二(三氟甲基)苯基]硼负离子；Ns = 硝基苯磺酰基

产率：65%～86%,
ee：93%～98%

当用铱代替铑作为中心金属时，1,2-二取代炔烃与磺酰亚胺同样可以发生高对映选择性的还原偶联反应得到烯丙胺衍生物。对于不对称炔烃，反应不仅获得了很好的对映选择性，同时还表现出了很好的区域选择性，主要得到较大取代基一端进行还原偶联产物[60]。

R = Me, Pr, iPr, TBSO(CH$_2$)$_2$
R$^1$ = Ph, 4-ClC$_6$H$_4$, 4-MeOC$_6$H$_4$, 3-MeOC$_6$H$_4$, 萘-2-基,
4-MeO$_2$CC$_6$H$_4$, PhCH＝CH, 呋喃-2-基, 噻吩-2-基,
Cy, iPr, cPr, 环戊基, Me, Bu, iBu
ArSO$_2$：PhSO$_2$, Ts

产率：64%～80%,
rr：10/1～＞99/1
ee：92%～99%

炔醛在铑催化氢化的条件下也可以顺利地发生分子内还原偶联反应得到环状烯丙醇衍生物。当以 ($R_a$)-18l 为配体时，一系列炔醛可以高对映选择性地转化为相应的还原偶联产物 [61]。

R = Ph, CH$_3$, cPr, H, CH$_3$(CH$_2$)$_5$
R$^1$ = H, Me, (CH$_2$)$_4$, (CH$_2$)$_5$
X = O, H$_2$
Y = BnN, 4-BrBnN, NTs, O

产率：63%～99%
ee：91%～99%

Krische 小组使用金属铑催化体系，实现了乙炔与醛、酮在氢气介入下的还原偶联反应，当用 ($R_a$)-MeO-Biphep(18a) 为配体时，可以实现对反应的立体控制，得到光学富集的还原偶联产物 [62]。

R = 邻苯二甲酰亚胺基    产率：85%；ee：88%

R = TBDPS—O—    产率：77%；ee：89%

### 3.1.3.5 烯丙基化/炔丙基化反应

羰基化合物及其衍生物的烯丙基化反应是构建碳碳键重要的反应之一，可合成烯丙醇或胺等重要的药物中间体。

Yamamoto 等人报道了 BINAP 与三氟甲磺酸银形成的手性配合物可以有效地催化烯丙基以及甲代烯丙基三丁锡烷对醛的不对称烯丙基化反应，且该反应体系对芳香醛和脂肪醛均有优异的反应性，可得到高对映选择性的烯丙基烷基化产物 [63]。

R = Ph, (E)-PhCH=CH, 萘-1-基, 呋喃-2-基
(E)-PrCH=CH, 2-MeC$_6$H$_4$, 4-MeOC$_6$H$_4$,
4-BrC$_6$H$_4$, PhCH$_2$CH$_2$
R$^1$ = H, Me

产率：47%～95%
ee：88%～97%

当使用 γ-取代烯丙基锡烷作为烯丙基化试剂时，反应表现出很好的区域选择性，仅得到 γ-烷基化产物，同时手性产物具有良好的对映选择性和非对映选择性[64]。

$$\text{RCHO} + R^1\text{CH=C}(R^1)\text{CH}_2\text{SnR}_3^2 \xrightarrow{(R_a)\text{-1a/AgOTf}} \text{产物}$$

R = Ph, (E)-PhCH=CH, 呋喃-2-基, 2-MeC$_6$H$_4$,
4-MeOC$_6$H$_4$, PhCH$_2$CH$_2$
R$^1$ = Me, CH$_2$=CH
R$^2$ = Me, Bu

产率：41%～73%
ee：58%～94%

Yamamoto 等人发现 AgOTf/($R_a$)-1a 催化体系不能催化烯丙基三甲氧基硅烷与醛的不对称烯丙基化反应。但是加入催化量的 KF 和冠醚 18-冠-6 后，反应不仅可以顺利地进行，而且相应的烯丙基化产物具有很好的对映选择性。在优化反应条件下，无论是芳香醛还是脂肪醛均能以高对映选择性得到手性烯丙醇产物[65]。

$$\text{CH}_2\text{=CHCH}_2\text{Si(OMe)}_3 + \text{RCHO} \xrightarrow[\text{THF, KF, 18-冠-6}]{\text{AgOTf/}(R_a)\text{-1a}} \text{产物}$$

R = Ph, (E)-PhCH=CH, 萘-1-基, 呋喃-2-基
4-MeOC$_6$H$_4$, 4-BrC$_6$H$_4$, 2-MeOC$_6$H$_4$,
Cy, PhCH$_2$CH$_2$

产率：57%～95%
ee：86%～97%

此外，在 AgF/($R_a$)-1a 的存在下，巴豆基三甲氧基硅烷同样可以作为烯丙基化试剂，高区域及立体选择地得到相应的 γ-烷基化合物。反应中巴豆基硅烷中双键构型对反应的立体选择性几乎没有影响，无论是使用单一构型还是使用混合物均表现出相似的反应性[66]。

$$\text{CH}_3\text{CH=CHCH}_2\text{Si(OMe)}_3 + \text{RCHO} \xrightarrow[\text{MeOH, -20℃(7h)→室温(17h)}]{(R_a)\text{-1a(摩尔分数6\%)}, \text{AgF(摩尔分数10\%)}} anti(1R, 2R) + syn(1R, 2S)$$

| E/Z | 产率/% | anti(ee/%)/syn(ee/%) |
|---|---|---|
| 83/17 | 77 | 92(96)/8(62) |
| <1/99 | 82 | 94(94)/6(60) |
| 45/55 | 99 | 93(94)/7(60) |

Yamamoto 小组报道，Ag(Ⅰ)/($R_a$)-19e 催化体系可以催化烯丙基三甲

氧基硅烷对酮的不对称烯丙基化反应得到重要的合成中间体高烯丙醇衍生物。反应具有很好的底物适用范围，无论是芳香酮还是脂肪酮都获得了很好的立体选择结果。

$$\text{Ph}\overset{O}{\underset{}{\|}}\text{Me} + \diagup\!\!\!\diagdown\text{Si(OMe)}_3 \xrightarrow[\text{THF, -78°C}]{\text{AgF/}(R_a)\text{-19e} \atop \text{MeOH}} \text{Ph}\overset{\text{Me OH}}{\underset{}{|}}\diagdown\!\!\!\diagup$$

产率：96%
ee：82%

此外，对于 $\gamma$-取代烯丙基硅烷区域专一地得到 $\gamma$-加成产物，反应不仅表现出很好的对映选择性，同时具有不错的非对映选择性[67]。

$$\text{Ph}\overset{O}{\underset{}{\|}}\text{Me} + \text{Me}\diagup\!\!\!\diagdown\text{Si(OMe)}_3 \xrightarrow[\text{MeOH}]{\text{AgF/}(R_a)\text{-19e}} \text{Ph}\overset{\text{Me OH}}{\underset{\text{Me}}{|}}\diagdown\!\!\!\diagup$$

(Z) 产率 95%, syn/anti 90/10, ee 93%
(E) 产率 60%, syn/anti 90/10, ee 95%

Kanai 和 Shibasaki 等发展了 Cu(Ⅱ)/($R_a$)-19d 催化体系催化烯基及苯基硅烷对醛的不对称烷基化反应，除了 $\alpha$-取代烯基硅烷只得到了中等程度的对映选择性之外，其他烯基硅烷无论与脂肪醛还是芳香醛反应可获得高 ee 值的烷基化产物[68]。

$$\text{R}\overset{O}{\underset{}{\|}}\text{H} + \text{R}^1\text{-SiY(OMe)}_2 \xrightarrow[\text{2) TBAF}]{\text{1) CuF}_2\cdot 2\text{H}_2\text{O/}(R_a)\text{-19d}} \text{R}\overset{\text{OH}}{\underset{}{|}}\text{R}^1$$

R = Ph, 4-ClC$_6$H$_4$, 4-MeC$_6$H$_4$, 4-MeOC$_6$H$_4$, 噻吩-2-基,
(E)-PhCH=CH, PhCH$_2$CMe$_2$, Cy
R$^1$ = CH$_2$=CH, CH$_3$(CH$_2$)$_3$CH=CH, CH$_3$(CH$_2$)$_2$C=CH$_2$, Ph
Y = OMe, Me

产率：48%～99%
ee：52%～99%

他们进一步将该催化体系应用于烯基或苯基硅烷对三氟甲基芳酮的不对称烃基化反应制备相应的光学富集含三氟甲基的叔醇，但是相对于醛的烷基化而言，反应的立体选择性有一定程度的下降[69]。

$$\text{R}\overset{O}{\underset{}{\|}}\text{H} + \text{R}^1\text{-SiY(OMe)}_2 \xrightarrow[\text{2) TBAF}]{\text{1) CuF}_2/(R_a)\text{-19d}} \text{R}\overset{\text{HO R}^1}{\underset{\text{CF}_3}{|}}$$

R = Ph, 4-ClC$_6$H$_4$, 4-MeC$_6$H$_4$, 4-BrC$_6$H$_4$, 萘-2-基
R$^1$ = CH$_2$=CH, CH$_3$(CH$_2$)$_3$CH=CH, Ph
Y = OMe, Me

产率：12%～100%
ee：49%～84%

Studer 等发现 Cu(Ⅰ) 或 Cu(Ⅱ)/($R_a$)-19e 催化体系在 1,4-环己二烯-3-

基三异丙氧基硅烷对醛的烷基化反应中可以实现对底物的去对称化，同时以中等至很好的非对映选择性以及很好的对映选择性得到光学活性1,3-环己二烯衍生物[70]。

$$\text{环己二烯-Si(O}i\text{Pr)}_3 + \text{ArCHO} \xrightarrow{\text{Cu(OTf)或Cu(OTf)}_2/(R_a)\text{-19e}} \text{环己二烯-CH(OH)Ar}$$

Ar = Ph, 4-MeC$_6$H$_4$, 4-$i$PrC$_6$H$_4$, 4-$i$BuC$_6$H$_4$, 4-MeOC$_6$H$_4$,
4-FC$_6$H$_4$, 4-BrC$_6$H$_4$, 4-ClC$_6$H$_4$, 2-MeC$_6$H$_4$, 呋喃-2-基,
2-MeOC$_6$H$_4$, 3-MeC$_6$H$_4$, 3-MeOC$_6$H$_4$, 3-BrC$_6$H$_4$,
5-Br-呋喃-2-基, 噻吩-2-基, 5-Br-噻吩-2-基, 萘-2-基

产率：44%～91%
syn/anti：2.5/1～29/1
ee：85%～94%

Kanai 和 Shibasaki 等人还发现 Cu(Ⅱ)/($R_a$)-19d 催化体系对烯基硼酸酯与醛的不对称烷基化反应同样有效。虽然反应的产率相对较低，但是他们发现添加剂二氟三苯基硅酸四丁基铵（TBAT）的加入可以显著提高催化活性。对于芳醛和 $\alpha,\beta$-不饱和醛底物而言，反应得到高产率、高 ee 值的烯基化产物，对于 $\alpha$-支链化脂肪醛在添加剂 TBAT 和 BF$_3$·Et$_2$O 的存在下同样可以得到满意的结果，但是对于支链脂肪醛和芳香酮则不能得到正常的加成产物[71]。

$$\text{RCHO} + \text{pinB-CH=CH-R}^1 \xrightarrow[\text{2) H}_2\text{O}]{\text{1) CuF}_2\cdot\text{2H}_2\text{O}/(R_a)\text{-19d, TBAT或TBAT/BF}_3\cdot\text{Et}_2\text{O}} \text{R-CH(OH)-CH=CH-R}^1$$

R = Ph, 4-ClC$_6$H$_4$, 4-MeC$_6$H$_4$, Cy, $c$Pr
R$^1$ = CH$_2$=CH, CH$_3$(CH$_2$)$_3$CH=CH

产率：87%～99%
ee：88%～98%

Ito 等人报道了手性 [Pd($\pi$-C$_3$H$_5$)(cod)]BF$_4$/($R_a$)-1a 配合物（C$_3$H$_5$为烯丙基）的催化下烯丙醇醋酸酯对 $\alpha$-乙酰氨基-$\beta$-羰基膦酸酯的不对称烯丙基取代反应，得到具有广泛生物活性的手性 $\alpha$-氨基膦酸酯衍生物[72]。将底物拓展为 $\alpha$-乙酰氨基-$\beta$-羰基酸酯时，取得了更好的对映选择结果，实验数据表明，烯丙醇醋酸酯 $\gamma$-位取代基的空间体积对反应对映选择性的影响明显，随着取代基体积的增大，产物的对映选择性显著提高，而 $\alpha$-乙酰氨基-$\beta$-羰基酸酯底物上的取代基空间体积对反应对映选择性的影响很小[72,73]。

| R | 产率/% | ee/% |
|---|---|---|
| —PO(OMe)$_2$ | 78 | 88 |
| —CO$_2$Me | 71 | 95 |

在该催化体系下，2-取代-1,3-二酮类底物同样能够顺利进行烯丙基化反应，无论是环状或是链状的1,3-二酮都能以高产率得到具有季碳手性中心的2,2-双取代-1,3-二酮衍生物[74]。

$R^1$ = Ph, 4-MeOC$_6$H$_4$
$R^2$ = Me
$R^1$-$R^2$ = (CH$_2$)$_3$, (CH$_2$)$_4$, (CH$_2$)$_5$, (CH$_2$)$_6$, (CH$_2$)$_2$CMe$_2$CH$_2$
$R^3$ = Me, Et
$R^4$ = Pr, Cy, Ph, 4-MeOC$_6$H$_4$, 4-CF$_3$C$_6$H$_4$

产率：65%～99%
ee：64%～89%

Takahashi 和 Lin 等考察了配体 **1a** 以及配体 **1a** 的 4,4'-位上三甲硅基取代基对钯催化的 α-乙酰氨基-β-羰基酸酯的烯丙基取代反应的影响，4,4'-位上引入三甲硅基得到的新配体 **3h** 与母体化合物 **1a** 相比具有更好的不对称诱导能力，产物的对映选择性均有一定程度的提高，但是反应的产率有所降低[75]。

| L* | R | $R^1$ | 产率/% | ee/% |
|---|---|---|---|---|
| ($R_a$)-BINAP(**1a**) | Me | H | 87 | 68 |
| ($R_a$)-TMS-BINAP(**3h**) | Me | H | 75 | 77 |
| ($R_a$)-BINAP(**1a**) | Me | Ph | 78 | 90 |
| ($R_a$)-TMS-BINAP(**3h**) | Me | Ph | 68 | 93 |
| ($R_a$)-BINAP(**1a**) | Ph | H | 93 | 72 |
| ($R_a$)-TMS-BINAP(**3h**) | Ph | H | 90 | 84 |

**3h**
($R_a$)-TMS-BINAP

在钯催化的不对称烯丙基化反应中，亲核试剂大多局限于丙二酸酯、β-酮酸酯等具有活泼亚甲基的化合物，而醛酮作为相应的亲核试剂则鲜有报道。Saicic 巧妙地利用仲胺与醛形成的烯胺对烯丙基钯中间体的亲核取代实现了以醛作为亲核试剂的分子内烯丙基化反应,在手性钯催化体系 Pd/($R_a$)-1a 的催化下,实现了对反应过程的立体控制,得到光学富集的环戊烷及四氢吡咯衍生物[76-77]。

X = C(CO$_2$Et)$_2$  产率:40%; ee:91%
X = NTs  产率:27%; ee:59%

2002 年 Taguchi 等利用氮亲核试剂的不对称烯丙基取代反应成功地实现了具有轴手性的光学活性 N-芳基-N-烯丙基酰胺衍生物的合成,虽然反应只获得了中等程度的对映选择性,但这是首例利用不对称催化的方法实现非双芳基的阻旋异构体的对映选择合成[77]。

| R | P配体 | 产率/% | 比例(S,R/S,S) |
|---|---|---|---|
| CH$_3$ | ($R_a$)-1a | 95 | 15.9 |
| CH$_3$ | ($S_a$)-1a | 95 | 5.2 |
| Ph | ($R_a$)-1a | 93 | 13.5 |
| Ph | ($S_a$)-1a | 92 | 3.2 |

Curran 小组使用 ($S_a$)-1a 为配体,烯丙基醋酸酯为烯丙基试剂,在低温反应条件下实现了 N-(2-叔丁基苯基)酰胺的不对称烯丙基取代反应,但由于低温反应导致反应活性降低,通常底物无法全部转化而导致产率降低[78]。

R = CH$_2$=CH, Me, Bn, Ph, tBu
产率:27%~97%
ee:12%~53%

Taguchi 等进一步使用光学活性乳酰胺及扁桃酰胺作为底物,通过不对称诱导的方法可以高非对映选择性地得到相应的 N-烯丙基化产物,通过进一步转化可以得到 ee 值 > 97% 的阻旋异构苯胺衍生物[79]。

Kobayashi 等人首次开发了使用氨水作为氮源进行钯催化的烯丙基胺化以制备伯胺,$NH_3$ 作为氮亲核试剂参与烯丙基取代反应,无论是链状还是环状的烯丙醇酯均能高选择性地得到相应的伯胺化合物。当采用 $(R_a)$-1a 作为配体时,可实现不对称的烯丙基取代反应,得到高对映选择性的手性伯胺[80]。

Ph⌒⌒CH(OAc)Ph  [PdCl(C$_3$H$_5$)]$_2$ / $(R_a)$-1a / aq NH$_3$/1,4-二氧六环 / 0.04mol/L, 室温, 18h → Ph⌒⌒CH(NH$_2$)Ph  产率:71%  ee:87%

Krische 等人报道了 [Ir(cod)Cl]$_2$ 与双膦配体 $(R_a)$-18l 以及间硝基苯甲酸原位产生的金属铱配合物在氢转移条件下可以实现醇或醛与醋酸烯丙酯的不对称烯丙基化反应,得到光学富集的高烯丙醇衍生物。其中添加剂间硝基苯甲酸是反应实现高转化率的关键,其与金属铱发生 C—H 插入得到的环状金属铱化合物是反应的活性催化剂[81]。

⌒OAc + HOCH$_2$R  [Ir(cod)Cl]$_2$/$(R_a)$-18l / 3-NO$_2$C$_6$H$_4$CO$_2$H → ⌒⌒CH(OH)R
R = (E)-PhC=CH, (E)-Me$_2$C=CH(CH$_2$)$_2$C(Me)=CH,
(E)-Me(CH$_2$)$_5$CH=CH, Ph(CH$_2$)$_3$, Me(CH$_2$)$_7$, iPr,
BnOCH$_2$C(Me$_2$), BnOCH$_2$, BnO(CH$_2$)$_2$, BnO(CH$_2$)$_3$
产率:63%~83%  ee:86%~95%

⌒OAc + O=CHR  [Ir(cod)Cl]$_2$/$(R_a)$-18l / 3-NO$_2$C$_6$H$_4$CO$_2$H → ⌒⌒CH(OH)R
R = (E)-PhC=CH, (E)-Me$_2$C=CH(CH$_2$)$_2$C(Me)=CH,
(E)-Me(CH$_2$)$_5$CH=CH, Ph(CH$_2$)$_3$, Me(CH$_2$)$_7$, iPr,
BnOCH$_2$C(Me$_2$), BnOCH$_2$, BnO(CH$_2$)$_2$, BnO(CH$_2$)$_3$
产率:41%~88%  ee:92%~97%

Krische 小组报道了 Ir/$(S_a)$-SegPhos(19a) 催化体系下实现醇或醛与 α-甲基取代烯丙基醋酸酯的不对称烯丙基化反应,高产率地得到相应的巴豆基化产物,反应获得了较好的非对映选择性和很好的对映选择性[82]。

$$\text{\raisebox{0pt}{}} \underset{R}{\overset{\text{OAc}}{\diagdown\!\!\!\!\diagup}}\!\!\!\!\underset{\text{Me}}{|} + \underset{R}{\overset{\text{OH}}{\diagdown\!\!\diagup}} \left(\text{或}\ \overset{O}{\underset{R}{\diagdown\!\!\diagup}}\right) \xrightarrow[3\text{-}O_2NC_6H_4CO_2H]{\text{Ir}/(S_a)\text{-}\mathbf{19a}} \underset{\text{Me}}{\overset{\text{OH}}{\diagdown\!\!\diagup}}\!\!R$$

R = Ph, 3-MeOC$_6$H$_4$, 4-MeOC$_6$H$_4$, 4-BrC$_6$H$_4$, 4-MeO$_2$CC$_6$H$_4$,
N-甲基-吲哚-2-基, PhCH=CH, Ph(CH$_2$)$_2$, BnO(CH$_2$)$_3$

醇，产率：61%～73%
anti/syn：5/1～8/1
ee：86%～97%
醛，产率：66%～80%
anti/syn：7/1～11/1
ee：96%～98%

Krische 等人用 [Ir(cod)Cl]$_2$ 与 ($R_a$)-**1a** 形成的手性铱配合物有效地催化乙酸烯丙酯与醇的不对称偶联反应，反应有着优异的普适性，在该体系下可高收率、高对映选择性地得到手性烯丙醇产物[83]。

$$\diagdown\!\!\!\!\diagup\text{OAc} + \underset{R}{\overset{\text{OH}}{\diagdown\!\!\diagup}} \xrightarrow[\substack{m\text{-}NO_2BzOH,\ Cs_2CO_3,\\ \text{THF, 100}^\circ\text{C}}]{[\text{Ir}(cod)Cl]_2/(R_a)\text{-}\mathbf{1a}} \underset{}{\overset{\text{OH}}{\diagdown\!\!\diagup}}\!\!\overset{*}{R}$$

产率：55%～80%
ee：90%～93%

R = p-NO$_2$Ph, Ph, p-MeOPh, p-(MeO$_2$C)Ph, p-BrPh,
3,5-Cl$_2$Ph, 胡椒基, o-MeOPh, N-甲基吲哚-2-基

Hayashi 等人发现了一种新型的烯基砜与芳基钛试剂的电离取代反应，在手性金属铑配合物 [Rh(OH)($S_a$)-**1a**]$_2$ 催化下苯基钛试剂可以与 $\alpha,\beta$-不饱和砜反应，得到相应的光学活性烯丙基取代芳烃衍生物[84]。

$$\underset{\text{SO}_2\text{Ph}}{\bigcirc\!\!\!\!\!=} + \text{ArTi}(O i\text{Pr})_3 \xrightarrow[\substack{(\text{Rh}摩尔分数3\%)\\ \text{THF, 40}^\circ\text{C, 12h}}]{[\text{Rh(OH)}(S_a)\text{-}\mathbf{1a}]_2} \underset{}{\bigcirc\!\!\!\!\!-}\!\!\text{Ar}$$

Ar = Ph：产率94%，ee＞99%
Ar = 4-MeOC$_6$H$_4$：产率99%，ee＞99.9%

$$\underset{\text{SO}_2\text{Ph}}{\overset{n\text{-}C_6H_{13}}{\diagdown\!\!\diagup}} \xrightarrow[\substack{[\text{Rh(OH)}(S_a)\text{-}\mathbf{1a}]_2\\ (\text{Rh}摩尔分数3\%)\\ \text{THF, 40}^\circ\text{C, 12h}}]{4\text{-MeOC}_6H_4\text{Ti}(Oi\text{Pr})_3} \underset{C_6H_4(4\text{-OMe})}{\overset{n\text{-}C_6H_{13}}{\diagdown\!\!\diagup}}$$

产率46%，ee99.2%

Tian 等人在钯催化剂和硼酸的存在下，催化一系列硝基乙酸盐与烯丙基伯胺进行烯丙基化反应，然后进行脱羧得到结构多样的硝基化合物，得到的手性产物具有中等至优异的产率以及优异对映选择性[85]。

$R^1$ = Me, Ph, 2-ClC$_6$H$_4$, 4-MeOC$_6$H$_4$, 萘-2-基, Cy
$R^2$ = H, Me
$R^3$ = Me, Et, Ph
$R^4$ = Me, Et, Cy, tBu, Bn
$R^5$ = NO$_2$, CN, COMe, SO$_2$Ph

产率：55%～92%；ee：88%～99%

$R^1$ = H, $R^2$ = Bn
$R^1$-$R^2$ = (CH$_2$CH$_2$)$_2$O

产率：80%；ee：94%
产率：55%；ee：94%

Hou 等人报道，Cu(Ⅰ)/($R_a$)-18l 催化体系能催化烯胺与 α-芳基取代炔丙醇醋酸酯之间的炔丙基代反应。反应具有较好的底物使用范围，无论是拉电子还是给电子基团取代的芳香族烯胺均获得了不错的对映选择结果[86]。

产率：77%；ee：85%

### 3.1.3.6 共轭加成及1,6加成反应

在共轭加成反应中，各种有机金属可作为碳亲核试剂参与到反应中，如有机硼试剂、有机硅试剂、格氏试剂、有机钛试剂以及有机锌试剂，其中有机硼试剂对空气、水较稳定，反应活性较低，能有效避免背景反应和其他副反应而成为其中报道最多的一类碳亲核试剂。

Hayashi 等人利用金属铑催化剂实现了芳基硼酸和烯基硼酸对烯酮的不对称共轭加成，在 Rh(acac)(C$_2$H$_4$)$_2$(acac= 乙酰丙酮) 与 ($S_a$)-1a 原位产生的手性金属铑配合物的存在下，无论是环酮还是直链或支链的烯酮均可得到高 ee 值的加成产物。其中催化剂前体对反应的影响十分明显，如果使用 Rh(acac)(CO)$_2$ 代替 Rh(acac)(C$_2$H$_4$)$_2$ 则会导致反应产率的极度降低以及反应选择性的明显下降[87,88]。

$R^1$ = iPr, $C_5H_{11}$    R = Ph, 4-MeC$_6$H$_4$, 4-CF$_3$C$_6$H$_4$, 3-MeOC$_6$H$_4$,    产率：51%~>99%
$R^2$ = Me          3-ClC$_6$H$_4$, (E)-1-庚烯基, (E)-tBuCH=CH    ee：91%~99%
$R^1$-$R^2$ = (CH$_2$)$_3$, (CH$_2$)$_4$, (CH$_2$)$_5$

Gong 等人使用 7,7-位双取代的手性 BINAP 衍生物 ($R_a$)-**7a** 代替 BINAP 作为配体在铑催化的芳基硼酸对 α,β-不饱和酮的共轭加成反应中也获得了极大成功。无论是环状烯酮还是非环状烯酮都可以获得很好的结果[89]。

产率：90%；ee：99%

Miyaura 等人报道了使用 Rh(acac)(C$_2$H$_4$)$_2$ 与 ($S_a$)-**1a** 原位生成的手性铑配合物作为催化剂，芳基硼酸为硼试剂的 α,β-不饱和酸酯的不对称共轭加成反应，得到了高达 97% 的对映选择性，但是产率稍微不理想。并同时发现使用苯基硼酸酯作为硼试剂可得到较好的结果[90]。

$R^1$ = Me, iPr, Ph
$R^2$ = Et, tBu, Bn, Cy, iPr, tBu    产率：26%~98%
Ar = 4-MeC$_6$H$_4$, 2-MeOC$_6$H$_4$, 4-MeOC$_6$H$_4$, 3-MeOC$_6$H$_4$    ee：77%~98%

1 产率：99%；ee：87%
2 产率：79%；ee：87%

Miyaura 等人对该反应进行了深入研究，2001 年该课题组报道了 α,β-不饱和酰胺不对称共轭加成，反应活性不如相应的酯，需要加入水溶性碱作为添加剂才能转化。底物局限于 β-甲基和伯烷基取代不饱和羧酸酰胺，如果使用 γ-位支链化或 β-芳基取代底物，则反应不能进行或者转化率极低。4-位拉电子基团取代基的硼酸反应活性高,4-位给电子基团取代基的硼酸反应反而会导致反应速率减慢、产率降低,并且会导致反应立体选择性下降[91]。

$$R^1\text{-CH=CH-C(O)NHR}^2 + ArB(OH)_2 \xrightarrow[K_2CO_3]{Rh(acac)(C_2H_4)_2/(S_a)\text{-1a}} R^1\text{-*CH(Ar)-CH}_2\text{-C(O)NHR}^2$$

$R^1$ = Me, $C_5H_{11}$, $i$Pr
$R^2$ = H, Ph, Cy, Bn
Ar = Ph, 4-MeC$_6$H$_4$, 4-MeOC$_6$H$_4$, 4-CF$_3$C$_6$H$_4$

产率:19%~89%
ee:77%~95%

Miyaura 等在此基础上重新研究了配体和碱在铑(Ⅰ)催化 1,4-加成芳基硼酸中对 $\alpha,\beta$-不饱和羰基化合物的影响,发现碱的加入对反应有明显的加速作用,可以降低反应的温度。当三乙胺作为添加剂时,不仅反应得到加速,而且反应产率和对映选择性均有一定程度的提高。当以 [Rh(nbd)$_2$]BF$_4$/($R_a$)-**1a**(nbd= 降冰片二烯)为催化剂、三乙胺为添加剂时,室温下或 50℃条件下,对 $\alpha,\beta$-不饱和酮、醛、酯和酰胺底物分别得到 ee 值高达 99%、92%、94% 和 92% 的对映选择性[92]。

产率:99%;ee:99%    产率:56%;ee:92%    产率:93%;ee:94%    产率:97%;ee:92%

Lukin 等人使用一种商品化的 Rh(nbd)$_2$BF$_4$ 和配体 ($S_a$)-**1a** 原位形成催化剂的方法可获得同样催化效果,同时催化剂的用量可降低到 1.5%(摩尔分数),加成试剂硼酸的用量也可以从以前的 1.5~3eq. 减少到仅 1.05~1.1eq.。这样大大地简化了反应的后处理过程,从而使反应在合成上更具有可行性[93]。

$$R^1\text{-CH=CH-C(O)R}^2 + ArB(OH)_2 \xrightarrow[Et_3N]{Rh(nbd)_2BF_4/(S_a)\text{-1a}} R^1\text{-*CH(Ar)-CH}_2\text{-C(O)R}^2$$

$R^1$ = Me, $C_5H_{11}$
$R^2$ = Me, EtO
$R^1$-$R^2$ = (CH$_2$)$_2$, (CH$_2$)$_3$
Ar = Ph, 4-BrC$_6$H$_4$, 4-MeOC$_6$H$_4$, 4-CF$_3$C$_6$H$_4$

产率:75%~85%
ee:92%~99%

由于某些有机硼酸的制备、分离和纯化较难进行，Hayashi 等人发现邻苯二酚硼烷（HBCat）与炔烃氢硼化后得到的烯基硼酸酯在上述催化体系下同样可以实现对烯酮的共轭加成，当加入 10eq. 的三乙胺作为添加剂时，不仅反应的产率得到大幅提高，而且反应的立体选择性也有少许提高[94]。

$$R\!-\!\!\!\!\!\!\!\!\!\underset{O}{\diagup}\!\!\!\!\!\!\!R^1 + R^2\!\!\!\!\!\!\!\underset{BCat}{\diagup}\!\!\!\!\!\!\!R^3 \xrightarrow[Et_3N]{Rh(acac)(C_2H_4)_2/(S_a)\text{-}1a} $$

R = iPr
R¹ = Me
R-R¹ = (CH₂)₂, (CH₂)₃
R² = Me, C₅H₁₁, Ph, tBu, CH₂OMe
R³ = H, Me

产率：68%～92%
ee：81%～95%

另外，Hayashi 小组使用溴代芳烃经丁基锂处理后再与硼酸酯作用原位生成了芳基硼酸锂盐，并发展了用其代替芳基硼酸作为亲核试剂的方法，同样获得了良好的对映选择性，从而避免了芳基硼酸的分离纯化过程。原来使用硼酸试剂由于硼酸的分解导致不能得到产物的底物，在此体系下也能得到高产率的加成产物。其中水的存在与否以及水的用量对反应有着决定性的影响[95]。

R = Me, iPr, C₅H₁₁
R¹ = Me
R-R¹ = (CH₂)₂, (CH₂)₃
Ar = 4-MeC₆H₄, 4-CF₃C₆H₄, 4-MeOC₆H₄,
3,5-Me₂-4-MeOC₆H₂, 萘-2-基

产率：75%～＞99%
ee：91%～99%

Hayashi 等人将底物扩展为 α,β-不饱和酸酯，对于链状 α,β-不饱和酸酯，同样使用芳基硼酸锂盐代替芳基硼酸作为亲核试剂，反应的产率得到了明显提高，此外酯基的体积越大越有利于反应选择性的提高。然而对于环状 α,β-不饱和酸酯情况正好相反，使用芳基硼酸试剂得到更高产率的产物，另外，环的大小对反应的产物影响十分显著，六元环底物能得到高产率和高 ee 值的产物，而五元环底物虽然产物的对映选择性很好，但反应的产率很不理想[96]。

R = Pr, iPr, Bu
R¹ = Me, Et, iPr, tBu
R-R¹ = (CH$_2$)$_2$, (CH$_2$)$_3$
Ar = Ph, 4-ClC$_6$H$_4$, 4-MeC$_6$H$_4$, 4-CF$_3$OC$_6$H$_4$,
3-MeOC$_6$H$_4$, 萘-2-基

产率: 64%~>99%
ee: 89%~98%

n = 0, 1
Ar = Ph, 4-ClC$_6$H$_4$, 4-MeC$_6$H$_4$, 4-CF$_3$OC$_6$H$_4$,
3-MeOC$_6$H$_4$, 萘-2-基

产率: 33%~95%
ee: 96%~98%

Hayashi 等人进一步地将该催化体系中芳基环硼氧烷代替芳基硼酸并应用到 α,β-不饱和膦酸酯的不对称共轭加成反应中，得到了高对映选择性且较高产率的手性膦酸酯衍生物，并且同样水是反应不可缺的助溶剂，但大量的水可导致催化活性降低，因此水的用量也是反应重要的影响因素。其中双键的构型决定了产物新产生手性中心的构型，对于 (Z)-构型底物而言，由于在反应条件下会缓慢向 (E)-构型转化，会导致反应选择性有少许下降，通过缩短反应时间，在牺牲转化率的前提下同样可获得很好的立体选择性，所得到的手性膦酸酯经过 Hornner-Witting 反应可以快速转化为相应的手性烯烃[97]。

产率: 94%; ee: 96%

Hayashi 等人报道了在手性膦-铑催化剂存在下，使用芳基硼试剂首次实现了对 5,6-二氢-2-吡啶酮的不对称共轭加成反应，得到了高对映选择性的 4-芳基-2-哌啶酮衍生物。通过在低温反应温度下使芳基 4-氟苯基环硼氧烷和 1eq. 水反应得到具有高对映选择性、高产率的 (R)-4-(4-氟苯基)-2-哌啶酮，是合成 3,4-二取代哌啶衍生物的重要合成中间体[98]。

产率: 82%; ee: 98%

Hayashi 报道了 Rh/($R_a$)-SegPhos(**19a**) 催化体系能有效地催化芳基硼酸对香豆素衍生物的不对称共轭加成制备相应的 4-芳基苯并吡喃-2-酮。该方法可以方便地用于泌尿系统药物 (*R*)-托特罗定的对映选择合成[99]。

R = 6-Me, H, 6-MeO$_2$C
Ar = Ph, 4'-MeC$_6$H$_4$, 3'-MeC$_6$H$_4$,
4'-ClC$_6$H$_4$, 4'-MeOC$_6$H$_4$

产率：45%～94%
ee：99.1%～99.8%

(*R*)-托特罗定

Hayashi 等人报道了手性膦-铑催化 α-取代硝基烯的 1,4-共轭加成反应，得到了具有高光学活性的硝基烷烃。对于个别对映选择性较低的产物，将反应介质换成 DMA/H$_2$O 或者 DMF/H$_2$O，产物的 ee 值会得到大幅度提高，其中，对于环状硝基烯而言，环的大小对反应的立体选择性影响很大，六元环和七元环非对映选择性较好，六元环主要得到 *cis*-选择性，而七元环主要表现出 *trans*-选择性。五元环底物则和开链底物相似，非对映选择性较差，主要是以 *trans*-产物为主[100]。

R = Pr, R$^1$ = Me
R-R$^1$ = (CH$_2$)$_3$, (CH$_2$)$_4$, (CH$_2$)$_5$
R$^2$ = Ph, 4-MeC$_6$H$_4$, 4-CF$_3$C$_6$H$_4$, 3-ClC$_6$H$_4$,
萘-2-基, (*E*)-*n*C$_5$H$_{11}$CH=CH

产率：33%～93%
*cis*/*trans*：17/83～88/12
ee：73%～99%

Konno 等人使用 [Rh(cod)$_2$]BF$_4$ 催化 β-三氟甲基取代的 α,β-不饱和酮可顺利地得到相应的加成产物。大多数底物可以获得很好的立体选择性，位阻较大的邻位取代芳基硼酸和烯基硼酸会导致反应产率和选择性下降，尤其是 2-氯取代苯硼酸几乎得到正常的加成产物。由于在催化体系中，(*Z*)-构型烯烃可以快速地转化为热力学稳定的 (*E*)-构型烯烃，无论使用何种构型均得到相同的结果,用二氯甲基替代三氟甲基,虽然也能高产率地得到加成产物,但反应的立体选择性明显下降。其他三氟甲基取代的活化烯烃除了 $\alpha,\beta$-不饱和酰胺之外,均未能取得令人满意的结果[101]。

$$F_3C\text{—CH=CH—C(O)Ph} + PhB(OH)_2 \xrightarrow[\text{甲苯/H}_2\text{O, 回流, 3h}]{[Rh(cod)_2]BF_4/(S_a)\text{-1a}} Ph\overset{CF_3}{\underset{*}{C}}H\text{—CH}_2\text{—C(O)Ph}$$

产率:90%;ee:90%

Genêt 等人报道了有机三氟硼酸钾在铑催化烯酮的不对称共轭加成反应中表现出比有机硼酸更好的反应活性,无论是环状还是链状烯酮都可以在短时间内得到高产率和高 ee 值的加成产物[102,103]。

$$R\text{—CH=CH—C(O)R}^1 + R^2BF_3K \xrightarrow{Rh(cod)_2PF_6/(S_a)\text{-1a}} R^*\text{—CH(R}^2\text{)—CH}_2\text{—C(O)R}^1$$

R = $C_5H_{11}$, $R^1$ = Me
R-$R^1$ = $(CH_2)_2$, $(CH_2)_3$, $(CH_2)_5$
$R^2$ = Ph, 4-MeOC$_6$H$_4$, 4-FC$_6$H$_4$, 3-ClC$_6$H$_4$, (E)-4-MeC$_6$H$_4$CH=CH

产率:70%~99%
ee:92%~98%

Corey 等人发现使用 [Rh(cod)$_2$]BF$_4$/($S_a$)-1a 催化异丙烯基三氟硼酸钾对环状烯酮的不对称加成反应中三乙胺添加剂的存在是反应成功的关键因素,不仅可以使反应在温室进行提高反应的化学选择性,同时可提高反应对映选择性。在此条件下,其他环状烯酮底物同样可以获得很好的结果,但是对于非环状的烯酮或烯醛底物虽然可以得到高产率的加成产物,但是反应没有立体选择性,得到的是消旋产物,其他烯基三氟硼酸钾同样可以得到相应的共轭加成产物,但 ee 值有所下降[104]。

产率:96%;ee:94%

在上述铑催化的有机硼试剂的共轭加成反应中,反应体系中需要加入水,这样就不能分离得到手性硼烯醇酯,并且有机硼酸在水中会部分水解,导致需要加入过量的硼酸以确保反应产率。Hayashi 等人开发了一种新催化体系，利用 9-芳基-9-硼双环 [3.3.1] 壬烷作为硼试剂在手性铑 [Rh(OMe)(cod)]$_2$ 和 ($S_a$)-**1a** 形成的配合物催化作用下首次实现了与烯酮的不对称 1,4-加成反应，得到了手性硼烯醇化物。反应具有很好的对映选择性，但是底物的普适性较窄，只有环己烯酮和环庚烯酮是适合的底物，环戊烯酮以及链状的烯酮则不能得到相应的产物。所得到的手性硼烯醇酯是非常有用的有机合成中间体，不用经过分离可通过一系列转化反应生成各种手性化合物[105]。

大多数铑催化的有机硼试剂对活化烯烃的共轭加成反应中所使用的是 $\beta$-取代的活化烯烃，而 Darses 和 Genêt 等报道了铑催化的有机硼酸钾试剂对 $\alpha$-取代活化烯烃 $\alpha$-乙酰氨基丙烯酸甲酯的共轭加成反应。在以往的 $\beta$-取代活化烯烃的加成反应中立体控制步骤在手性金属催化有机硼试剂向双键转移过程；而当采用 $\alpha$-取代活化烯烃作为底物时，反应的第一步加成时并没有产生新的手性中心，反应的立体控制步骤在淬灭加成后所形成的烯醇金属盐中间体，即对映选择性的质子化过程。研究过程中发现质子化试剂是决定反应选择性的关键因素，当采用愈创木酚（2-MeOC$_6$H$_4$OH）为质子化试剂时获得了比较满意的对映选择性结果。

进一步研究表明 N 上的取代基以及酯基的体积大小对反应的对映选择性的影响比较明显。其中乙酰基以及叔丁氧羰基是较好的氮上保护基，对映选择性随着酯基体积的增大呈上升的趋势。当乙酰基或叔丁氧羰基保护的 $\alpha$-氨基丙烯酸异丙酯为底物时，反应的选择性可以得到进一步提高[106,107]。

$$\underset{CO_2Me}{\overset{NHAc}{\diagup}} + RBF_3K \xrightarrow[2-MeOC_6H_4OH]{[Rh(cod)_2]BF_4/(R_a)\text{-}\mathbf{1a}} R\underset{CO_2Me}{\overset{NHAc}{\diagup}}*$$

R = Ph, 4-MeOC$_6$H$_4$, 4-FC$_6$H$_4$, 4-BrC$_6$H$_4$,
2-Naph, 噻吩-2-基, (E)-4-MeC$_6$H$_4$CH=CH

产率：68%～96%
ee：81%～89.5%

在铑催化的不对称共轭加成反应中，有机硅试剂作为反应原料也有不少例子。Inouet 等报道了一种 [Rh(cod)$_2$(MeCN)$_2$]BF$_4$ 与 ($S_a$)-**1a** 形成的手性铑配合物可以催化三烷氧基硅烷对 $\alpha,\beta$-不饱和酮的共轭加成，无论是芳基还是烯基硅烷都能得到加成产物，同时获得很好的对映选择性。就对映选择性而言，环状烯酮效果好于非环状烯酮。在非环状烯酮底物中，立体选择性随 $\beta$-位取代基空间体积的增大而增加，而加成效率则正好相反，初步研究表明，该催化体系对不饱和羧酸酯、酰胺等底物同样有效[108]。

$$R^1\diagdown\diagup\overset{R^2}{\underset{O}{\diagdown}} + R-Si(OR')_3 \xrightarrow[\text{二氧六环}/H_2O]{[Rh(cod)_2(MeCN)_2]BF_4/(S_a)\text{-}\mathbf{1a}} R^1\underset{R}{\overset{R^2}{\diagdown\diagup\diagdown}}_O$$

R = Ph, p-ClC$_6$H$_4$, p-MeOC$_6$H$_4$, CH$_2$=CH,
(E)-PhCH=CH, (Z)-PhCH=CH
R' = Me, Et
R$^1$ = Me, Pr, iPr
R$^2$ = Me, Ph, OMe, NH$_2$
R$^1$-R$^2$ = (CH$_2$)$_2$, (CH$_2$)$_3$

产率：54%～93%
ee：75%～98%

Inoue 等人在上述研究的基础上，进一步考察了芳基三烷氧基硅烷与 $\alpha,\beta$-不饱和酯和酰胺的 1,4-加成反应。实验结果表明，环状酯的立体选择性明显高于链状酯，非环状酯中 $\beta$-取代基和酯基的体积大小对反应的产率以及反应选择性影响很大，随着取代基和酯基体积的增大，立体选择性同样是逐渐增加，反应产率则反之。该规律同样存在不饱和酰胺的加成反应中，另外酰胺底物中，氮上双取代的选择性低于相应的单取代和未取代的底物[109]。

$$\underset{R^1}{\overset{O}{\diagup}}\!\!\!\diagdown_{OR^2} + Ar\!-\!Si(OR)_3 \xrightarrow[\text{二氧六环/H}_2\text{O}]{[Rh(cod)_2(MeCN)_2]BF_4/(S_a)\text{-}\mathbf{1a}} \underset{R^1}{\overset{Ar}{\diagup}}\!\!\!\underset{}{\overset{O}{\diagdown}}_{OR^2}$$

$R^1$ = Me, $n$Pr, $i$Pr  Ar = Ph, 4-MeO-C$_6$H$_4$, 4-Cl-C$_6$H$_4$  产率：28%~93%
$R^2$ = Me, Et, $i$Pr   R = Me, Et                               ee：84%~99%
$R^1$-$R^2$ = (CH$_2$)$_2$

$$\underset{R^1}{\overset{O}{\diagup}}\!\!\!\diagdown_{N{R^2 \atop R^3}} + Ar\!-\!Si(OR)_3 \xrightarrow[\text{二氧六环/H}_2\text{O}]{[Rh(cod)_2(MeCN)_2]BF_4/(S_a)\text{-}\mathbf{1a}} \underset{R^1}{\overset{Ar}{\diagup}}\!\!\!\underset{}{\overset{O}{\diagdown}}_{N{R^2 \atop R^3}}$$

$R^1$ = Me, $n$Pr, $i$Pr        Ar = Ph, 4-MeO-C$_6$H$_4$, 4-Cl-C$_6$H$_4$   产率：30%~75%
$R^2$ = H, Me, Et, $i$Pr,   R = Me, Et                                ee：72%~92%
$R^3$ = H, Me, PhCH$_2$

  Hayashi 等人发展了串联的铑催化氢硅烷对炔烃的加成和加成产物烯基硅烷对烯酮的不对称共轭加成"一锅法"制备手性 $\beta$-烯基取代酮的方法。反应过程中，在氢硅烷加成之后，必须加入水分解过量的氢硅烷以避免将底物烯酮还原从而降低共轭加成产物的产率。虽然三乙氧基氢硅烷对炔烃的加成存在区域选择性，但由于其中支链产物反应活性差对后续的加成产物的分离并没有造成影响[110]。

$$R\!-\!\!\equiv\!\!\!- + (EtO)_3SiH \xrightarrow[\substack{2)\ H_2O \\ 3)\ \underset{R^1}{\overset{O}{\diagup}}\!\!\!\diagdown_{R^2}}]{1)\ Rh/(S_a)\text{-}\mathbf{1a}} \underset{R^1}{\overset{R}{\diagup}}\!\!\!\diagdown\!\!\!\diagdown_{R^2}^{O}$$

R = Ph, 4-MeOC$_6$H$_4$, 2-MeC$_6$H$_4$, 萘-1-基,
  萘-2-基, C$_6$H$_{13}$, Et$_3$Si                      产率：65%~89%
$R^1$ = $i$Pr                                          (Z)/(E)：91/9~100/0
$R^2$ = Me                                             ee：78%~98%
$R^1$-$R^2$ = (CH$_2$)$_2$, (CH$_2$)$_3$, (CH$_2$)$_4$

  Hayashi 等人通过使用有机锌试剂作为亲核试剂，铑与 ($R_a$)-$\mathbf{1a}$ 作为催化体系，通过 1,4-加成反应制备 2-芳基-4 哌啶酮，并具有非常好的对映选择性。在探索研究的过程中发现，使用有机硼试剂时，虽能得到高 ee 值的加成产物，但是反应的转化率中等，当使用活性较高的有机钛试剂时则会存在 1,2-和 1,4-加成竞争反应。使用有机锌试剂反应不仅高产率地得到 1,4-加成产物，并且获得非常高的对映选择性。无论是吸电子还是给电子基团取代的苯基氯化锌均能得到几乎光学纯的目标产物。其中给电子基团取代的苯基氯化锌需要使用更多剂量的催化剂，而烷基和烯

基锌试剂在该体系下不能得到加成产物。

Ar = Ph, 4-PhC$_6$H$_4$, 4-MeOC$_6$H$_4$, 4-FC$_6$H$_4$,
3,5-Me$_2$C$_6$H$_3$, 2-MeC$_6$H$_4$

产率: 87%～100%
ee: 99%～99.5%

这种1,4-加成方法可用于制备速激肽拮抗剂的关键中间体，即 **1** 与 4-氟-2-甲基苯基氯化锌的反应。催化量的 [RhCl((*R*)-**1a**)]$_2$，然后除去苄氧基羰基，可高收率地得到高对映选择性的手性化合物 **3**（两步收率 73%，ee 值 97%）。

反应过程中会形成手性烯醇中间体 **4**，如果向反应中加入亲电试剂，则可发生"一锅法"官能化反应，如加入烯丙基溴以反式形式得到 *R*-Ming Bao 产物 **5** 的单一非对映异构体，具有高收率（产率 83%）。若使用新戊酰氯作为亲电子试剂，可以以优异的收率得到 *O*-酰化产物 **6**（产率 97%）。

此外，有机锌试剂对链状或环状的 α,β-不饱和烯酮底物同样有效，以几乎定量的产率和高对映选择性得到手性产物。

[反应式:环己烯酮 + PhZnCl → [RhCl((R)-1a)]₂ (Rh摩尔分数3%) / THF, 20°C → H₂O → 3-苯基环己酮]

[反应式:iPr-CH=CH-C(O)Me + PhZnCl → [RhCl((R)-1a)]₂ (Rh摩尔分数3%) / THF, 20°C → H₂O → iPr-CH(Ph)-CH₂-C(O)Me]

进一步将底物拓展为 2,3-二氢-4-喹啉酮同样有效,为 2-芳基-2,3-二氢-4-喹啉酮类抗肿瘤药物的合成提供了便利的方法[111,112]。

[反应式:4-喹啉酮(N-CO₂Bn, R取代) + ArZnCl → [RhCl(C₂H₄)₂]₂ (Rh摩尔分数7.5%), ($R_a$)-1a (摩尔分数8.2%), Me₃SiCl (3.0eq.), THF, 20°C, 20h, 10% HCl aq → 2-芳基-2,3-二氢-4-喹啉酮]

R = 6-Cl, 5,7-(MeO)₂, 6,7-OCH₂O
Ar = Ph, 2-MeC₆H₄, 3,5-Me₂C₆H₃, 4-MeOC₆H₄, 4-FC₆H₄, 萘-2-基
产率:72%~100%
ee:86%~99%

Charette 等人报道了一种使用二烷基锌试剂在铜催化下的共轭加成反应,在 (CuOTf)₂Tol 与 ($R_a$)-1a 原位形成的手性铜配合物存在下,二乙基锌可以对乙烯基吡啶基-2-砜发生对映选择性共轭加成。β-芳基取代和β-烷基取代不饱和砜都取得了很好的立体选择性。一般来说,反应介质四氢呋喃与苯相比往往可以给出更高的产率和相对较低的对映选择性。反应仅仅局限于伯烷基形成的二烷基锌试剂,如果使用 α-支链化的锌试剂会导致反应产率和产物 ee 值的急剧下降[113]。

[反应式:R-CH=CH-SO₂Py + Et₂Zn → (CuOTf)₂Tol/($R_a$)-1a, THF或PhH, 60°C → R-CH(Et)-CH₂-SO₂Py]

R = Ph, 4-MeOC₆H₄, 4-CF₃C₆H₄, 萘-2-基, iPr, Me
THF 产率:67%~93%;ee:84%~93%
PhH 产率:55%~93%;ee:88%~98%

由于 2-杂环芳基硼酸的不稳定性,易于在过渡金属催化过程中发生分解从而限制了其在不对称共轭加成中的应用。Martin 等报道了 2-杂环芳基钛及锌试剂在 Rh/($R_a$)-MeO-Biphep(18a) 催化体系的存在下可以顺利地与一系列烯酮、不饱和内酯及内酰胺发生共轭加成,从而为该类光学活性含 2-杂环芳基羰基化合物的合成提供了有效的方法[114]。

Tol-BINAP：产率90%, ee 90%
MeO-Biphep：产率90%, ee 98%

Hayashi 等人报道了铑催化的芳基三异丙氧基钛与烯酮的共轭加成反应，发现加成反应中间体手性烯醇钛虽不能分离，但通过异丙醇锂处理后再加入三甲基氯硅烷则可分离得到相应的烯醇硅醚，无论是环状还是链状烯酮均可得到高 ee 值的烯醇硅醚。该化合物在有机合成中具有较为广泛的应用。

分离得到的甲硅烷基烯醇醚是非常有用的合成中间体，可易于转化成各种富含对映异构体的化合物。不经分离的手性烯醇钛中间体可与烷基卤代化合物发生烷基化反应，与酰氯发生氧酰基化反应，与醛发生羟醛缩合等反应制备其他各种手性物质[115]。

Oestreich 等人利用铑催化活化硼-硅键，形成的亲核性硅对 $\alpha,\beta$-不饱和羰基化合物进行共轭加成实现了不对称新型硅-碳键的形成，对于环状烯酮或环状不饱和酯底物，反应具有优异的对映选择性，但是部分底物由于在催化体系下形成相应的还原产物而导致加成产物的产率很低。对于链状底物 $\alpha,\beta$-不饱和酮不能实现转化，不饱和酸酯衍生物则同时得到还原产物和加成产物，其中双键的构型对反应的立体选择性影响很大，与环状底物构型相同的 (Z)-构型底物可以获得近乎完美的立体控制结果，而 (E)-构型底物则只获得很低的对映选择性[116]。

Loh 等人报道了在手性金属铜配合物的催化下，格氏试剂可以和 $\alpha,\beta$-不饱和羧酸酯发生不对称共轭加成反应。芳基取代不饱和羧酸酯的反应活性较低，通常需要更多的催化剂用量。除了甲基格氏试剂之外，其余格氏试剂得到高 ee 值的加成产物。反应中配体 **1b** 与催化剂前体 CuI 的配比是反应能否获得好的立体选择性的关键因素。除了可以通过改变配体的构型来实现对产物构型的控制之外，通过改变底物双键的几何构型也可以达到同样的目的[117]。

为了提高甲基格氏试剂共轭加成的产率，Loh 小组进一步对反应条件进行优化，当反应温度提升至-20℃，催化剂用量为 2%（摩尔分数）时，反应的产率得到显著提高而不影响反应的对映选择性。在此反应条件下，一系列脂肪族 α,β-不饱和羧酸酯与甲基格氏试剂反应均可得到令人满意的结果[118]。

$$\text{R}\diagup\!\!\!\diagup\text{CO}_2\text{Me} + \text{MeMgBr} \xrightarrow{\text{CuI}/(S_a)\text{-1b}} \text{R}^*\text{CH(Me)CH}_2\text{CO}_2\text{Me}$$

R = Et, Pr, iPr, Bu, PhCH$_2$CH$_2$, Bn

产率：50%～86%
ee：95%～>99%

Feringa 等人使用同样的催化体系研究了甲基格氏试剂对 α,β-不饱和羧酸硫醇酯的不对称加成，与 α,β-不饱和羧酸酯相比，在铜催化的该类底物的共轭加成反应中，甲基格氏试剂同样可以获得高产率和高 ee 的加成产物。对于苯基格氏试剂，尽管可以顺利反应，但反应得到的是外消旋产物。当肉桂酸衍生物为底物时，苯环上的取代基对反应活性影响很大，4-位拉电子基团的引入有利于反应的进行，而给电子基团的引入则会导致反应活性的降低[119]。

格氏试剂对 α,β-不饱和吡啶-2-基砜的不对称共轭加成在手性铜配合物的催化下也能顺利实现，催化剂前体对反应的选择性有一定的影响，其中效果最好的是 CuCl 与 ($R_a$)-1b 形成的手性铜配合物。从反应的对映选择性来看，脂肪族 α,β-不饱和砜明显优于芳香族 α,β-不饱和砜，前者与脂肪族格氏试剂的加成获得了很好的结果，而后者只能得到中等程度 ee 值的加成产物。苯基格氏试剂虽然也能得到高产率的加成产物，但产物没有光学活性[120]。

$$\text{C}_5\text{H}_{11}\diagup\!\!\!\diagup\text{SO}_2\text{Py} + \text{EtMgBr} \xrightarrow{\text{CuCl}/(R_a)\text{-1b}} \text{C}_5\text{H}_{11}\text{CH(Et)CH}_2\text{SO}_2\text{Py}$$

产率：97%
ee：93%

$$\text{R}^1\diagup\!\!\!\diagup\text{C(O)SR}^2 + \text{RMgBr} \xrightarrow{\text{CuI}/(S_a)\text{-1b}} \text{R}^{1*}\text{CH(R)CH}_2\text{C(O)SR}^2$$

R = Me, Et, iPr, Bu, iBu, Ph
R$^1$ = Ph, 4-ClC$_6$H$_4$, 4-MeC$_6$H$_4$, 4-MeOC$_6$H$_4$, Me, iPr, Pent, CH$_2$OTBDPS
R$^2$ = Me, Et

产率：15%～94%
ee：0%～99%

3 手性双膦配体

在铑催化的端炔对 $\alpha,\beta$-不饱和酮的共轭加成反应中，由于炔基铑中间体与炔烃的反应活性往往高于其与烯酮的反应活性，因此主要得到的是炔烃的二聚产物而不是相应的共轭加成产物。Hayashi 小组报道了通过改变在硅基乙炔硅原子上取代基以及双膦配体磷原子上取代基的体积大小可以避免二聚反应的发生，从而专一地生成共轭加成产物。在 Rh/($R_a$)-19d 催化体系的催化下，三异丙硅基乙炔与烯酮化学专一地得到了相应的共轭加成产物。无论是环状还是链状烯酮都获得了很好的立体选择性[121]。

R = Ph, 4-MeOC$_6$H$_4$, 呋喃-2-基, (E)-PhCH=CH, Et, Me
R$^1$ = Me, Et, C$_5$H$_{11}$

产率：54%～99%
ee：88%～97%

进一步将该催化体系扩展至端炔对 $\alpha,\beta$-不饱和醛的共轭加成同样获得了成功。反应的化学选择性很大程度上决定于反应介质，例如在 1,4-二氧六环中同时得到共轭加成和双加成产物，而在甲醇介质中则化学专一地得到共轭加成产物。反应表现出很好的立体选择性和底物适用范围，一系列烯醛均可以得到高 ee 值的加成产物[122]。

产率：93%
ee：96%

### 3.1.3.7 羰基的 $\alpha$-芳/烷基化

Buchwald 等人报道了首例钯催化的酮烯醇盐的不对称芳基化反应，但是反应的对映选择性具有很大程度的底物结构依赖性。例如对于 2-甲基-$\alpha$-四氢萘酮反应获得了较好的对映选择性；对于 $\alpha'$-苯亚甲基-$\alpha$-甲基环己酮反应只得到很低的立体控制结果，但是对于相应的环戊酮底物而言，反应则表现出了很好的对映选择性[123]。

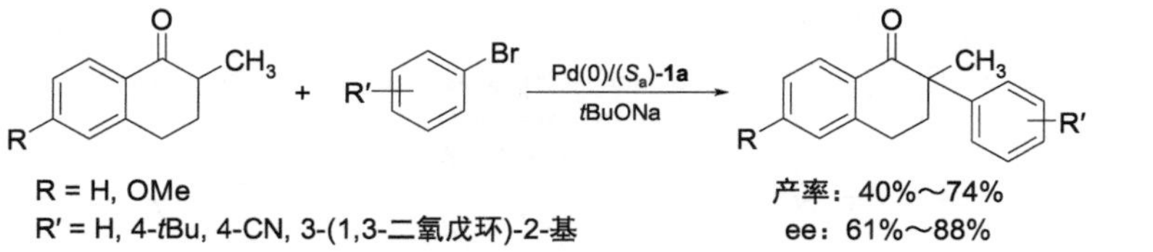

Pd(0)/($S_a$)-1a
tBuONa
R = H, OMe
R′ = H, 4-tBu, 4-CN, 3-(1,3-二氧戊环)-2-基
产率：40%～74%
ee：61%～88%


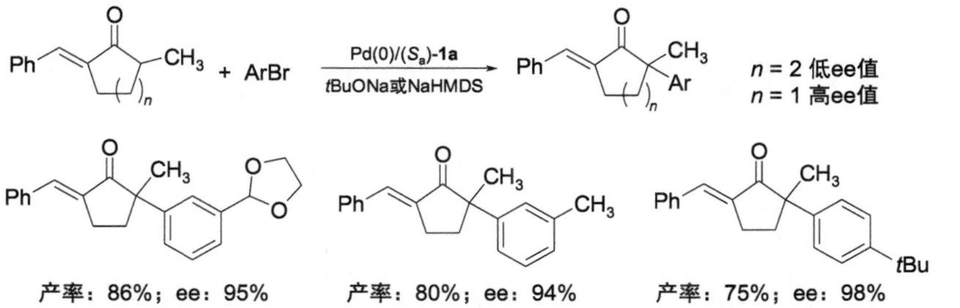

Pd(0)/($S_a$)-1a
tBuONa或NaHMDS
n = 2 低ee值
n = 1 高ee值
产率：86%；ee：95%
产率：80%；ee：94%
产率：75%；ee：98%


Buchwald 等人使用 Ni/($S_a$)-**1a** 催化体系实现了对 2-取代 $\gamma$-丁内酯的不对称 $\alpha$-芳基化反应，反应同样获得了很好的立体选择性。反应中发现催化量 $ZnBr_2$ 的加入可以明显加速反应，尽管 2-位取代基对反应的对映选择性影响并不明显，但对反应活性(产率)影响比较显著，随着取代基体积的增大，反应速率减慢，同时需要在较高反应温度下进行反应[124]。

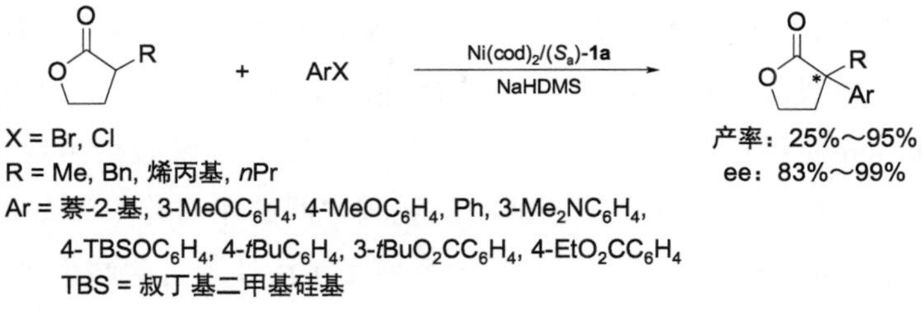

Ni(cod)$_2$/($S_a$)-1a
NaHDMS
X = Br, Cl
R = Me, Bn, 烯丙基, nPr
Ar = 萘-2-基, 3-MeOC$_6$H$_4$, 4-MeOC$_6$H$_4$, Ph, 3-Me$_2$NC$_6$H$_4$,
4-TBSOC$_6$H$_4$, 4-tBuC$_6$H$_4$, 3-tBuO$_2$CC$_6$H$_4$, 4-EtO$_2$CC$_6$H$_4$
TBS = 叔丁基二甲基硅基
产率：25%～95%
ee：83%～99%


Breit 等人报道了使用 Rh/ 光氧化还原助催化体系开发了胺的区域化芳基化，使得胺与炔烃和丙二烯偶联。其中使用改进的 Rh/Ir 双催化体系，线性选择性可以仅用炔烃和丙二烯两者获得，以良好至极好的产率和区域选择性合成有用的直链高烯丙基胺，补充了传统的过渡金属催化的烯丙基化反应[125]。

Ar = Ph, 4-MePh, 4-OMePh, 4-CF$_3$Ph, 4-BrPh, 4-FPh,
3,4-Cl$_2$Ph, 3-BrPh, 3-FPh, 3-MePh, 3-BrPh
R = H, 5-Cl, 7-Cl

产率：14%~96%；L：B：75：25~95：5

R$^1$ = H, Me
R$^2$ = Me, Cl(CH$_2$)$_3$, Br(CH$_2$)$_2$, MeO$_2$C(CH$_2$)$_2$,
CN(CH$_2$)$_2$, PhS(CH$_2$)$_2$, PhthN(CH$_2$)$_2$, Cy

产率：60%~94%；L：B：81：19~95：5
E/Z：69：31~95：5

### 3.1.3.8　Friedel-Crafts 芳/烷基化反应

1999 年，Johannsen 利用 CuPF$_6$/(R$_a$)-**1b** 催化体系催化的吲哚及吡咯与 N-磺酰基乙醛酸酯亚胺之间的 Friedel-Crafts 烷基化反应实现了光学活性杂环 α-氨基酸衍生物的不对称合成。虽然 5-位吸电子基团取代吲哚反应活性有所下降，但提高反应温度后同样可以得到相应的烷基化产物。对于 N-甲基吡咯则得到大约 1：1 的 2-位和 3-位烷基化产物的混合物，但其中 3-位产物得到了较好的对映选择性。2-乙酰基吡咯虽然反应产率不是很高，但可以高对映选择性地得到专一的 4-烷基化产物[126]。Jørgensen 使用相同的催化体系实现了 N,N-二烷基苯胺衍生物及 1,3-二甲氧基苯及 N-烷氧羰基保护乙醛酸酯亚胺之间的不对称 Friedel-Crafts 烷基化反应，高对映选择性地得到了一系列光学活性苯甘氨酸衍生物[127]。

R = H, OMe, NO$_2$, CO$_2$Me, Br

产率：67%~89%
ee：78%~97%

Umani-Ronchi 发展了另一种手性钯催化体系用于催化吲哚与 α, β-不饱和羧酸衍生物之间的不对称 Michael 类型 Friedel-Crafts 烷基化反应。在 PdCl$_2$(MeCN)$_2$、($S_a$)-**1b** 与 AgSbF$_6$ 原位产生的离子型钯配合物的催化下，一系列（取代）吲哚与 $\alpha,\beta$-不饱和丁酸硫酯反应可以得到具有较高 ee 值的烷基化产物[128]。

R = H, 2-Me, 5-MeO, 2-Ph, 5-Bn, 2-Me-7-Br
R$^1$ = H, Me

产率：20%～80%
ee：70%～86%

Michelet 发展了一种 Au(Ⅰ)/($R_a$)-**18h** 催化体系催化的 1,6-烯炔的串联 Friedel-Crafts 烷基化/环化异构化反应，可以实现光学活性的环戊烷或四氢吡咯烷衍生物的不对称合成。如果在炔端引入芳基，在相同的反应条件下通过分子内的反应可以实现多环体系的构建[129]。

R = H, Me；R$^1$ = Ph, Me, 3,4-(OCH$_2$O)C$_6$H$_3$
Z = C(CO$_2$Me)$_2$, C(CO$_2$iPr)$_2$, C(CO$_2$Bn)$_2$, C(SO$_2$Ph)$_2$, O
Ar—H = 1,3,5-三甲氧基苯, 1-甲基吲哚, 吡咯,
1-甲基-2-苯基吲哚, 1,3-二甲氧基苯,
1,3,5-三甲氧基-2-溴苯

产率：37%～99%
ee：53%～98%

### 3.1.3.9 Mannich 反应

Mannich 反应是有机合成中一类重要的碳-碳键形成反应，可以方便地实现 $\beta$-氨基羰基类化合物的合成，Lectka 等人报道了手性双膦配体 ($R_a$)-**1a** 与 AgSbF$_6$ 形成的手性银配合物同样可以催化烯醇硅醚和 N-磺酰基乙醛酸酯亚胺之间的不对称 Mannich 反应，但是反应条件比较苛刻，为了抑制背景反应的发生以获得满意的立体控制结果，反应往往需要在-80℃下进行。当使用 ($R_a$)-**1a** 与 CuClO$_4$ 形成的手性铜配合物作催化剂时，反应条件却十分温和，反应不仅可以在零摄氏度下进行，而且反应的对映选择性也有大幅度提高[130]。Lectka 课题组在此基础上又拓展了相关不对称 Mannich 反应的研究[131,132]。

R = Ph, 4-MeOC$_6$H$_4$, 4-FC$_6$H$_4$, 4-ClC$_6$H$_4$,
4-CF$_3$C$_6$H$_4$, tBu

AgSbF$_6$/($R_a$)-**1a**, −80℃, 产率为70%~95%, ee值为61%~90%
CuClO$_4$/($R_a$)-**1a**, 室温, 产率为65%~93%, ee值为89%~98%

2009 年，Shibasaki 小组在催化剂 CuOAc/($R_a$)-**19d** 和添加剂 (EtO)$_2$Si(OAc)$_2$ 的作用下，也开发了 N-次膦酰基芳香酮亚胺与烯酮硅缩醛的不对称 Mannich 反应。反应可以顺利地得到光学活性的加成产物 β-氨基酸酯。该催化体系还实现了外消旋 α,α-双取代氰基乙酸对 N-二苯基次膦酰基醛亚胺的脱羧 Mannich 反应，无论是芳香醛亚胺还是脂肪醛亚胺均获得了很好的对映选择性[133]。

R = Ph, 4-ClC$_6$H$_4$, 4-MeOC$_6$H$_4$, 萘-2-基,
呋喃-2-基, 噻吩-3-基, 环己烯-1-基
R$^1$ = Me, Et
Xy = 3,4-二甲苯基

产率: 29%~92%
ee: 87%~97%

Kanai 和 Shibasaki 等人巧妙地利用铜催化的氢硼烷对 α,β-不饱和羧酸酯还原过程中产生的高浓度烯醇盐实现了对 N-二苯基次膦酰酮亚胺的亲核加成，即所谓的还原 Mannich 反应，并在此基础上使用 CuOAc/($R_a$)-**19e** 催化体系实现了 α,β-不饱和酸酯对酮亚胺的不对称还原 Mannich 反应，并取得了中等至很好的非对映选择性和很好的对映选择性[134]。

R = Ph, 4-ClC$_6$H$_4$, 4-MeOC$_6$H$_4$, 萘-2-基,
1-环己烯基
R$^1$ = H, Me, CO$_2$Et

产率: 47%~95%
dr: 3/1~30/1
ee: 82%~93%

### 3.1.3.10 烯醇硅醚的质子化反应

前手性烯醇盐的不对称质子化反应是获取光学活性 α-取代羰基化合物的有效方法，但是对于大多数不对称质子化过程往往需要至少化学

计量的手性酸作为质子源。1997 年，Nakai 等人报道了 [$PdCl_2$(($R_a$)-**1a**)]/$AgPF_6$ 催化体系可以催化水对烯醇硅醚的不对称质子化。其中碱性添加剂二异丙基胺是决定反应立体选择性的关键因素，不加二异丙基胺得到的几乎是外消旋产物。而加入二异丙基胺可以很好地实现对反应的立体控制 [135]。

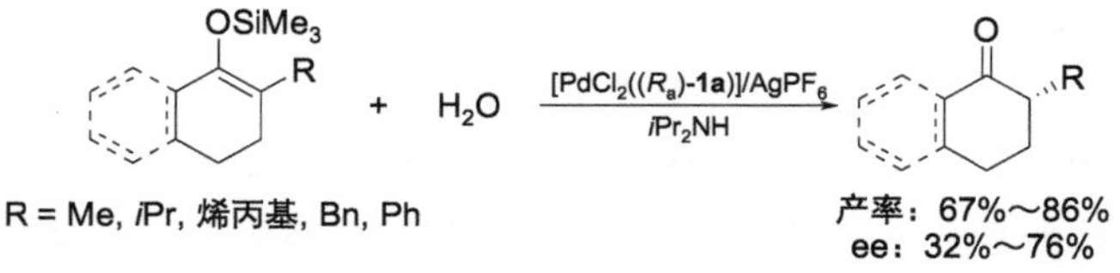

OSiMe3
R
+ H2O
[PdCl2((Ra)-1a)]/AgPF6
iPr2NH
O
R
R = Me, iPr, 烯丙基, Bn, Ph
产率：67%～86%
ee：32%～76%


Yanagisawa 等人报道了在醛的不对称烯丙基化反应和 Aldol 反应中表现出很好催化活性的 AgF/($R_a$)-**1a** 催化体系同样可以催化甲醇对烯醇硅醚的不对称质子化反应，2-芳基取代环己酮衍生的烯醇硅醚是一类最佳反应底物，可以得到几乎光学纯的 2-芳基环己酮，而 2-烷基取代的底物对映选择相对较低 [136]。

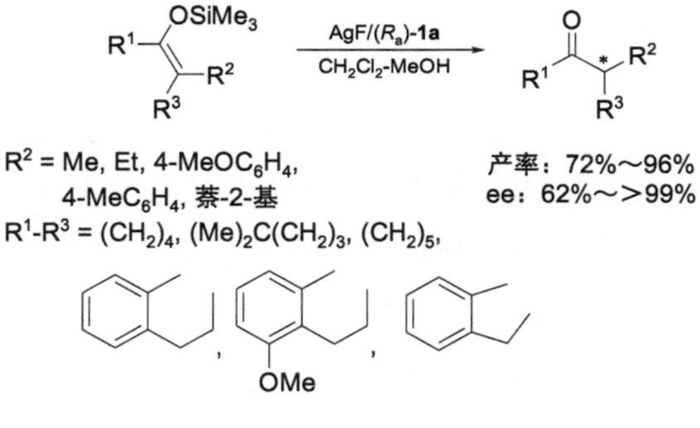

OSiMe3
R1
R2
R3
AgF/(Ra)-1a
CH2Cl2-MeOH
O
R1
*
R2
R3
R2 = Me, Et, 4-MeOC6H4,
4-MeC6H4, 萘-2-基
R1-R3 = (CH2)4, (Me)2C(CH2)3, (CH2)5,
OMe
产率：72%～96%
ee：62%～>99%


### 3.1.3.11 Heck 反应

将 Heck 反应应用于催化不对称合成是一个非常重要的研究领域，Shibasaki 等人首次报道了钯催化前手性烯基碘化物发生分子内 Heck 反应合成顺式萘烷衍生物，反应的对映选择性可达 46%[137]。

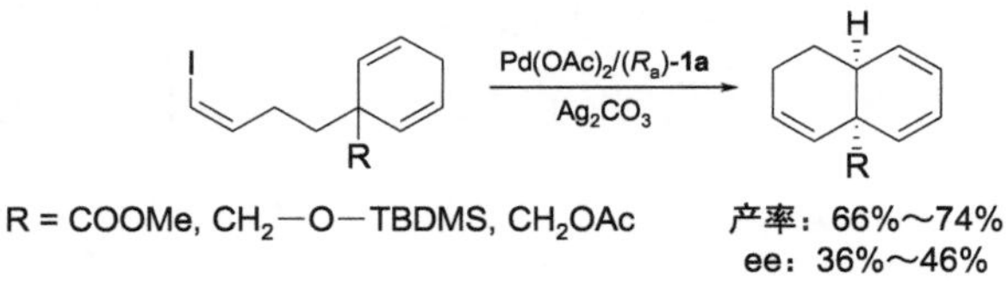

I
R
Pd(OAc)2/(Ra)-1a
Ag2CO3
H
R
R = COOMe, CH2—O—TBDMS, CH2OAc
产率：66%～74%
ee：36%～46%

用 Pd(OAc)$_2$/($R_a$)-**1a** 催化体系，Hayashi 实现了 2,3-二氢呋喃和芳基三氟甲磺酸酯的分子间不对称 Heck 反应。反应主要得到 2-芳基-2,3-二氢呋喃以及少量 2-芳基-2,5-二氢呋喃，其中主要产物具有很好的 ee 值[138]。

Ar = Ph, 4-ClC$_6$H$_4$, 3-ClC$_6$H$_4$, 4-AcC$_6$H$_4$,
4-NCC$_6$H$_4$, 4-MeOC$_6$H$_4$, 萘-2-基

产率：42%～86%
A/B：71/29～89/11
ee：73%～93%

Gelman 等人报道了 Pd(OAc)$_2$/($R_a$)-MeO-Biphep(**18a**) 催化体系在 Cu(OAc)$_2$ 作为氧化剂存在下，可以催化芳基硼酸对 2,3-二氢呋喃的不对称 Heck 反应，最终异构化得到热力学更稳定的 2-芳基-2,3-二氢呋喃。但是对于个别邻位取代的苯硼酸底物而言，反应的对映选择性急剧下降，例如 2-甲基苯硼酸得到的几乎是外消旋的产物[139]。

Ar = Ph, 4-MeC$_6$H$_4$, 3-MeC$_6$H$_4$, 2-MeC$_6$H$_4$,
4-CF$_3$C$_6$H$_4$, 4-ClC$_6$H$_4$, 3-ClC$_6$H$_4$,
2-ClC$_6$H$_4$, 4-FC$_6$H$_4$, 萘-1-基

产率：36%～74%
ee：17%～86%

Overman 等人进一步利用该催化体系成功地实现了天然产物毒扁豆碱(physostigmine)的两个对映体的全合成，使用 ($S_a$)-**1a** 为配体得到左旋异构体，使用 ($R_a$)-**1a** 为配体则得到右旋异构体[140]。

(-)-毒扁豆碱

Lassaletta 等人开发了一种 Pd(dba)$_2$/**1e** 催化的杂二芳基磺酸盐与富电子烯烃的不对称 Heck 反应[141]，用于合成高对映选择性和非对映选择性的轴手性化合物。

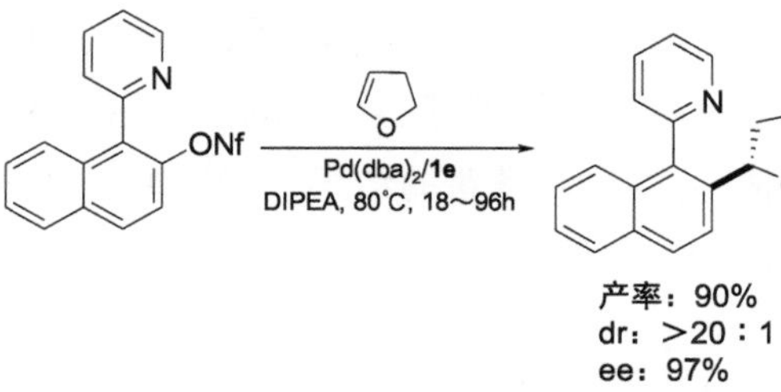
N
ONf
O
Pd(dba)₂/1e
DIPEA, 80°C, 18～96h
N
O
产率：90%
dr：>20：1
ee：97%

### 3.1.3.12 Buchwald-Hartwig 交叉偶联反应

催化不对称合成光学活性螺旋体在不对称催化研究领域具有一定的挑战性。Sasai 等人巧妙地利用 $Pd(OAc)_2/(S_a)$-**1a** 催化不对称 *N*-芳基化实现了对映选择性高达 70 % 手性螺环衍生物的合成 [142]。

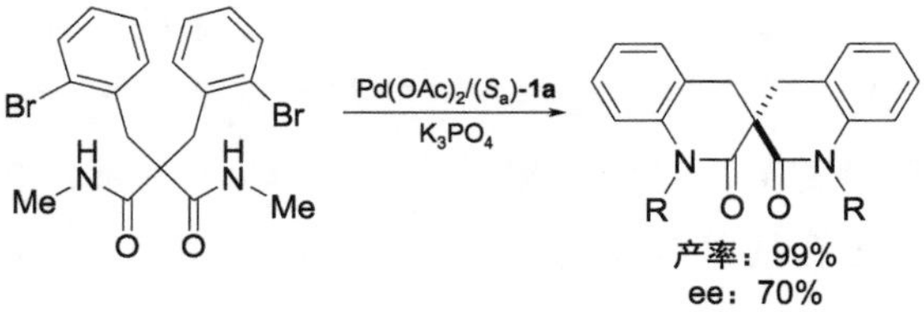
Br
Br
H
N
Me
H
N
Me
O
O
Pd(OAc)₂/(Sₐ)-1a
K₃PO₄
N
R
O
O
N
R
产率：99%
ee：70%

### 3.1.3.13 Aldol 反应及还原 Aldol 反应

手性双膦配体涉及的 Aldol 反应及还原 Aldol 反应使用到的催化剂主要是与 Pd、Cu、Ag 和 Rh 形成的络合物。

Shibasaki 等人报道了 $PdCl_2$ 与 $(R_a)$-**1a** 所形成的钯配合物与 AgOTf 作用产生的离子型手性钯配合物可以催化烯醇硅醚和醛之间的不对称 Aldol 反应，经酸处理得到相应的 $\beta$-羟基酮 [143]。Kiyooka 等人同样研究了手性离子钯催化配合物催化的 Aldol 反应，使用 $PdCl_2/(R_a)$-**1a** 和 $AgSbF_6$ 在无水 DMF 中原位产生的钯配合物为催化剂，他们发现 3Å 分子筛的加入可以显著地提高反应的产率，并且催化剂用量可以降低至 1%(摩尔分数) [144]。

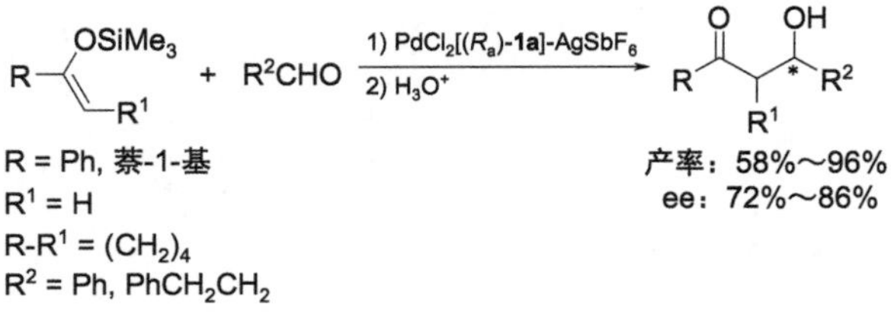
OSiMe₃
R
R¹
+ R²CHO
1) PdCl₂[(Rₐ)-1a]-AgSbF₆
2) H₃O⁺
O
OH
R
R²
R¹
R = Ph, 萘-1-基
R¹ = H
R-R¹ = (CH₂)₄
R² = Ph, PhCH₂CH₂
产率：58%～96%
ee：72%～86%

Carreira 报道了由 ($S_a$)-**1b**、Cu(OTf)$_2$ 和 (Bu$_4$N)Ph$_3$SiF$_2$ 原位络合产生的手性铜配合物可以有效地催化二烯醇硅醚和醛的不对称 Aldol 反应。对于芳香醛和 α,β-不饱和醛可以得到满意的结果，脂肪醛虽然也能得到高对映选择结果，但是反应的产率很低（< 40%）。

RCHO + [二烯醇硅醚] $\xrightarrow{\text{Cu(OTf)}_2/(S_a)\text{-}\mathbf{1b}/(\text{Bu}_4\text{N})\text{Ph}_3\text{SiF}_2}$ [产物]

R = Ph, 4-MeOC$_6$H$_4$, 萘-2-基, 噻吩-2-基
呋喃-2-基, 2-MeOC$_6$H$_4$, CH=CH,
PhCH=CH, CH$_3$CH=CH, Me$_2$C=CH,
PhCH=CMe

产率：48%～98%
ee：65%～95%

Yamamoto 小组使用手性银配合物 AgOTf/($R_a$)-**1b**，在反应的过程中加入催化量三烷基甲氧基锡和 2eq. 的甲醇，实现了烯醇三氯乙酸酯与醛的不对称 Aldol 反应。反应的过程中，三烷基甲氧基锡将烯醇三氯乙酸酯转化为相应的烯醇锡盐，进而和醛反应得到 Aldol 加成产物，其中甲醇的作用是再生出三烷基甲氧基锡从而使反应进行完全。三烷基甲氧基锡中的烷基对反应有一定影响，三甲基甲氧基锡的结果优于三丁基甲氧基锡[145]。

RCHO + [环己烯OCOCl$_3$] $\xrightarrow[\text{Me}_3\text{SnOMe, MeOH}]{\text{AgOTf/}(R_a)\text{-}\mathbf{1b}}$ [产物]

R = Ph, 4-MeOC$_6$H$_4$, 2,3-(OCH$_2$O)C$_6$H$_3$,
萘-1-基, PhCH=CH, PrCH=CH, PhCH$_2$CH$_2$

产率：29%～86%
syn/anti：4/96～24/76
ee：77%～96%

使用 AgOTf/($R_a$)-**1b** 催化体系，Yamamoto 实现了烯醇三烷基锡烷与亚硝基苯之间的不对称 O-Nitroso Aldol 反应。反应获得了很好的选择性，不仅可以高选择性地得到相应的 α-氨氧基酮，而且产物具有很高的 ee 值[146]。

R = Me, Bu
R$^1$ = Et
R$^2$ = H, Ph
R$^1$-R$^2$ = (CH$_2$)$_3$, (CH$_2$)$_4$, (CH$_2$)$_5$,
CMe$_2$(CH$_2$)$_3$, (CH$_2$)$_2$CMe$_2$CH$_2$
R$^3$ = Me

产率：92%～96%
A/B：81/19～>99/1
ee：82%～97%

在 Aldol 反应中，羰基化合物往往集中在醛，而相应的酮的 Aldol 反应研究相对较少。Shibasaki 等报道了由 CuF·3PPh$_3$·2EtOH 和 (EtO)$_3$SiF 组成的催化体系可以有效地催化烯酮硅缩醛和酮之间的 Aldol 反应，得到高产率的加成产物。当反应体系中加入手性膦配体 ($S_a$)-1a 时，可以实现对映选择酮的 Aldol 反应。反应中底物烯酮硅缩醛的构型对反应的立体控制几乎没有影响，无论是 (Z)-构型还是 (E)-构型的底物都得到了相近的产率和对映选择性[147]。

(Z) 产率 95%, ee 80%
(E) 产率 99%, ee 82%

Morken 等人报道了在手性铑配合物和还原试剂氢硅烷的存在下，醛和丙烯酸酯可以发生串联的共轭还原 Aldol 反应得到 β-羟基酸酯衍生物，避免了提前准备烯醇盐或烯醇硅醚的过程，极大地简化了反应操作，反应获得了不错的对映选择性，但非对映选择性并不理想[148]。

R = Me$_2$C=CH, MeCH=CH, MeCH=C(Me), 环己烯-1-基

产率：48%～82%
syn/anti: 1.8/1～5.1/1
ee: 45%～88%

由硅基保护的 Aldol 加成产物在合成上具有广泛用途，Morken 等通过改变还原试剂氢硅烷的结构发现，Me$_2$iPrSiH 不仅可以高化学选择性得到加成产物硅醚，而且比以前使用 Et$_2$MeSiH 获得的硅醚更加稳定，可以分离得到硅醚产物，从而提高了该还原 Aldol 反应在有机合成，尤其是复杂天然产物合成方面的应用可能[149]。

R = Pr, iPr, Ph, Cy

产率：57%～79%
A/B: >20/1
syn/anti: 3/1～5/1
ee: 79%～88%

Krische 等使用 [Rh(cod)Cl]$_2$ 与 ($R_a$)-**1a** 原位形成的手性铑配合物为催化剂，实现了经由同时含有烯酮和甲基酮结构单元分子的分子内串联不对称共轭加成-Aldol 反应制备相应的光学活性环戊烷及环己烷衍生物。反应具有很好的非对映选择性，几乎得到单一的非对映异构体，同时获得了很好的对映选择性[150]。

### 3.1.3.14 Pauson-Khand 反应

所谓的 Pauson-Khand 反应就是指烯烃、炔烃以及一氧化碳在过渡金属的催化下发生的 [2+2+1] 环化构建环戊烯酮衍生物的过程，由于环戊烯酮衍生物不仅是重要生物活性物质，同时也是重要的合成砌块，因此该反应受到广泛关注。

Gibson 等人使用 Co$_4$(CO)$_{12}$ 为钴源，($S_a$)-**1a** 为配体对钴催化的不对称 Pauson-Khand 反应进行了研究，得到了很好的对映选择性，并通过 $^{31}$P NMR 试验以及 X 射线单晶衍射证实反应过程中配体 ($S_a$)-**1a** 首先形成一端配位的双核金属钴配合物，在其催化下烯、炔和一氧化碳发生环化反应，整个催化过程发生在另一个未被膦配位的金属钴中心上[151]。

Consiglio 等报道了 Co(0)/($R_a$)-MeO-Biphep(**18a**) 体系能有效催化不对称 Pauson-Khand 反应，其中 Co(0) 源既可以是 Co$_2$(CO)$_8$，也可以由 CoCl$_2$ 在还原剂 Zn 存在下原位还原产生。但是就催化活性和立体选择性而言，以 Co(CO)$_8$ 为 Co(0) 源的催化体系获得了更好的结果[152]。

金属铑配合物既可以催化醛的脱羰基化反应，同时又能催化烯炔的 Pauson-Khand 反应，Chung 等报道了一种利用 [Rh(cod)Cl]$_2$/($S_a$)-**1a** 作为催化剂的水相中不对称 Pauson-Khand 反应，反应介质为水/1,4-二氧六环(体积比 =1∶1)。他们发现，表面活性剂的加入对反应有一定的影响，当加入阴离子型表面活性剂十二烷基磺酸钠(SDS)为添加剂时反应会明显加速，并且对反应的对映选择性几乎没有影响，在此条件下一系列烯炔底物可以顺利地发生分子内 Pauson-Khand 反应得到很高 ee 值的环化产物。反而加入阳离子型表面活性剂十六烷基三甲基溴化铵(CTAB)时反应明显减速，同时对映选择性也有所降低[153]。

TsN⟶[Rh(cod)Cl]$_2$/($S_a$)-**1a**, SDS, CO, H$_2$O/二氧六环⟶ TsN 产物

产率：86%
ee：93%

Jeong 等人发展了另一种 [RhCl(CO)$_2$]/($S_a$)-**1a** 催化体系应用于分子内不对称 Pauson-Khand 反应，并且发现反应中 CO 的压力对反应影响比较明显，CO 压力增大有利于提高反应的产率，但是会造成反应立体选择性降低，一般来说，炔端芳基取代的底物通常比烷基取代的底物具有更高的产率，但产物的 ee 值相对低于相应的烷基取代的产物[154]。

X⟶[RhCl(CO)$_2$]/($S_a$)-**1a**, AgOTf, CO, THF, 回流⟶ 产物

X = C(CO$_2$Me)$_2$, C(CO$_2$Et)$_2$, C(CO$_2$iPr)$_2$,
CH$_2$, O, NTs
R = Me, Ph, C$_4$H$_9$

产率：40%～99%
ee：22%～96%

Kwong 等人发现在 [Rh(cod)Cl]$_2$ 和双膦配体 ($S_a$)-**1e** 的存在下，甲酸酯作为一氧化碳源，可以实现串联的甲酸酯脱羰基反应和烯炔的不对称 Pauson-Khand 反应，获得了较为满意的对映选择性结果。不足之处是部分底物的产率很低，尤其是位阻较大的炔端 2-取代苯基底物只得到 11% 的分离产率，究其原因可能是由于位阻妨碍了炔烃单元与金属中心的配位[155]。

R = Me, Et, Ph, 4-MeC$_6$H$_4$,
3-MeOC$_6$H$_4$, 4-MeOC$_6$H$_4$,
4-ClC$_6$H$_4$, 4-FC$_6$H$_4$,
2-MeC$_6$H$_4$, 噻吩-2-基
X = C(CO$_2$Et)$_2$, O, NTs

产率：11%~65%
ee：65%~94%

Shibata 等人巧妙地利用铑配合物的这一特性，实现了铑催化剂的以醛作为一氧化碳源的 Pauson-Khand 反应，从而避免了有毒一氧化碳气体的使用，当采用 [Rh(cod)Cl]$_2$/($S_a$)-**1b** 为催化剂时，肉桂醛为一氧化碳源时，一系列烯炔底物可以发生高立体选择的分子内不对称 Pauson-Khand 反应得到相应的环戊烯酮的衍生物[156]。

Shibata 等报道铱同样可以催化 Pauson-Khand 反应，当采用 [Ir(cod)Cl]$_2$/($S_a$)-**1b** 为催化剂时，得到了具有很高 ee 值的环戊烯酮衍生物，不足之处是当烯端引入取代基时会导致反应产率的明显下降。该催化体系不仅对分子内反应有效，而且对于分子间反应同样有效，如 1-苯基丙炔和降冰片烯在同样的条件下可以高立体选择地（包括非对映选择性及对映选择性）得到相应的环化产物，但是反应的产率不是很理想[157]。

R = Me, Ph(CH$_2$)$_3$, 4-MeOC$_6$H$_4$, Ph
R$^1$ = H, Me
X = C(CO$_2$Et)$_2$, O, NTs

产率：30%~80%
ee：88%~98%

产率：32%(A+B), A/B：>10/1
ee：93%(A)

在上述金属铱催化的不对称 Pauson-Khand 反应中，在烯端引入取代基会导致反应产率的大幅下降，但是采取降低反应体系中一氧化碳的分压至 0.2atm（约 20kPa）时不仅可以有效地克服这一缺点，而且还提高了反应的选择性：当 R$^1$=Me、R=Ph 时，产率和 ee 值分别从 30% 和 88% 提高

到 86% 和 93%；当 R¹ 为烯丙基、R 为苯基时，产率和 ee 分别从 22% 和 86% 增加至 62% 和 94%。此外，利用醛作为一氧化碳源同样可以实现不对称 Pauson-Khand 反应，虽然反应产率有所下降，但几乎不影响反应的选择性[158]。

R = Me, Ph, 4-MeOC$_6$H$_4$, 4-ClC$_6$H$_4$
R¹ = H, Me, 烯丙基
X = C(CO$_2$Et)$_2$, O, NTs

产率：22%～89%
ee：84%～98%

### 3.1.3.15 Diels-Alder 反应

不对称 Diels-Alder 反应及杂 Diels-Alder 反应是合成手性环状化合物有效方法之一。Ghosh 等人报道了手性 Pt 和 Pd 双膦配合物在催化环戊二烯和酰基-2-噁唑啉酮的不对称 Diels-Alder 反应及其在对映选择性中的拮抗离子效应。结果显示，手性铂配合物的效果好于相应的钯配合物，并且拮抗离子不仅对催化活性同时对不对称诱导能力有显著影响，其中 $ClO_4^-$ 和 $SbF_6^-$ 的离子型铂配合物能获得更好的结果[159]。

R = H, Me, CO$_2$Et

| | R | 产率/% | endo/exo | ee/% |
|---|---|---|---|---|
| X = $ClO_4^-$ | H | 93 | 98 : 2 | 97 |
| | Me | 51 | 91 : 9 | 90 |
| | CO$_2$Et | 75 | 90 : 10 | 90 |
| X = $SbF_6^-$ | H | 99 | 97 : 3 | 98 |
| | Me | 74 | 92 : 8 | 93 |
| | CO$_2$Et | 84 | 89 : 11 | 90 |

Oi 等人使用手性阳离子钯(Ⅱ)和铂(Ⅱ)配合物为催化剂对未活化的共轭二烯与芳甲酰甲醛实现了不对称杂 Diels-Alder 反应。得到相应的3,6-二氢吡喃衍生物。反应的对映选择性受双烯结构影响较大，而与芳甲酰基苯环上取代基性质关系不大。其中 2,3-二甲基-1,3-丁二烯和环己二烯获得了很高 ee 值的环加成产物。

手性钯配合物催化的基于乙醛酸(酯)的杂 Diels-Alder 反应也得到正常加成产物，但是与芳甲酰基甲醛的反应不同，除了环己二烯之外，其余双烯底物同时还伴随着几乎等量的 ene 加成产物的形成[160]。

产率：67%
ee：99%

产率：34%　　产率：34%
ee：95%　　　ee：57%

Jørgensen 等报道了 $CuClO_4/(R_a)$-**1b** 催化体系能催化亚胺与共轭烯烃之间的杂 Diels-Alder 反应得到相应的含氮杂环化合物，在该催化体系下，活化的乙醛酸酯磺酰亚胺得到了较好的结果，对映选择性可高达 95%。而普通的磺酰亚胺虽然也能得到相应的环化产物，但是产物只有中等程度的 ee 值(46%)[161]。

产率：87%　产率：83%　产率：85%　产率：52%　产率：64%
ee：79%　　ee：94%　　ee：83%　　ee：95%　　ee：65%

在杂 Diels-Alder 反应中，除了醛、亚胺之外，偶氮化合物也可以作为底物，Yamamoto 等报道了在 AgOTf 与 $(R_a)$-**1a** 形成的手性银配体物催化下实现了 2-偶氮吡啶与一系列官能团化的硅氧基双烯顺利地发生高对映选择的环化反应得到相应的环加成反应产物，环化产物可以进一步方便地转化为手性二胺[162]。

研究发现手性 Cu(Ⅰ)/($S_a$)-**19e** 催化体系则能有效地催化非环状双烯体与 6-甲基-2-亚硝基吡啶之间的不对称 Nitoso-Diels-Alder 反应[163]。

Nishimura 等人在阳离子铱与 ($R_a$)-**1a** 形成的配合物的催化下,通过 C-H 活化实现了芳香酮亚胺与炔烃的不对称 [3+2] 环化反应,同时获得了具有高对映选择性的螺氨基茚衍生物,有趣的是添加苯甲酸却导致该产物的立体构型发生翻转[164]。

### 3.1.3.16 [2+2+2] 环加成反应

Tanaka 等人利用金属铑的催化作用成功实现了 1,6-二炔与异硫氰酸

酯的 [2+2+2] 环加成反应，得到相应的双环硫代吡喃亚胺，同时当使用 ($R_a$)-**1a** 为配体时，可实现对映选择环化反应，得到具有中等对映选择性的环化产物 [165]。

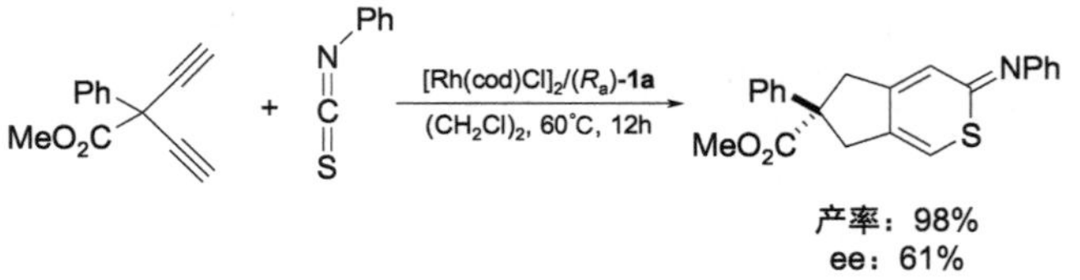

Ph
MeO2C
+
N
Ph
C
S
[Rh(cod)Cl]2/(Ra)-1a
(CH2Cl)2, 60°C, 12h
NPh
S
产率：98%
ee：61%


过渡金属催化的炔与异氰酸酯的 [2+2+2] 环化反应是构筑 2-吡啶酮的有效方法。Tanaka 等报道，Rh(Ⅰ)$^+$/($R_a$)-**19d** 催化体系能有效地催化 1,6-二炔与异氰酸酯之间的环化反应得到具有轴手性的 2-吡啶酮衍生物 [166]。

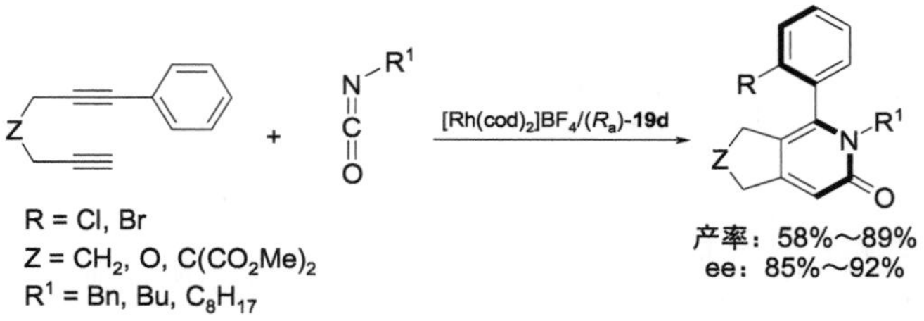

Z
+
N
R1
C
O
[Rh(cod)2]BF4/(Ra)-19d
R
N
R1
O
R = Cl, Br
Z = CH2, O, C(CO2Me)2
R1 = Bn, Bu, C8H17
产率：58%～89%
ee：85%～92%


Tanaka 等人使用铑催化 1,6- 二炔和 $\alpha$-脱氢氨基酸酯的分子间不对称 [2+2+2] 环化反应得到高产率及高对映选择性 $\alpha,\alpha$-双取代手性氨基酸衍生物。实验表明无论是对称还是不对称的二炔均获得了优异的对映选择性，对于不对称二炔反应的区域选择性很大程度上取决于底物结构 [167]。

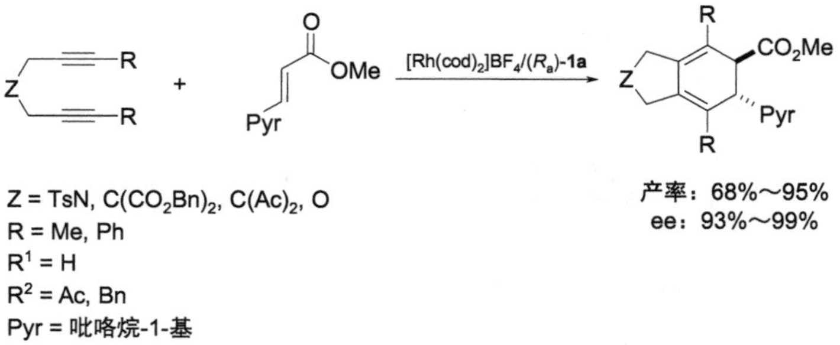

Z
R
R
+
O
OMe
Pyr
[Rh(cod)2]BF4/(Ra)-1a
CO2Me
Pyr
Z = TsN, C(CO2Bn)2, C(Ac)2, O
R = Me, Ph
R1 = H
R2 = Ac, Bn
Pyr = 吡咯烷-1-基
产率：68%～95%
ee：93%～99%

[Rh(cod)₂]BF₄/(Rₐ)-**1a**

R = Ph; R¹ = Me  产率：73%；ee：99%
R = CO₂Me; R¹ = Me  产率：67%；ee：99%

Tanaka 等人报道了手性铑配合物 [Rh(($S_a$)-H$_8$-BINAP)]BF$_4$ 在炔端 2-取代苯基取代的酯基桥连 1,6-二炔与炔烃之间的不对称 [2+2+2] 环化反应表现出了不错的催化活性和不对称诱导能力，为具有轴手性的光学活性苯并呋喃酮的不对称合成提供了新途径。其中不对称炔烃获得了中等至较好的区域选择性和较好的对映选择性，而对称炔烃则得到了很好的对映选择结果[168]。

Ar = 2-MeC₆H₄, 2-CF₃C₆H₄, 2-ClC₆H₄, 萘-1-基
R = H, AcOCH₂, HOCH₂
R¹ = AcOCH₂, HOCH₂, HOCH₂C≡C

产率：45%～96%
A : B = 64/36～90/10
ee：73%～＞99%

Tanaka 等人报道了手性铑配合物 [Rh(($S_a$)-H$_8$-BINAP)]BF$_4$ 催化体系应用于 2-取代苯基取代的(高)炔丙醇及其酯与丁炔二酸二酯之间的不对称 [2+2+2] 环化反应同样获得了很好的对映选择结果，提供了制备光学活性联苯型化合物的有效方法[169]。

Ar = 2-MeC₆H₄, 2-EtC₆H₄, 2-ClC₆H₄, 萘-1-基
R = HOCH₂, AcO, EtCOO
E = COOMe

产率：61%～89%
ee：84%～95%

Tanaka 小组利用 Rh/($R_a$)-H$_8$-BINAP 催化体系还发展了另一种通过分子内三炔的不对称 [2+2+2] 环化反应，实现了平面间环芳烷的催化对映选择性合成。由于反应同时伴随非手性的邻位环化产物的形成，因而反应的产率并不尽如人意，但是值得注意的是，反应获得了很好的对映选择性。

他们进一步研究发现，当使用 $(S_a)$-H$_8$-Xyl-BINAP(**12b**) 代替 $(R_a)$-H$_8$-BINAP(**12a**) 配体时，该催化体系同样适用于 1,12-二炔与丁炔二酸二叔丁酯的分子间环化反应，以 ee 值为 92% 的立体选择性和 15% 的产率得到相应的 [9] 间环芳烷，但不足之处是反应仍然伴随副产物的形成，目标化合物的产率很低[170]。

在 Rh(Ⅰ)/$(R_a)$-SegPhos(**19a**) 催化体系的催化下，1,6-二炔与炔端大位阻基团取代的 N,N-二取代丙炔酰胺可以发生不对称 [2+2+2] 环化反应制备光学活性 2,6-二取代苯甲酰胺衍生物。反应表现出了非常完美的对映选择性，对于所测试底物均得到了光学纯的环化产物[171]。

Shibata 等人巧妙地利用 [Rh(Ⅰ)]/$(R_a)$-SegPhos(**19a**) 催化的炔端含邻位取代芳基的 1,6-烯炔与炔烃的不对称 [2+2+2] 环化反应，实现了同时含有中心手性和轴手性的化合物的合成。其中炔组分中大位阻取代基的存在是反应获得很好非对映选择性的关键因素。当把炔组分中的叔丁氧羰基变为甲氧羰基或甲氧甲基时，反应的非对映选择性急剧下降[172]。

R = 2-MeC$_6$H$_4$, 联苯-2-基, 萘-1-基
Z = C(CO$_2$Me), NTs, O

产率：85%~98%
dr：>20/1
ee：98%~>99%

2016 年 Tanaka 小组实现了铑与 ($S_a$)-H$_8$-BINAP 络合物催化的邻位取代苯二炔与腈的对映选择性 [2+2+2] 环加成反应，成功地获得了高收率和高 ee 值的轴向手性 3-(2-卤代苯基)吡啶。并考察了邻位取代基对区域和对映选择性的影响。发现当邻位取代基为甲氧基和甲氧基羰基时，轴向手性 3-芳基吡啶是主要产物，而对映选择性明显降低。另外，当邻位取代基为烷基时，得到非手性 6-芳基吡啶[173]。

R$^1$ = H, Me
R$^3$ = CO$_2$Et, CO$_2$Me, COMe, CH$_2$CN, P(O)(OEt)$_2$
Z = [C(CO$_2$Et)$_2$]$_2$, [C(CO$_2$Me)$_2$]$_2$, NTs

| R$^2$ | 手性产物产率及ee值/% | 非手性产物产率/% |
|---|---|---|
| Br, Cl, OMe CO$_2$Me | 21%~97%；9%~99% | |
| Me, CF$_3$, CH$_2$OMe | | 85%~98% |

Roglans 等人实现了阳离子铑配合物 [Rh(cod)$_2$]BF$_4$ 与 ($R_a$)-1a 催化二炔和 Morita-Baylis-Hillman(MBH) 加合物的 [2+2+2] 环加成反应，并通过动力学拆分，得到具有高对映选择性(84%~97%)的含有两个相邻手性中心(分别为季碳和叔碳)的产物[174]。

产率：51%
ee：94%

Shibata 等人报道了手性离子型铑配合物可以催化 1,6-烯炔和炔烃之间的不对称 [2+2+2] 环化反应得到 1,3-己二烯衍生物。反应获得了很好的对映选择性结果。对于不对称的炔烃底物，虽然反应的区域选择性不是很好，但是所有异构体都有很高的 ee 值[175]。

R = H, Me, Ph
R$^1$ = Me, Ph
R$^2$ = CH$_2$OMe, CH$_2$OH, H
Z = NTs, C(CO$_2$Me)$_2$, O

产率：50%～96%
ee：88%～98%

此外，Shibata 等人还研究了手性铑配合物催化的分子内 1,4-二烯炔的不对称 [2+2+2] 环化反应，发现其中 1,4-二烯单元中 2-位取代基的存在与否决定了反应的化学选择性，当 2-位有取代基时反应得到的是桥环化合物，而当 2-位没有取代基时则得到的是稠环化合物[176]。

R = Me, Bu, H, Ph, BnOCH$_2$
R$^1$ = Me, Ph
Z = NTs, C(CO$_2$Bn)$_2$, O

产率：40%～83%
ee：88%～99%

R = Me, H, Ph, Ph(CH$_2$)$_3$
Z = NTs, C(CO$_2$Bn)$_2$, O

产率：55%～91%
ee：90%～99%

Tanaka 等人报道了使用阳离子 Rh/($R_a$)-1a 络合物催化 1,6-烯炔与环丙亚基乙酰胺的对映选择性 [2+2+1] 环加成反应，得到具有高对映选择性的双环(环戊-2-烯-1-亚基)乙酰胺[177]。

(R)-BINAP    产率：69%；ee：94%
(R)-Tol-BINAP    产率：68%；ee：97%

Shibata 等人报道了手性铑配合物 [Rh(cod)(($S_a$)-**1e**)]$BF_4$ 催化的 1,6-二炔和具有环外双键的环状化合物之间的 [2+2+2] 环化反应,高对映选择性地得到了手性螺环化合物。其中烯酮类底物反应活性较高,反应在较低温度下进行,但是反应的选择性有所下降。杂原子连接的二炔由于更容易自身聚合,往往需要加入过量的烯烃以抑制自身聚合副反应。对于不对称的二炔底物,反应不仅获得了很好的区域选择性,同时还获得了很好的对映选择性[178]。

产率:84%
ee:99%

### 3.1.3.17 1,3-偶极环加成反应

甲亚胺叶立德与烯烃的环加成反应是构建含有吡咯烷环结构活性化合物的有效方法之一。Komatsu 等人使用 ($R_a$)-**1a** 与 Cu(OTf)$_2$ 原位形成的配合物作为催化剂,在 N-亚烷基甘氨酸酯与活化烯烃间的 1,3-偶极加成反应中,获得了高非对映选择性和对映选择性的四氢吡咯衍生物。对于马来酰亚胺类亲偶极体而言,反应表现出很好的非对映选择性和较好的对映选择性,对于非环状亲偶极体而言,尽管反应同样获得很好的对映选择性,但是非对映选择性不理想[179]。

R = Ph, 4-MeOC$_6$H$_4$, 4-O$_2$NC$_6$H$_4$, 4-ClC$_6$H$_4$

产率:71%~83%
exo/endo:>95/5
ee(exo):62%~87%

Toste 等报道了 Au(Ⅰ)/($S_a$)-**19c** 催化体系能有效地催化吖内酯与一系列活化烯烃之间的 1,3-偶极环加成反应,反应获得了很好的立体选择性。此外,对于丙烯酸衍生物类活化烯烃,金催化反应同样具有不同区域选择性[180]。

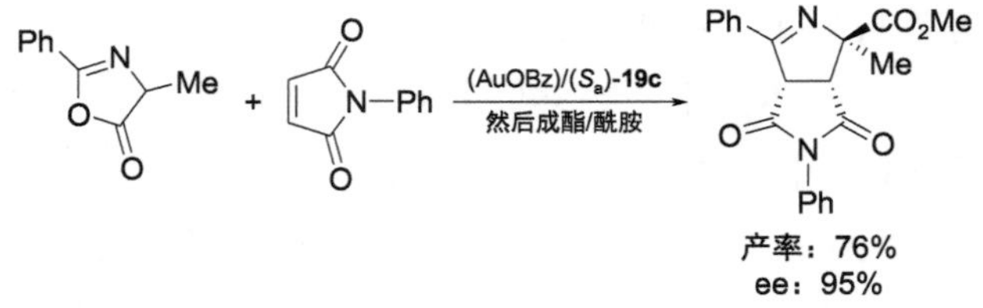
Ph
N
Me
O
O
+
O
N—Ph
O
(AuOBz)/(S_a)-19c
然后成酯/酰胺
Ph
N
CO2Me
Me
O
N
O
Ph
产率：76%
ee：95%

## 3.1.3.18 ene 反应

Lewis 酸催化的 ene 反应是一类重要的 C—C 键形成反应，可以方便地构筑 1,3-双官能团体系，在有机合成中具有广泛用途。

Zhang 等人报道了一种手性铑配合物催化的分子内炔和烯烃的 ene 反应。在 $[Rh(cod)Cl]_2/(S_a)$-**1a** 催化下酯基桥连的 1,6-烯炔实现了分子内炔和烯烃的 ene 反应，高产率地得到 $\alpha$-亚甲基-$\gamma$-丁内酯衍生物，反应具有很好的立体选择性，得到的产物几乎是光学纯的，利用该策略可实现天然产物咪唑类生物碱 (+)-毛果芸香碱 [+(pilocarpine)] 的全合成 [181]。

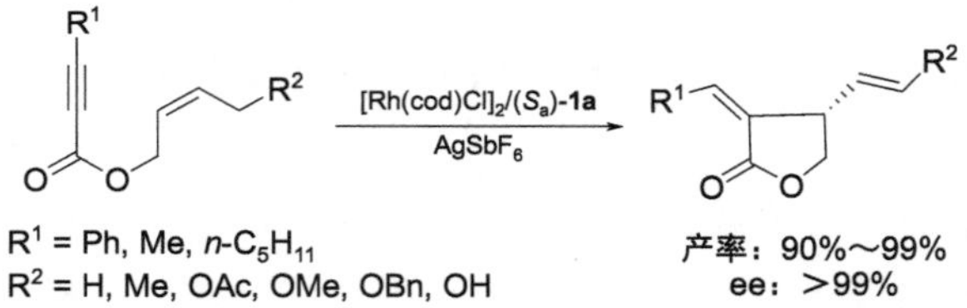
R1
R2
[Rh(cod)Cl]2/(S_a)-1a
AgSbF6
R1
R2
O
O
O
O
R1 = Ph, Me, n-C5H11
R2 = H, Me, OAc, OMe, OBn, OH
产率：90%～99%
ee：>99%

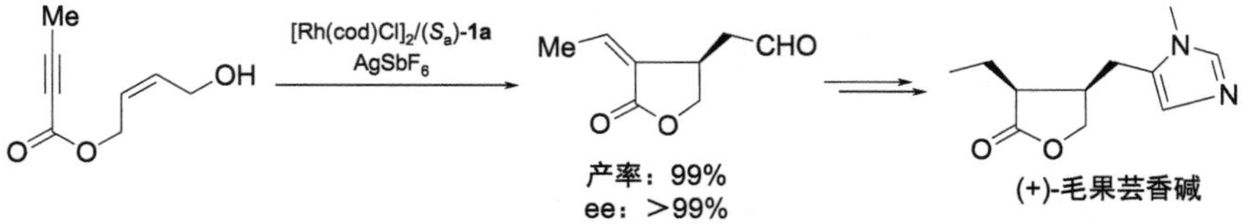
Me
OH
[Rh(cod)Cl]2/(S_a)-1a
AgSbF6
Me
CHO
N
N
产率：99%
ee：>99%
(+)-毛果芸香碱

此外，醚桥连的 1,6-烯炔 [182]，酰胺键桥联的 1,6-烯炔 [183] 和杂原子氮桥连的 1,6-烯炔 [184] 在同样的催化体系下可以高产率和高立体选择性地得到相应的光学构型产物。

Mikami 等报道了另一例基于三氟丙酮酸乙酯的羰基 -ene 反应，所用催化剂是配体 SegPhos(**19a**) 形成的手性钯配合物。该催化体系表现出了很好的底物适用范围，无论是反应活性较低的单取代烯烃还是二取代、三取代烯烃都能在 1h 内得到相应的加成产物，并且获得很好的立体选择性[185]。

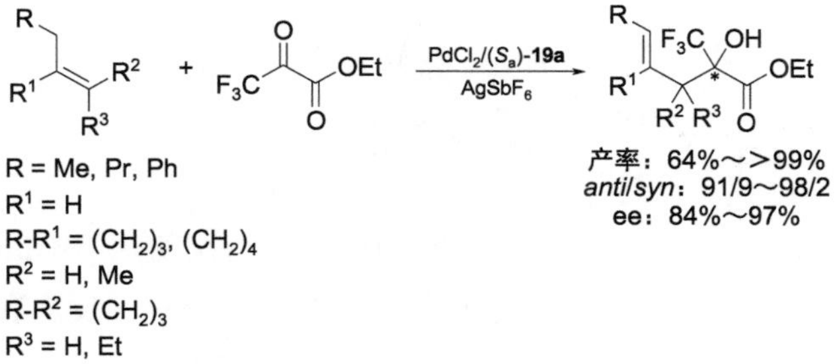

PdCl$_2$/($S_a$)-19a
AgSbF$_6$
R = Me, Pr, Ph
R$^1$ = H
R-R$^1$ = (CH$_2$)$_3$, (CH$_2$)$_4$
R$^2$ = H, Me
R-R$^2$ = (CH$_2$)$_3$
R$^3$ = H, Et
产率：64%～>99%
anti/syn：91/9～98/2
ee：84%～97%


Lectka 等利用手性铜配合物 $CuClO_4$ • $(MeCN)_2$/($R_a$)-**1b** 催化的 *N*-乙醛酸亚胺磺酸酯的 ene 反应实现了光学活性非天然氨基酸衍生物的合成。无论是链状的、环状的还是含有杂原子的 ene 底物均可获得高 ee 加成产物[186]。使用六氟磷酸铜代替高氯酸亚铜作为催化剂且催化剂用量降低至 1%(摩尔分数)时，甚至降低至 0.1%，基本不影响反应的对映选择性[187]。

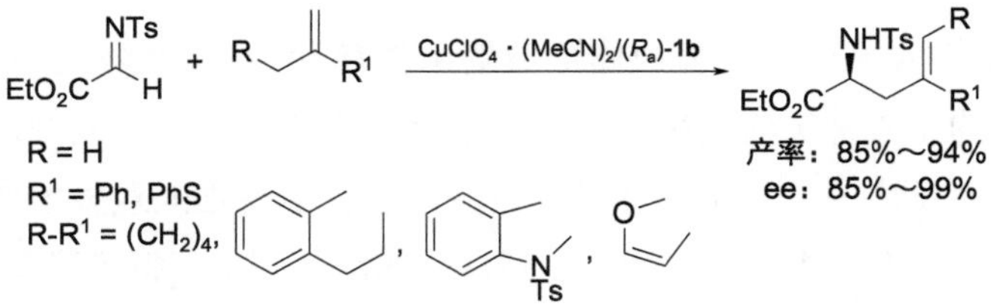

CuClO$_4$ · (MeCN)$_2$/($R_a$)-1b
R = H
R$^1$ = Ph, PhS
R-R$^1$ = (CH$_2$)$_4$,
产率：85%～94%
ee：85%～99%


### 3.1.3.19　分子内氢酰化反应

双膦配体形成的手性铑配合物能够催化酮醛分子的不对称氢酰化得到具有光学活性的环状分子。Morehead 等人报道了 2-乙烯基苯甲醛系列化合物在铑催化下发生的分子内氢酰化反应，实现了光学活性 2,3-二氢-1-茚酮衍生物的不对称合成。在离子手性铑配合物 [Rh($R_a$-BINAP)(nbd)]$ClO_4$ 的催化下，一系列 2-烯基取代苯甲醛高产率高对映选择性地得到相应的茚酮衍生物，除三甲基硅基取代底物立体选择性稍低之外。在反应过程中为了有效地抑制二聚副反应的发生，通常采用缓慢滴加的方式控制底物浓度[188]。

R = Me, Et, Ph, 萘-2-基, CH₂CH₂OH, SiMe₃, CF₃, COOEt

产率：89%～98%
ee：70%～99%

Sato 等人为解决催化剂回收再利用问题，首次使用 [BBIm][NTf₂] 作为反应介质进行上述催化剂催化的分子内氢酰化反应，反应在离子液体中进行时，同样可高产率及高立体选择性得到相应的氢酰化产物，并且催化剂可回收再利用至少 5 次而不影响催化活性和对映选择性[189]。

| 循环 | 1 | 2 | 3 | 4 | 5 |
|---|---|---|---|---|---|
| 产率/% | >99 | >99 | >99 | 99 | 99 |
| ee/% | 99 | 99 | 99 | 99 | 99 |

Dong 等人报道了双膦配体形成的手性铑配合物能够催化酮醛的分子内酮羰基的不对称氢酰化关环得到光学活性内酯衍生物。其中双膦配体的碱性强弱对反应的化学选择性影响十分明显，碱性强的双膦配体有利于抑制脱羰基反应得到相应的内酯，而碱性弱的配体则主要得到脱羰基副产物。在配体 ($R_a$)-19d 的存在下，反应不仅获得了很好的化学选择性，高产率地得到相应的氢酰化产物，而且在反应过程中实现了完美的立体控制[190]。

R = Ph, 4-ClC₆H₄, 萘-2-基
Me, Bu, Bn, *i*Pr, *t*Bu

产率：85%～99%
ee：99%～>99%

### 3.1.3.20 环异构化反应

烯基环丙烷(VCP)存在于天然和人工生物活性化合物中，是重要的有机合成中间体，因此引起了大量研究者的关注。Murakami 等发现芳基硼酸与 3-甲氧基-1-丙炔形成的 3-甲氧基-1-烯丙基铑中间体与烯烃反应可

方便地实现烯基环丙烷的构建，当使用手性配体 **1a** 形成的铑配合物为催化剂时，可实现手性烯基环丙烷衍生物的不对称合成[191]。

Toste 等人发展了一种金配合物催化的 1,7-或 1,8-烯炔的环化异构化构筑双环 [5.1.0] 及 [6.1.0] 衍生物的有效途径。在 **1e** 与 AuCl 形成的手性金配合物催化下，一系列具有 1,2-亚苯基结构单元桥连的 1,7-或 1,8-烯炔高对映选择性地得到相应的环化异构化产物[192]。

$R^1$ = Me, Et, 丙烯基, BnCH$_2$,
$R^2$ = H, Me
$n$ = 1, 2

Mikami 等人发现 Pd(Ⅱ)/($S_a$)-SegPhos(**19a**) 催化体系可以催化氧桥连的 1,6-烯炔的环化异构化反应得到相应的光学活性四氢呋喃衍生物。反应体系中催化剂前体以及溶剂的极性对反应的影响比较明显，在非极性溶剂氯代苯中 Pd(OTf)$_2$ 是较好的催化剂前体，而在极性溶剂 DMSO 中则使用 [(MeCN)$_4$Pd](BF$_4$)$_2$ 为催化剂前体可以获得更好的结果[193]。

| Pd(Ⅱ) | 产率/% | ee/% |
|---|---|---|
| Pd(OTf)$_2$ | >99 | >99 |
| [(MeCN)$_4$Pd](BF$_4$)$_2$ | >99 | 90 |

### 3.1.3.21 环化二聚反应

($R_a$)-BINAP(**1a**)、AgBF$_4$ 与 [Rh(cod)Cl]$_2$ 原位产生的手性离子型铑配合物可用于不对称催化二聚反应，Tam 等人报道了在该催化体系下，氧杂及氮杂苯并二烯环发生二聚反应一步产生萘并 [1,2-b]-呋喃环衍生物和吡咯环衍生物，具有优异的产率和优异的对映选择性[194]。

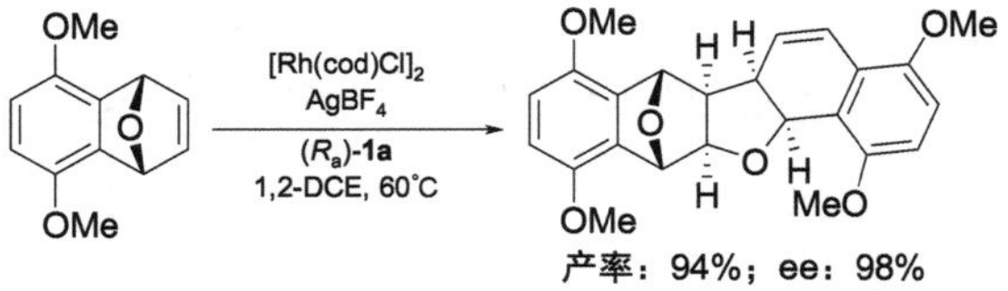

产率：94%；ee：98%

Hayashi 小组发展了另一种 [RhCl((Rₐ)-1a)]₂/NaBArF 催化体系，该催化体系表现出了更强的催化活性和不对称诱导能力，仅需要 1%（摩尔分数）剂量即可获得高达 99% 的立体控制结果。此外，该催化体系还可催化氧杂苯并二烯与丁炔二酸甲酯之间的交叉环化二聚反应，同样以高产率和高 ee 值得到相应的二氢呋喃衍生物[195]。

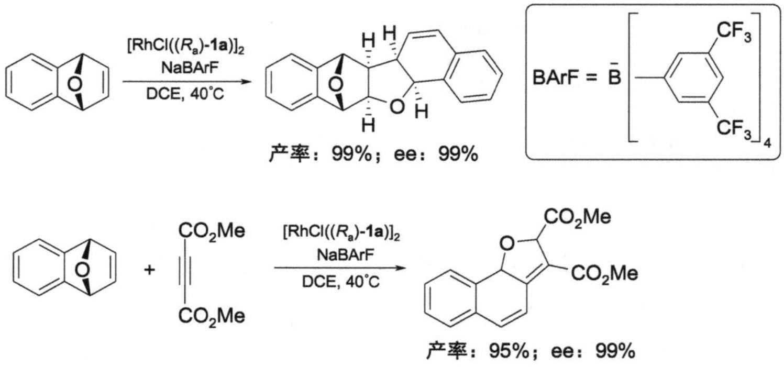

产率：99%；ee：99%

产率：95%；ee：99%

### 3.1.3.22 其他环化反应

Chegondi 等人报道了在 Rh 催化下的炔烃-环己二烯酮实现了高度区域选择性和对映选择性地还原环化，高产率地得到顺式氢苄基呋喃和顺式羟吲哚。同时还研究了 1,3-二炔-环己二烯酮的去对称化，其中 Rh 络合物与环己二烯酮环的分子内配位诱导了排他性的区域选择性。并通过氢-氘交叉实验的机理研究表明，氢活化是串联还原环化的速率决定步骤[196]。

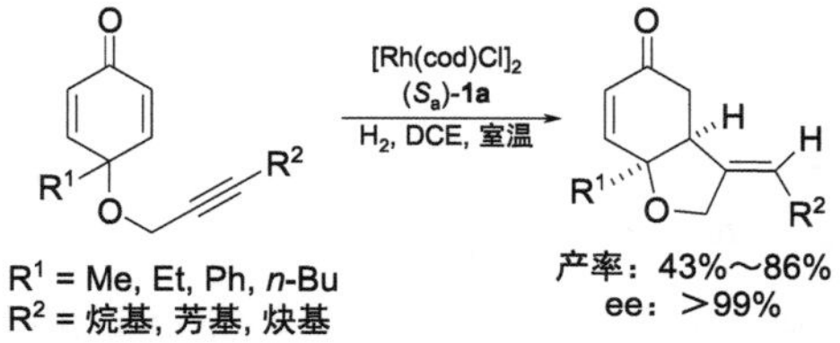

R¹ = Me, Et, Ph, n-Bu
R² = 烷基, 芳基, 炔基

产率：43%～86%
ee：>99%

Zhang 等人报道了一种通过钌/**1a** 催化体系催化连续脱氢环化和 N-烷基化一锅合成烷基氨基喹啉和萘啶，同时保持了中等到优秀的收率，该反应不需要添加外部氢源和卤化试剂[197]。

Akiyama 等人报道了 ($R_a$)-**1b** 与 [Cu(MeCN)$_4$]BF$_4$ 或 [Cu(MeCN)$_4$]PF$_6$ 原位产生的手性铜配合物可以催化 1-甲氧基取代丙二烯基硅烷及锗烷与 N-磺酰基乙醛酸酯亚胺之间的 [2+2] 环化反应得到四元氮杂环衍生物，环化产物经酸处理可以定量地转化为相应的官能团密集的 $\alpha,\beta$-不饱和-$\beta'$-氨基酰基硅（锗）烷。反应的对映选择性很大程度上取决于硅烷上取代基的空间体积，一般来说，随着取代基体积的增大，反应的立体选择性呈逐步下降的趋势[198]。

Lu 等使用配体 ($R_a$)-**1b** 与 Pd(CH$_3$CN)$_4$(BF$_4$)$_2$ 原位形成的离子型手性钯配合物为催化剂实现了 2-甲酰基苯硼酸与联烯酸酯之间的串联环化反应。反应获得了很好的非对映选择性和较好的对映选择性，可以方便地实现具有环外双键的手性二取代二氢茚结构单元的构建[199]。

Hayashi 小组报道了在 Rh/($R_a$)-SegPhos(**19a**) 催化体系中，炔端三烷基硅基或锗基取代的炔基芳基酮与硼酸酯反应可以环化得到 3,3′-双取代-1-茚酮衍生物。与芳基硼酸酯相比，烷基硼酸酯在产率和立体选择性方面均有所下降[200]。

## 3.1.3.23 开环反应

Yang 等人报道了 [Ir(cod)Cl]$_2$/($S_a$)-**1a** 催化体系可有效地催化仲胺对 $N$-Boc 保护氮杂苯并降冰片二烯的不对称开环反应制备相应的光学活性 1,2-二胺衍生物。其中脂肪族仲胺 4-取代哌嗪表现出了比芳香族仲胺 $N$-甲基苯胺更好的反应活性[201]。

R = 4-MeOC$_6$H$_4$, 4-MeC$_6$H$_4$, 4-FC$_6$H$_4$, 4-ClC$_6$H$_4$, 4-AcC$_6$H$_4$, 4-CF$_3$C$_6$H$_4$, 4-NO$_2$C$_6$H$_4$, 2-MeOC$_6$H$_4$, 2-MeC$_6$H$_4$, 2-FC$_6$H$_4$, 2-ClC$_6$H$_4$, 2-NCC$_6$H$_4$, 3-MeC$_6$H$_4$, 3-CF$_3$C$_6$H$_4$, 3,4-Cl$_2$C$_6$H$_3$, 2,4-F$_2$C$_6$H$_3$, 3,4-Me$_2$C$_6$H$_3$, 2,5-Me$_2$C$_6$H$_3$, 2,4-Me$_2$C$_6$H$_3$, 2,3-Me$_2$C$_6$H$_3$, Ph$_2$CH, Bn, EtO$_2$C

产率：55%～86%
ee：65%～87%

Nishibayashi 等报道了在 Cu(Ⅰ)/($R_a$)-**18h** 催化体系中，芳胺对炔基取代的外消旋环氧烷发生不对称开环反应可以提高产率地得到光学活性炔基取代氨基醇衍生物。其中芳基取代环氧烷底物获得了很好的对映选择性。虽然烷基取代的底物同样能发生反应，但是产率和立体选择性均有所下降[202]。

R = Ph, 4-ClC$_6$H$_4$, 4-BrC$_6$H$_4$, 4-MeC$_6$H$_4$, 4-PhC$_6$H$_4$, 4-FC$_6$H$_4$, 2-Naph, Me, $t$Bu
R$^1$ = CF$_3$, CO$_2$Me

产率：80%～98%
ee：54%～94%

Murakami 等发展了一种铑催化的硼酸酯取代环丁酮的分子内加成-开环制备具有季碳手性中心的苯并环戊酮衍生物的方法。其中双膦配体 ($S_a$)-SegPhos(**19a**) 获得了最好的对映选择结果。该策略在萜类天然产物 (−)-α-Herbertenol 的全合成中得到了成功应用 [203]。

R = Et, *i*Pr, BnO(CH$_2$)$_3$
R$^1$ = H, Me

产率：81%~97%
ee：79%~95%

(−)-α-Herbertenol

### 3.1.3.24 丙二烯化反应

尽管光学活性丙二烯衍生物在有机合成上占有重要地位，但其不对称催化合成仍然是不对称催化中最富有挑战性的研究课题之一。Hayashi 小组报道了 Pd/($R_a$)-**1a** 催化体系可以催化溴代共轭二烯与亲核试剂之间的 $S_N2'$ 类型亲核取代，一步实现光学富集的丙二烯衍生物的合成。研究表明，在催化体系中，添加剂二亚苄基丙酮(DBA)的存在是反应获得令人满意的对映选择性的关键因素 [204]。

R = Ph, 二茂铁基, *t*Bu, *n*-C$_8$H$_{17}$
Nu = C(NHAc)(COOEt)$_2$, C(COOMe)$_2$Me

产率：34%~98%
ee：41%~89%

在 Rh/($R_a$)-SegPhos(**19a**) 催化体系以及三甲基氯硅烷的存在下，芳基钛试剂可以与 β-炔基环己烯酮和环戊烯酮发生 1,6-加成得到光学活性的联烯基取代烯醇硅醚。其中三甲基氯硅烷的存在是反应成功的必要条件，否则反应只得到 1,2-加成产物 [205]。

Ar = Ph, 4-FC$_6$H$_4$, 4-MeC$_6$H$_4$
R = Bu, Cy, *t*Bu, 4-MeOC$_6$H$_4$
*n* = 0, 1

产率：60%~>99%

产率：44%~91%
ee：26%~92%

### 3.1.3.25 卤代反应

Gustafson 等人在催化剂控制的苯酚区域选择性氯化反应中证明了不同的催化剂结构可改变产物的区域选择性,其中使用硫化膦 **14** 为催化剂时得到对氯苯酚,而使用双硫脲催化剂时则克服了亲电苯酚氯化反应固有的对位选择性,通过标准的亲电氯化作用产生了不易获得的邻氯苯酚[206]。

R = 2-Ph, 2-$t$Bu, 2-CN, 2-I, 3-F, 3-Cl, 3-Br, 3-I, 3-$t$Bu, 3-$CO_2$Me

产率:53%~89%

Deng 等人使用 BINAP(S)(BINAP 一硫化物)**15a** 催化 $N$-溴代琥珀酰亚胺(NBS)对烯丙基苯胺的对映选择性溴化。得到了收率高达 90%,ee 值高达 87% 的手性 2-溴甲基吲哚啉产物[207]。

R = Ts, Ns, Ms, Bs, $p$-ClC$_6$H$_4$SO$_2$, $p$-MeOC$_6$H$_4$SO$_2$, PhSO$_2$
R$^1$ = H, 4-MeO, 5-MeO, 4-Me, 4-$t$Bu, 4-Ph, 4-F, 5-F, 4-Cl, 5-Cl, 4-Br, 4-NC, 4-CF$_3$, 4-NO$_2$, 3,5-Me$_2$, 4,6-Me$_2$
R$^2$ = H, Me, Ph, 4'-MeOC$_6$H$_4$, 4'-CF$_3$C$_6$H$_4$

产率:39%~90%
ee:0~87%

### 3.1.3.26 其他反应

具有联萘骨架的双膦配体除了在上述不对称反应中获得了成功应用之外,在许多其他类型不对称反应中也获得了很好的立体控制结果。

Mikami 等报道了双膦配体 ($S_a$)-**1a** 形成的手性离子型钯配合物在不对称 Suzuki-Miraura 反应中表现出了很好的催化活性,在数小时内即可高产率地得到相应的偶联产物,并且获得中等程度及较好的对映选择性[208]。

[反应式: 萘基溴 + 萘基硼酸, Pd(MeCN)₄SbF₆, (S_a)-1a, Ba(OH)₂; R = OMe, OiPr, OBn, NMe₂; 产率：61%～99%；ee：49%～70%]

在铑催化的有机金属试剂对活化烯烃的不对称1,4-加成反应中，有机金属试剂往往局限于芳基或烯基金属试剂，而炔基金属试剂的报道相对较少，并且底物适用范围较窄。Hayashi小组巧妙地利用铑催化的炔基烯丙醇的重排反应得到相应的 $\beta$-炔基酮类化合物，相当于间接地实现了炔基金属试剂对 $\alpha,\beta$-不饱和酮的不对称共轭加成反应[209]。

[反应式: 炔基烯丙醇底物 + [Rh(OH)(cod)]₂/(R_a)-1a → β-炔基酮；产率：88%；ee：94%]

Zhao等人使用 Rh/(R_a)-1a 催化体系对映选择性地氧化还原二级烯丙醇得到一系列具有高产率和良好至高水平的对映选择性具有 $\alpha$ 或 $\beta$ 立体中心的酮[210]。

[反应式: 1,3,3-三苯基烯丙醇 + [Rh(cod)]₂BF₄/(R_a)-1a, Ag₂CO₃, 甲苯, 24℃, MS 4Å, 20h → 产物；产率：89%；ee：84%]

Zhou等人报道了在 (R_a)-1a 与 PdCl₂ 催化下的外消旋 $\alpha$-甲基苄基溴与有机铝试剂对映选择性不对称交叉偶联反应，所得到的产物具有优异的立体选择性，ee值高达99%[211]。

[反应式: PhCH(Me)Br + 对甲基苯乙烯基Al(iBu)₂, PdCl₂/(R_a)-1a, THF, 35℃ → 产物；产率：54%～68%；ee：81%～99%]

Miura 等人用酰胺和胺进行 Pd/($R_a$)-**1a** 催化体系催化的仲苄基碳酸酯的不对称苄基取代，以高产率和高对映体比率形成相应的光学活性苄胺。同样可适用于苯酚亲核试剂，从而提供具有可接受产率和对映选择性的手性醚[212]。

$$\text{OBoc} \atop R^1 \quad R^2 \quad + \quad \text{Ts-NH-Me} \xrightarrow[K_2CO_3, \text{MeCN}, 60{}^\circ\text{C}, 6h]{[\text{CpPd}(\eta^3\text{-C}_3\text{H}_5)]/(R_a)\text{-1a}} \quad \text{Ts-N(Me)} \atop R^1 \quad R^2$$

$R^1$ = 2-萘基, 1-萘基, 6-MeO-2-萘基, 菲基, 2-苯并噻吩基

$R^2$ = Me, Ph, 4-MeOC$_6$H$_4$, 4-CF$_3$C$_6$H$_4$, 4-ClC$_6$H$_4$, 3-MeOC$_6$H$_4$, 2-MeC$_6$H$_4$

产率：34%～99%
er：96∶4～52∶48

Kotani 小组在手性双（氧化膦）($S_a$)-**17d** 作为路易斯碱催化剂，用四氯化硅活化羧酸以原位形成相应的双（三氯甲硅烷基）亚乙基二醇酯，随后将其与醛或酮进行醛醇缩合反应以生成 ee 值高达 92% 的高对映选择性的 $\beta$-羟基羧酸[213]。

$$\text{Br(CH}_2\text{)}_3\text{COOH} + \text{PhCHO} \xrightarrow[\text{CH}_2\text{Cl}_2, -60{}^\circ\text{C}, 24h]{\text{SiCl}_4/i\text{Pr}_2\text{NBu}/(S_a)\text{-17d}} \text{Ph-CH(OH)-CH(COOH)-(CH}_2\text{)}_2\text{Br}$$

产率：77%
ee：92%
dr：9:1

Consiglio 等报道了 Pd/($S_a$)-MeO-Biphep(**18a**) 催化体系能催化苯乙烯的不对称双烷氧羰基化得到光学活性苯基丁二酸甲酯，但是反应的转化率、化学选择性以及对映选择性与催化剂前体有关。当以 Pd(acac)$_2$/($S_a$)-**18a**/TsOH（1∶1∶2）为催化剂组合时，苯乙烯转化率为 48%，化学选择性为 49%，对映选择性为 93%；而使用 Pd(OSO$_2$CF$_3$)$_2$·2H$_2$O/($S_a$)-**18a**（1∶1）组合则可以将转化率和化学选择性分别提高到 69% 和 78%，并且获得了相近的立体选择性（ee 值 92%）[214]。

## 3.2 手性螺环双膦配体

2003 年,van Leeuwen 等报道了首例具有螺二色烷骨架的螺环双膦配体 SPANPhos(**36**)[215]。但由于制备这类螺环双膦配体的光学异构体非常困难,影响了这类螺环骨架配体的发展及应用。直到 2012 年,他们才将该类手性螺环双膦配体应用到钯催化 α-氰基苯乙酸乙酯的不对称氟化反应中,并给出了 ee 值为 93% 的对映选择性[216]。

2012 年,丁奎岭等基于他们发展的 α,α′-二-(2-羟基芳亚甲基)环酮的不对称催化氢化/缩酮化反应构筑手性 2,2′-螺二色烷结构,并设计合成了具有螺二色烷骨架的手性螺环双膦配体 SKP(**37**)。该手性螺环双膦配体 SKP 的钯催化剂 Pd-($R_a$,$R$,$R$)-**37b** 对 Morita-Baylis-Hillman(MBH) 加成产物类型的烯丙基醋酸酯与芳胺的不对称烯丙基胺化反应具有很高的催化活性和对映选择性[217,218]。反应的收率达 67%~96%,ee 值可达 91%~97%。催化剂 Pd-($R_a$,$R$,$R$)-**37b** 的化学选择性好,活性也非常高,转化数可达 4750。该不对称烯丙基氨基化反应已成功应用于降胆固醇新药依折麦(ezetimibe)的不对称合成。

最近，周剑等采用手性螺二色烷双膦配体 SKP 的的金催化剂 Au-($S_a$,$S$,$S$)-**37a** 实现了 3-重氮吲哚酮与烯烃的不对称环丙烷化反应[219]。反应给出了较高的收率 (44% ～ 86%)、很高的非对映选择性 (dr > 20∶1) 和对映选择性 (ee 值最高达 94%)。反应的底物适应范围也较广，反式和顺式的 1,2-二取代烯烃都能给出很好的结果。

早在 1998 年，蒋耀忠等人就研究具有螺 [4.4] 壬烷骨架的螺环亚磷酸酯配体 **38**，并成功应用于铑催化芳基乙烯的不对称氢甲酰化反应，得到了较好的区域选择性（最高达到 97:3），对映选择性可达到 69 % ee [220,221]。2004 年，陈新滋等[222]发展了磷氮键相连的手性螺环次亚膦酰胺配体 SpiroNP (**39**)。该配体在铑催化 α- 脱氢氨基酸酯的不对称氢化反应中表现优异，ee 值可达 94%～ > 99%。然而，手性螺环配体 SpiroBIP(**40**) 却表现出略低的对映选择性[223]。

手性螺环双膦-钌-双胺催化剂 ($S_a$,$R$,$R$)-**41a** 在简单酮的不对称催化氢化反应中表现出了优异的催化活性 (转化数 TON 高达 100000) 和对映选择性 (ee 值高达 > 99%)[224]。

手性螺环双膦 (chiral spiro diphosphines, SDP) 配体 (S)-**42b** 在钯催化丙二酸酯类化合物对 1,3-二苯基烯丙基醋酸酯的不对称烯丙基取代反应中也给出了很好的结果，ee 值达到 99.1%[225]。

SDP 配体 (R)-**42a** 还被成功用于铑催化 1,6-烯炔与硅烷试剂的不对称硅氢化环化反应，获得了 41%～93% 的收率和 89%～99.5% 的 ee 值[226]。

周其林等发现，手性螺环双膦-钌-双胺催化剂对消旋 α-取代醛和酮的不对称催化氢化反应非常有效，可为多样性手性醇的不对称合成提供新的动态动力学拆分合成方法，并在手性药物和生物活性天然产物分子的不对称全合成中得到了应用。例如，在消旋 α-烷基取代芳基乙醛的不对称催化氢化中，手性螺环双膦配体与 1,2-环己二胺 (DACH) 组合的手性螺环双膦-钌-双胺催化剂 ($S_a$,R,R)-**41c**，可以获得 ee 值为 78%～96% 的对映选择性。这是首次基于动态动力学拆分过程实现从醛的不对称催化氢化合成手性伯醇[227,228]。

在环状和链状的消旋 α-氨基取代脂肪酮的不对称催化氢化反应中，同样可以实现动态动力学拆分，手性螺环双膦-钌-双胺催化剂 ($S_a$,R,R)-**41b** 可给出最好的对映选择性结果，相应的产物 β-氨基醇的 ee 值高达 99.9%，cis/anti 选择性高达 99:1 反应的转化数可达 30000[229]。

($S_a$,$R$,$R$)-**41a** [Ar = 3,5-Me$_2$C$_6$H$_3$, R = C$_6$H$_5$]
($S_a$,$R$,$R$)-**41b** [Ar = C$_6$H$_5$, R = C$_6$H$_5$]
($S_a$,$R$,$R$)-**41c** [Ar = 4-MeO-3,5-Me$_2$C$_6$H$_2$, R-R = (CH$_2$)$_4$]

$n$ = 1～3；R$^1$ = H, 烷基
R$^2$ = 烷基, 芳基

(1MPa)

($S_a$,$R$,$R$)-**41a**
或($S_a$,$R$,$R$)-**41b**, 摩尔分数0.1%
$t$BuOK, $i$PrOH, 室温, 0.5~72h

ee：91%～99.9%
cis/trans高达99：1
TON高达30000

另外，该催化剂对动态动力学拆分消旋 $\alpha$,$\alpha'$-双取代环酮的不对称化反应氢也非常有效。以 ($S_a$,$R$,$R$)-**41a** 为催化剂对系列 $\alpha$-芳基-$\alpha'$-(2-乙氧基-2-氧代乙基)环酮的不对称氢化中，可以获得很高的 cis,cis-选择性和 ee 值为 75%～99.9% 的对映选择性[230]。

(5MPa)

($S_a$,$R$,$R$)-**41a** 摩尔分数0.1%
$t$BuOK, $i$PrOH, 室温, 50h

$n$ = 1～3

ee：75%～99.9%
cis, cis-选择性 >99：1

周其林等人在发现手性螺二氢茚骨架的双膦配体具备优异的手性诱导性能后，进一步在螺二氢茚骨架上进行改造，引入另一苯并环使之成为刚性更强、二面角更大(接近 90°)的螺二芴骨架，得到的手性螺环双膦配体 SFDP(**43**) 表现出更强的刚性[231,232]。实验表明，基于螺二芴骨架的手性螺环双膦配体 ($R$)-**43a** 和 ($R$)-**43b** 均在醋酸钌催化的肉桂酸和巴豆酸类 $\alpha$,$\beta$- 不饱和羧酸的不对称氢化反应中表现出很好的手性诱导能力和催化性能，获得了 ee 值为 97% 的对映选择性，底物/催化剂之比可高达 10000。

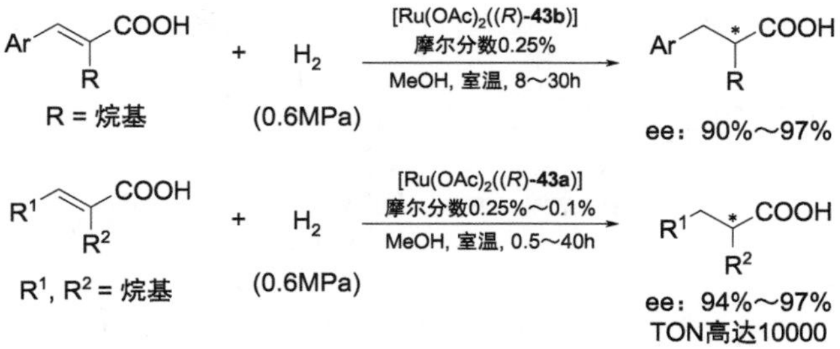
[Ru(OAc)2((R)-43b)]
摩尔分数0.25%
MeOH, 室温, 8～30h
Ar
COOH
R
R = 烷基
H2
(0.6MPa)
ee：90%～97%
[Ru(OAc)2((R)-43a)]
摩尔分数0.25%～0.1%
MeOH, 室温, 0.5～40h
R1, R2 = 烷基
(0.6MPa)
ee：94%～97%
TON高达10000

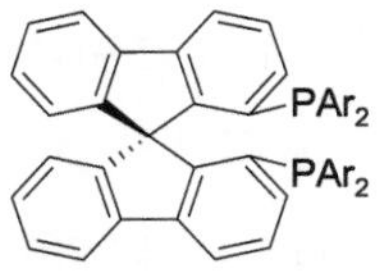
PAr2
PAr2

(*R*)-**43a** [Ar = 3,5-$(CH_3)_2C_6H_3$]
(*R*)-**43b** [Ar = 3,4,5-$(CH_3)_3C_6H_2$]

2007 年，徐立进、范青华、陈新滋等人报道了基于螺二氢茚骨架的手性螺环亚膦酸酯配体 SDPO(**44**)[233]，并发现该配体的铱络合物在喹啉化合物的不对称催化氢化反应中表现出优异的催化活性和对映选择性。例如，在 2-位烷基取代的喹啉不对称氢化反应中，催化剂 Ir-(*R*)-**44** 取得了 ee 值为 92% 的对映选择性，底物 / 催化剂之比可高达 5000。

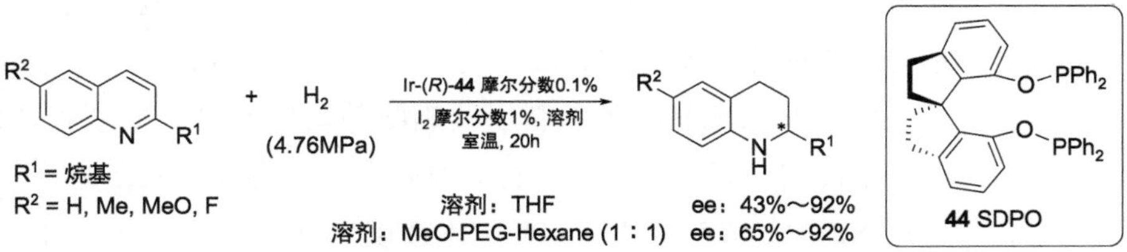
Ir-(R)-44 摩尔分数0.1%
I2 摩尔分数1%, 溶剂
室温, 20h
H2
(4.76MPa)
R1 = 烷基
R2 = H, Me, MeO, F
溶剂：THF
ee：43%～92%
溶剂：MeO-PEG-Hexane (1 ∶ 1)
ee：65%～92%
PPh2
PPh2
44 SDPO

# 3.3
# 碳手性双膦配体

1992 年，Burk 等在以 Et-DuPhos/Rh 体系催化氢化亚胺时，发现底物的 N 原子上羰基的存在对反应起着决定作用，羰基能活化底物，提高反应的活性[234]。

2015 年，Zhou 课题组以 $NiBr_2(DME)/(R,R)$-Me-DuPhos 为催化体系，$HCO_2H/Et_3N$（体积比为 5∶2）作为氢源的条件，80°C 下实现了不饱和烯烃的不对称转移氢化反应。该镍催化体系底物适用范围比较广，适用于含羧酸酯基、酰氨基和氰基的不饱和烯烃类底物[235]。

运用相同的催化体系 $NiBr_2(DME)/(R,R)$-Me-DuPhos 在水和 $AlCl_3$ 存在的条件下，以 N,N-二甲基甲酰胺 (DMF) 作为氢源，水和 $AlCl_3$ 有助于 DMF 分解生成甲酸根离子，相当于甲酸的等价类似物，成功实现了 $\alpha,\beta$-不饱和羧酸酯的不对称转移氢化。底物兼容性强，含有不同吸电子和给电子取代基的苯基、吡啶基以及噻吩基团的底物都能顺利地进行反应，以非常高的产率、良好到优秀的对映选择性得到还原产物[236]。Chirik 等采用 $Ni(OAc)_2/(S,S)$-Me-DuPhos 体系以 $H_2$ 作为氢源，$Bu_4NI$ 存在的条件下，可以催化前手性 $\alpha,\beta$-不饱和羧酸酯的对映选择性氢化，取得了良好到优秀的对映选择性（ee值为 73%～96%）[237]。

手性 DuPhos 双膦配体衍生物在相应双键氢化还原中的应用，例如，$Ni(OAc)_2/(S,S)$-Ph-BPE 体系成功地实现了环状磺酰亚胺酯的不对称氢化反应，以高产率和高对映选择性获得光学纯的环状磺酰胺酯产物（ee值为 83%～>99%[238]），以及催化 $\alpha,\beta$-不饱和磺酰亚胺的化学选择性氢化反应，高化学选择性和高对映选择性地获得了一系列保留烯烃 C=C 双键的手性烯丙基胺化合物[239]。

1995 年，Achiwa 等将 BCMP 的铱催化剂用于 2,3,3-三甲基-3H-吲哚的不对称氢化，结果发现，添加物对反应的影响很大，能提高反应的对映选择性。以 BiI$_3$ 为添加物时，获得了 ee 值最高 91% 的对映选择性[240]。此后，Zhang 课题组以 BICP 的铱催化剂，加入邻苯二甲酰亚胺作为添加物，将反应的 ee 值提高到 95 %[241]。

N-烷基亚胺稳定性差、易于分解，且往往存在 Z/E 异构体。因此，N-烷基亚胺的不对称氢化反应挑战性较大。大多数的 Ru、Rh、Ir、Ti 等催化剂仅得到中等的对映选择性。直到 1989 年，Bakos 等将磺酰化的 BDPP/铑催化剂用于 N-苄基亚胺的氢化，得到了 ee 值最高 96% 的对映选择性[242]。进一步的研究表明，配体磺酰化的程度决定着反应的对映选择性[243]。

非对称杂化的膦-氨基膦配体 HW-Phos 应用于苯并噁嗪的不对称氢化，获得了 ee 值最高 95% 的对映选择性，反应的 TON 高达 5000。

# 3.4 二茂铁双膦配体

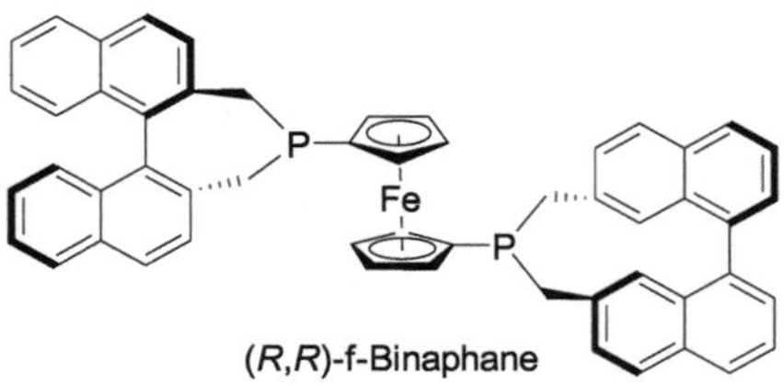

(R,R)-f-Binaphane

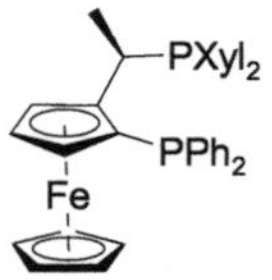

Xyliphos

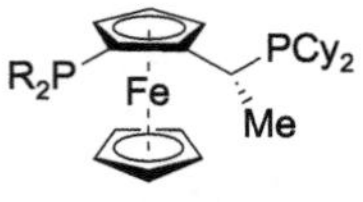

R = 3,5-Me₂-4-MeO-C₆H₂

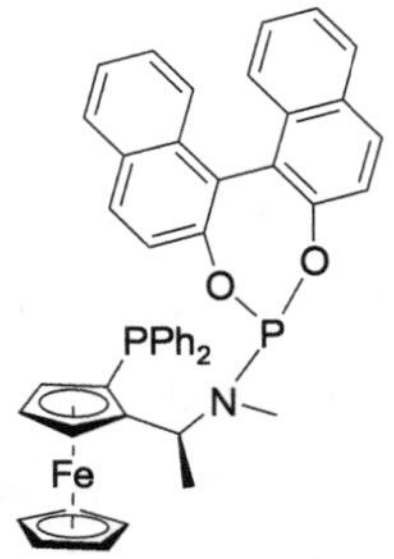

**45a** ($S_c,R_p,S_a$)
**45b** ($S_c,R_p,R_a$)

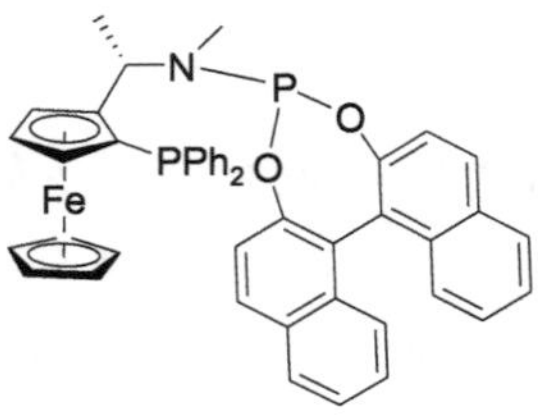

**45c** ($S_c,S_p,R_a$)
**45d** ($S_c,S_p,S_a$)

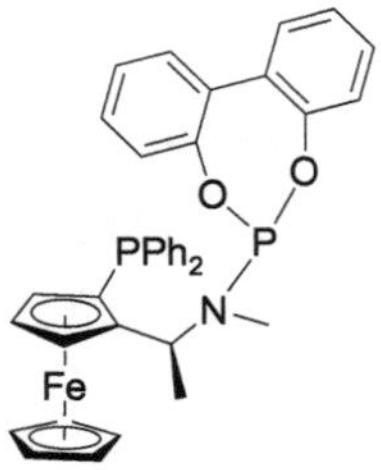

**45e** ($S_c,R_p$)

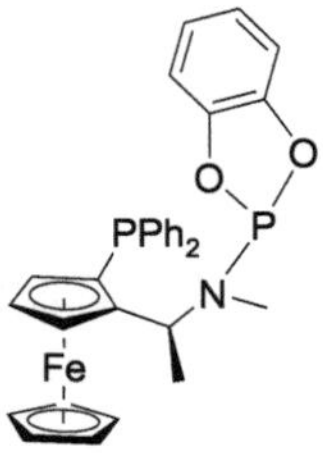

**45f** ($S_c,R_p$)

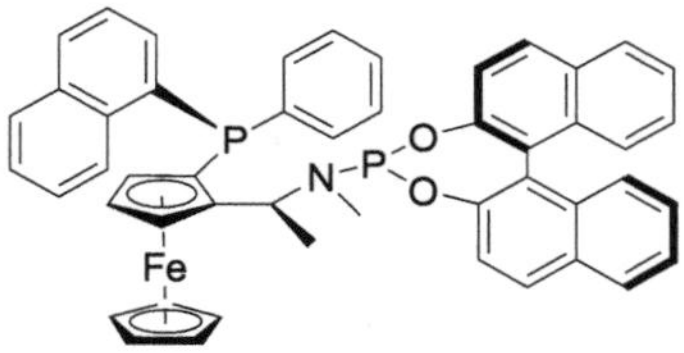

($S_c,R_{Fc},S_a,R_p$)-PPFA

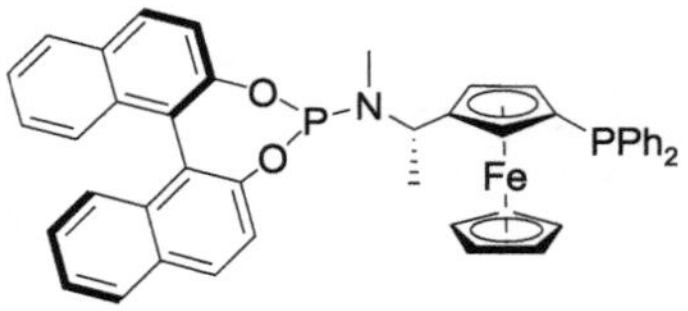

($S_c,R_p,R_a$)-PPFA

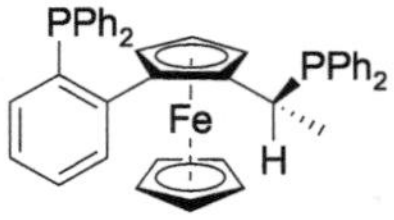

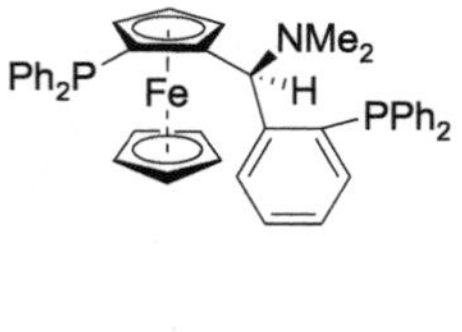

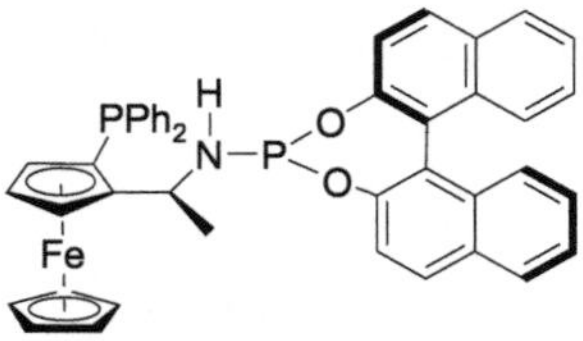

($S_c,R_p,S_a$)-PPFA

  BoPhoz 配体是最早应用于不对称反应的膦配体，该类配体合成简便，且很容易进行结构修饰，并在多类不饱和 C=C 双键催化不对称氢化反应中显示出优异的催化活性和立体选择性。具有极强的空气稳定性，放置三年后与铑离子络合催化氢化 α-脱氢氨基酸和衣康酸衍生物，依然可以获得高于 95% 的 ee 值。

张绪穆等报道的 Ir-(R,R)f-Binaphane 催化 N-芳基烷基芳基亚胺的氢化，得到了 ee 值为 96% ~ 99% 的对映选择性还原产物 [244]。

Xyliphos/铱配合物在催化 N-芳基亚胺的氢化中也表现出超高的活性，反应的底物/催化剂之比 S/C > 1000000，氢化产物的 ee 值为 79%。这一超高效催化剂已被成功用于年产量万吨的手性除草剂"金都尔"[(S)-异丙甲草胺, (S)-metolachlor] 的工业化生产 [245]。

基于二茂铁和 BINOL 衍生的膦-亚膦酰胺酯配体在不对称氢化反应中也能进行完美的对映选择性控制 [246]，以二茂铁为骨架的双膦配体除了经典的氢化反应外，也能有效地对其他多种反应进行不对称转化。

将二茂铁衍生的膦-亚磷酰胺酯配体 PPFAPhos 用于 Ag 催化亚甲胺叶立德与马来酸二甲酯的不对称 [3+2] 环加成反应。结果表明使用含有磷手性中心的配体 $(S_c,R_{Fc},S_a,R_p)$-PPFA 时，可获得最佳的非对映选择性(endo/exo 最高为 99/1)和对映选择性(ee 值最高 99%)[247]。

## 3.5 磷手性双膦配体

催化不对称氢化(AH)是制备单一对映体化合物最可靠的方法之一。其操作简单、收率高,在工业中占主导地位。多年来,这种转化在反应物结构和催化剂效率方面得到了长足发展。功能化烯烃的氢化主要由手性双膦配体在 Rh 或 Ru 催化剂存在的情况下进行的。P-立体中心双膦配体早期应用于 L-DOPA(一种用于治疗帕金森病的药物)的工业生产,而大多数双膦配体的设计是通过在配体主链上引入立体碳中心而不是在 P 原子上,从而失去了靠近金属中心的立体碳中心所能提供的优势。主要原因是很难制备在 P 原子上具有手性的磷化氢。随着新的、直接的方法的发展,P-立体中心双膦配体成功地应用于催化不对称氢化过程。

Diéguez 等成功合成了一些 P-立体中心双膦配体,用于 Rh 催化的烯烃的不对称氢化反应。这些催化剂对 α-脱氢氨基酸衍生物和 α-烯酰胺有优秀的对映体控制(ee 值高达 99%),对更具挑战性的 β-类似物(ee 值高达 80%)具有良好的结果,同时还利用该催化体系合成几种有价值的药理活性化合物前体[248]。

(R,R)-DIPAMP  (R,R)-BenzP  (S)-Binspine  ZhangPhos

DuanPhos  BIBOP  (S)-TCFP  MaxPhos

(S,S,S,S)-WingPhos  (S,S)-tBu-BisP  TangPhos  (R,R)-QuinoxP

Tang 等报道通过使用 WingPhos 或 PFBO-BIBOP 作为配体，可实现高度对映选择性的铑催化芳基环硼氧烷与 N-未保护的酮亚胺的加成，得到一系列手性具有优异的 ee 值和产率的产物。该方法能够有效地对映选择性合成 Cipargamin[249]。此外，Rh/WingPhos 催化体系还可应用于一些医药关键中间体的合成。例如，西洛多辛的关键中间体[250]。

Tang 等发现使用 ArcPhos 对环状四取代烯酰胺的不对称氢化反应可以获得优异的产率和对映选择性。克级规模试验证明催化剂 Rh/ArcPhos 最高的转化数可以达到 10000，具有很好的实用前景[251]，值得注意的是，该体系作为关键合成步骤可实现 (R)-托法替尼的克级规模合成。

$$\underset{R}{\text{NHAc}} \xrightarrow[\text{MeOH, H}_2(3.45\text{MPa})]{[\text{Rh(nbd)}]\text{SbF}_6/\text{ArcPhos}} \underset{R}{\text{NHAc}}$$

产率：95%～99%
ee：高达99%
TON高达10000

ArcPhos

2006 年，Imamoto 等以 P-手性的双膦配体 (S,S)-tBu-BisP 的铱配合物为催化剂，在常压 (101 kPa) 下，研究了 N-芳基亚胺的氢化反应，得到了最高 99% 的对映选择性。底物的电子效应对反应立体选择性的影响十分明显。同样底物在 DuanPhos/铱的催化下也能获得满意的产率和对映选择性[252]。对于 N-苯基取代的苯乙酮衍生的底物 (Ar′ = Ph)，当 Ar 为 4-MeOC$_6$H$_4$ 时，反应的 ee 值仅为 69%。对于取代的 N-苯基苯乙酮衍生的底物 (Ar=Ph)，当 Ar′为 4-CF$_3$C$_6$H$_4$ 时，得到了 ee 值最高 99% 的对映选择性[253]。

$$\underset{\text{Ar}}{\overset{}{\underset{N}{\parallel}}}\text{Ar}' \xrightarrow[\text{H}_2(101\text{kPa}), \text{CH}_2\text{Cl}_2]{\text{配体} \atop [\text{Ir(cod)}_2]\text{BArF}} \underset{\text{Ar}}{\overset{*}{\underset{H}{\text{N}}}}\text{Ar}'$$

配体为(S,S)-tBu-BisP, ee：69%～99%
配体为(S,R)-DuanPhos, ee：89%～98%

张绪穆课题组发现 TangPhos 的铑配合物能高效地催化 α-亚胺酯的氢化，提供了一种制备手性 α-氨基酸衍生物的高效方法[254]，以及 N-磺酰基亚胺氢化转化方法[255]。

$$\underset{R}{\overset{\text{NPMP}}{\underset{\text{CO}_2\text{R}'}{\parallel}}} \xrightarrow[\text{H}_2(5.0\text{MPa}), \text{CH}_2\text{Cl}_2, 50°\text{C}]{\text{TangPhos} \atop [\text{Rh(cod)}_2]\text{BF}_4} \underset{R}{\overset{\text{NHPMP}}{\underset{\text{CO}_2\text{R}'}{*}}}$$

ee：高达91%

张万斌等利用 Ni/(R,R)-QuinoxP 体系实现链状酮和芳基酮的叔丁基磺酰酮亚胺的不对称催化氢化反应，可高对映选择性地得到一系列手性叔丁基磺酰胺产物底物普适性好，最高可以获得 ee 值 99% 的对映选择性。对于苯并环类磺酰亚胺类底物的不对称氢化，该类催化剂的活性非常高，底物和催化剂用量的比例能达到 10500，是镍催化均相不对称氢化反应的重大突破，使其有望应用于工业应用[256]。

## 3.6 膦-亚磷酸酯及膦-亚磷酰胺酯

1993 年，Takaya 和 Nozaki 等发展了含有两个手性联萘单元的膦-亚磷酸酯杂合配体 (R,S)-BinaPhos。首次实现了苯乙烯的高效高对映选择性不对称氢甲酰化。该反应催化剂用量低，反应条件温和，具有良好的区域选择性和优异的对映选择性（b/l = 7.3, ee 值 94%），因此该工作被认为是不对称氢甲酰化领域中具有里程碑意义的研究之一[257]。Nozaki 等将 Rh(Ⅰ)/(R,S)-BinaPhos 催化体系应用于 2-乙烯基噻吩的不对称氢甲酰化反应中，获得了很好的区域选择性和对映选择性（b/l = 16, ee 值 93%）。在此基础上他们将该反应体系用于 2-苯酰基-5-乙烯基噻吩的不对称氢甲酰化反应中，最后进行氧化，以 84% 的 ee 值得到非甾体类消炎药 (S)-噻洛芬酸[258]。随后 Rh(Ⅰ)/(R,S)-BinaPhos 催化体系在 2,5-二氢呋喃、2,5-二氢吡咯[259] 和非对称 1,2-二取代烯烃的不对称氢甲酰化方面都有相应的报道[260]。

2005 年，Leitner 等首次将手性膦-亚磷酸酯配体 (R,S)BinaPhos 用于芳香烯烃与频哪醇 H-磷酸酯的氢膦酰化反应，在钯催化下可以高选择性地得到马氏规则的产物，但相应的对映选择性较低[261]。

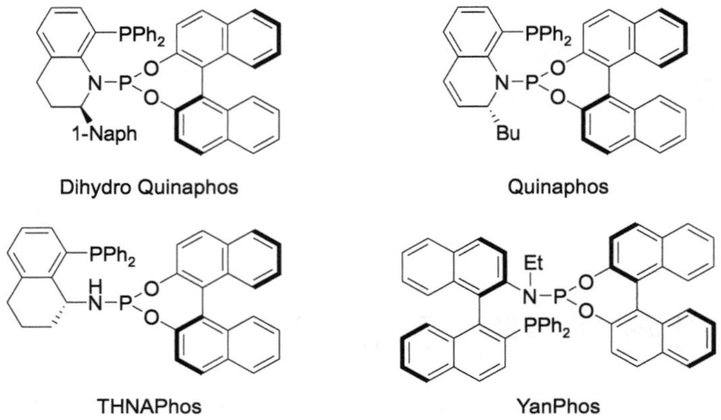

Dihydro Quinaphos

Quinaphos

THNAPhos

YanPhos

  膦-亚磷酰胺酯作为一类双膦配体，其常用于经典的不饱和双键的不对称氢化反应中[262-266]。例如，Leitner 等将 Quinaphos/Rh 配合物用于催化烯酰胺和衣康酸二甲酯的不对称催化氢化反应，得到了 97.8% 和 98.8% 的对映选择性，其中 ($R_c,R_a$) 为匹配构型。当他们采用含有柔性骨架 1,2,3,4-四氢喹啉衍生的配体时，却发现 ($S_c,R_a$) 为匹配构型，Dihydro Quinaphos/Rh 催化双键的不对称氢化反应，可得到 ee 值 > 99% 的产物，该反应的转化频率（TOF）高达 20000 $h^{-1}$[267]。根据上述介绍可看出，传统的氢化反应一般用二氯甲烷、甲苯等有毒溶剂。

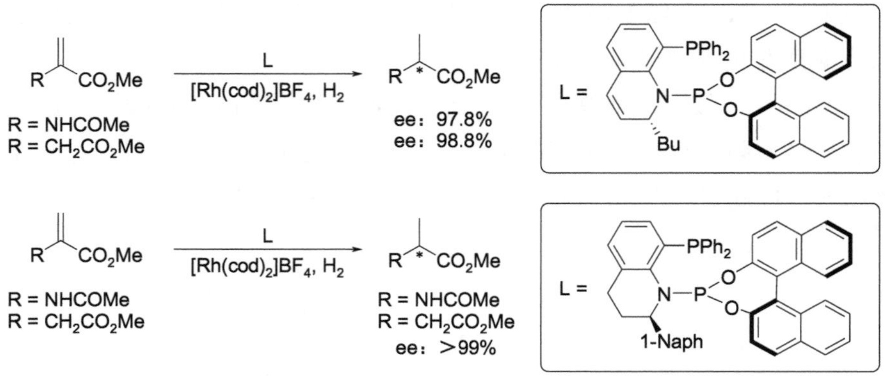

  以绿色化学基本准则为契机，Bakos 等尝试使用对环境友好的有机碳酸酯作为溶剂，在 Rh/ 膦-亚磷酰胺酯的催化反应后，经过简单的过滤、洗涤纯化便可得到合成治疗帕金森病的药物 L-DOPSA（产率 71%，ee 值 99.6%）[268]。

Zheng 等分别以四氢萘胺和 1-萘胺为原料合成了配体 THNAPhos 和配体 (S)-HY-Phos。在氢化反应过程中，(S)-HY-Phos 虽然比 THNAPhos 少了 1 个碳手性中心，但在烯醇膦酸酯的氢化反应中，依然可以得到最高 99% 的对映选择性。在配体合成成本上 (S)-HY-Phos 更具优势[269,270]。

类似结构的膦-亚磷酰胺酯在不同双键进行不对称氢化的例子也有相应的报道，如采用铱/膦-亚磷酰胺酯催化体系可对烷基芳基亚胺、二烷基亚胺、α-亚胺酯和喹啉等底物进行氢化，均能实现完全转化且得到很好的对映选择性 (ee 值最高 99%)[271,272]。Ru/膦-亚磷酰胺酯催化体系也能实现良好的对映选择性[273]。

铑/膦-亚磷酰胺酯催化体系在催化不对称氢甲酰化反应中也得到很好的应用[267,274,275]。以 YanPhos 配体 (R,S) 为例，其与铑形成的配合物对苯乙烯衍生物的不对称氢甲酰化反应表现出很强的对映选择性控制[276,277]。

Reek 等将 IndolPhos 配体与烯丙基氯化钯原位生成的催化剂用于单取代和二取代烯丙醇乙酸酯的烯丙基烷基化反应。对于 1,3-二苯基-2-丙烯醇乙酸酯 ee 值最高可达 90%。对于单取代的肉桂醇乙酸酯,虽然其区域选择性较低 ($b/l$ = 14:86),仍可得到较好的对映选择性 (ee 值 81%)[278,279]。

## 参考文献

[1] Miyashita A, Yasuda A, Takaya H, et al. Synthesis of 2,2′-bis(diphenylphosphino)-1,1′-binaphthyl (BINAP), an atropisomeric chiral bis(triaryl)phosphine, and its use in the rhodium(Ⅰ)-catalyzed asymmetric hydrogenation of α-(acylamino)acrylic acids. J. Am. Chem. Soc., 1980, 102: 7932-7934.

[2] Cai D, Payack J F, Bender D R, et al. Synthesis of chiral 2,2′-bis(diphenylphosphino)-1,1′-binaphthyl (BINAP) via a novel Nickel-catalyzed Phosphine Insertion. J. Org. Chem., 1994, 59: 7180-7181.

[3] Ager D J, Laneman S A Convenient and direct preparation of tertiary phosphines via nickel-catalysed cross-coupling. Chem. Commun., 1997, 2359-2360.

[4] Noyori R, Ohkuma T, Kitamura M, et al. Asymmetric hydrogenation of β-keto carboxylic esters. A practical, purely chemical access to β-hydroxy esters in high enantiomeric purity. J. Am. Chem. Soc., 1987, 109: 5856-5858.

[5] Kitamura M, Ohkuma T, Takaya H, et al. A practical asymmetric synthesis of carnitine. Tetrahedron Lett., 1988, 29: 1555-1556.

[6] Huang H L, Liu L T, Chen SF, et al. The synthesis of a chiral fluoxetine intermediate by catalytic enantioselective hydrogenation of benzoylacetamide. Tetrahedron: Asymmetry, 1998, 9: 1637-1640.

[7] 李月明, 范清华, 陈新滋: 不对称有机反应. 北京: 化学化学工业出版社, 2005, 11-29.

[8] Jeulin S, Paule S D de, Ratovelomanana-Vidal V, et al. Difluorphos, an electron-poor diphosphane: a good match between electronic and steric features. Angew. Chem. Int. Ed., 2004, 43: 320-325.

[9] Ohkuma T, Koizumi M, Muñiz K, et al. trans-RuH($\eta^1$-BH$_4$)(binap)(1,2-diamine): a catalyst for asymmetric hydrogenation of simple ketones under base-free conditions. J. Am. Chem. Soc., 2002, 124: 6508-6509.

[10] Mikami K, Aikawa K, Kainuma S, et al. Enantioselective catalysis of carbonyl-ene and Friedel-Crafts reactions with trifluoropyruvate by 'naked' palladium(Ⅱ) complexes with SEGPHOS ligands. Tetrahedron: Asymmetry, 2004, 15: 3885-3889.

[11] Flowers B J, Gautreau-Service R, Jessop P G. β-hydroxycarboxylic acids from simple ketones by

carboxylation and asymmetric hydrogenation. Adv. Synth. Catal., 2008, 350, 2947-2958.
[12] Starodubtseva E V, Turova O V, Vinogradov M G, et al. A convenient route to chiral γ-lactones via asymmetric hydrogenation of γ-ketoesters using the $RuCl_3$–BINAP–HCl catalytic system. Tetrahedron, 2008, 64: 11713-11717.
[13] Turova O V, Starodubtseva E V, Vinogradov M G, et al. A concise synthesis of highly enantiomerically enriched 2-alkylparaconic acid esters via ruthenium-catalyzed asymmetric hydrogenation of acylsuccinates. Tetrahedron: Asymmetry, 2009, 20: 2121-2124.
[14] Blanc D, Ratovelomanana-Viadal V, Marinetti A, et al. Enantioselective synthesis of *anti* 1,3-diols *via* Ru(II)-catalyzed hydrogenations. Synlett, 1999, 480-482.
[15] Meng Q, Sun Y, Ratovelomanana-Vidal V, et al. $CeCl_3·7H_2O$: an effective additive in Ru-catalyzed enantioselective hydrogenation of aromatic α-ketoesters. J. Org. Chem., 2008, 73: 3842-3847.
[16] Wan X, Meng Q, Zhang H, et al. An efficient synthesis of chiral β-hydroxy sulfones via Ru-catalyzed enantioselective hydrogenation in the presence of iodine. Org. Lett., 2007, 9: 5613-5616.
[17] Clarke M .L, France M B, Knight F R, et al. Enantio- and diastereoselective hydrogenation of a fluorinated diketone. synlett, 2007, 1739-1741.
[18] Donate P M, Frederico D, Silva R da, et al. Asymmetric synthesis of γ-butyrolactones by enantioselective hydrogenation of butenolides. Tetrahedron: Asymmetry, 2003, 14: 3253-3256.
[19] Maligres P E, Krska S W, Humphrey G R. Enantioselective hydrogenation of α-aryloxy α, β-unsaturated acids. Asymmetric synthesis of α-aryloxycarboxylic acids. Org. Lett., 2004, 6: 3147-3150.
[20] Uemura T, Zhang X, Matsumura K, et al. Highly efficient enantioselective synthesis of optically active carboxylic acids by $Ru(OCOCH_3)_2[(S)-H_8$-BINAP]. J. Org. Chem., 1996, 61: 5510-5516.
[21] Kita Y, Hida S, Higashihara K, et al. Chloride-bridged dinuclear rhodium(iii) complexes bearing chiral diphosphine ligands: Catalyst precursors for asymmetric hydrogenation of simple olefins. Angewandte Chemie, 2016, 55: 8299-8303.
[22] Yuan S, Gao G, Wang L, et al. The combination of asymmetric hydrogenation of olefins and direct reductive amination. Nature Communications, 2020, 11: 621.
[23] Scrivanti A, Bovo S, Ciappa A, et al. The asymmetric hydrogenation of 2-phenethylacrylic acid as the key step for the enantioselective synthesis of Citralis Nitrile. Tetrahedron. Lett., 2006, 47: 9261-9265.
[24] Wang X B, Wang D W, Lu S M,et al. Highly enantioselective Ir-catalyzed hydrogenation of exocyclic enamines. Tetrahedron: Asymmetry, 2009, 20: 1040-1045.
[25] Wu S, He M, Zhang X. Synthesis of ortho-phenyl substituted MeO-BIPHEP ligand and its application in Rh-catalyzed asymmetric hydrogenation. Tetrahedron: Asymmetry, 2004, 15: 2177-2180.
[26] Tang W, Chi Y, Zhang X. An *ortho*-substituted BIPHEP ligand and its applications in Rh-catalyzed hydrogenation of cyclic enamides. Org. Lett., 2002, 4: 1695-1698.
[27] Abe H, Amii H, Uneyama K. Pd-catalyzed asymmetric hydrogenation of α-fluorinated iminoesters in fluorinated alcohol: a new and catalytic enantioselective synthesis of fluoro α-amino acid Derivatives. Org. Lett., 2001, 3: 313-315.
[28] Wang Y Q, Zhou Y G. Highly enantioselective Pd-catalyzed asymmetric hydrogenation of N-diphenylphosphinyl ketimines. Synlett, 2006, 1189-1192.
[29] Wang W B, Lu S M, Yang P Y, et al. Highly enantioselective iridium-catalyzed hydrogenation of heteroaromatic compounds, quinolines. J. Am. Chem. Soc., 2003, 125: 10536-10537.
[30] Yang P Y, Zhou Y G. The enantioselective total synthesis of alkaloid (−)-galipeine. Tetrahedron: Asymmetry, 2004, 15: 1145-1149.
[31] Wang X B, Zeng W, Zhou Y G. Iridium-catalyzed asymmetric hydrogenation of pyridine derivatives, 7,8-dihydro-quinolin-5(6H)-ones. Tetrahedron Lett., 2008, 49: 4922-4924.
[32] Chen M W, Duan Y, Chen Q A, et al. Enantioselective Pd-catalyzed hydrogenation of fluorinated imines: facile access to chiral fluorinated amines. Org. Lett., 2010, 12: 5075.
[33] Zhou X Y, Bao M, Zhou Y G. Palladium-catalyzed asymmetric hydrogenation of simple ketimines using a Brønsted acid as activator. Adv. Synth. Catal., 2011, 353: 84.
[34] Bunlaksananusorn T, Rampf F. A facile one-pot synthesis of chiral β-amino esters. Synlett, 2005, 2682-

2684.

[35] Gao K, Wu B, Yu C.-B, et al. Iridium catalyzed asymmetric hydrogenation of cyclic imines of benzodiazepinones and benzodiazepines. Org. Lett., 2012, 14: 3890-3893.

[36] Berhal F, Wu Z, Zhang Z, et al. Enantioselective synthesis of 1-aryl-tetrahydroisoquinolines through iridium catalyzed asymmetric hydrogenation. Org. Lett., 2012, 14: 3308-3311.

[37] Charette A B, Giroux A. Asymmetric hydrogenation of $N$-tosylimines catalyzed by BINAP-ruthenium(II) complexes Tetrahedron Lett., 1996, 37: 6669-6672.

[38] Xu H, Yang P, Chuanprasit P, et al. Nickel-catalyzed asymmetric transfer hydrogenation of hydrazones and other ketimines. Angew. Chem. Int. Ed., 2015, 54: 5112.

[39] Ren Y, Xu X, Sun K, et al. A new and effective method for providing optically active monosubstituted malononitriles: selective reduction of $\alpha,\beta$-unsaturated dinitriles catalyzed by copper hydride complexes. Tetrahedron: Asymmetry, 2005, 16: 4010-4014.

[40] Appella D H, Moritani Y, Shintani R, et al. Asymmetric conjugate reduction of $\alpha, \beta$-unsaturated esters using a chiral phosphine-copper catalyst. J. Am. Chem. Soc., 1999, 121:9473-9474.

[41] Moritani Y, Appella D H, Jurkauskas V, et al. Synthesis of $\beta$-alkyl cyclopentanones in high enantiomeric excess via copper-catalyzed asymmetric conjugate reduction. J. Am. Chem. Soc., 2000, 122: 6797-6798.

[42] Hughes G, Kimura M, Buchwald S L. Catalytic enantioselective conjugate reduction of lactones and lactams. J. Am. Chem. Soc., 2003, 125: 11253-11258.

[43] Llamas T, Arrayás R G, Carretero J C. Catalytic asymmetric conjugate reduction of $\beta,\beta$-disubstituted $\alpha,\beta$-unsaturated sulfones. Angew. Chem. Int. Ed., 2007, 46: 3329-3332.

[44] Deng J, Hu X P, Huang J D, et al. Enantioselective synthesis of $\beta$-aryl-$\gamma$-amino acid derivatives via Cu-catalyzed asymmetric 1,4-reductions of $\gamma$-phthalimido-substituted $\alpha,\beta$-unsaturated carboxylic Acid Esters. J. Org. Chem., 2008. 73: 6022-6024.

[45] Lipshutz B H, Servesko J M, Petersen T B, et al. Asymmetric 1,4-reductions of hindered â-substituted cycloalkenones using catalytic SEGPHOS−ligated CuH. Org. Lett., 2004, 6: 1273-1275.

[46] Lipshutz B H, Lower A, Kucejko R J, et al. Applications of Asymmetric Hydrosilylations Mediated by Catalytic (DTBM-SEGPHOS)CuH. Org. Lett., 2006, 8: 2969-2972.

[47] Lipshutz B H, Caires C C, Kuipers P, et al. Tweaking copper hydride (CuH) for synthetic gain. A practical, one-pot conversion of dialkyl ketones to reduced trialkylsilyl ether derivatives. Org. Lett., 2003, 5(17): 3085-3088.

[48] Sirol S, Courmarcel J, Mostefai N, et al. Tweaking efficient enantioselective hydrosilylation of ketones catalyzed by air stable copper fluoride-phosphine complexes. Org. Lett., 2001, 3: 4111-4113.

[49] Kantam M K, Laha S, Yadav J. Tweaking asymmetric hydrosilylation of prochiral ketones catalyzed by nanocrystalline copper(II) Oxide. Adv. Synth. Catal., 2007, 349: 1797-1802.

[50] Issenhuth J T, Dagorne S, Bellemin-Laponnaz S. Efficient enantioselective hydrosilylation of aryl ketones catalyzed by a chiral BINAP-copper(I) Catalyst-Phenyl(methyl)silane System. Adv. Synth. Catal., 2006, 348: 1991-1994.

[51] Dorta R, Egli P, Zürcher F, et al. The [IrCl(Diphosphine)]$_2$/fluoride system. Developing catalytic asymmetric olefin hydroamination. J. Am. Chem. Soc., 1997, 119: 10857-10858.

[52] Kawatsura M, Hartwig J F. Palladium-catalyzed intermolecular hydroamination of vinylarenes using arylamines. J. Am. Chem. Soc., 2000, 122: 9546-9547.

[53] Hu A, Ogasawara M, Sakamoto T, et al. Palladium-catalyzed intermolecular asymmetric hydroamination with 4,4′-disubstituted BINAP and SEGPHOS. Adv. Synth. Catal., 2006, 348: 2051-2056.

[54] Zhang Z, Bender C F, Widenhoefer R A. Gold(I)-catalyzed enantioselective hydroamination of $N$-allenyl carbamates. Org. Lett., 2007, 9: 2887-2889.

[55] Zhang Z, Widenhoefer R A. Gold(I)-catalyzed intramolecular enantioselective Hydroalkoxylation of allenes. Angew. Chem. Int. Ed., 2007, 46: 283-285.

[56] Yang X H, Davison R T, Dong V M. Catalytic hydrothiolation: regio- and enantioselective coupling of

thiols and dienes. J. Am. Chem. Soc., 2018, 140: 10443-10446.

[57] Rubina M, Rubin M, Gevorgyan V. Catalytic enantioselective hydroboration of cyclopropenes. J. Am. Chem. Soc., 2003, 125: 7198-7199.

[58] Huddleston R R, Jang H Y, Krische M J. First catalytic reductive coupling of 1,3-diynes to carbonyl partners: a new regio- and enantioselective C—C bond forming hydrogenation. J. Am. Chem. Soc., 2003, 125: 11488-11489.

[59] Skucas E, Kong J R, Krische M J. Enantioselective reductive coupling of acetylene to N-arylsulfonyl imines via rhodium catalyzed C-C bond-forming hydrogenation: (Z)-dienyl allylic amine. J. Am. Chem. Soc., 2007, 129: 7242-7243.

[60] Ngai M Y, Barchuk A, Krische M J. Enantioselective iridium-catalyzed imine vinylation: optically enriched allylic amines via alkyne-imine reductive coupling mediated by hydrogen. J. Am. Chem. Soc., 2007, 129: 12644-12645.

[61] Rhee J U, Krische M J. Highly enantioselective reductive cyclization of acetylenic aldehydes via rhodium catalyzed asymmetric hydrogenation. J. Am. Chem. Soc., 2006, 128: 10674-10675.

[62] Kong J R, Krische M J. Catalytic carbonyl Z-dienylation via multicomponent reductive coupling of acetylene to aldehydes and α-ketoesters mediated by hydrogen: carbonyl insertion into cationic rhodacyclopentadienes. J. Am. Chem. Soc., 2006, 128: 16040-16041.

[63] Yanagisawa A, Nakashima H, Ishiba A, et al. Catalytic asymmetric allylation of aldehydes using a chiral silver( I ) complex. J. Am. Chem. Soc., 1996, 118: 4723-4724.

[64] Akira Y, Hiroshi N, Yoshinari N, et al. Asymmetric addition of allylic stannanes to aldehydes catalyzed by BINAP·Ag( I ) Complex. Chem. Soc. Jpn., 2001, 74: 1129-1137.

[65] Wadamoto M, Ozasa N, Yanagisawa A, et al. BINAP/AgOTf/KF/18-Crown-6 as new bifunctional catalysts for asymmetric sakurai—Hosomi allylation and mukaiyama aldol reaction. J. Org. Chem., 2003, 68: 5593-5601.

[66] Yanagisawa A, Kageyama H, Nakatsuka Y, et al. Enantioselective addition of allylic trimethoxysilanes to aldehydes catalyzed by p-Tol-BINA AgF. Angew. Chem. Int. Ed., 1999, 38: 3701-3703.

[67] Wadamoto M, Yamamoto H. Silver-catalyzed asymmetric sakurai—hosomi allylation of ketones. J. Am. Chem. Soc., 2005, 127: 14556-14557.

[68] Tomita D, Wada R, Kanai M, et al. Enantioselective alkenylation and phenylation catalyzed by a chiral CuF complex. J. Am. Chem. Soc., 2005, 127: 4138-4139.

[69] Motoki R, Tomita D, Kanai, M, et al. Catalytic enantioselective alkenylation and phenylation of trifluoromethyl ketone. Tetrahedron Lett, 2006, 47: 8083-8086.

[70] Umeda R, Studer A. Desymmetrization of 1,4-cyclohexadienyltriisopropoxysilane using copper catalysis. Org. Lett., 2007, 9: 2175-2178.

[71] Tomita D, Kanai M, Shibasaki M. Nucleophilic activation of alkenyl and aryl boronates by a chiral $Cu^IF$ complex: catalytic enantioselective alkenylation and arylation of aldehydes. Chem. Asian J., 2006, 1: 161-166.

[72] Kuwano R, Nishio R, Ito Y. Enantioselective construction of quaternary α-carbon centers on α-amino phosphonates via catalytic asymmetric allylation. Org. Lett., 1999, 1: 837-839.

[73] Kuwano R, Ito Y. Catalytic Asymmetric allylation of prochiral nucleophiles, α-acetamido-β-ketoesters. J. Am. Chem. Soc., 1999, 121: 3236-3237.

[74] Kuwano R, Uchida K, Ito, Y. asymmetric allylation of unsymmetrical 1,3-diketones using a BINAP–palladium catalyst. Org. Lett., 2003, 5: 2177-2179.

[75] Ogasawara M, Ngo H L, Sakamoto T, et al. Applications of 4,4'-$(Me_3Si)_2$-BINAP in transition-metal-catalyzed asymmetric carbon-carbon bond-forming reactions. Org. Lett., 2005, 7: 2881-2884.

[76] Bihelovic F, Matovic R, Vulovic B, et al. Organocatalyzed cyclizations of π-allylpalladium complexes: a new method for the construction of five- and six-membered rings. Org. Lett., 2007, 9: 5063-5066.

[77] Kitagawa O, Kohriyama M, Taguchi T. Catalytic asymmetric synthesis of optically active atropisomeric anilides through enantioselective N-allylation with chiral Pd-tol-BINAP catalyst. J. Org. Chem., 2002, 67: 8682-8684.

[78] Terauchi J, Curran D P. N-allylation of anilides with chiral palladium catalysts: the first catalytic asymmetric synthesis of axially chiral anilides. Tetrahedron: Aymmetry, 2003, 14: 587-592.

[79] Kitagawa O, Takahashi M, Kohriyama M, et al. Synthesis of optically active atropisomeric anilide derivatives through diastereoselective N-allylation with a chiral Pd−π-allyl catalyst. J. Org. Chem., 2003, 68: 9851-9853.

[80] Nagano T, Kobayashi S. Palladium-catalyzed allylic amination using aqueous ammonia for the synthesis of primary amines. J. Am. Chem. Soc., 2009, 131: 4200-4201

[81] Kim I S, Ngai M Y, Krische M J. Enantioselective iridium-catalyzed carbonyl allylation from the alcohol or aldehyde oxidation level via transfer hydrogenative coupling of allyl acetate: departure from chirally modified allyl metal reagents in carbonyl addition. J. Am. Chem. Soc., 2008, 130: 14891-14899.

[82] Kim I S, Han S B, Krische M J. anti-diastereo- and enantioselective carbonyl crotylation from the alcohol or aldehyde oxidation level employing a cyclometallated iridium catalyst: α-methyl allyl acetate as a surrogate to preformed crotylmetal reagents. J. Am. Chem. Soc., 2009, 131: 2514-2520.

[83] Kim I S, Ngai M Y, Krische M J. Enantioselective iridium-catalyzed carbonyl allylation from the alcohol or aldehyde oxidation level using allyl acetate as an allyl metal surrogate. J. Am. Chem. Soc., 2008, 130: 6340-6341.

[84] Yoshida K, Hayashi T. A New cine-substitution of alkenyl sulfones with aryltitanium reagents catalyzed by rhodium: mechanistic studies and catalytic asymmetric synthesis of allylarenes. J. Am. Chem. Soc., 2003, 125: 2872-2873.

[85] Xu J K, Wang Y, Gu Y, et al. Palladium-catalyzed stereospecific allylation of nitroacetates with enantioenriched primary allylic amines. Adv. Synth. Catal., 2016, 358: 1854-1858.

[86] Fang P, Hou X L. Asymmetric copper-catalyzed propargylic substitution reaction of propargylic acetates with enamines. Org. Lett., 2009, 11: 4612-4615.

[87] Takaya Y, Ogasawara M, Hayashi T, et al. Rhodium-catalyzed asymmetric 1,4-addition of aryl- and alkenylboronic acids to enones. J. Am. Chem. Soc., 1998, 120: 5579-5580.

[88] Takaya Y, Ogasawara M, Hayashi T. Asymmetric 1,4-addition of phenylboronic acid to 2-cyclohexenone catalyzed by Rh( I )/binap complexes. Chirality, 2000, 12: 469-471.

[89] Yuan W C, Cun L F, Mi A Q, et al. A class of readily available optically pure 7,7'-disubstituted BINAPs for asymmetric catalysis. Tetrahedron, 2009, 65: 4130-4141.

[90] Sakuma S, Sakai M, Itooka R, et al. Asymmetric conjugate 1,4-addition of arylboronic acids to α,β-unsaturated esters catalyzed by rhodium( I )/(S)-binap. J. Org. Chem., 2000, 65: 5951-5955.

[91] Sakuma S, Miyaura N. Rhodium( I )-catalyzed asymmetric 1,4-addition of arylboronic acids to α,β-unsaturated amides. J. Org. Chem., 2001, 66: 8944-8946.

[92] Itooka R, Iguchi Y, Miyaura N. Rhodium-catalyzed 1,4-addition of arylboronic acids to α,β-unsaturated carbonyl compounds: large accelerating effects of bases and ligands. J. Org. Chem., 2003, 68: 6000-6004.

[93] Lukin K, Zhang Q, Leanna M R. Practical method for asymmetric addition of arylboronic acids to α,β-unsaturated carbonyl compounds utilizing an in situ prepared rhodium catalyst. J. Org. Chem., 2009, 74: 929-931.

[94] Takaya Y, Ogasawara M, Hayashi T. Rhodium-catalyzed asymmetric 1,4-addition of 2-alkenyl-1,3,2-benzodioxaboroles to α,β-unsaturated ketones. Tetrahedron, 1998, 39: 8479-8482.

[95] Takaya Y, Ogasawara M, Hayashi T. Rhodium-catalyzed asymmetric 1,4-addition of arylboron compounds generated in situ from aryl bromides. Tetrahedron Lett, 1999, 40: 6957-6961.

[96] Takaya Y, Senda T, Kurushima H, et al. Rhodium-catalyzed asymmetric 1,4-addition of arylboron reagents to α,β-unsaturated esters. Tetrahedron:Asymmetry, 1999, 10: 4047-4056.

[97] Hayashi T, Senda T, Takaya Y, et al. Rhodium-catalyzed asymmetric 1,4-addition to 1-alkenylphosphonates. J. Am. Chem. Soc., 1999, 121: 11591-11592.

[98] Senda T, Ogasawara M, Hayashi T. Rhodium-catalyzed asymmetric 1,4-addition of organoboron reagents to 5,6-dihydro-2(1H)-pyridinones. Asymmetric synthesis of 4-aryl-2-piperidinones. J. Org.

Chem., 2001, 66: 6852-6856.
[99] Chen G, Tokunaga N, Hayashi T. Rhodium-catalyzed asymmetric 1,4-addition of arylboronic acids to coumarins: asymmetric synthesis of (R)-tolterodine. Org. Lett., 2005, 7: 2285-2288.
[100] Hayashi T, Senda T, Ogasawara M. Rhodium-catalyzed asymmetric conjugate Addition of organoboronic acids to nitroalkenes. J. Am. Chem. Soc., 2000, 122: 10716-10717.
[101] Konno T, Tanaka T, Miyabe T, et al. A first high enantiocontrol of an asymmetric tertiary carbon center attached with a fluoroalkyl group via Rh(I)-catalyzed conjugate addition reaction. Tetrahedron. Lett., 2008, 49: 2106-2110.
[102] Pucheault M, Darses S, Genêt J P. Potassium organotrifluoroborates: new partners in catalytic enantioselective conjugate additions to enones. Tetrahedron Lett, 2002, 43: 6155-6157.
[103] Pucheault M, Darses S, Genêt J P. Potassium organotrifluoroborates in rhodium-catalyzed asymmetric 1,4-additions to enones. Eur. J. Org. Chem., 2002: 3552-3557.
[104] Lalic G, Corey E J. Enantioselective rhodium(I)-triethylamine catalyzed addition of potassium isopropenyl trifluoroborate to enones. Tetrahedron Letters, 2008, 49: 4894-4896.
[105] Yoshida K, Ogasawara M, Hayashi T. Generation of chiral boron enolates by rhodium-catalyzed asymmetric 1,4-addition of 9-aryl-9-borabicyclo[3.3.1]nonanes (B-Ar-9BBN) to $\alpha,\beta$-unsaturated ketones. J. Org. Chem., 2003, 68: 1901-1905.
[106] Navarre L, Darses S, Genêt J P. Tandem 1,4-addition/enantioselective protonation catalyzed by rhodium complexes: efficient access to $\alpha$-amino acids. Angew. Chem. Int. Ed., 2004, 116: 737-741.
[107] Navarre L, Maritinez R, Genêt J P, et al. Access to enantioenriched $\alpha$-amino esters via rhodium-catalyzed 1,4-addition/enantioselective protonation. J. Am. Chem. Soc., 2008, 130: 6159–6169.
[108] Oi S, Taira A, Honma Y, et al. Asymmetric 1,4-addition of organosiloxanes to $\alpha,\beta$-unsaturated carbonyl compounds catalyzed by a chiral rhodium complex. Org. Lett., 2003, 5: 97-99.
[109] Oi S, Taira A Honma Y, et al. Asymmetric 1,4-addition of aryltrialkoxysilanes to $\alpha,\beta$-unsaturated esters and amides catalyzed by a chiral rhodium complex. Tetrahedron: Asymmetry, 2006, 17: 598-602.
[110] Otomaru Y, Hayashi T. Rhodium-catalyzed asymmetric 1,4-addition of alkenylsilanes generated by hydrosilylation of alkynes: a one-pot procedure where a rhodium/(S)-binap complex catalyzes the two successive reactions. Tetrahedron: Asymmetry, 2004,15: 2647-2651.
[111] Shintani R, Tokunaga N, Doi H, et al. A new entry of nucleophiles in rhodium-catalyzed asymmetric 1,4-addition reactions: addition of organozinc reagents for the synthesis of 2-aryl-4-piperidones. J. Am. Chem. Soc., 2004, 126: 6240-6241.
[112] Shintani R, Yamagami T, Kimura T, et al. Asymmetric synthesis of 2-aryl-2,3-dihydro-4-quinolones by rhodium-catalyzed 1,4-addition of arylzinc reagents in the presence of chlorotrimethylsilane. Org. Lett., 2005, 7: 5317-5319.
[113] Desrosiers J N, Bechara W S, Charette A B. Catalytic enantioselective addition of diorganozinc reagents to vinyl sulfones. Org. Lett., 2008, 10: 2315-2318.
[114] Smith A J, Abbott L K, Martin S F. Enantioselective conjugate addition employing 2-heteroaryl titanates and zinc reagents. Org. Lett., 2009, 11: 4200-4203.
[115] Hayashi T, Tokunaga N, Yoshida K, et al. Rhodium-catalyzed asymmetric 1,4-addition of aryltitanium reagents generating chiral titanium enolates: isolation as silyl enol ethers. J. Am. Chem. Soc., 2002, 124: 12102-12103.
[116] Walter C, Fröhlich R, Oestreich M. Rhodium(I)-catalyzed enantioselective 1,4-addition of nucleophilic silicon. Tetrahedron, 2009, 65: 5513–5520.
[117] Wang S Y, Ji S J, Loh T P. Cu(I) Tol-BINAP-catalyzed enantioselective michael reactions of grignard reagents and unsaturated esters. J. Am. Chem. Soc., 2007, 129: 276-277.
[118] Wang S Y, Lum T K, Ji S J, et al. Highly efficient copper(I) iodide-tolyl-BINAP-catalyzed asymmetric conjugate addition of methylmagnesium bromide to $\alpha,\beta$-unsaturated esters. Adv. Synth. Catal., 2008, 350: 673-677.
[119] Ruiz B M, Geurts K, Fernández-Ibáñez M Á, et al. Highly versatile enantioselective conjugate

addition of grignard reagents to α,β-unsaturated thioesters. Org. Lett., 2007, 9: 5123-5126.
[120] Bos P H, Minnaard A J, Feringa B L. Catalytic asymmetric conjugate addition of grignard reagents to α,β-unsaturated sulfones. Org. Lett., 2008, 10: 4219-4222.
[121] Nishimura T, Guo X X, Uchiyama N, et al. Steric tuning of silylacetylenes and chiral phosphine ligands for rhodium-catalyzed asymmetric conjugate alkynylation of enones. J. Am. Chem. Soc., 2008, 130: 1576-1577.
[122] Nishimura T, Sawano T, Hayashi T. Asymmetric synthesis of β-alkynyl aldehydes by rhodium-catalyzed conjugate alkynylation. Angew. Chem. Int. Ed., 2009, 48:8057-8059.
[123] Åhman J, Wolfe J P, Troutman M V, et al. Asymmetric arylation of ketone enolates. J. Am. Chem. Soc., 1998, 120: 1918-1919.
[124] Spielvogel D J, Buchwald S L. Nickel-BINAP catalyzed enantioselective α-arylation of α-substituted γ-butyrolactones. J. Am. Chem. Soc., 2002, 124: 3500- 3501.
[125] Zheng J, Breit B. Regiodivergent hydroaminoalkylation of alkynes and allenes by a combined rhodium and photoredox catalytic system. Angew. Chem. Int. Ed., 2019, 58: 3392-3397.
[126] Johannsen M. An enantioselective synthesis of heteroaromatic N-tosyl α-amino acids. Chem. Commun., 1999: 2233-2234.
[127] Saaby S, Fang X, Gathergood N, et al. Formation of optically active aromatic α-amino acids by catalytic enantioselective addition of imines to aromatic compounds. Angew. Chem. Int. Ed., 2000, 39: 4114-4116.
[128] Bandini M, Melloni A, Tommasi S, et al. Designing new α,β-unsaturated thioesters for the catalytic, enantioselective Friedel-Crafts alkylation of indoles. Chim. Acta., 2003, 86: 3753-3763.
[129] Chao C M, Vitale M R, Toullec P Y, et al. Asymmetric gold-catalyzed hydroarylation/cyclization reactions. Chem. Eur. J., 2009, 15: 1319-1323.
[130] Ferraris D, Young B, Dudding T, et al. Catalytic, enantioselective alkylation of α-imino esters using late transition metal phosphine complexes as catalysts. J. Am. Chem. Soc., 1998, 120: 4548-4549.
[131] Ferraris D, Young B, Cox C, et al. Catalytic, enantioselective alkylation of α-imino esters: the synthesis of nonnatural α-amino acid derivatives. J. Am. Chem. Soc., 2002, 124: 67-77.
[132] Ferraris D, Young B, Cox C, et al. Diastereo- and enantioselective alkylation of α-imino esters with enol silanes catalyzed by (R)-Tol-BINAP-CuClO$_4$,(MeCN)$_2$. J. Org. Chem., 1998, 63: 6090-6091.
[133] Yin L, Kanai M, Shibasaki M. Nucleophile generation via decarboxylation: asymmetric construction of contiguous trisubstituted and quaternary stereocenters through a Cu(Ⅰ)-catalyzed decarboxylative mannich-type reaction. J. Am. Chem.Soc., 2009, 131: 9610-9611.
[134] Du Y, Xu L W, Shimizu Y, et al. Asymmetric reductive mannich reaction to ketimines catalyzed by a cu(Ⅰ) complex. J. Am. Chem. Soc., 2008, 130: 16146-16147.
[135] Sugiura M, Nakai T. Asymmetric catalytic protonation of silyl enol ethers with chiral palladium complexes. Angew. Chem. Int. Ed., 1997, 36: 2366-2368.
[136] Yanagisawa A, Touge T, Arai T. Asymmetric protonation of silyl enolates catalyzed by chiral phosphine-silver(Ⅰ) complexes. Pure Appl. Chem., 78: 519–523.
[137] Sato Y, Sodeoka M, Shibasaki M. Catalytic asymmetric C-C bond formation: asymmetric synthesis of cis-decalin derivatives by palladium-catalyzed cyclization of prochiral alkenyl iodides. J. Org. Chem.,1989, 54: 4738-4739.
[138] Ozawa F, Kubo A, Hayashi T. Catalytic asymmetric arylation of 2,3-dihydrofuran with aryl triflates. J. Am. Chem. Soc.,1991, 113: 1417-1419.
[139] Penn L, Shpruhman A, Gelman D. Enantio- and regioselective Heck-type reaction of arylboronic acids with 2,3-dihydrofuran. J. Org. Chem., 2007, 72: 3875-3879.
[140] Ashimori A, Matsuura T, Overman L E, et al. Catalytic asymmetric synthesis of either enantiomer of physostigmine. Formation of quaternary carbon centers with high enantioselection by intramolecular Heck reactions of (Z)-2-butenanilides. J. Org. Chem., 1993, 58: 6949-6951.
[141] Carmona J A, Hornillos V, Ramírez-López P, et al. Dynamic kinetic asymmetric heck reaction for the simultaneous generation of central and axial chirality. J. Am. Chem. Soc., 2018, 140: 11067-11075.

[142] Takenaka K, Itoh N, Sasai H. Enantioselective synthesis of $C_2$-symmetric spirobilactams via Pd-catalyzed intramolecular double N-arylation. Org. Lett., 2009: 11, 1483-1486.

[143] Sodeoka M, Ohrai K, Shibasaki M. Catalytic asymmetric aldol reaction via chiral Pd(II) enolate in Wet DMF. J. Org. Chem., 1995, 60: 2648-2649.

[144] Kiyooka S, Hosokawa S, Tsukasa S. Dicationic (BINAP)palladium-catalyzed enantioselective aldol reaction of aldehydes with a silyl enol ether: a simplified practical procedure. Tetrahedron Letters, 2006, 47: 3959-3962.

[145] Yanagisawa A, Matsumoto Y, Asakawa K, et al. Asymmetric aldol reaction of enol trichloroacetate catalyzed by tin methoxide and BINAP·silver(I) complex. Tetrahedron, 2002, 58: 8331-8339.

[146] Momiyama N, Yamamoto H. Catalytic enantioselective synthesis of α-aminooxy and α-hydroxy ketone using nitrosobenzene. J. Am. Chem. Soc., 2003, 125: 6038-6039.

[147] Oisaki K, Suto Y, Kanai M, et al. A new method for the catalytic aldol reaction to ketones. J. Am. Chem. Soc., 2003, 125: 5644-5645.

[148] Russell A E, Fuller N O, Taylor S J, et al. Investigation of the Rh-catalyzed asymmetric reductive aldol reaction. Expanded scope based on reaction analysis. Org. Lett., 2004, 6: 2309-2312.

[149] Fuller N O, Morken J P. Direct formation of synthetically useful silyl-protected aldol adducts via the asymmetric reductive aldol reaction. Synlett, 2005: 1459-1461.

[150] Cauble D F, Gipson J D, Krische M. Diastereo- and enantioselective catalytic carbometallative aldol cycloreduction: tandem conjugate addition-aldol cyclization. J. Am. Chem. Soc., 2003, 125: 1110-1111.

[151] Gibson S E, Lewis S E, Loch J A, et al. Identification of an asymmetric Pauson-Khand precatalyst. Organometallics, 2003, 22: 5382-5384.

[152] Schmid T M, Consiglio G. Asymmetric cyclocarbonylation of 1,6-enynes with cobalt catalysts. Tetrahedron: Asymmetry, 2004, 15: 2205-2208.

[153] Suh W H, Choi M, Lee S I, et al. Rh(I)-catalyzed asymmetric intramolecular Pauson-Khand reaction in aqueous media. Synthesis, 2003, 14: 2169-2172.

[154] Jeong N, Sung B K, Choi Y K. Rhodium(I)-catalyzed asymmetric intramolecular Pauson-Khand-type reaction. J. Am. Chem. Soc., 2000, 122: 6771-6772.

[155] Lee H W, Chan A S C, Kwong F Y. Formate as a CO surrogate for cascade processes: Rh-catalyzed cooperative decarbonylation and asymmetric Pauson-Khand-type cyclization reactions. Chem. Commun., 2007: 2633-2635.

[156] Shibata T, Toshida N, Takagi K. Rhodium complex-catalyzed Pauson-Khand-type reaction with aldehydes as a CO source. J. Org. Chem., 2002, 67: 7446-7450.

[157] Shibata T, Takagi K. Iridium-chiral diphosphine complex catalyzed highly enantioselective Pauson-Khand-type reaction. J. Am. Chem. Soc., 2000, 122, 40: 9852-9853.

[158] Shibata T, Toshiba N, Yamasaki M, et al. Iridium-catalyzed enantioselective Pauson–Khand-type reaction of 1,6-enynes. Tetrahedron, 2005, 61: 9974-9979.

[159] Ghosh A K, Matsuda H. Counterions of BINAP-Pt(II) and -Pd(II) complexes: novel catalysts for highly enantioselective Diels – Alder Reaction. Org. Lett., 1999, 1: 2157-2159.

[160] Oi S, Terada E, Ohuchi K, et al. Asymmetric hetero Diels – Alder reaction catalyzed by chiral cationic palladium(II) and platinum(II) complexes. J. Org. Chem., 1999, 64: 8660-8667.

[161] Yao S, Saaby S, Hazell R G, et al. Catalytic enantioselective Aza-Diels-Alder reactions of imines-an approach to optically active nonproteinogenic α-amino acids. Chem. Eur. J., 2000, 6: 2435-2448.

[162] Kawasaki M, Yamamoto H. Catalytic enantioselective Hetero-Diels-Alder reactions of an azo compound. J. Am. Chem. Soc., 2006, 128: 16482-16483.

[163] Yamamoto Y, Tamamoto H. Catalytic asymmetric Nitroso-Diels–Alder reaction with acyclic dienes. Angew. Chem. Int. Ed., 2005, 44: 7082-7085.

[164] Nagamoto M, Yamauchi D, Nishimura T. Iridium-catalyzed asymmetric [3+2] annulation of aromatic ketimines with alkynes via C-H activation: unexpected inversion of the enantioselectivity induced by protic acids. Chem. Commun., 2016, 52: 5876-5879.

[165] Tanaka K, Wada A, Noguchi K. Rhodium-catalyzed [2+2+2] cycloaddition of 1,6-diynes with isothiocyanates and carbon disulfide. Org. Lett., 2006, 8:907-909.

[166] Shen Z, Khan H A, Dong V M. Rh-catalyzed carbonyl hydroacylation: an enantioselective approach to lactones. J. Am. Chem. Soc., 2008, 130: 2916-2917.

[167] Tanaka K, Takahashi M, Imase H, et al. Enantioselective synthesis of $\alpha,\alpha$-disubstituted $\alpha$-amino acids by Rh-catalyzed [2+2+2] cycloaddition of 1,6-diynes with protected dehydroamino acid. Tetrahedron, 2008, 64:6289-6293.

[168] Tanaka K, Nishida G, Wada A, et al. Enantioselective synthesis of axially chiral phthalides through cationic [RhI($H_8$-binap)]-catalyzed cross alkyne cyclotrimerization. Angew. Chem. Int. Ed., 2004, 43: 6510-6512.

[169] Tanaka K, Nishida G, Ogino M, et al. Enantioselective synthesis of axially chiral biaryls through rhodium-catalyzed complete intermolecular cross-cyclotrimerization of internal alkynes. Org. Lett., 2005, 7:3119-3121.

[170] Tanaka K, Sagae H, Toyoda K, et al. Enantioselective synthesis of planar-chiral metacyclophanes through cationic Rh(I)/modified-BINAP-catalyzed inter- and intramolecular alkyne cyclotrimerizations. Tetrahedron, 2008, 64: 831-846.

[171] Suda T, Noguchi K, Hirano M, et al. Enantioselective synthesis of $N,N$-dialkylbenzamides with aryl-carbonyl axial chirality by rhodium-catalyzed [2+2+2] cycloaddition. Chem. Eur. J., 2008, 14: 6593-6596.

[172] Shibata T, Otomo M, Tahara Y, et al. Highly diastereo- and enantioselective construction of both central and axial chiralities by Rh-catalyzed [2 + 2 + 2] cycloaddition. Org. Biomol. Chem., 2008, 6: 4296-4298.

[173] Kashima K, Teraoka K, Uekusa H, et al. Rhodium-catalyzed atroposelective [2+2+2] cycloaddition of ortho-substituted phenyl diynes with nitriles: effect of ortho substituents on regio- and enantioselectivity. Org. Lett., 2016, 18: 2170-2173.

[174] Fernández M, Parera M, Parella T, et al. Rhodium-catalyzed [2+2+2] cycloadditions of diynes with Morita-Baylis-Hillman adducts: a stereoselective entry to densely functionalized cyclohexadiene scaffolds. Adv. Synth. Catal., 2016, 358:1848-1853.

[175] Shibata T.; Arai, Y.; Tahara, Y. Enantioselective construction of quaternary carbon centers by catalytic [2+2+2] cycloaddition of 1,6-enynes and alkynes. Org. Lett., 2005, 7: 4955-4957.

[176] Shibata T, Tahara Y. Enantioselective intramolecular [2+2+2] cycloaddition of 1,4-diene-ynes: a new approach to the Construction of quaternary carbon stereocenters. J. Am. Chem. Soc., 2006, 128: 11766-11767.

[177] Suzuki S, Nishigaki S, Shibata T, et al. Rhodium-catalyzed enantioselective [2+2+1] cycloaddition of 1,6-enynes with cyclopropylideneacetamides. Org. Lett., 2018, 20: 7461-7465.

[178] Tsuchikama K, Kuwata Y, Shibata T. Highly Enantioselective construction of a chiral spirocyclic structure by the [2+2+2] cycloaddition of diynes and exo-methylene cyclic compounds. J. Am. Chem. Soc., 2006, 128: 13686-13687.

[179] Oderaotoshi Y, Cheng W, Fujitomi S, et al. Exo- and enantioselective cycloaddition of azomethine ylides generated from $N$-alkylidene glycine esters using chiral phosphine-copper complexes. Org. Lett., 2003, 5: 5043-5046.

[180] Melhado A D, Luparia M, Toste F D. Au(I)-catalyzed enantioselective 1,3-dipolar cycloadditions of münchnones with electron-deficient alkenes. J. Am. Chem. Soc., 2007, 129: 12638-12639.

[181] Lei A, He M, Zhang X. Highly enantioselective syntheses of functionalized $\alpha$-methylene-$\gamma$-butyrolactones via Rh(I)-catalyzed intramolecular alder ene reaction: application to formal synthesis of (+)-pilocarpine. J. Am. Chem. Soc., 2002, 124: 8198-8199.

[182] Lei A, Waldkirch J P, He M, et al. Highly enantioselective cycloisomerization of enynes catalyzed by rhodium for the preparation of functionalized lactams. Angew. Chem. Int. Ed., 2002, 41: 4526-4529.

[183] Lei A, Waldkirch J P, He M, et al. Highly enantioselective cycloisomerization of enynes catalyzed by rhodium for the preparation of functionalized lactams. Angew. Chem. Int. Ed., 2002, 41: 4526-4529.

[184] Nicolaou K C, Li A, Ellery S P, et al. Rhodium-catalyzed asymmetric enyne cycloisomerization of terminal alkynes abd formal total synthesis of (−) platensimycin. Angew. Chem. Int. Ed., 2009, 48: 6293-6295.

[185] Mikami K, Kawakami Y, Akiyama K, et al. Enantioselective catalysis of ketoester-ene reaction of silyl enol ether to construct quaternary carbons by chiral dicationic palladium(II) complexes. J. Am. Chem. Soc., 2007, 129: 12950-12951.

[186] Ferraris D, Young B, Cox C, et al. Diastereo- and enantioselective alkylation of $\alpha$-imino esters with enol silanes catalyzed by (R)-Tol-BINAP-CuClO$_4$,(MeCN)$_2$. J. Org. Chem., 1998, 63: 6090-6091.

[187] Yao S, Fang X, Jørgensen K A. Catalytic enantioselective ene reactions of imines: a simple approach for the formation of optically active $\alpha$-amino acids. Chem. Commun., 1998: 2547-2548.

[188] Kundu K, McCullagh J V, Morehead A T. Hydroacylation of 2-vinyl benzaldehyde systems: an efficient method for the synthesis of chiral 3-substituted indanones. J. Am. Chem. Soc., 2005, 127: 16042-16043.

[189] Oonishi Y, Ogura J, Sato Y. Rh(I)-catalyzed intramolecular hydroacylation in ionic liquids. Tetrahedron. Letters, 2007, 48: 7505-7507.

[190] Shen Z, Khan H A, Dong V M. Rh-catalyzed carbonyl hydroacylation: an enantioselective approach to lactones. J. Am. Chem. Soc., 2008, 130: 2916-2917.

[191] Miura T, Sasaki T, Harumashi T, et al. Vinylcyclopropanation of olefins via 3-methoxy-1-propenylrhodium(I). J. Am. Chem. Soc., 2006, 128, 8: 2516-2517.

[192] Watson I D G, Ritter S, Toste F D. Asymmetric synthesis of medium-sized rings by intramolecular au(I)-catalyzed cyclopropanation. J. Am. Chem. Soc., 2009, 131: 2056-2057.

[193] Hatano M, Terada M, Mikami K. Highly enantioselective palladium-catalyzed ene-type cyclization of a 1,6-enyne. Angew. Chem. Int. Ed., 2001, 40: 249-253.

[194] Allen A, Marquand P L, Burton R, et al. Rhodium-catalyzed asymmetric cyclodimerization of oxabenzonorbornadienes and azabenzonorbornadienes: scope and limitations. J. Org. Chem., 2007, 72: 7849-7857.

[195] Allen A, Marquand P L, Burton R, et al. Rhodium-catalyzed asymmetric cyclodimerization of oxabenzonorbornadienes and azabenzonorbornadienes: scope and limitations. J. Am. Chem. Soc., 2007, 129: 1492-1493.

[196] Gollapelli K K, Donikela S, Manjula N, V et al. Rhodium-catalyzed highly regio- and enantioselective reductive cyclization of alkyne-tethered cyclohexadienones. ACS Catal., 2018, 8: 1440-1447.

[197] Lv W, Xiong B, Jiang H, et al. Synthesis of 2-alkylaminoquinolines and 1,8-naphthyridines by successive ruthenium-catalyzed dehydrogenative annulation and N-alkylation processes. Adv. Synth. Catal., 2017, 359: 1202-1207.

[198] Akiyama T, Daidouji K, Fuchibe K. Cu(I)-catalyzed enantioselective [2 + 2] cycloaddition of 1-methoxyallenylsilane with $\alpha$-imino ester: chiral synthesis of $\alpha,\beta$-unsaturated acylsilanes. Org. Lett., 2003, 5: 3691-3693.

[199] Yu X, Lu X. Cationic palladium complex catalyzed diastereo- and enantioselective tandem annulation of 2-formylarylboronic acids with allenoates. Org. Lett., 2009, 11: 4366-4369.

[200] Shintani R, Takatsu K, Hayashi T. Rhodium-catalyzed asymmetric synthesis of 3,3-disubstituted 1-indanones. Angew. Chem. Int. Ed., 2007, 46: 3735-3737.

[201] Yang D, Long Y, Wang H, et al. Iridium-catalyzed asymmetric ring-opening reactions of N-Boc-azabenzonorbornadiene with secondary amine nucleophiles. Org. Lett., 2008, 10: 4723-4726.

[202] Hattori G, Yoshida A, Miyake Y, et al. Enantio selective ring-opening reactions of racemic ethynyl epoxides via copper-allenylidene intermediates: efficient approach to chiral β-amino alcohols. J. Org. Chem., 2009, 74: 603-7607.

[203] Matsuda T, Shigeno M, Makino M, et al. Enantioselective C − C bond cleavage creating chiral quaternary carbon centers. Org. Lett., 2006, 8: 3379-3381.

[204] Ogasawara M, Ikeda H, Nagano T, et al. Palladium-catalyzed asymmetric synthesis of axially chiral allenes: a synergistic effect of dibenzalacetone on high enantioselectivity. J. Am. Chem. Soc., 2001,

123: 2089-209.

[205] Hayashi T, Tokunaga N, Inoue K. Rhodium-catalyzed asymmetric 1,6-addition of aryltitanates to enynones giving axially chiral allenes. Org. Lett., 2004, 6: 305-307.

[206] Maddox S M, Dinh A N, Armenta F, et al. The catalyst-controlled regiodivergent chlorination of phenols. Org. Lett., 2016, 18: 5476-5479.

[207] Yu S N, Li Y L, Deng J. Enantioselective synthesis of 2-bromomethyl indolines via BINAP(S)-catalyzed bromoaminocyclization of allyl aniline. Adv. Synth. Catal., 2017, 359: 2499-2508.

[208] Mikami K, Miyamoto T, Hatano M. A highly efficient asymmetric Suzuki–Miyaura coupling reaction catalyzed by cationic chiral palladium(ii) complexes. Chem. Commun., 2004: 2082-2083.

[209] Nishimura T, Katoh T, Takatsu K, et al. Rhodium-catalyzed asymmetric rearrangement of alkynyl alkenyl carbinols: synthetic equivalent to asymmetric conjugate alkynylation of enones. J. Am. Chem. Soc., 2007, 129: 14158-14159.

[210] Liu T L, Ng T W, Zhao Y. Rhodium-catalyzed enantioselective isomerization of secondary allylic alcohols. J. Am. Chem. Soc., 2017, 139: 3643-3646.

[211] Fang H, Yang Z, Zhang L, et al. Transmetal-catalyzed enantioselective cross-coupling reaction of racemic secondary benzylic bromides with organoaluminum reagents. Org. Lett., 2016, 18: 6022-6025.

[212] Najib A, Hirano K, Miura M. Palladium-catalyzed asymmetric benzylic substitution of secondary benzyl carbonates with nitrogen and oxygen nucleophiles. Org. Lett., 2017, 19: 2438-2441.

[213] Kotani S, Yoshiwara Y, Ogasawara M, et al. Catalytic enantioselective aldol reactions of unprotected carboxylic acids under phosphine oxide catalysis. Angew. Chem. Int. Ed., 2018, 57: 15877-15881.

[214] Nefkens S C A, Sperrle D C M, Consiglio G. Palladium-catalyzed enantioselective bis-alkoxycarbonylation of olefins. Angew. Chem. Int. Ed., 1993, 32: 1719-1720.

[215] Freixa Z, Beentjes M S, Batema G D, et al. SPANphos: A C2-Symmetric trans-coordinating diphosphane ligand. Angew. Chem. Int. Ed., 2003, 42: 1284.

[216] Jacquet O, Clément N D, Blanco C, et al. SPANphos ligands in palladium-catalyzed asymmetric fluorination. Eur. J. Org. Chem., 2012: 4844-4852.

[217] Wang X, Meng F, Wang Y, et al. Aromatic spiroketal bisphosphine ligands: palladium-catalyzed asymmetric allylic amination of racemic Morita–Baylis–Hillman adducts. Angew. Chem. Int. Ed., 2012, 51: 9276-9282.

[218] Wang X, Guo P, Wang X, et al. Practical asymmetric catalytic synthesis of spiroketals and chiral diphosphine ligands. Adv. Synth. Catal., 2013, 355: 2900-2907.

[219] Cao Z -Y, Wang X, Tan C, et al. Highly stereoselective olefin cyclopropanation of diazooxindoles catalyzed by a C2-symmetric spiroketal bisphosphine/Au( I ) complex. J. Am. Chem. Soc., 2013: 135, 8197.

[220] Jiang Y, Xue S, Li Z, et al. Asymmetric hydroformylation of styrene catalyzed by chiral spiro diphosphite–rhodium( I ) complex.Tetrahedron: Asymmetry, 1998, 9: 3185-3189.

[221] Jiang Y, Xue S, Yu K, et al. Asymmetric hydroformylation catalyzed by rhodium( I ) complexes of novel chiral spiro ligands.J. Organomet. Chem., 1999, 586:159.

[222] Lin C W, Lin C C, Lam L F L, et al. Synthesis of a novel spiro bisphosphinamidite ligand for highly enantioselective hydrogenation. Tetrahedron Lett. 2004, 45: 7379-7381.

[223] Guo Z Q, Guan X Y, Chen Z Y, Synthesis of a novel spiro bisphosphinite ligand and its application in Rh-catalyzed asymmetric hydrogenation. Tetrahedron: Asymmetry 2006, 17: 468-373.

[224] Xie J-H, Wang L X, Fu Y, et al. Synthesis of spiro diphosphines and their application in asymmetric hydrogenation of ketones. J. Am. Chem. Soc., 2003, 125: 4404.

[225] Xie J H, Duan H F, Fan B M, et al. Application of SDP ligands for Pd-catalyzed allylic alkylation. Adv. Synth. Catal., 2004, 346: 625-632.

[226] Fan B M, Xie J H, Li S, et al. Highly enantioselective hydrosilylation/cyclization of 1,6-enynes catalyzed by rhodium( I ) complexes of spiro diphosphines. Angew. Chem. Int. Ed., 2007, 46: 1275.

[227] Zhou Z T, Xie J H, Zhou Q L. Enantioselective synthesis of chiral β-aryloxy alcohols by asymmetric

hydrogenation of α-aryloxy aldehydes via dynamic kinetic resolution. Adv. Synth. Catal., 2009: 351-363.

[228] Xie J H, Zhou Z T, Kong W L, et al. Ru-catalyzed asymmetric hydrogenation of racemic aldehydes via dynamic kinetic resolution: efficient synthesis of optically active primary alcohols. J. Am. Chem. Soc., 2007, 129: 1868.

[229] Liu S, Xie J H, Wang L X, et al. Dynamic kinetic resolution allows a highly enantioselective synthesis of cis-α-aminocycloalkanols by ruthenium-catalyzed asymmetric hydrogenation. Angew. Chem. Int. Ed., 2007, 46: 7506.

[230] Liu C, Xie J H, Li Y L, et al. Asymmetric hydrogenation of α, α'-disubstituted cycloketones through dynamic kinetic resolution: an efficient construction of chiral diols with three contiguous stereocenters. Angew. Chem. Int. Ed. 2013, 52: 593.

[231] Cheng X, Zhang Q, Xie J H, et al. Highly rigid diphosphane ligands with a large dihedral angle based on a chiral spirobifluorene backbone. Angew. Chem. Int. Ed., 2005, 44: 11.

[232] Cheng X, Xie J H, Li S, et al. Asymmetric hydrogenation of α,β-unsaturated carboxylic acids catalyzed by ruthenium(II) complexes of spirobifluorene diphosphine (SFDP) ligands. Adv. Synth. Catal., 2006, 348: 1271.

[233] Tang W J, Zhu S F, Xu L J, et al. Asymmetric hydrogenation of quinolines with high substrate/catalyst ratio. Chem. Commun., 2007: 613.

[234] Burk M, Feaster J E. Enantioselective hydrogenation of the C:N group: a catalytic asymmetric reductive amination procedure. J. Am. Chem. Soc., 1992, 114: 6266.

[235] Guo S Y, Yang P, Zhou, Nickel-catalyzed asymmetric transfer hydrogenation of conjugated olefins. J. Chem. Commun., 2015, 51: 12115.

[236] Guo S, Zhou J. N,N-dimethylformamide as hydride source in nickel-catalyzed asymmetric hydrogenation of α,β-unsaturated esters. Org. Lett., 2016, 18: 5344.

[237] Shevlin M, Friedfeld M R, Sheng H, et al. Nickel-catalyzed asymmetric alkene hydrogenation of α,β-unsaturated esters: high-throughput experimentation-enabled reaction discovery, optimization, and mechanistic elucidation. J. Am. Chem. Soc., 2016, 138: 3562.

[238] Liu Y, Yi Z, Tan X, et al. Nickel-catalyzed asymmetric hydrogenation of cyclic sulfamidate imines: efficient synthesis of chiral cyclic sulfamidates. Science, 2019, 19: 63.

[239] Zhao X, Zhang F, Liu K, et al. Nickel-catalyzed chemoselective asymmetric hydrogenation of α,β-unsaturated ketoimines: an efficient approach to chiral allylic amines. Org. Lett., 2019, 21: 8966.

[240] Morimoto T, Nakajima N, Achiwa K. Asymmetric hydrogenation of cyclic ketimines catalyzed by iridium(I)-complexes of BCPM and its analog1. Synlett., 1995: 748.

[241] Zhu G, Zhang X. Additive effects in Ir−BICP catalyzed asymmetric hydrogenation of imines. Tetrahedron: Asymmetry, 1998, 9: 2415.

[242] Bakos J, Orosz A, Heil B, et al. Rhodium(I)–sulfonated-bdpp catalysed asymmetric hydrogenation of imines in aqueous–organic two-phase solvent systems. J. Chem. Soc., Chem. Commun., 1991: 1684.

[243] Lensink C, de Vries J G. Improving enantioselectivity by using a mono-sulphonated diphosphine as ligand for homogeneous imine hydrogenation. Tetrahedron: Asymmetry, 1992, 3: 235.

[244] Xiao D, Zhang X. Highly enantioselective hydrogenation of acyclic imines catalyzed by Ir–f-binaphane complexes. Angew. Chem. Int. Ed., 2001, 40: 3425.

[245] Blaser H U, Malan C, Pugin B, et al Selective hydrogenation for fine chemicals: recent trends and new developments. Adv. Synth. Catal., 2003, 345: 103.

[246] Hu X P, Zheng Z. Unsymmetrical hybrid ferrocene-based phosphine-phosphoramidites: a new class of practical ligands for Rh-catalyzed asymmetric hydrogenation. Org. Lett., 2004, 6: 3585.

[247] Yu S, Hu X, Deng J, et al. Enantioselective Ag(I)-catalyzed [3+2] cycloaddition of azomethine ylides using a chiral ferrocene-based phosphine–phosphoramidite ligand having a stereogenic P-center. Tetrahedron: Asymmetry, 2009, 20: 621-625.

[248] Maria B. PStereogenic NPhosphine-phosphite ligands for the Rh catalyzed hydrogenation of olefins. J.

Org. Chem., 2020, 85: 4730-4739.

[249] Zhu J, Huang L, Dong W, et al. Enantioselective rhodium-catalyzed addition of arylboroxines to N-unprotected ketimines: efficient synthesis of cipargamin. Angew. Chem. Int. Ed., 2019, 58: 16119.

[250] Liu G, Liu X, Cai Z, et al. Design of phosphorus ligands with deep chiral pockets: practical synthesis of chiral β-arylamines by asymmetric hydrogenation. Angew. Chem. Int. Ed., 2013, 52: 4235-4238.

[251] Li C, Wan F, Chen Y, et al. Stereoelectronic effects in ligand design: enantioselective rhodium-catalyzed hydrogenation of aliphatic cyclic tetrasubstituted enamides and concise synthesis of (R)-tofacitinib. Angew. Chem. Int. Ed., 2019, 58: 13573.

[252] Imamoto T, Iwadate N, Yoshida K. Enantioselective hydrogenation of acyclic aromatic N-aryl imines catalyzed by an iridium complex of (S,S)-1,2-bis(tert-butylmethylphosphino)ethane. Org. Lett., 2006, 8: 2289.

[253] Li W, Hou G, Chang M, et al. Highly efficient and enantioselective iridium-catalyzed asymmetric hydrogenation of N-arylimines. Adv. Synth. Catal., 2009, 351: 3123.

[254] Yang Q, Shang G, Gao W, et al. AHighly Enantioselective, Pd–TangPhos-catalyzed hydrogenation of N-tosylimines. Angew. Chem. Int. Ed., 2006, 45: 3832.

[255] Li W Hou G Chang M et al. Highly efficient and enantioselective iridium-catalyzed asymmetric hydrogenation of N-arylimines. Adv. Synth. Catal., 2009, 351: 3123.

[256] Li B, Chen J, Zhang Z, et al. Nickel-catalyzed asymmetric hydrogenation of N-sulfonyl imines. Angew. Chem. Int. Ed., 2019, 58: 7329.

[257] Sakai N, Mano S, Nozaki K, et al. Highly enantioselective hydroformylation of olefins catalyzed by new phosphine phosphite-rhodium(Ⅰ) complexes. J. Am. Chem. Soc., 1993, 115: 7033-7034.

[258] Tanaka R, Nakano K, Nozaki K. Synthesis of α-heteroarylpropanoic acid via asymmetric hydroformylation catalyzed by Rh(Ⅰ)-(R,S)-BINAPHOS and the subsequent oxidation. J. Org. Chem., 2007, 72: 8671-8676.

[259] Horiuchi T, Ohta T, Shirakawa E, et al. Asymmetric hydroformylation of heterocyclic olefins catalyzed by chiral phosphine-phosphite-Rh(Ⅰ) complexes J. Org. Chem., 1997, 62: 4285.

[260] Sakai N, Nozaki K, Takaya H. Asymmetric hydroformylation of 1,2-disubstituted olefins catalysed by chiral phosphinephosphite-rhodium(Ⅰ) complexes. J. Chem. Soc, Chem. Commun., 1994: 395-396.

[261] Shulyupin M O, Franciò G, Beletskaya I P, et al. Regio- and enantioselective catalytic hydrophosphorylation of vinylarenes. Adv. Synth. Catal., 2005, 347: 667.

[262] Qiu M, Wang D Y, Hu X P, et al. Asymmetric synthesis of chiral Roche ester and its derivatives via Rh-catalyzed enantioselective hydrogenation with chiral phosphine-phosphoramidite ligands. Tetrahedron: Asymmetry, 2009, 20: 210.

[263] Vargas S, Rubio M, Suárez A, et al. Asymmetric hydrogenation of imines catalyzed by iridium complexes with phosphine–phosphite ligands: importance of backbone flexibility. Tetrahedron Lett., 2005, 46: 2049.

[264] Vargas S, Rubio M, Suárez A, et al. Iridium complexes with phosphine – phosphite ligands. structural aspects and application in the catalytic asymmetric hydrogenation of N-aryl imines. Organometallics, 2006, 25: 961.

[265] Huang J D, Hu X P, Duan Z C, et al. Readily available phosphine – phosphoramidite ligands for highly efficient Rh-catalyzed enantioselective hydrogenations. Org. Lett., 2006, 8: 4367.

[266] Zhou X M, Huang J D, Luo L B, et al. Chiral 1-phenylethylamine-derived phosphine-phosphoramidite ligands for highly enantioselective Rh-catalyzed hydrogenation of β-(acylamino)acrylates: significant effect of substituents on 3,3′-positions of binaphthyl moiety. Org. Biomol. Chem. 2010, 8: 2320.

[267] Franciò C, Faraone F, Leitner W. Asymmetric catalysis with chiral phosphane/phosphoramidite ligands derived from quinoline (QUINAPHOS). Angew. Chem. Int. Ed., 2000, 39: 1428.

[268] Balogh S, Farkas G, Madarász J, et al. Asymmetric hydrogenation of C[double bond, length as m-dash]C double bonds using Rh-complex under homogeneous, heterogeneous and continuous mode conditions. Green Chem., 2012, 14: 1146.

[269] Wang D Y, Hu X P, Huang J D, et al. Highly enantioselective synthesis of a-hydroxy phosphonic acid

derivatives by Rh-catalyzed asymmetric hydrogenation with phosphine–phosphoramidite ligands. Angew. Chem. Int. Ed., 2007, 46: 7810.

[270] Yu S B, Huang J D, Wang D Y, et al. Novel chiral phosphine-phosphoramidite ligands derived from 1-naphthylamine for highly efficient Rh-catalyzed asymmetric hydrogenation. Tetrahedron: Asymmetry., 2008, 19: 1862.

[271] Hou C J, Wang Y H, Zheng Z, et al. Chiral phosphine–phosphoramidite ligands for highly efficient Ir-catalyzed asymmetric hydrogenation of sterically hindered N-arylimines. Org. Lett., 2012, 14: 3554.

[272] Eggenstein M, Thomas A, Theuerkauf J, et al. Highly efficient and versatile phosphine-phosphoramidite ligands for asymmetric hydrogenation. Adv. Synth. Catal., 2009, 351: 725.

[273] Burk S, Franciò G, Leitner W. Ruthenium-catalysed asymmetric hydrogenation of ketones using QUINAPHOS as the ligand. Chem. Commun., 2005: 3460.

[274] Wassenaar J, de Bruin B, Reek J N H. Rhodium-catalyzed asymmetric hydroformylation with taddol-based Indol Phos ligands. Organometallics, 2010, 29: 2767.

[275] Chikkali S H, Bellini R, de Bruin B, et al. Highly selective asymmetric Rh-catalyzed hydroformylation of heterocyclic olefins. J. Am. Chem. Soc., 2012, 134: 6607.

[276] Yan Y, Zhang X. A hybrid phosphorus ligand for highly enantioselective asymmetric hydroformylation. Rhodium-catalyzed asymmetric hydroformylation with taddol-based IndolPhos ligands. J. Am. Chem. Soc., 2006, 128: 7198.

[277] Zhang X, Cao B, Yu S, et al. Rhodium-catalyzed asymmetric hydroformylation of N-allylamides: highly enantioselective approach to $\beta_2$-amino aldehydes. Angew. Chem. Int. Ed., 2010, 49: 4047.

[278] Wassenaar J, van Zutphen S, Mora G, et al. INDOLPhosphole and INDOLPhos palladium-allyl complexes in asymmetric allylic alkylations. Organometallics, 2009, 28: 2724.

[279] Wassenaar J, Jansen E, van Zeist W J, et al. Catalyst selection based on intermediate stability measured by mass spectrometry. Nat. Chem., 2010, 2: 417.

PH☉SPHORUS 磷科学前沿与技术丛书　　　　手性膦配体合成及应用

# 4

# 手性膦-杂原子双齿配体

4.1 手性膦-杂原子双齿配体的合成
4.2 手性膦-杂原子双齿配体的应用

Synthesis and Application of Chiral Phosphine Ligands

磷原子与其他具有配位能力的氮等杂原子形成的手性膦-杂原子双齿配体，是极具发展潜力的一类膦配体，这方面的研究工作十分活跃，本章将简要汇总手性膦-杂原子双齿配体的研究进展。

# 4.1 手性膦-杂原子双齿配体的合成

## 4.1.1 膦/氮配体

由于磷和氮均属于强配位原子，且此类配体容易经手性池特别是氨基酸等天然手性化合物衍生合成，因此膦/氮双齿配体是膦/杂原子配体中研究得最早，同时也是最为重要的一类手性膦-杂原子双齿配体。

### 4.1.1.1 膦/噁唑啉配体

在1993年，Pfaltz[1]、Helmchen[2]和Williams[3]等几乎同时报道了苯基膦/噁唑啉(PHOX)配体的合成(图4-1)及其在钯催化的不对称烯丙基取代反应中应用的研究。此类配体已在众多的不对称催化反应中发挥着重要作用，属于典型的优势结构配体之一。

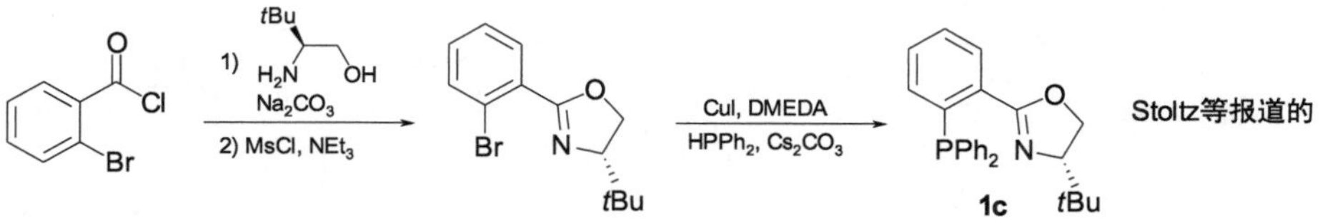

图 4-1 苯基膦/噁唑啉配体 1 的合成

  Pfaltz 等人从邻溴苯乙腈出发，依次通过溴锂交换、芳基锂对氯化磷的亲核取代反应引入二苯基膦，2-氰基苯基膦，接着在氯化锌催化下与氨基醇关环形成噁唑啉得到 PHOX 配体。或者先由苯腈与氨基醇反应形成苯基噁唑啉，再通过邻位锂化、芳基锂对氯化磷的亲核进攻形成 C—P 键。Helmchen 等则利用芳基镁对氯化磷的亲核进攻引入二芳基膦。Williams 等从邻氟苯乙腈出发，首先构筑 2-氟苯基噁唑啉，再通过二苯基膦钾对氟代苯环的亲核取代合成相应 PHOX 配体。另外，Stoltz 等[4]利用碘化亚铜/N,N'-二甲基乙二胺催化仲膦与 2-溴代噁唑啉间的 Ullmann 偶联反应构建 C—P 键，该方法尤其适用于缺电子的溴代芳烃(图 4-1)。

  手性 PHOX 配体从逆合成分析上可视为单膦化合物与噁唑啉通过桥连组装而成的杂化配体。在 Pfaltz、Helmchen 和 Williams 等所开展的 PHOX 配体应用研究工作基础上，人们通过改变磷原子上以及噁唑啉环上的取代基、磷原子与噁唑啉间的桥连片段即骨架很容易实现手性膦/噁唑啉配体的结构修饰与调控，从而开发出结构多样的配体，如图 4-2 所示。

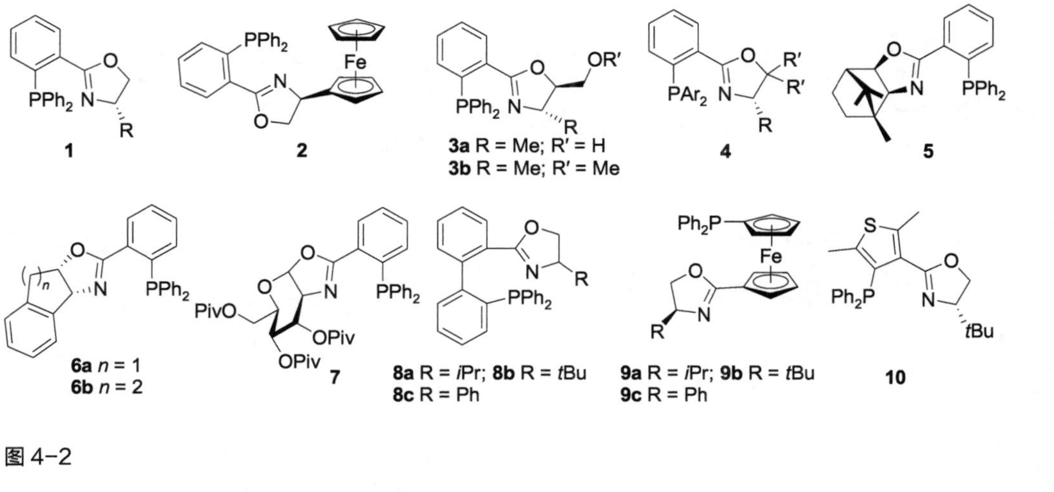

图 4-2

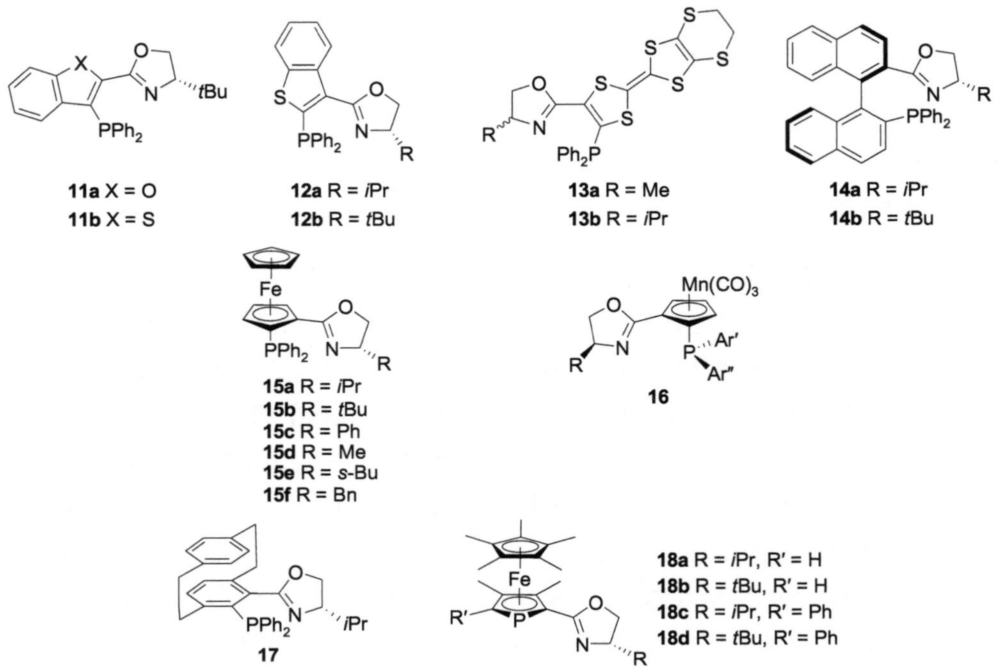
11a X = O
11b X = S
12a R = iPr
12b R = tBu
13a R = Me
13b R = iPr
14a R = iPr
14b R = tBu
15a R = iPr
15b R = tBu
15c R = Ph
15d R = Me
15e R = s-Bu
15f R = Bn
16
17
18a R = iPr, R′ = H
18b R = tBu, R′ = H
18c R = iPr, R′ = Ph
18d R = tBu, R′ = Ph

图 4-2 （杂）芳基手性膦 / 噁唑啉配体

针对噁唑啉环的结构调控最直接的办法就是改变手性 C4 上的取代基。与使用从天然手性氨基酸衍生的手性 $\beta$-氨基醇不同，Patti 等 [5] 制备了二茂铁取代的手性 $\beta$-氨基醇来构建噁唑啉环，成功实现了配体 **2** 的合成（图 4-3）。

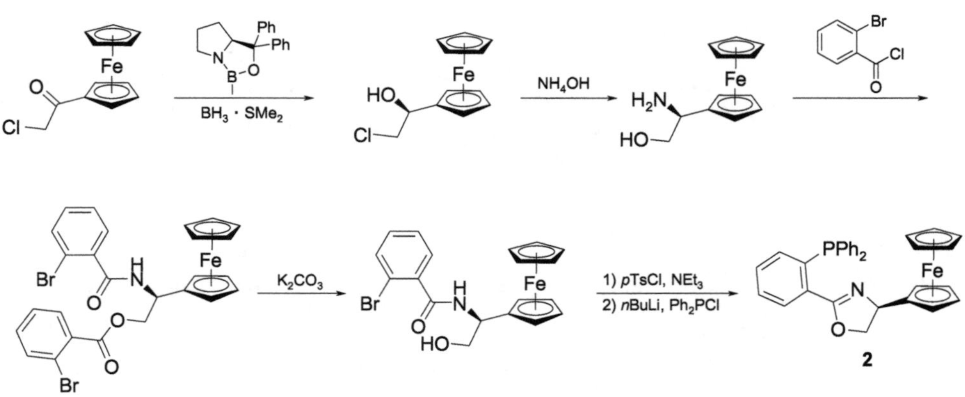
BH3 · SMe2
NH4OH
K2CO3
1) pTsCl, NEt3
2) nBuLi, Ph2PCl
2

图 4-3 配体 2 的合成

进一步改进噁唑啉环结构还有两种策略，一是在噁唑啉环的 C-5 位上引入合适的取代基，如 Moberg 等[6]和 Pfaltz 等[7]分别在噁唑啉环的 C-5 位上引入一个或两个取代基合成了配体 **3** 和 **4**。另一种策略是通过构建并环增加刚性和位阻，Helmchen 等[8]以 (2S,3R)-3-羟基冰片-2-胺、(1R,2S)-1-氨基-2-茚醇、(1R,2S)-四氢萘-1,2-二醇为原料，分别合成了 **5** 和 **6**。Kunz 等[9]合成了葡萄糖胺衍生的手性膦/噁唑啉配体 **7**。苯基骨架可继续拓展成其他(杂)芳环，如 Zhang 等[10]从联苯酚出发合成了构型相对灵活的联苯骨架手性膦/噁唑啉配体 **8**。Ikeda 等[11]合成了二茂铁骨架手性膦/噁唑啉配体 **9**，膦与噁唑啉分别位于不同的环戊二烯单元上。Tietze[12]和 Guiry[13] 课题组分别独立报道了(苯并)噻吩基、苯并呋喃基手性膦/噁唑啉配体 **10**、**11**、**12** 的合成。Avarvari 等[14]合成了具有氧化还原活性的四硫富瓦烯衍生的手性膦/噁唑啉配体 **13**。

简单的亚芳基 PHOX 类型配体合成路线简单，但手性诱导作用仅由手性噁唑啉结构控制，引入其他手性元素便成为后续配体设计、合成研究工作的重要方向。Ikeda 等[15]和 Hayashi 等[16]从消旋联萘酚出发，经双酯化、钯催化单羧化及磺酸酯基催化磷化形成膦氧化物中间体，此中间体按照两种略为不同的合成路线继续转化并经过关键的非对映体拆分步骤后得到轴手性的膦/噁唑啉配体 **14**(图 4-4)。

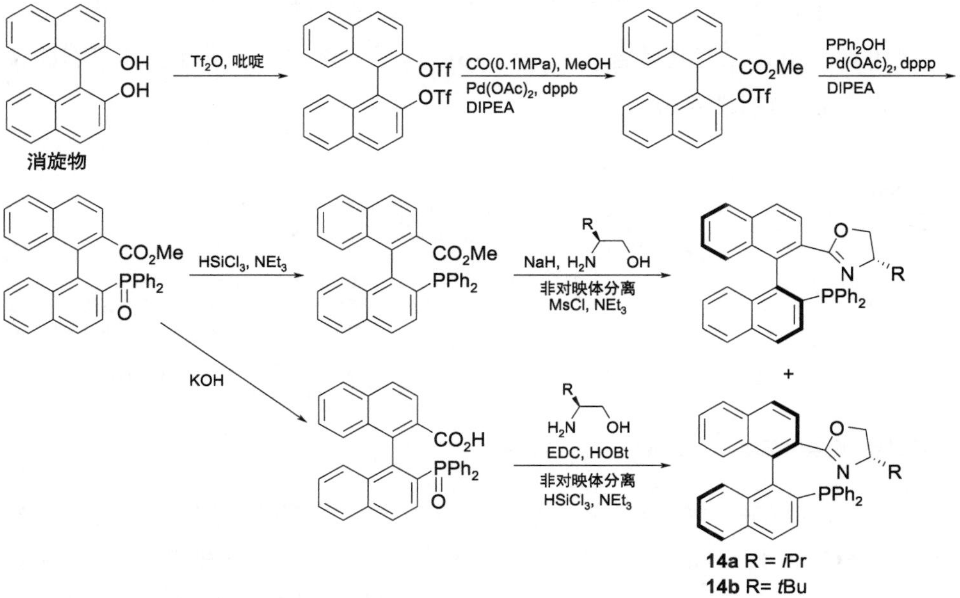

图 4-4　联萘骨架膦/噁唑啉配体 **14** 的合成

在含二茂铁骨架的平面手性膦/噁唑啉配体合成中，通常先由氰基二茂铁[17]或二茂铁甲酰氯[18]与手性氨基醇反应引入噁唑啉环，进一步与适当的烷基锂试剂反应。在手性噁唑啉的定向作用下，邻位锂化能够以高非对映选择性进行从而引入平面手性，随后芳基锂对氯化膦的亲核取代反应引入有机膦，得到相应的平面手性的膦/噁唑啉配体 15。光学纯的平面手性膦/噁唑啉配体最终经重结晶后获得（图 4-5）。

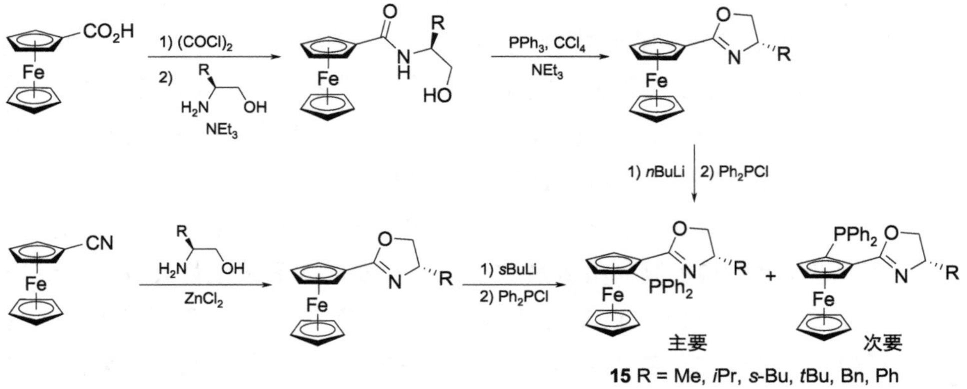

图 4-5　平面手性膦/噁唑啉配体 15 的合成

Helmchem 等[19]从三羰基锰环戊二烯基甲酰氯出发，依次构建噁唑啉环、引入二芳基膦合成了配体 16，其不仅含有平面手性，还具有磷及碳中心手性（图 4-6）。

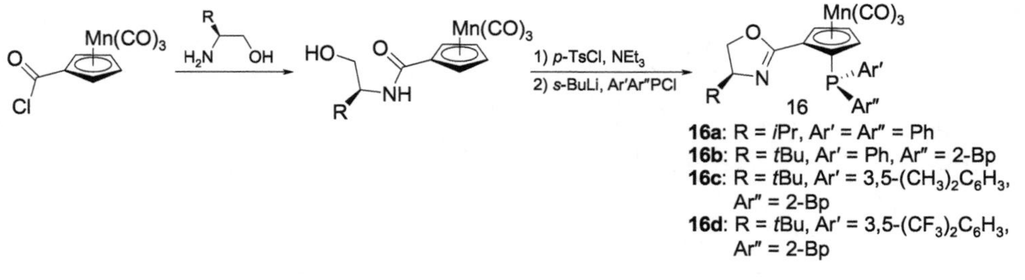

图 4-6　手性膦/噁唑啉配体 16 的合成

Bolm 等[20]报道了邻位取代的 [2.2] 对环芳烷骨架的平面手性膦/噁唑啉配体 17。Fu 等[21]报道了一种新颖结构的磷杂二茂铁骨架膦/噁唑啉配体 18 的合成（图 4-7）。作者巧妙地利用氨基醇双负离子与消旋的三氟甲酮化物反应经酰胺中间体顺利关环成噁唑啉。非对映体经色谱分离得到光学纯的平面手性磷杂二茂铁骨架化合物。基于类似合成路线，作者使用 (1$R$,2$S$)-1-氨基-2-茚醇还合成了并环结构的手性膦/噁唑啉配体。在这类配体中，作为配位原子的磷处于环戊二烯环上，为 $sp^2$ 杂化，通过 M → P$\pi$ 反馈键稳定金属配合物。

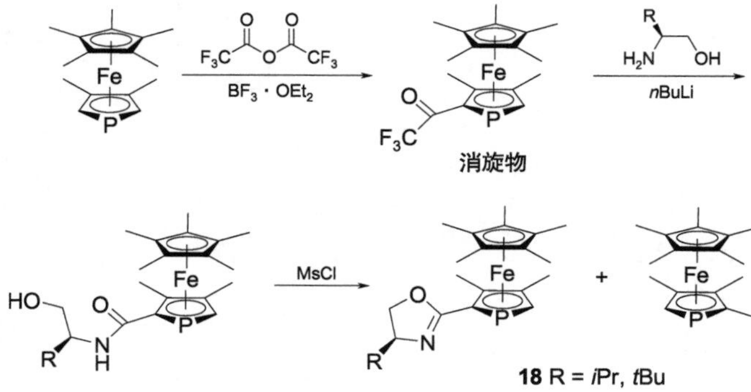

图 4-7　磷杂二茂铁骨架膦/噁唑啉配体 18 的合成

除了上述芳基骨架的手性膦/噁唑啉配体外，各种烷基骨架的膦/噁唑啉配体也被报道出来，如图 4-8 所示。

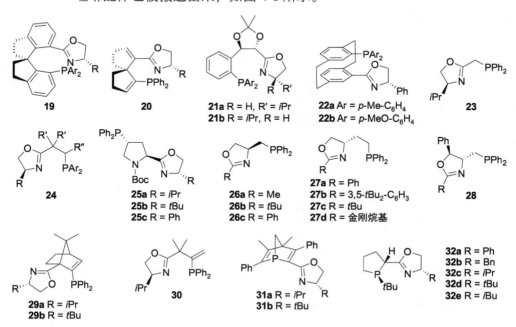

图 4-8　烷基和杂环骨架膦/噁唑啉配体

周其林等[22]从手性螺二酚出发合成了螺环骨架的膦/噁唑啉配体 **19**，而对应的氰基化合物不能与氨基醇反应形成噁唑啉（图 4-9）。

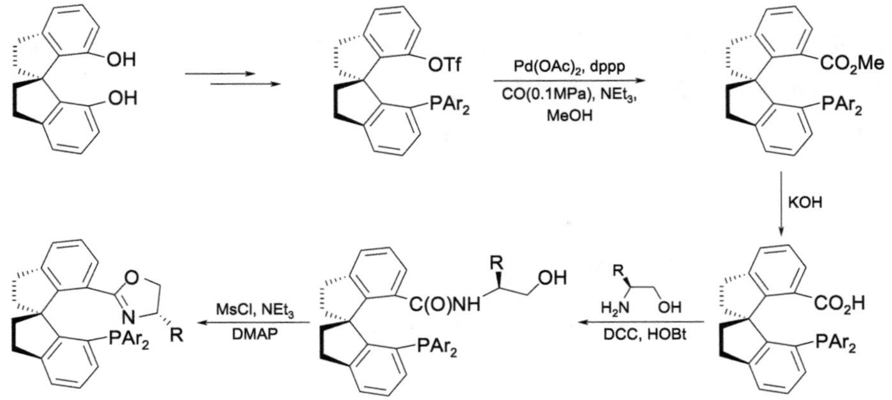

**19a** R = *i*Pr, Ar = Ph
**19b** R = Ph, Ar = Ph
**19c** R = Bn, Ar = Ph
**19d** R = Bn, Ar = 3,5-Me$_2$-C$_6$H$_3$

图 4-9　螺环骨架的膦/噁唑啉配体 **19** 的合成

丁奎岭等[23]从消旋的螺[4,4]壬烷-1,4-二酮出发合成了螺环骨架的膦/噁唑啉配体 **20**（图 4-10）。

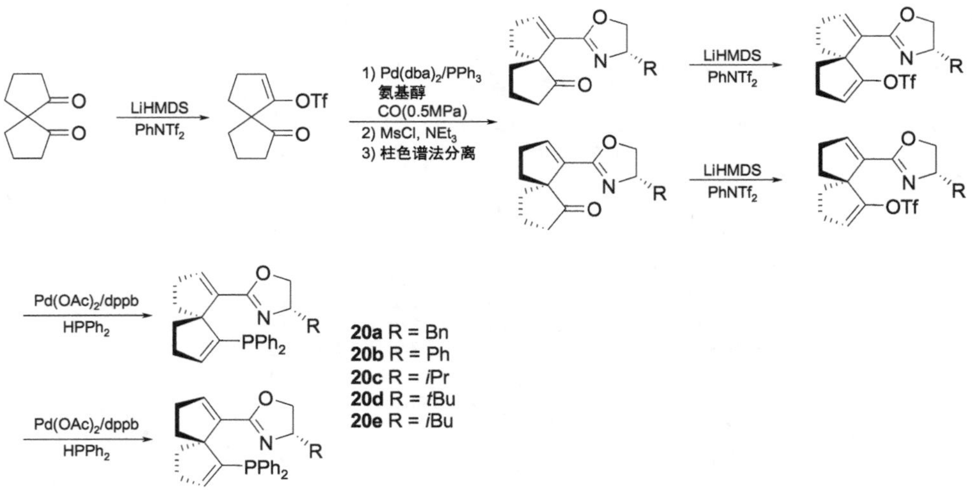

**20a** R = Bn
**20b** R = Ph
**20c** R = *i*Pr
**20d** R = *t*Bu
**20e** R = *i*Bu

图 4-10　螺环骨架的膦/噁唑啉配体 **20** 的合成

Morken 等[24]从香豆素出发经开环、Sharpless 不对称双羟化、缩酮保护、催化构建碳-磷键、环化形成噁唑啉、膦氧化物还原等多步反应合成了系列手性膦/噁唑啉配体 **21**（图 4-11）。

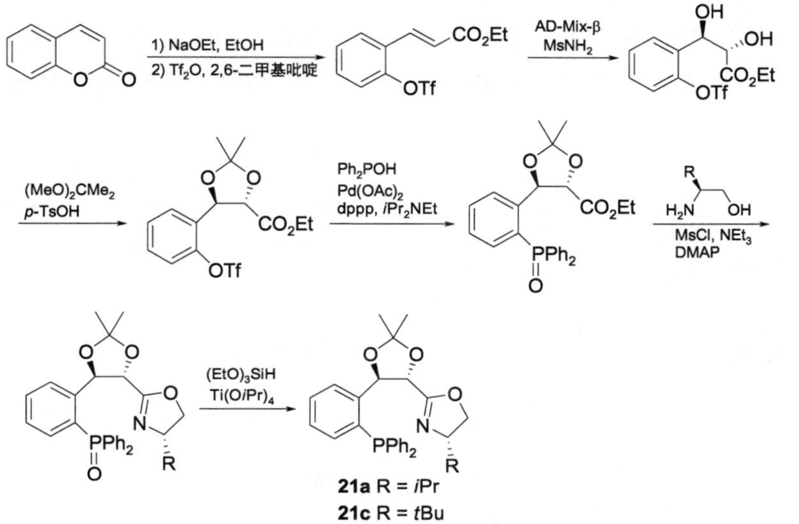

**图 4-11 膦/噁唑啉配体 21 的合成**

侯雪龙等[25]和 Pelter 等[26]报道了 [2.2] 对环芳烷骨架的平面手性膦/噁唑啉配体 **22**，磷原子和噁唑啉分别位于不同的苯环上。在 1993 年，Helmchem 等[2]报道了首例烷基桥连的膦/噁唑啉配体 **23**。Pfaltz 等[27]以商品化试剂 3-氯特戊酰氯为原料，先与手性氨基醇反应，酰胺可通过重结晶或蒸馏提纯。进一步关环蒸馏提纯得相应噁唑啉化合物。最后与二芳基磷钾发生亲核取代得到粗产物，过滤得膦/噁唑啉配体 **24**（图 4-12）。产物总收率达 20% 以上，无需柱色谱分离，适合大规模制备。

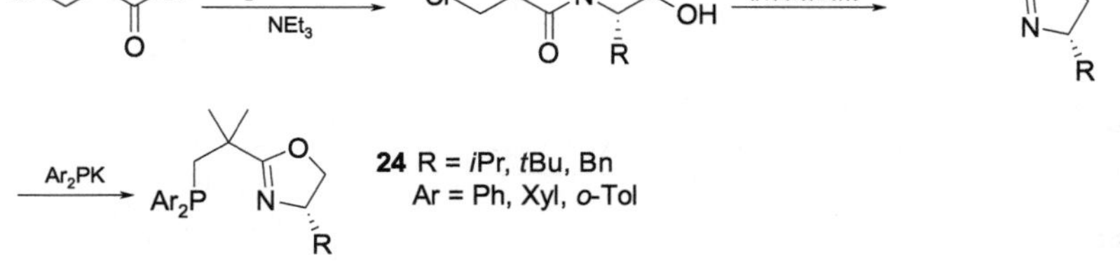

**图 4-12 膦/噁唑啉配体 24 的合成**

Gilbertson 等[28]从磷取代羧酸出发合成了类似结构的配体 **24**。鉴于脯氨酸骨架配体在不对称催化反应中往往表现出很高的立体选择性，该作者[29]从商品化的 Boc 保护的反式 4-羟基脯氨酸出发，先与氨基酸酯或氨基醇形成酰胺，二醇与甲基磺酰氯形成双酯，其中伯烷基磺酸酯先环化形成噁唑啉，仲烷基磺酸酯与二苯基膦钠发生亲核取代，随后形成膦硫化物进行纯化。雷尼镍还原脱硫得到最终配体 **25**（图 4-13）。

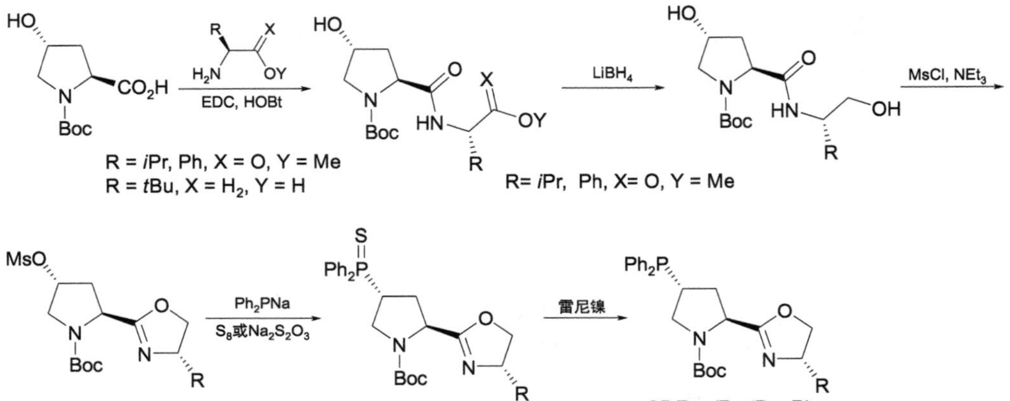

**25** R = *i*Pr, *t*Bu, Ph

图 4-13 膦/噁唑啉配体 **25** 的合成

Burgess 等[30]基于组合方式合成了另一类膦/噁唑啉配体 **26**（图 4-14），此类配体磷原子通过噁唑啉 C-4 上的烷基相连，与金属络合时形成的是

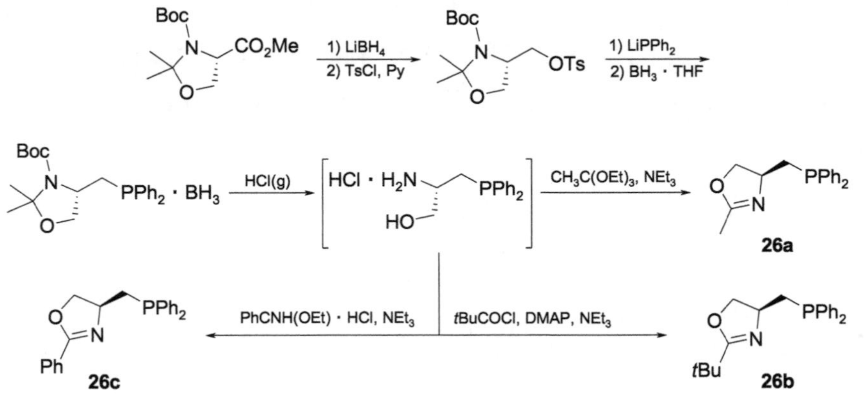

**26a** **26b** **26c**

图 4-14 膦/噁唑啉配体 **26** 的合成

非平面的五元螯合环，不同于常见的 PHOX 配体。作者发展了三种方法制备目标手性配体，C-2 上的取代基多达 13 种。在此基础上，作者又以天冬氨酸为原料合成了第二代配体 **27**[31]，其与金属络合时形成的是六元环。James 等 [32] 在噁唑啉环 C-5 上引入取代基对配体 **26** 结构进行修饰合成了配体 **28**。Gilbertson 等 [33] 通过对酮基膦酸进行系列转化可合成烯基膦/噁唑啉配体 **29** 及其类似结构配体 **30**。以磷杂环戊二烯为原料，经过与苯基炔丙酯 [4+2] 环加成、噁唑啉结构单元引入等步骤合成了配体 **31**[34]。

张绪穆等 [35] 从磷杂环戊烷硫化膦出发，通过非对映选择性羧化、酰胺合成、环化、还原脱硫等反应合成了含环膦结构的配体 **32**（图 4-15）。该配体同时具有多个碳手性中心及一个电负性磷手性中心。

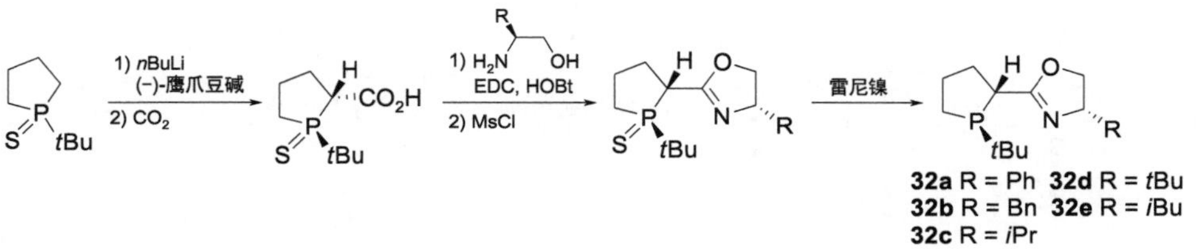

图 4-15 杂环膦/噁唑啉配体 **32** 的合成

当用 P—O/P—N 替代 P—C 键时可以对膦配体的立体性质和电子效应进行修饰，从而调节手性催化剂的不对称诱导能力，如图 4-16 所示。

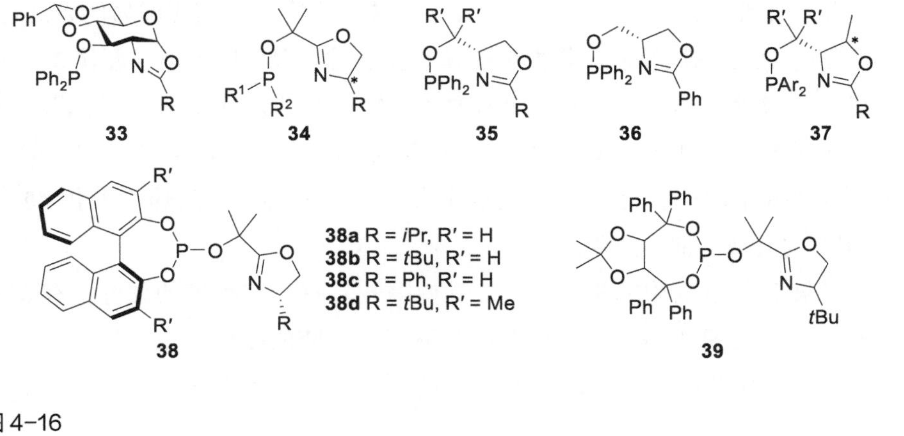

图 4-16

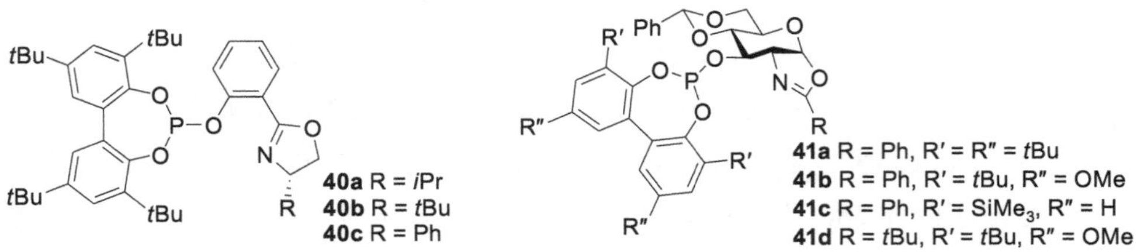

**图 4-16** 亚磷酸酯及亚次膦酸酯配体

在 1999 年，Uemura 等[36]从氨基葡萄糖出发，通过对羟基的选择性保护及转化合成了亚次膦酸酯/噁唑啉配体 **33**（图 4-17）。

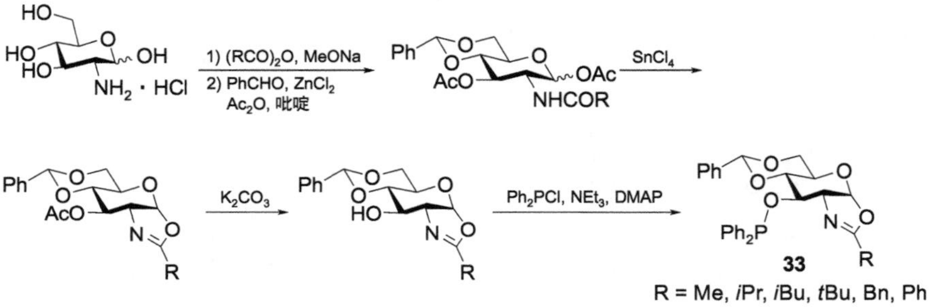

**图 4-17** 亚次膦酸酯/噁唑啉配体 33 的合成

Pfaltz 等[37]用简单的商品化氨基醇和 2-羟基异丁酸为起始原料只需两步反应就实现了一类亚次膦酸酯/噁唑啉配体 **34** 的合成。具有相似骨架的配体 **35**、**36** 分别由 Pfaltz 等[38]与 Richards 等[39]独立报道，他们以丝氨酸甲酯盐酸盐为原料通过三步或四步反应完成了相应配体的合成（图 4-18）。Pfaltz 等[40]从苏氨酸甲酯盐酸盐通过类似反应合成了配体 **37**，该配体的噁唑啉环 C-5 上的一个氢原子被甲基取代。从不同的羟基取代的噁唑啉及环状亚磷酰氯出发，Pfaltz 等合成了分别由联萘酚和 TADDOL 衍生的配体 **38**[41]、**39**[42]。Pàmies 等报道了亚磷酸酯/噁唑啉配体 **40**[43]、**41**[44] 的合成与催化应用。

胺类化合物中的极性 N—H 键使得胺具有一定程度的酸性，在碱存在下胺可与氯化膦反应形成亚次膦酰胺或亚磷酰胺酯。上述反应为合成各种含噁唑啉的膦/氮配体（图 4-19）提供了切实可行的方法。

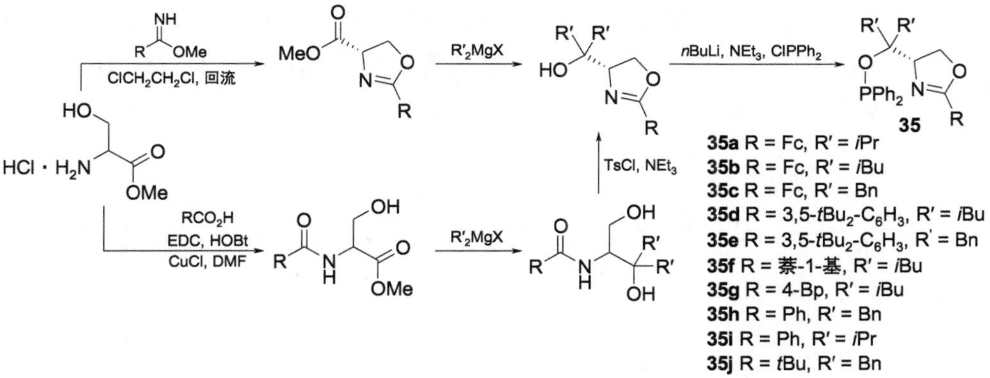

图 4-18 亚次膦酸酯/噁唑啉配体 35 的合成

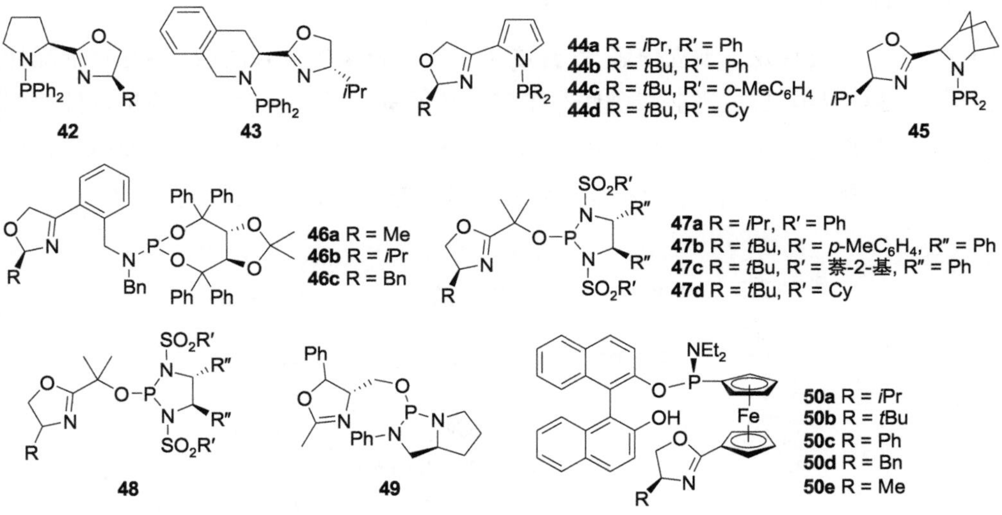

图 4-19 其他膦/噁唑啉配体

  Gilbertson 等[45]合成了由脯氨酸苄酯衍生的亚次膦酰胺/噁唑啉配体 **42**（图 4-20）。Agbossou-Niedercorn 等[46]随后采用类似路径合成了 **42** 的非对映体及由四氢异喹啉衍生的亚次膦酰胺/噁唑啉配体 **43**。Cozzi、Pfaltz 及其合作者[47]从 2-氰基吡咯出发合成了配体 **44**。研究表明，温和、弱碱条件下形成的 C—P 键可有效抑制其水解。Andersson 等[48]报道了氮杂降冰片烷骨架的配体 **45**。在 2007 年，Guiry 等[49]合成了配体 **46**，此类双齿配体可与金属形成八元环络合物。Pfaltz 等[50]从手性环己二胺和 1,2-二苯基乙二胺出发，合成相应的双(对甲苯磺酰氨基)亚磷酰氯中间体化合物，再与相应的含羟基的噁唑啉生成目标配体 **48**。Bondarev 等 [51] 利用 (S)-N-苯基-2-氨甲基吡咯烷衍生的亚磷酰氯合成了配体 **49**。

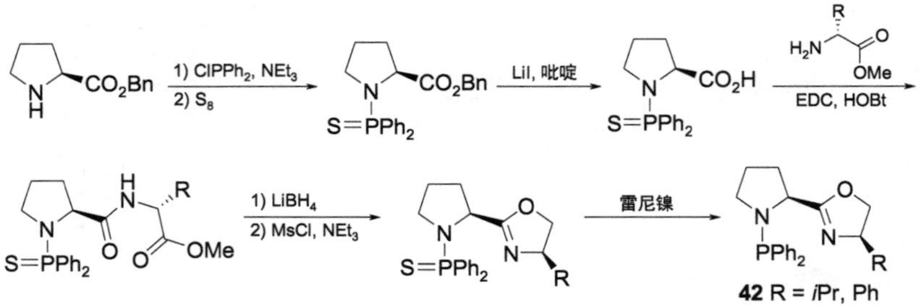

**图 4-20　亚次膦酰胺/噁唑啉配体 42 的合成**

侯雪龙、戴立信等 [52] 合成了二茂铁骨架的亚膦酰胺酯/噁唑啉配体 **50**（图 4-21）。当亚膦酰胺与不同构型的联萘酚反应时形成的是可柱色谱分离的非对映异构体，因而形成了一个磷手性中心。此类型配体看似结构复杂，但却能以克级规模制备，且对空气十分稳定。

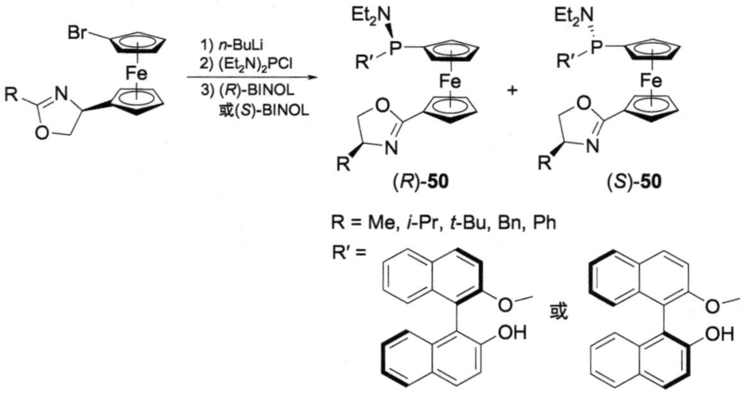

**图 4-21　亚膦酰胺酯/噁唑啉配体 50 的合成**

### 4.1.1.2　膦/咪唑啉配体

相对于结构多样的膦/噁唑啉配体而言，膦/咪唑啉配体的种类和数量有限，已经报道的部分配体结构如图 4-22 所示。

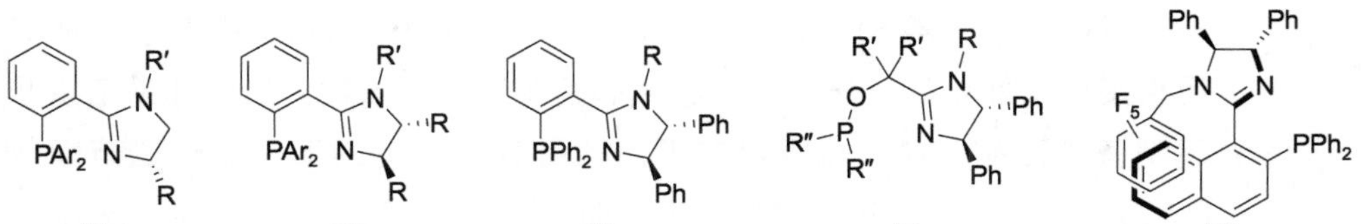

图 4-22 代表性的膦/咪唑啉配体

Pfaltz 等[53]从易于制备的 β-羟基苯甲酰胺出发，首先构筑咪唑啉环，再通过邻位锂化、与二芳基氯化磷发生亲核取代反应生成配体 **51**（图 4-23）。

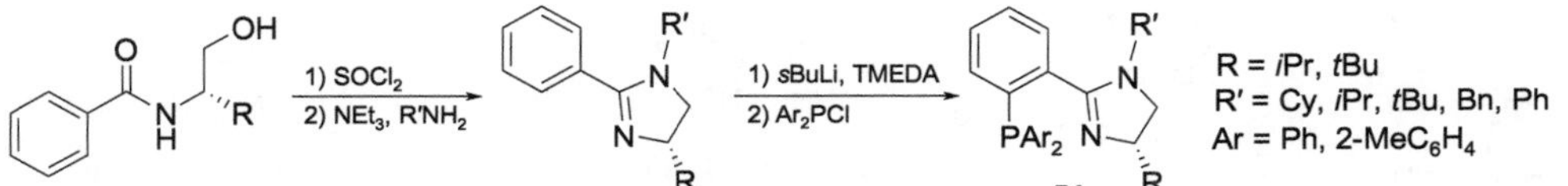

图 4-23 膦/咪唑啉配体 **51** 的合成

Busacca 等[54]以卤代亚氨酸乙酯为原料，与手性二胺缩合形成咪唑啉环，当 X=F 时，先与二芳基膦金属化合物发生亲核取代引入二芳基膦，再保护仲胺；当 X=I 时，先保护仲胺，再通过钯催化 C—P 键形成反应合成目标膦/咪唑啉 **52**（图 4-24）。作者对比研究了两种合成路线，认为后者更具有普适性。

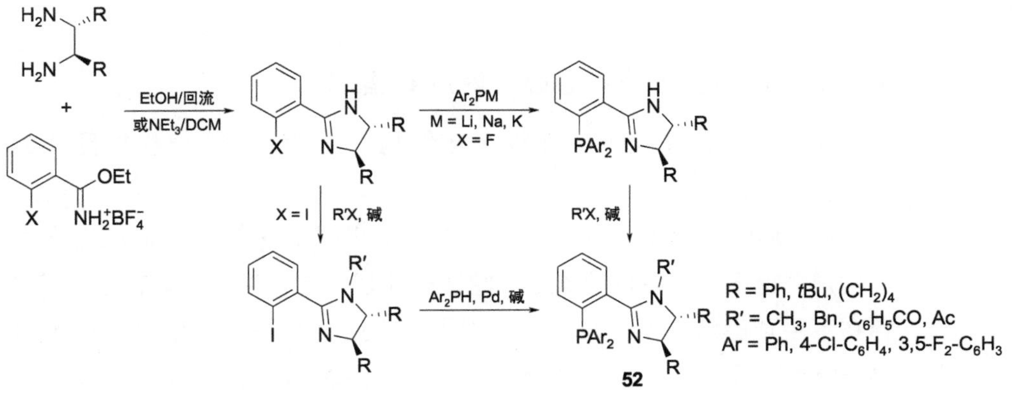

图 4-24 膦/咪唑啉配体 **52** 的合成

Claver 等[55] 报道了配体 **53** 的合成，其结构中的氨基没有经过保护而直接用于催化反应研究。作者从邻卤苯甲酰氯出发，先以几乎定量产率制备二硫酯。在氧化汞存在下，利用 HgS 沉淀的形成驱动二硫酯与 (1R,2R)-二苯基乙二胺的缩合反应从而以高收率得到咪唑啉。Pfaltz 等[56] 还在专利中报道了以 2-羟甲基咪唑啉为原料与各种亚磷酰氯反应制备配体 **54**。

Guiry 等[57] 利用 (1S,2S)-1,2-二苯基乙二胺与 2-甲氧基-萘甲醛缩合后再引入五氟苯基和二苯基膦合成了膦/咪唑啉配体 **55**(图 4-25)，光学纯化合物是通过非对映异构体分级结晶获得的。

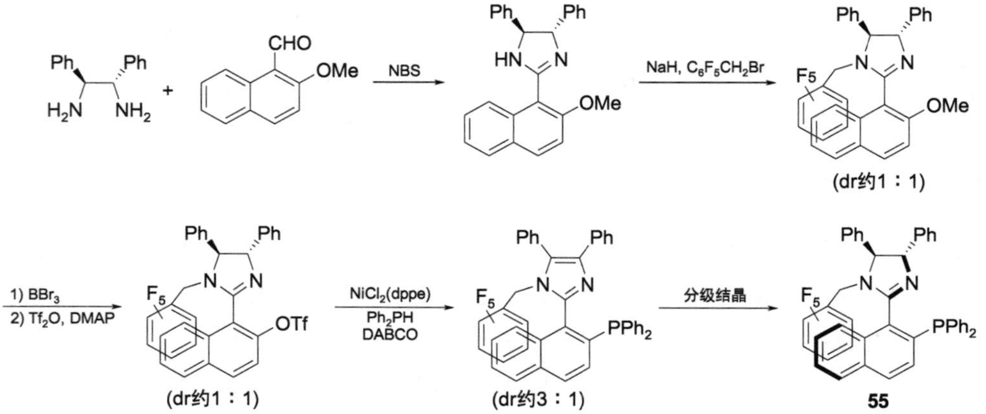

图 4-25 手性膦/咪唑啉配体 **55** 的合成

### 4.1.1.3 膦/噁唑、膦/噻唑、膦/咪唑配体

含有噁唑、噻唑、咪唑等杂环结构的膦/氮配体结构如图 4-26 所示。

在 2004 年，Andersson 等[58] 从偶氮双甲酮出发，在 $Rh_2(OAc)_4$ 催化下构建噁唑环，随后将酮羰基对映选择性还原成手性醇。醇在正丁基锂存在下生成醇氧负离子与二芳基氯化磷反应形成 P—O 键从而完成配体 **56** 的合成(图 4-27)。考虑到配体中的 P—O 键容易发生水解，作者没有对配体进行分离，而是过滤处理后直接与金属形成稳定的络合物。

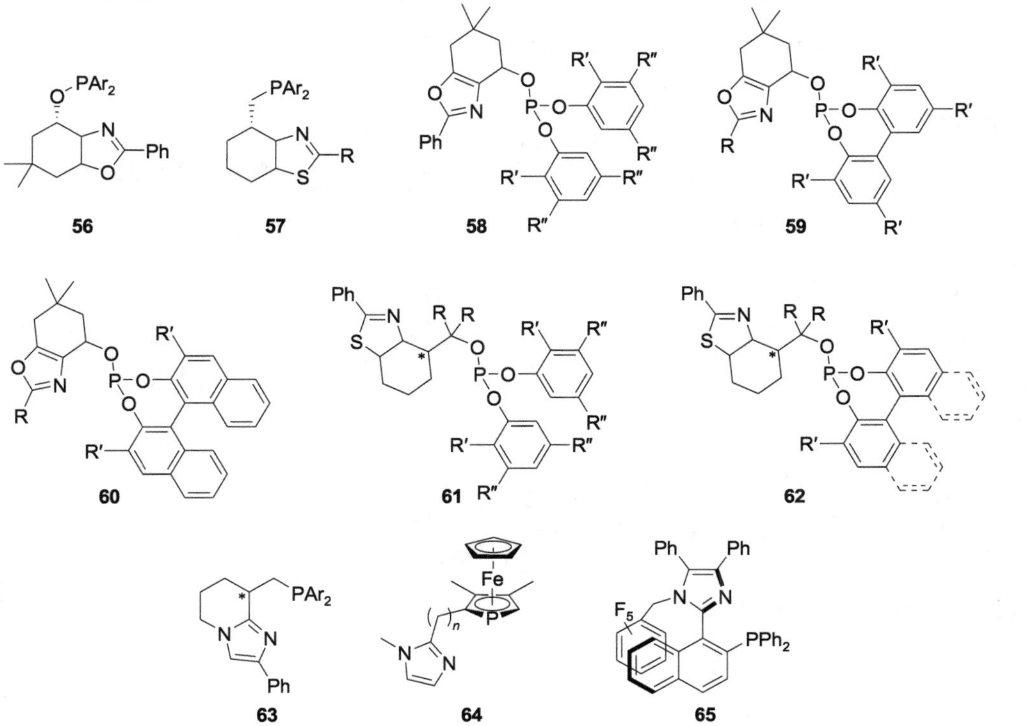

图 4-26 膦/其他氮杂环配体

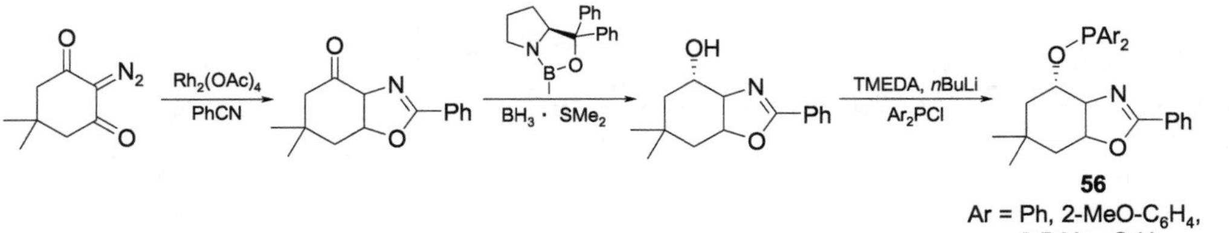

图 4-27 膦/噁唑配体 56 的合成

在上述研究基础上，该课题组[59]进一步将配体从噁唑环拓展到噻唑结构，从而得到了配体 57（图 4-28）。相比于膦/噁唑配体，膦/噻唑配体更易于制备和结构调控，且分子中的 C—P 键更加牢固，因而配体的稳定性得到很大程度的提高。Andersson 与 Diéguez 等[60]合作开发了一系列膦/噁唑配体 58～60、膦/噻唑配体 61 和 62。Andersson 等[61]从 2-氨基烟酸乙酯出发合成了手性膦/咪唑配体 **63**(图 4-29)。Ganter 等[62]报道了含有磷杂环戊二烯单元的膦/咪唑配体 **64**。

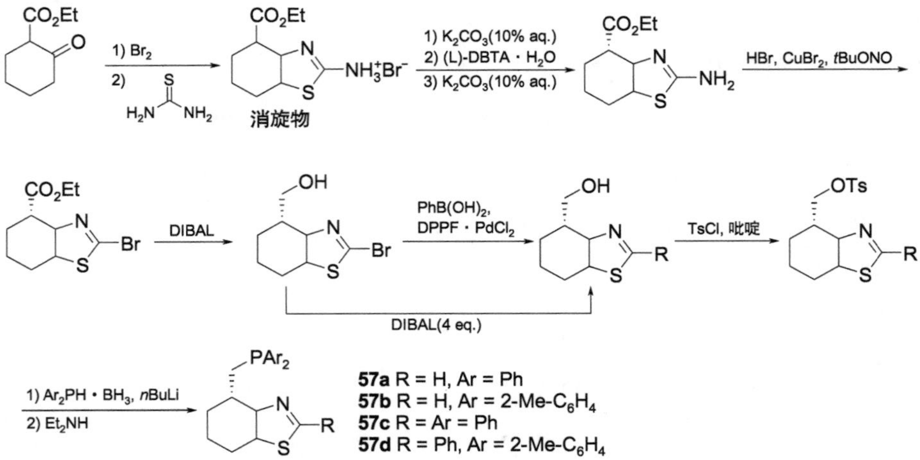

图 4-28 膦/噻唑配体 **57** 的合成

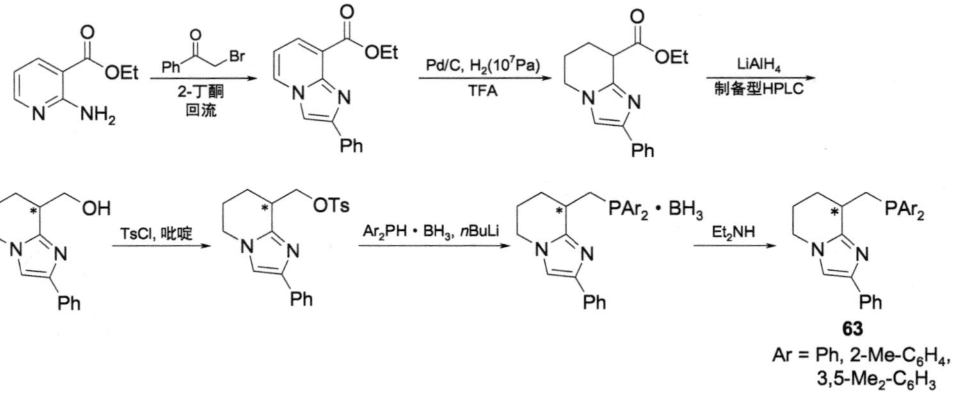

图 4-29 膦/咪唑配体 **63** 的合成

Aponick 等[63]报道了一类新颖结构的轴手性 P,N 配体 **65**(图 4-30),与含吲哚的联芳基 P,N 配体不同,得益于缺电子的五氟苯基与富电子的萘环间的 π-π 相互作用,此类膦/咪唑配体结构稳定,不会发生消旋。

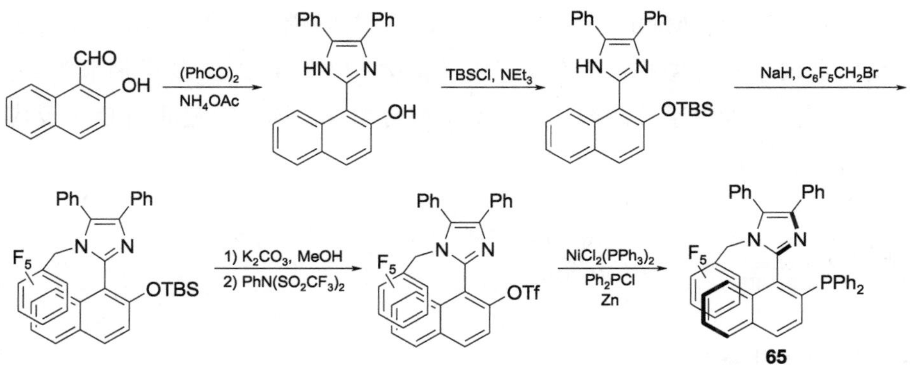

图 4-30 膦/咪唑配体 65 的合成

### 4.1.1.4 膦/吡啶相关配体

膦/吡啶及其类似物如喹啉、异喹啉、菲啶等构成了另一大类重要的手性膦/氮双齿配体（图 4-31）。

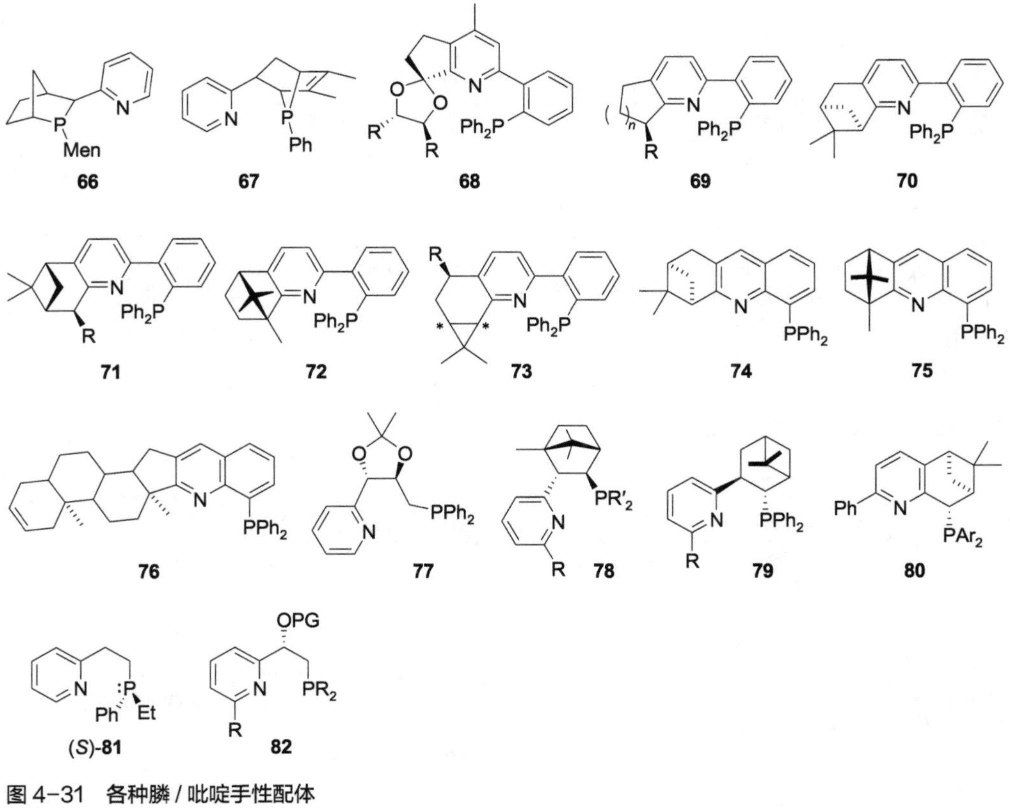

图 4-31 各种膦/吡啶手性配体

在1992年，Mathey等[64]利用环戊二烯与膦烯络合物间的不对称D-A反应合成了双环结构的手性膦/吡啶配体 **66**（图 4-32）。基于相似策略，Leung等[65]通过手性磷杂环戊二烯钯络合物与乙烯基吡啶间的D-A反应合成了配体 **67**。

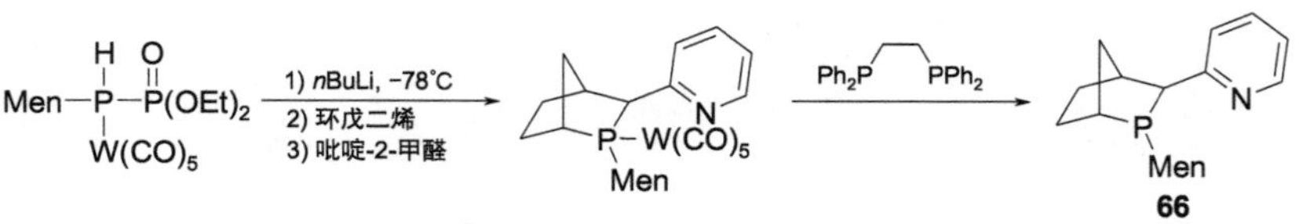

图 4-32 手性膦/吡啶配体 **66** 的合成

Wilson等[66]利用 2-氯吡啶酮与不同手性二醇的缩合反应合成手性膦/吡啶配体 **68**（图 4-33）。Ito、Katsuki等[67]从相应的手性 2-氯代吡啶出发，依次通过 Suzuki 偶联、酚羟基酯化、C-P 偶联反应合成了配体 **69**（图 4-34）。

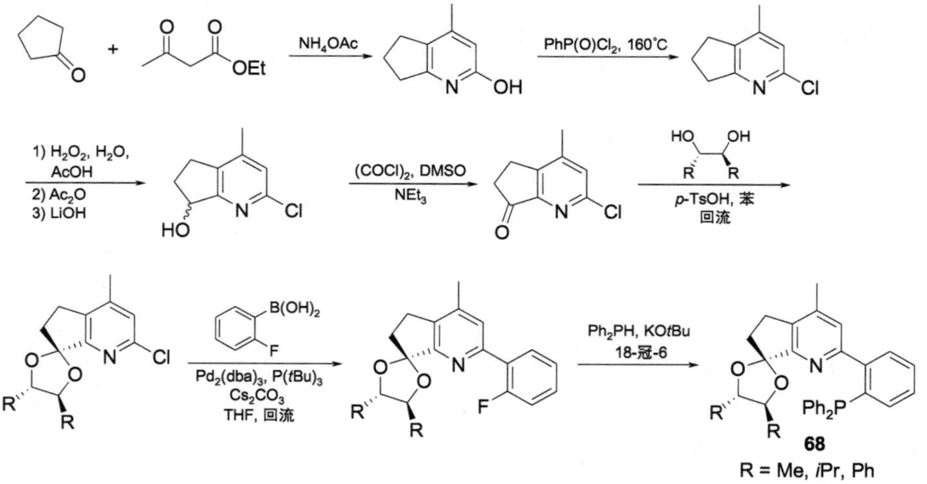

图 4-33 手性膦/吡啶配体 **68** 的合成

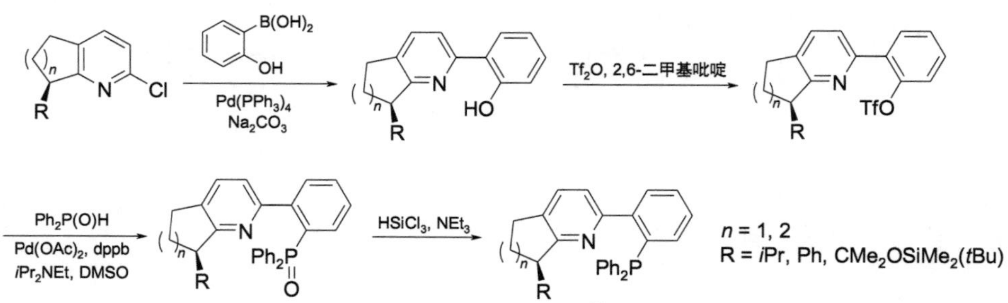

图 4-34 手性膦/吡啶配体 **69** 的合成

Chelucci[68] 和 Malkov、Kocovsky[69] 课题组分别独立报道了类似结构配体 **70** ~ **73** 的合成。两者均使用手性亚甲基酮与吡啶碘盐发生 Krohnke 关环反应得到四氢喹啉衍生物，但在 C—P 键形成反应上前者（图 4-35）通过三氟甲磺酸苯酯与二苯基膦氧化物催化偶联引入二苯基膦，后者（图 4-36）利用二苯基膦钾对苯环上氟原子的取代反应实现。

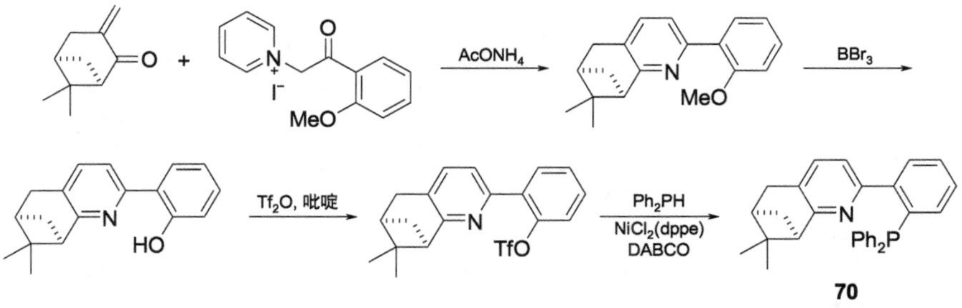

**图 4-35** 手性膦/吡啶配体 **70** 的合成

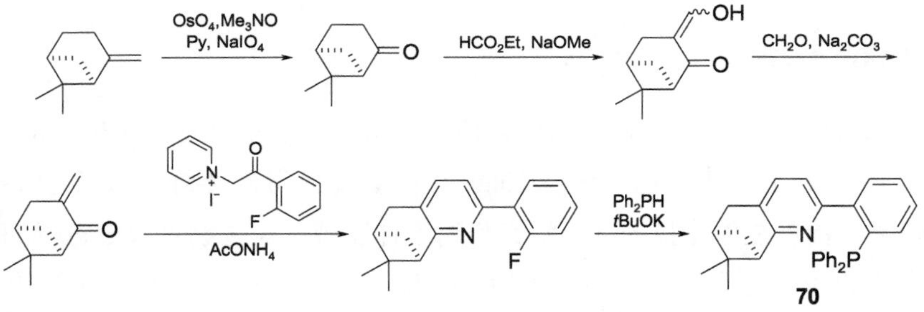

**图 4-36** 手性膦/吡啶配体 **70** 的合成

Chelucci 等[70] 随后还报道了手性配体 **74**（图 4-37）、**75**、**76**。关键中间体四氢吖啶衍生物由 2-氨基-3-氟苯甲醛与相应手性酮通过缩合反应获得，再与二苯基膦锂发生取代形成目标产物。

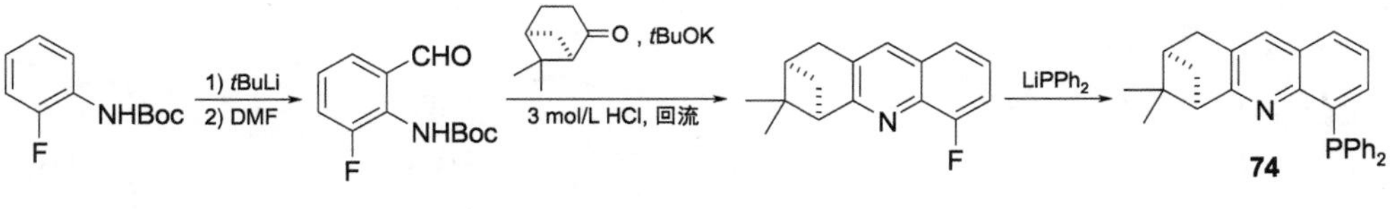

**图 4-37** 手性膦/吡啶配体 **74** 的合成

在 1996 年，Chelucci 等[71] 从 L-酒石酸衍生的二醇出发，通过选择性转化羟基为氰基,再利用钴催化的腈与乙炔环加成反应构筑吡啶环,继而引入二苯基膦合成了含有手性烷基骨架的 1,6-N,P 配体 **77**(图 4-38)。

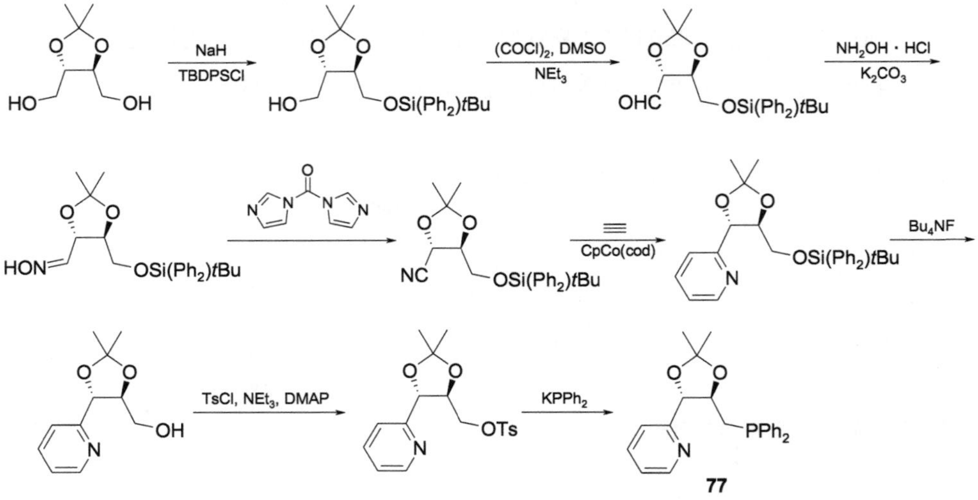

**图 4-38** 手性膦/吡啶配体 **77** 的合成

Knochel 等[72]从 D-樟脑或 (R)-诺蒎酮制得相应的烯基三氟甲磺酸酯,与 2-吡啶或 2-喹啉基溴化锌进行 Negishi 偶联反应得到 2-烯基吡啶或 2-烯基喹啉。接着在催化量的叔丁醇钾作用下仲膦或其氧化物对双键进行加成,生成的膦氧化物经还原制得手性 N,P-配体 **78**(图 4-39)、**79**。Li 等[73]通过杂原子定向烷基 C—H 直接锂化、亲核进攻相应的氯化磷合成了高非对映选择性的配体 **80**,其与金属络合后形成的是五元金属环。

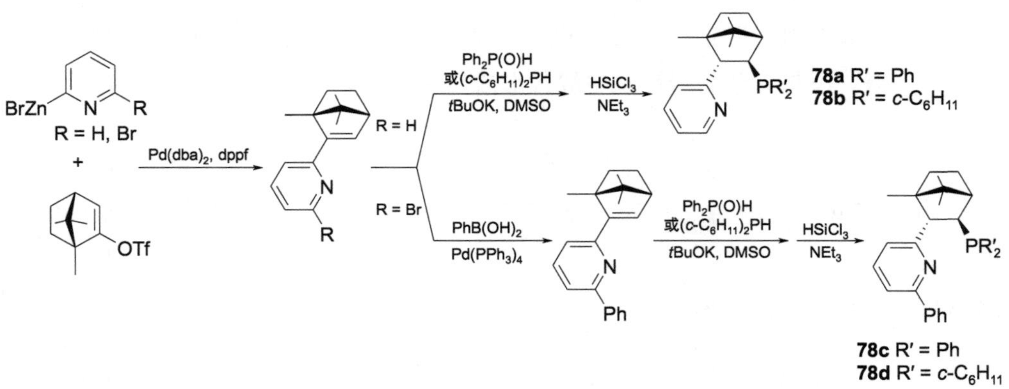

**图 4-39** 手性膦/吡啶配体 **78** 的合成

Leung 等[74]利用手性二乙腈钯络合物促进消旋的乙基苯基膦与 2-乙烯基吡啶发生不对称氢膦化反应，通过分级结晶得到一对非对映异构体钯络合物。两种络合物分别依次用浓盐酸、氰化钾水溶液处理可释放出游离吡啶膦配体 **81**（图 4-40）。

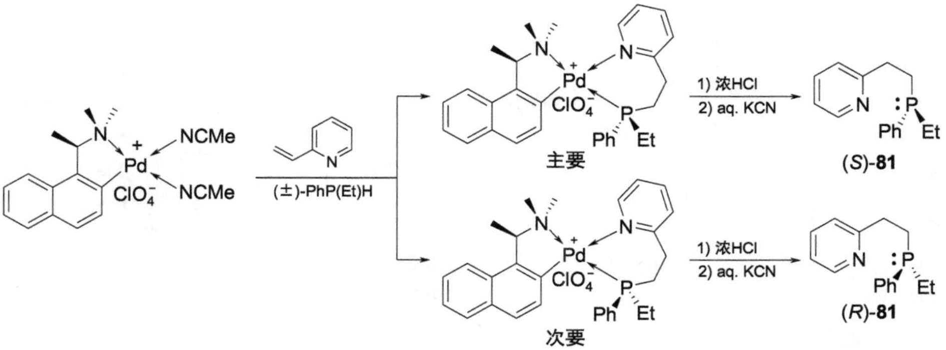

图 4-40　手性膦/吡啶配体 **81** 的合成

Pfaltz 等[75]以 2-吡啶甲酸乙酯及 2-乙烯基喹啉为原料，分别利用不对称羰基还原和 Sharpless 双羟基化反应合成关键的手性仲醇中间体，以不同的硅试剂保护羟基后再脱去硼烷得相应的手性配体 **82**（图 4-41）。

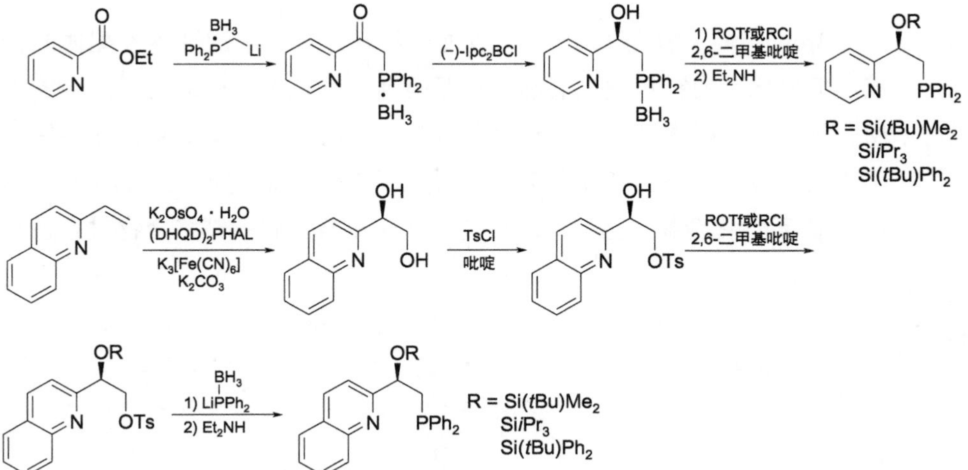

图 4-41　手性膦/吡啶配体 **82** 的合成

除上述含磷中心手性、碳中心手性的膦/吡啶配体外，人们还合成了具有轴手性或平面手性的膦/吡啶配体，如图 4-42 所示。

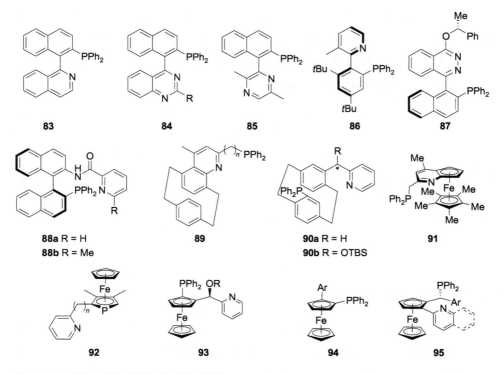

图 4-42 轴手性或平面手性的膦/吡啶配体

Brown 等[76]曾在 1992 年报道过一种联芳基骨架的轴手性配体的合成，但此化合物容易发生消旋，从而限制了其在不对称催化反应中的应用。随后，该课题组从 1-氯异喹啉与 2-甲氧基-1-萘硼酸的钯催化偶联反应出发，经五步操作将甲氧基转化成二苯基膦，制得消旋配体 **83**[77]（图 4-43）。手性化合物是通过 (R)-N,N-二甲基-α-萘乙胺的钯络合物对其进行

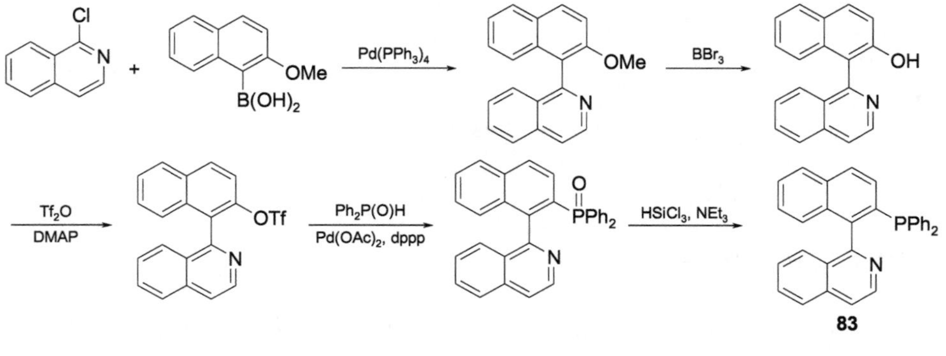

图 4-43 轴手性膦/吡啶配体 83 的合成

拆分获得，并在此基础上合成了菲啶/膦配体[78]及不同二芳基膦取代的喹啉/膦双齿配体[79]。Guiry 等使用氯代喹唑啉、吡嗪与 2-甲氧基-1-萘硼酸偶联合成了配体 84[80]、85[81]。作者发现在引入二苯基膦时，镍较钯催化剂在 C—P 形成反应中效率更高。

Chan 等[82]以三芳基膦为膦化试剂，从相应的芳基三氟甲烷磺酸酯通过 Pd(OAc)$_2$ 催化形成 C—P 键引入二苯基膦，制得消旋化合物。使用 (R)-N,N-二甲基-α-萘乙胺的钯络合物对其拆分从而实现轴手性配体 86 的合成（图 4-44）。作者曾尝试以光学活性芳基三氟甲烷磺酸酯为原料通过催化膦化直接合成手性配体，但由于手性原料在高温反应条件下易发生消旋没能取得成功。

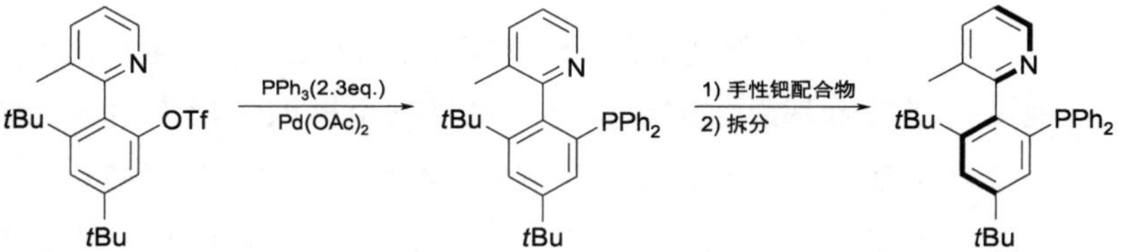

图 4-44 轴手性膦/吡啶配体 86 的合成

为避免使用计量的手性钯络合物拆分消旋配体这一步骤，Carreira 等[83]使用 1,4-二氯酞嗪与 2-萘酚偶联形成联芳基骨架，再与手性醇或胺进行亲核取代引入手性中心，目标手性配体 87（图 4-45）可通过柱色谱分离非对映异构体获得。

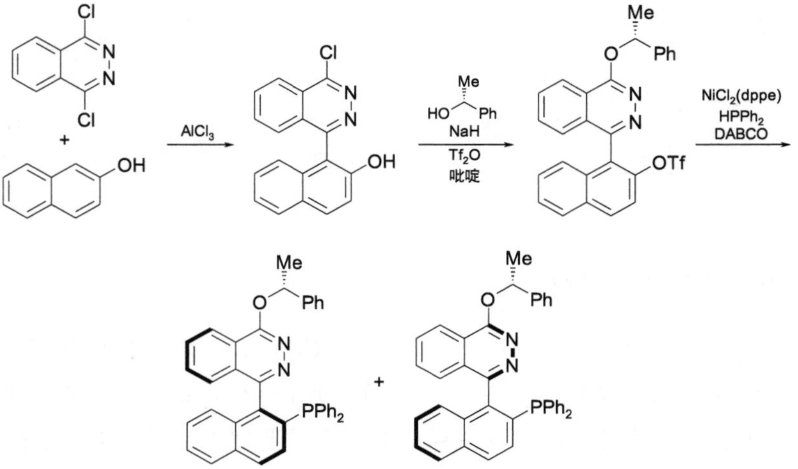

图 4-45 轴手性膦/酞嗪配体 87 的合成

张绪穆等[84]从轴手性化合物 NOBIN 出发先合成已知中间体氨基膦，再与吡啶-2-甲酸脱水形成酰胺，从而合成了一类具有很大的咬合角的配体 88（图 4-46）。作者认为酰胺桥链赋予联芳基配体较强的刚性，这对于配体更有效地发挥手性诱导作用至关重要。

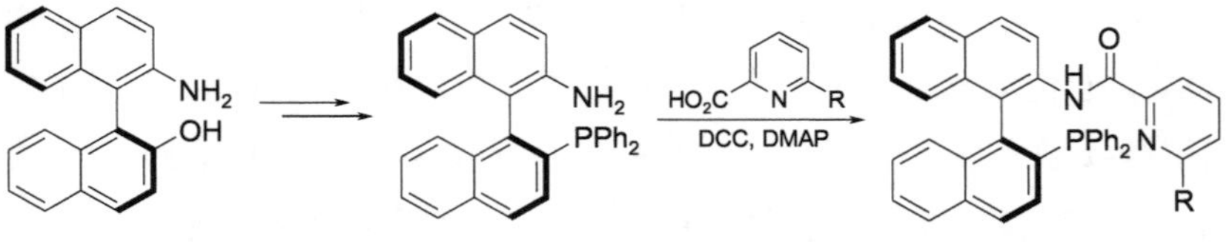

图 4-46 轴手性膦/吡啶配体 88 的合成

Ruzziconi 等[85]从平面手性的环芳烷骨架喹啉衍生物出发，通过相应锂试剂与二苯基氯化膦或溴甲基喹啉的亲核取代反应引入二苯基膦，进一步反应合成了不同碳链长度的膦/喹啉配体 89（图 4-47）。姜标等[86]用光学纯的溴代环芳烷经溴锂交换、锂试剂进攻吡啶或喹啉-2-甲醛、非对映异构体分离、羟基成醚保护合成了配体 90。

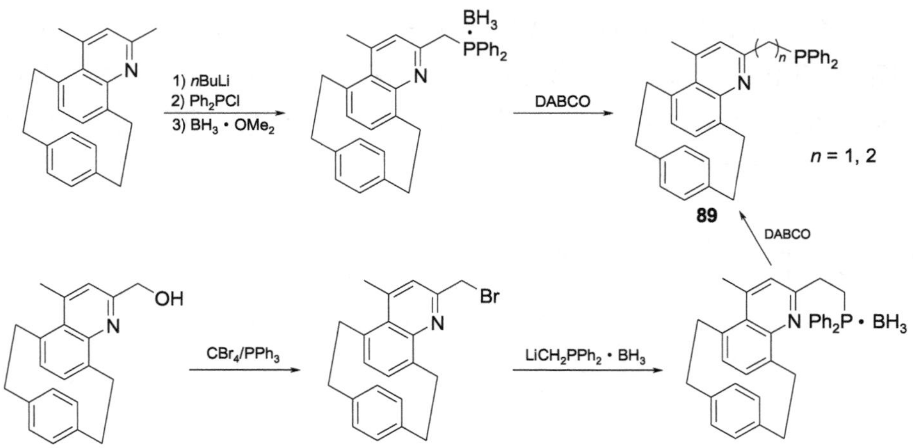

图 4-47 平面手性膦/吡啶配体 89 的合成

Fu 等[87]从五甲基环戊二烯锂（$C_5Me_5Li$）及 2-氯吡啶并环戊二烯锂出发合成 2-氯吡啶并二茂铁，通过 Kumada 偶联引入甲基，将其直接锂化后与二苯基氯化膦发生亲核取代生成吡啶膦。平面手性配体 91 由手性 HPLC 拆分得到（图 4-48）。

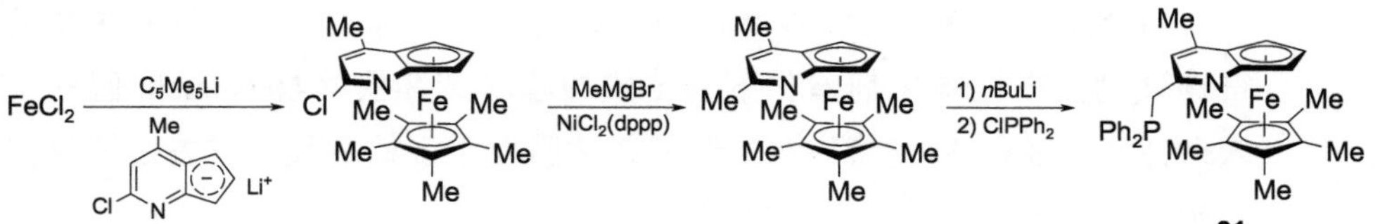

91

图 4-48　平面手性膦 / 吡啶配体 **91** 的合成

Ganter 等 [88] 使用含吡啶的锂试剂对手性 2-甲酰基-3,4-甲基磷杂二茂铁进行加成，再对仲醇脱氧即得相应的膦 / 吡啶配体 **92**（图 4-49）。

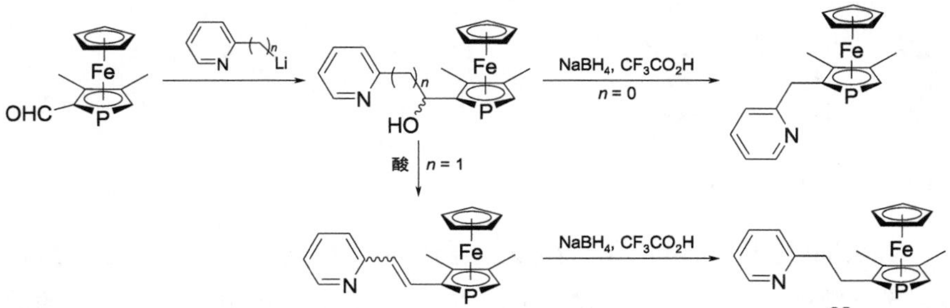

92

图 4-49　平面手性膦 / 吡啶配体 **92** 的合成

Knochel 等 [89] 从手性二茂铁亚砜出发，通过定向邻位锂化、膦化、硫化得到平面手性膦 / 硫化合物。通过亚砜-锂交换进攻吡啶-2-甲醛引入羟基吡啶，羟基转化为醚后可实现非对映异构体的分离，脱硫后得相应的膦 / 吡啶配体 **93**（图 4-50）。

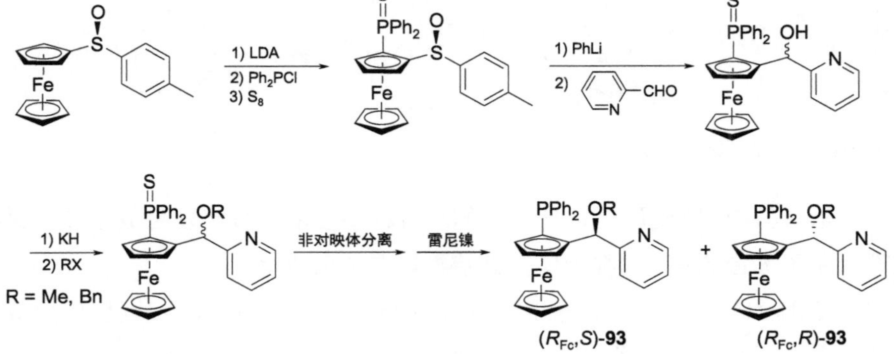

($R_{Fc}$,$S$)-**93**　　($R_{Fc}$,$R$)-**93**

图 4-50　平面手性膦 / 吡啶配体 **93** 的合成

同样是从手性二茂铁亚砜出发,经过锂化、与溴化锌进行金属交换得锌试剂,在钯催化剂催化下与1-碘异喹啉或8-溴喹啉Negishi偶联引入杂芳环。亚砜-锂交换后与二苯基氯化膦发生亲核取代、脱去硼烷保护基制得最终手性配体 **94**[90](图4-51)。

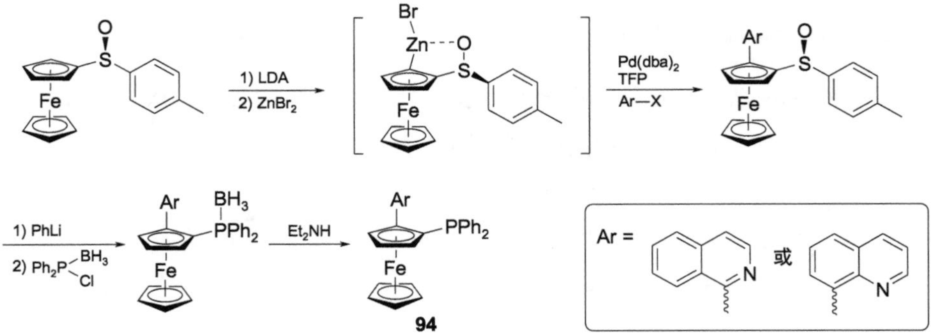

**图4-51** 平面手性膦/吡啶配体 **94** 的合成

Knochel等[91]还从二茂铁出发,通过傅-克酰化、不对称还原、醚化得到光学纯的α-甲氧基衍生物,利用Negishi偶联反应引入杂芳环,随后取代甲氧基引入二苯基膦得到另一类配体 **95**(图4-52)。为方便后续操作需对其进行原位保护。

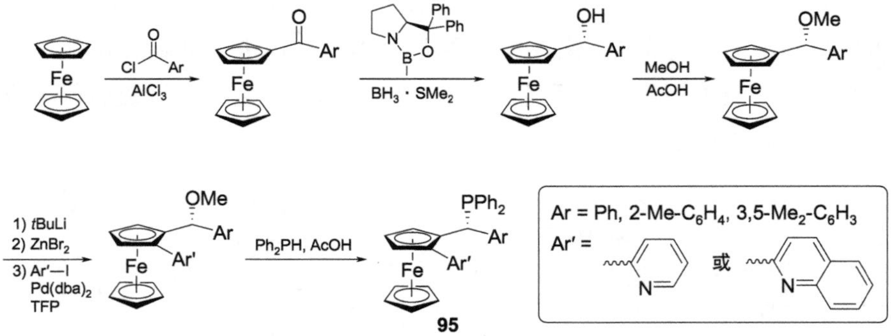

**图4-52** 平面手性膦/吡啶配体 **95** 的合成

除了叔膦/吡啶配体以外,人们从各种手性醇或胺出发合成了氧、氮等杂原子键合的膦/吡啶配体,其结构如图4-53所示。

在1994年,Faraone等[92]从(-)-薄荷酮出发,与2-吡啶锂反应形成手性醇,再与二苯基氯化膦反应制得配体 **96**。Goldfuss等[93]以(-)-小茴香酮为原料采用类似路线合成了配体 **97**。周永贵等[94]拆分2-吡啶基环

己醇合成了配体 **98**。Pfaltz 等[75]通过还原各种商品化的吡啶酮或用 2-吡啶锂进攻各种醛形成了含不同取代基的消旋仲醇，经手性色谱柱分离得到光学纯化合物。手性仲醇与二烷基氯化膦或 *N*,*N*-二乙基氨基二芳基膦反应合成了系列膦/吡啶配体 **99**。为了提高配体的刚性，随后 Pfaltz[95]和周永贵[96]课题组还报道了双环结构的配体 **100** 的合成，从而可以在环尺寸、磷原子及吡啶环上取代基等方面调节催化剂结构(图 4-54)。Li 等[97]报道了其类似结构配体 **101** 的合成。

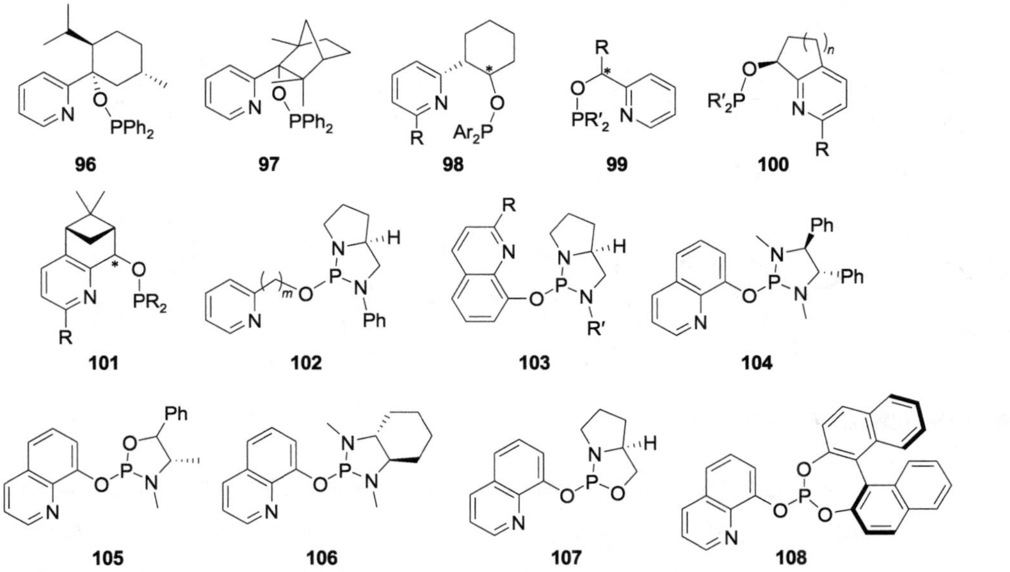

图 4-53 其他手性膦/吡啶配体

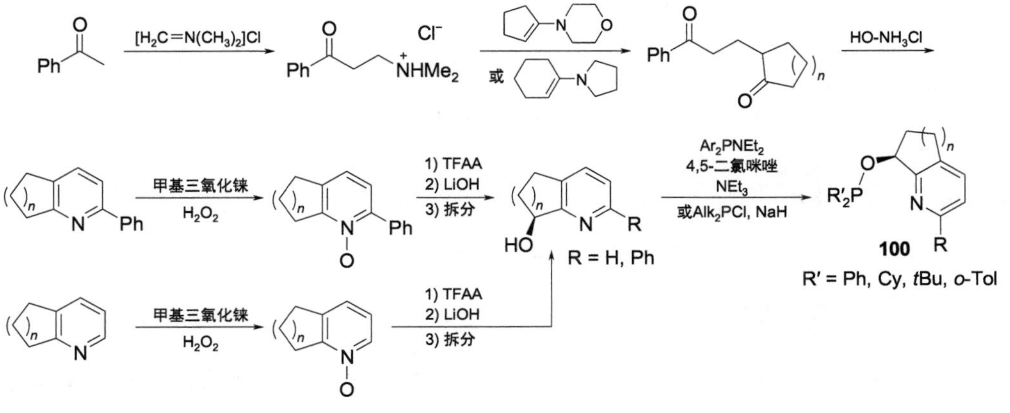

图 4-54 双环手性膦/吡啶配体 **100** 的合成

Buono 等[98]利用 (S)-(+)-2-(苯胺甲基)吡咯烷、三(二甲氨基)膦、2-羟基吡啶、2-羟甲基吡啶、8-羟基喹啉间的交换反应一步合成了配体 **102**、**103**。如使用其他手性二胺、氨基醇、二醇还能合成配体 **104~108**。

### 4.1.1.5 膦/胺、膦/亚胺配体

氨基膦配体相对较少，可能与氨基上氮原子的配位能力较弱有关。已经报道的结构如图 4-55 所示。

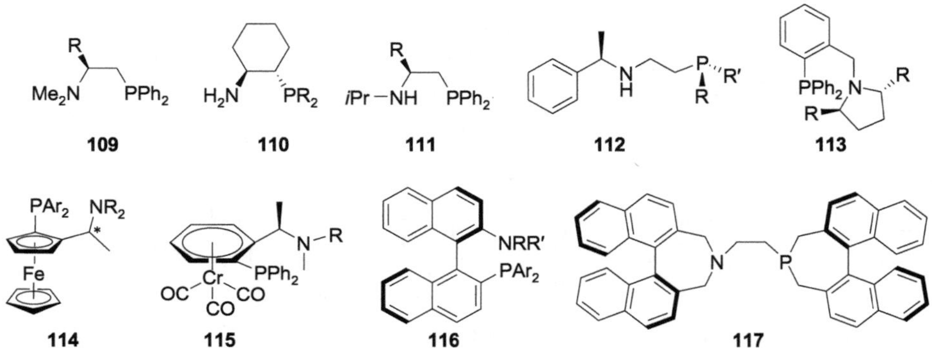

图 4-55 代表性的手性氨基膦配体

Kumada 等[99]从手性氨基酸合成了 β-氨基膦配体 **109**（图 4-56）。Yudin 等[100]报道了一种利用仲膦在酸性条件下亲核进攻氮杂环丙烷的开环反应构建 β-氨基膦配体的方法，手性氨基膦 **110** 可通过酒石酸拆分获得。

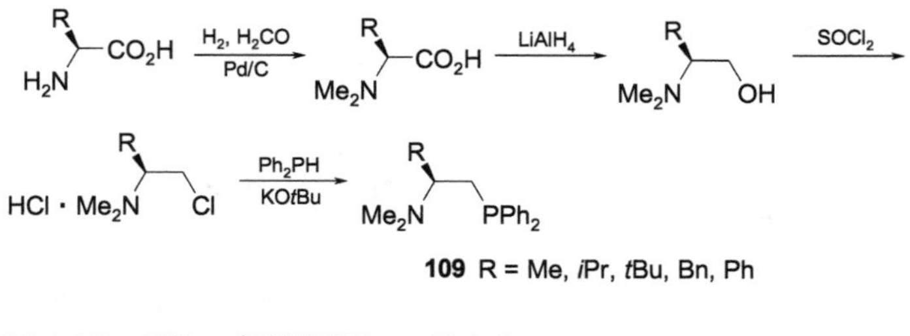

图 4-56 手性 β-氨基膦配体 **109** 的合成

Hilmersson 等[101]利用二苯基膦钾进攻手性环状磺酰胺合成了 β-氨基膦配体 **111**（图 4-57）。

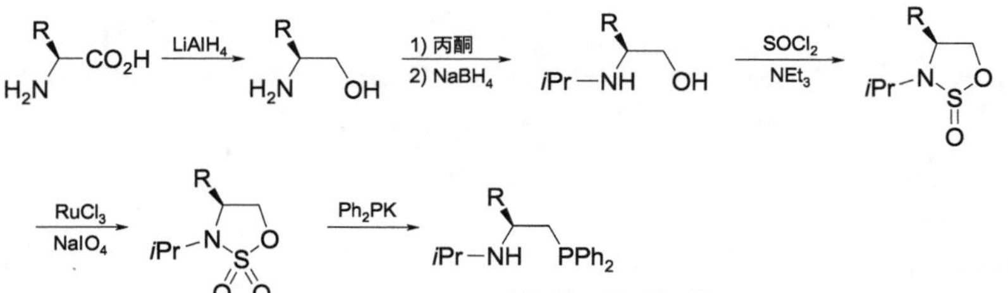


**111** R = *i*Pr, Bn, Ph

图 4-57　手性 $\beta$-氨基膦配体 **111** 的合成

Pietrusiewicz 和 Mortreux 等利用 $\alpha$-苯乙胺与手性乙烯基膦氧化物反应制得 $\beta$-氨基膦氧化物，还原后得相应配体 **112**[102]（图 4-58）。

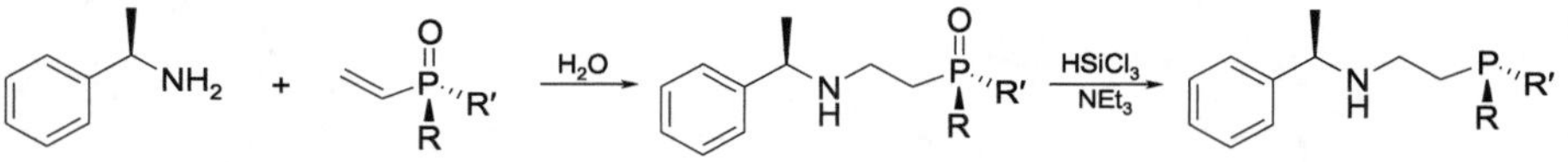


**112** R = *t*Bu, R′ = Ph
R = Ph, R′ = Me

图 4-58　P 手性 $\beta$-氨基膦配体 **112** 的合成

Guiry 等 [103] 通过邻溴苄胺与手性环状硫酸酯缩合生成四氢吡咯，溴锂交换后与二苯基氯化膦反应生成配体 **113**（图 4-59）。

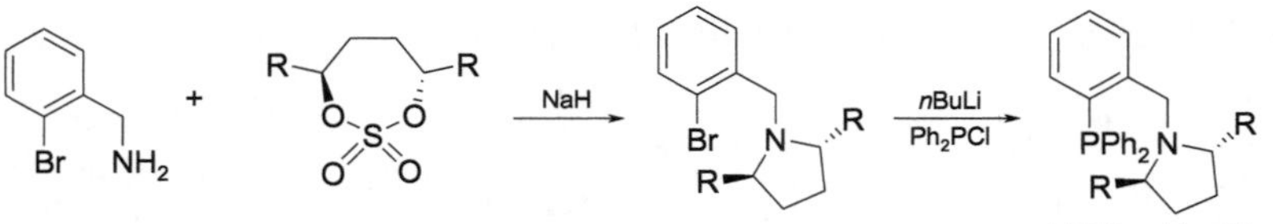


**113** R = Me, Et

图 4-59　手性氨基膦配体 **113** 的合成

在 1974 年，Kumada 等 [104] 从光学活性 Ugi 胺出发通过邻位锂化、亲核进攻二甲基或二苯基氯化膦合成了具有平面手性和碳中心手性的膦 / 胺配体 **114**。随后含不同取代基的二茂铁骨架膦 / 胺配体被报道。Uemura 和 Hayashi 等 [105] 报道了单芳烃三羰基铬络合物骨架的平面手性配体 **115**。Kocovsky 等 [106] 从轴手性 NOBIN 出发合成了配体 **116**（图 4-60）。

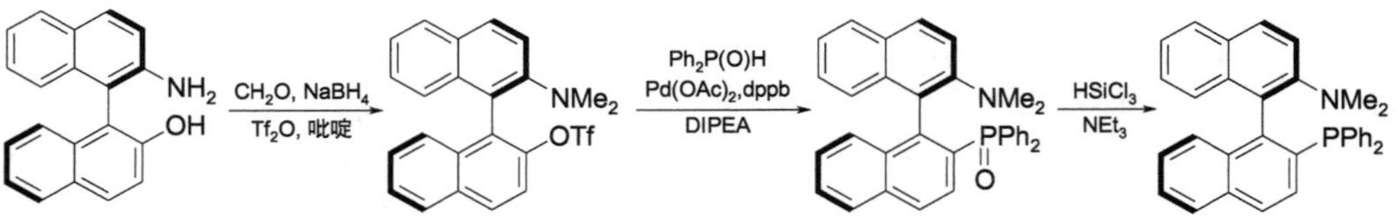


116

图 4-60 轴手性氨基膦配体 **116** 的合成

Moberg 等[107] 用氨乙基膦依次与两分子的 2,2′-二溴甲基-1,1′-联萘进行取代环化反应合成了配体及其非对映体 **117**（图 4-61）。

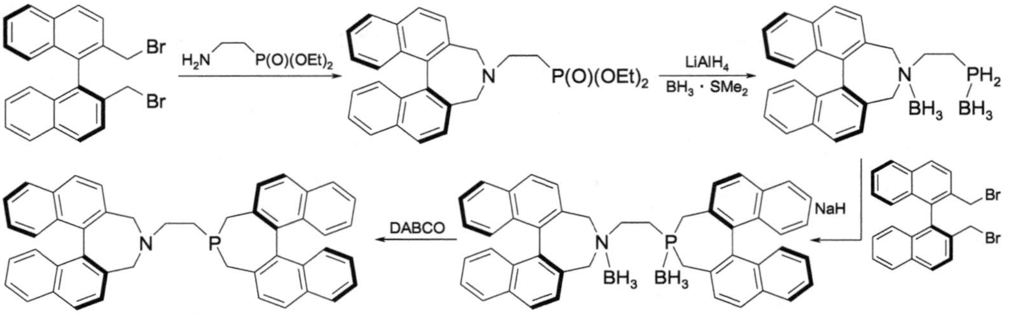


117

图 4-61 轴手性氨基膦配体 **117** 的合成

sp$^2$ 杂化的氮原子较 sp$^3$ 杂化的氮原子具有更高的电负性，因而与金属形成的键也更强。代表性的结构如图 4-62 所示。

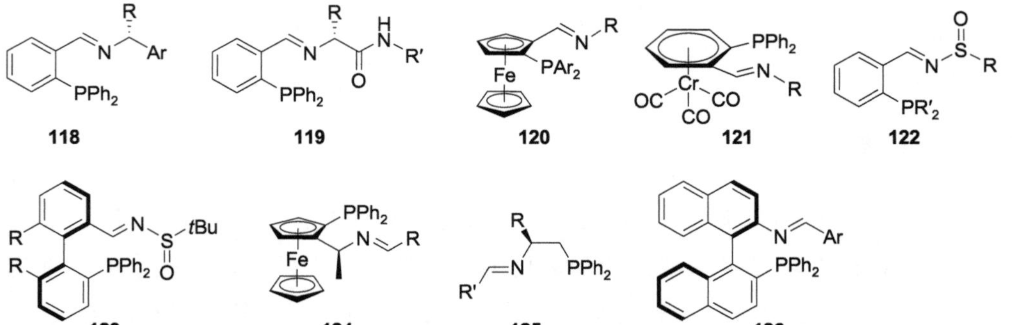

118 119 120 121 122 123 124 125 126

图 4-62 代表性的手性膦 / 亚胺配体

早在 1984 年，Brunner 等[108] 用 2-二苯基膦基苯甲醛与 $\alpha$-苯乙胺反应合成了配体 **118**。Hashimoto 课题组[109] 报道了类似结构的配体。

Hoveyda 等[110]基于肽合成了膦/亚胺配体 **119**。Chung 等[111]用二苯基膦基二茂铁甲醛与伯胺缩合合成了仅具有平面手性的配体 **120**。同时该课题组[112]还合成了平面手性的单芳烃三羰基铬配合物骨架的配体 **121**。Ellman 等[113]用叔丁基亚磺酰胺与 2-二芳基膦基甲醛缩合合成了膦/亚胺配体 **122**。秦勇等[114]将消旋的 2-二苯基膦基-2'-醛基联苯与叔丁基亚磺酰胺反应，分离非对映异构体得到轴手性膦/亚胺配体 **123**。利用氨基膦与醛或酮反应也可以方便地制备膦/亚胺配体。Hayashi 等[115]率先从配体 **114** 合成了膦/亚胺配体 **124**（图 4-63）。

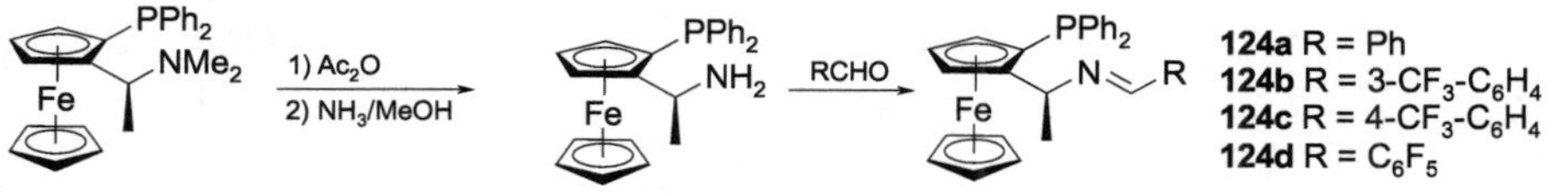

图 4-63 平面手性膦/亚胺配体 **124** 的合成

Achiwa 和 Morimoto 等[116]从天然手性氨基酸出发先合成 β-氨基膦，进而与醛或缩醛反应生成膦/亚胺配体 **125**（图 4-64）。Shi 等[117]合成了联萘骨架的膦/亚胺配体 **126**。

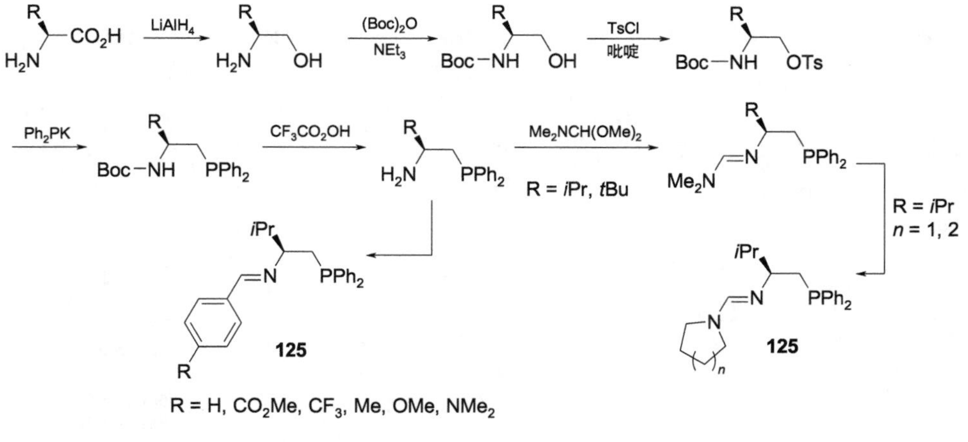

图 4-64 氨基酸衍生的手性膦/亚胺配体 **125** 的合成

## 4.1.2 膦/硫配体

与膦配体相比，硫配体具有较少的 σ-给体和较小的 σ-受体性质，膦/硫配体（图 4-65）中两个配位原子在电性上的显著差异有可能提高特定反应的区域和立体选择性。

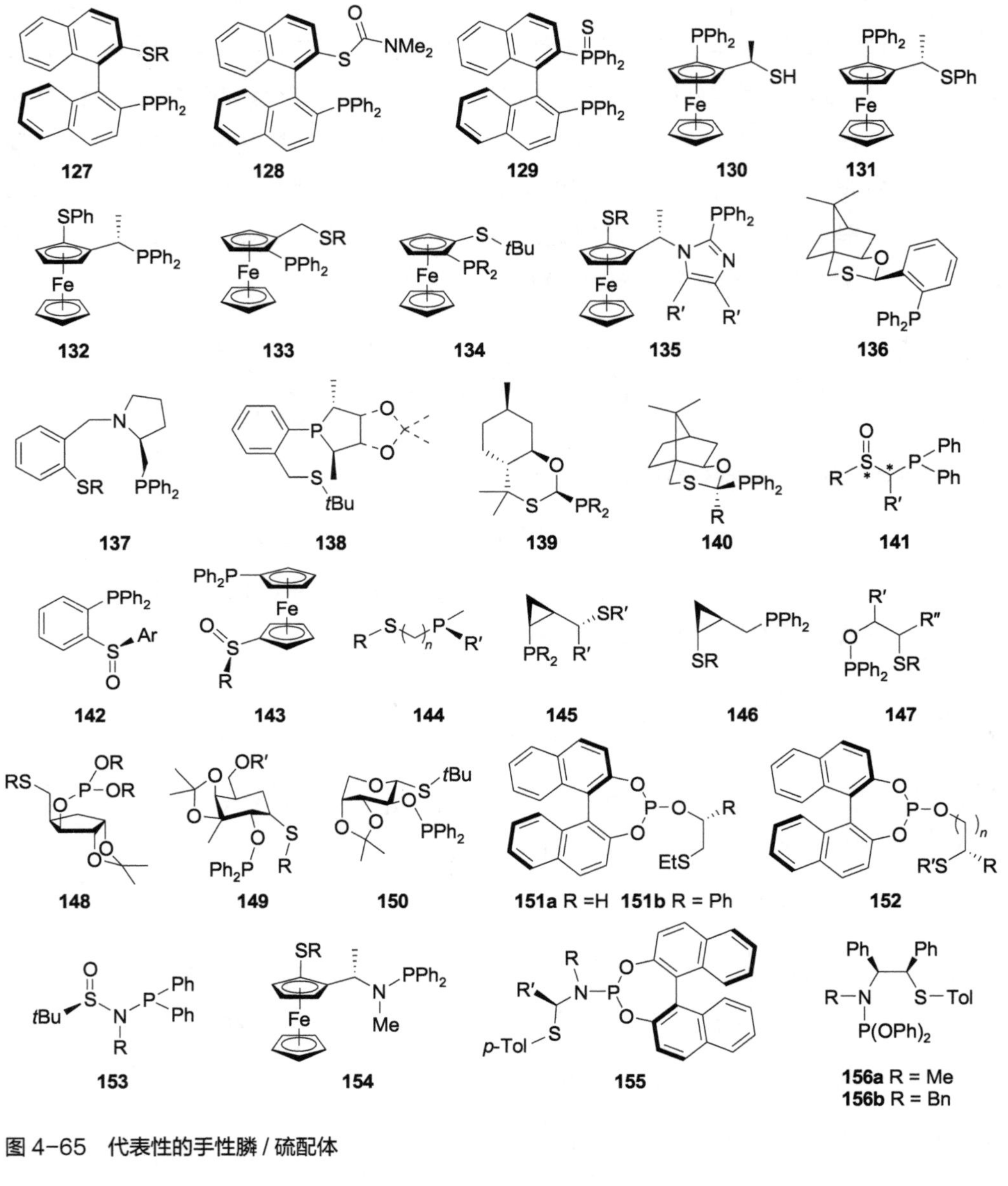

**图 4-65　代表性的手性膦/硫配体**

Gladiali 等[118]从 (R)-联萘酚出发分别经催化 C—P 键形成、Newmatm-Kwart 热重排引入磷、硫杂原子合成了配体 **127**（图 4-66）。利用相似反应路线，施敏等[119]合成了配体 **127** 和 **128**。Willis 等[120]直接硫化手性双膦化合物合成配体 **129**。

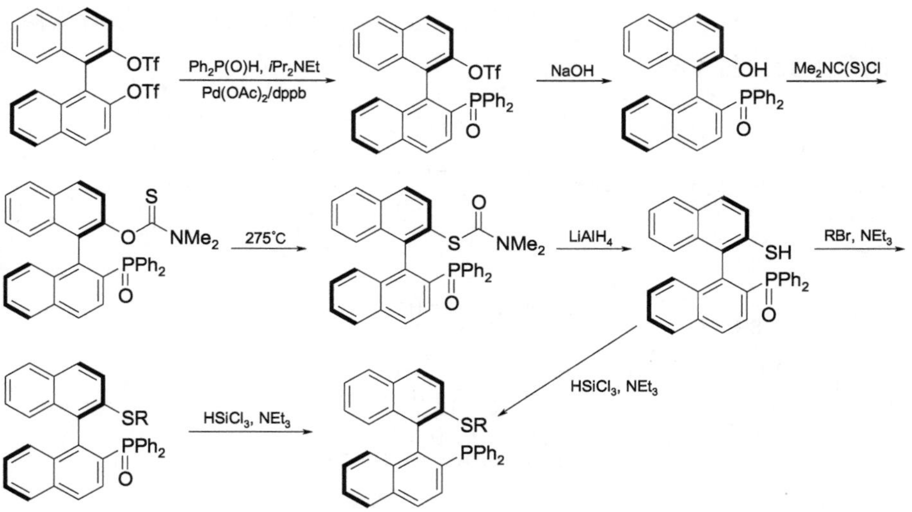

图 4-66 轴手性膦/硫配体 127 的合成

Togni 等[121] 从 (R)-Ugi 胺出发合成二茂铁骨架膦/硫配体 130（图 4-67）。侯雪龙、戴立信等[122] 从 (S)-Ugi 胺出发合成了磷和硫取代基位置发生互换的配体 131 和 132。Manoury 等[123] 从二苯基膦基二茂铁甲醛出发合成了仅有平面手性的膦/硫配体 133。

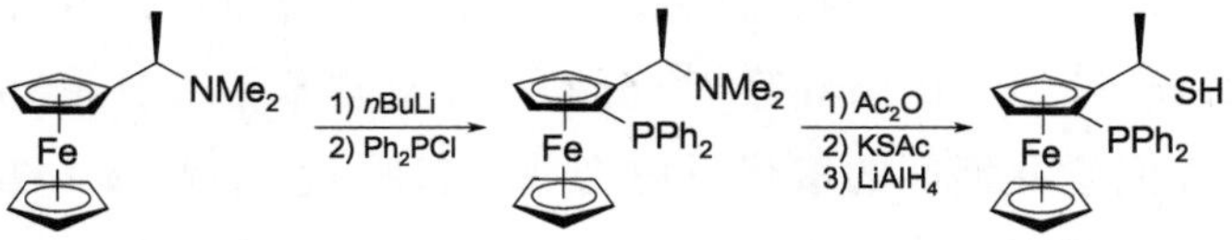

图 4-67 平面手性膦/硫醇配体 130 的合成

Carretero 等[124] 从叔丁基二茂铁亚砜出发，通过非对映选择性锂化、二取代氯化膦捕获锂中间体引入膦取代基，还原得到仅有平面手性的膦/硫配体 134（图 4-68）。

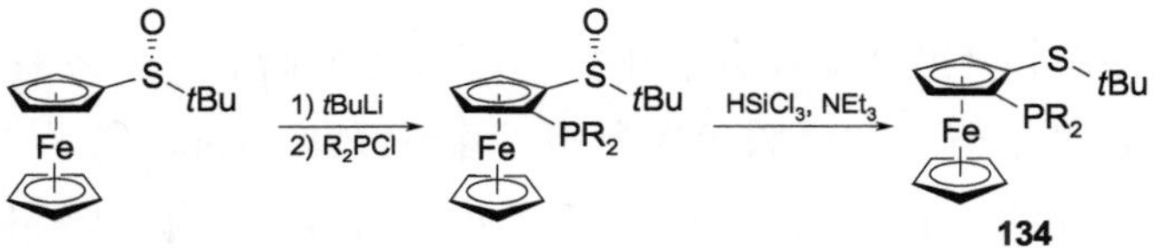

R = Ph, 4-F-$C_6H_4$, 4-$CF_3$-$C_6H_4$, 呋喃-2-基, Cy

图 4-68 平面手性膦/硫醚配体 134 的合成

陈新滋等[125]从 (S)-Ugi 胺通过三步反应合成了系列二茂铁骨架膦/硫配体 **135**（图 4-69）。

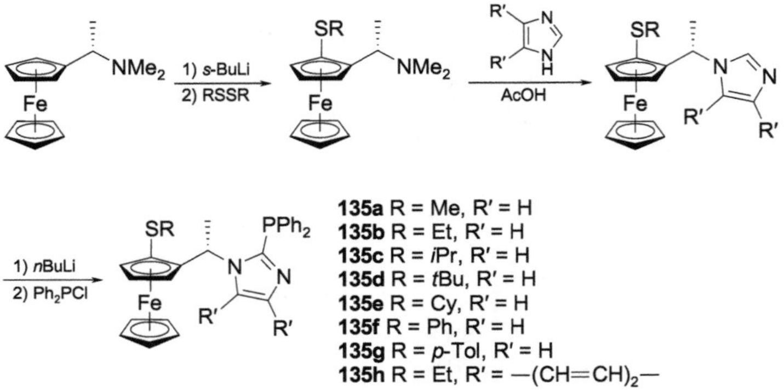

**图 4-69** 平面手性膦/硫醚配体 135 的合成

Nakano 等[126]用天然手性源化合物衍生的 $\gamma$-羟基硫醇与邻二苯基膦基甲醛反应合成了膦/硫配体 **136**。Hiroi 等[127]从邻溴苯甲酸和邻巯基苯甲酸出发合成了含吡咯烷骨架的膦/硫配体 **137**。Rajanbabu 等[128]从邻溴苄溴出发依次引入硫、磷杂原子合成了配体 **138**。Verdaguer 等[129]从长叶薄荷酮及 D-樟脑磺酸出发分别合成了配体 **139** 和 **140**。Verdaguer 等[130]通过对手性亚砜 $\alpha$-位 C—H 直接锂化、二苯基氯化膦捕获锂中间体形成配体 **141**。Hiroi 等[131]和 Toru 等[132]分别从邻氟碘苯和二茂铁出发利用芳基锂与手性亚磺酸酯间的取代反应合成了叔膦/亚砜配体 **142** 和 **143**。Imamoto 等[133]合成了仅有磷手性中心的叔膦/硫醚配体 **144**。Molander 等[134]通过手性环丙烷醇中碘、羟基的转化合成了环丙烷骨架的叔膦/硫醚配体 **145** 和 **146**，二者的区别是磷和硫取代基位置发生了互换。Evans 等[135]分别从手性 $\alpha$-溴代酰亚胺、环状亚磺酸酯、环己烯氧化物出发合成相应的 $\beta$-羟基硫醚，引入二苯基膦后形成手性膦/硫配体 **147**（图 4-70）。

Ruiz 等[136]从呋喃木糖出发先将分子中的伯醇转化成硫醚，仲羟基与苯酚、联苯酚、联萘酚衍生的氯膦反应制得了配体 **148**。Khiar 等[137]分别从 D-半乳糖五乙酸酯和 D-阿拉伯糖四乙酸酯出发合成了配体 **149** 和 **150**。Pregosin 和 Albinati 等[138]在碱性条件下用 $\beta$-羟基硫醚与联萘酚衍生的氯膦进行反应合成了配体 **151**。Takemoto 等 [139] 通过相似合成路线制备了配体 **152**。Verdaguer 等 [140] 直接从 (R)-叔丁基亚磺酰胺出发合成了配体 **153**。陈新滋等 [141] 从 (S)-Ugi 胺合成了二茂铁骨架配体 **154**。Crevisy 和 Mauduit 等 [142] 将商品化的 β-氨基醇转化为 β-氨基硫醚后与三氯化磷、手性联萘酚反应合成了配体 **155**。肖文精等 [143] 合成了类似结构的配体 **156**。

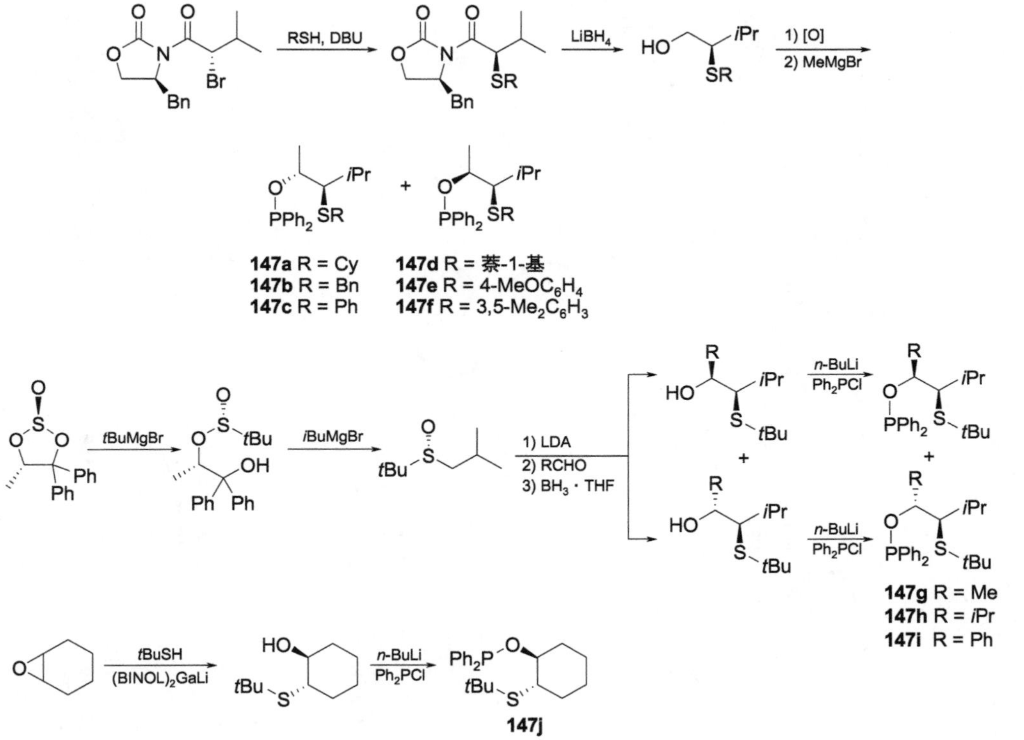

图 4-70　手性膦/硫醚配体 147 的合成

## 4.1.3　膦/氧配体

磷原子具有较强的配位能力，氧的配位稳定性较差，因此膦/氧配体属于半易变配体，部分手性膦/氧配体结构如图 4-71 所示。

Kumada 等 [144] 从 (R)-Ugi 胺出发，首先通过非对映选择性锂化、亲核进攻二苯基氯化膦引入二苯膦基，再经醋酸酐酯化、水解或醇解合成二茂铁骨架的膦/氧配体 **157**（图 4-72）。

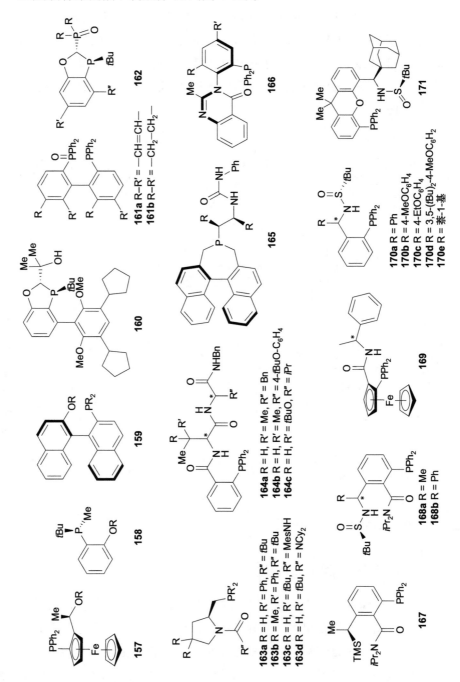

图 4-71　代表性的手性膦/氧配体

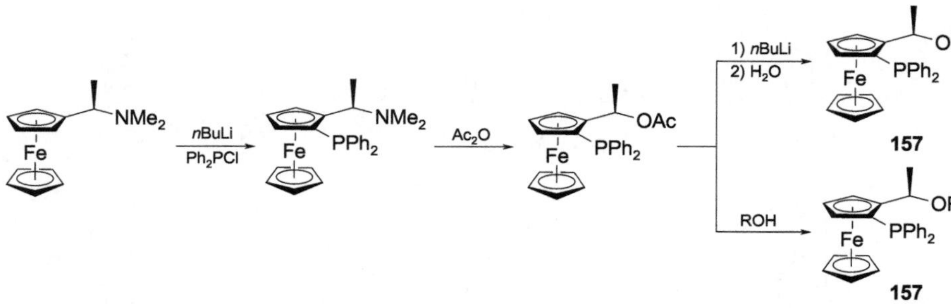

图 4-72 平面手性膦/氧配体 157 的合成

Imamoto 等[145]从叔丁基二氯化膦出发依次引入芳基和甲基合成了含有磷中心手性膦/氧配体 158。Morgans 等[146]、Hayashi 等[147]、Ito 等[148]从手性联萘酚出发采用同一反应路线合成了联萘骨架的膦/酚配体 159（图 4-73）。此配体中羟基烷基化得相应膦/醚配体。

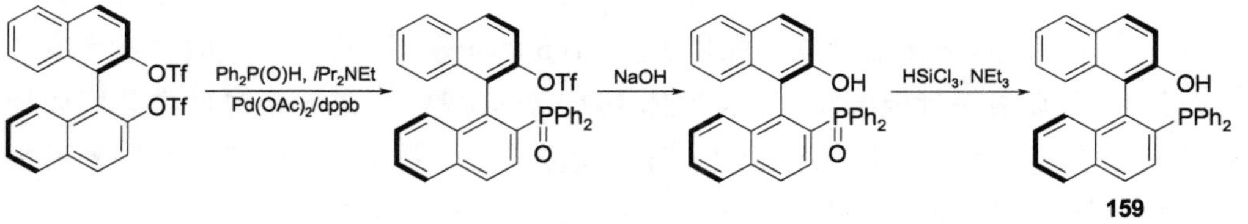

图 4-73 轴手性膦/酚配体 159 的合成

汤文军等[149]经多步反应合成了含有磷手性中心的膦/醇配体 160（图 4-74）。

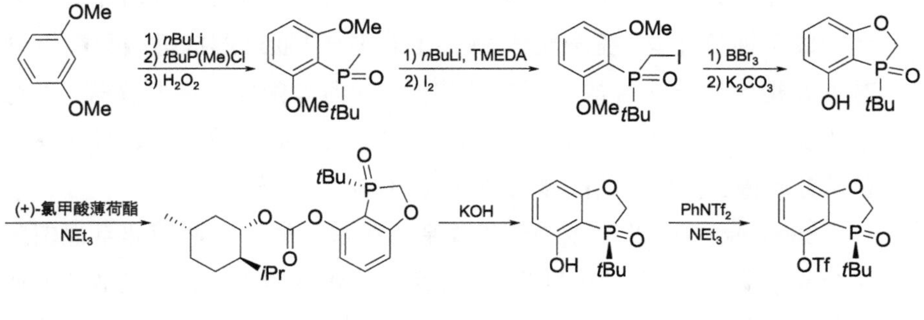

图 4-74

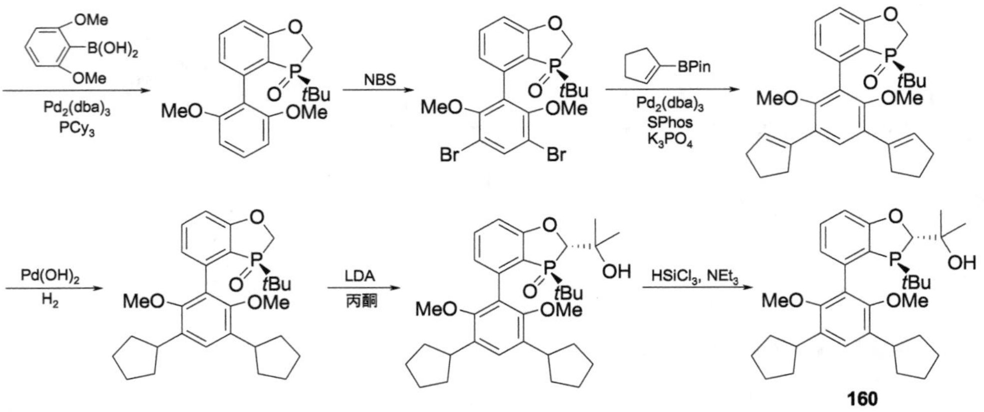

图 4-74 手性膦/醇配体 160 的合成

Gladiali 等[150]以消旋联萘酚为原料，在合成相应消旋 BINAP 的单氧化物后经拆分得到配体 161。Grushin[151]利用手性双膦配体如 BINAP 的选择性氧化制得目标产物 161a。汤文军等[152]通过三氯硅烷选择性还原 1,3-二膦氧化物合成了配体 162。Tomioka 等[153]从 L-脯氨酸出发合成了含吡咯烷骨架的膦/酰胺配体 163。Pfaltz 等[154]进一步对此类型配体结构进行扩展，以各种酰氯、异腈酸酯、二烷基氨基甲酰氯与 L-脯氨酸衍生的氨基膦化合物反应合成了类似结构配体。Hoveyda 等[155]利用 2-二苯基膦基苯甲酸与氨基酸形成的二肽反应合成了膦/三酰胺配体 164。Reek 等[156]以手性氨基膦与苯基异腈酸酯反应合成了膦/脲配体 165。Dai 和 Virgil 等[157]通过 2-N-乙酰氨基苯甲酸与 2-氨基苯基膦缩合形成喹唑啉酮骨架，再对其拆分得轴手性配体 166。Clayden 等[158]利用 (−)-金雀花碱/仲丁基锂定向锂化乙基 α-C-H 并引入三甲基硅，重结晶分离非对映异构体合成了轴手性膦/酰胺配体 167。Dai 等[159]通过形成樟脑酸酯拆分非对映异构体的方法合成了相应结构的膦/酰胺配体。徐利文等[160]通过格氏试剂对叔丁基亚磺酰亚胺的不对称亲核反应形成非对映异构体，柱色谱分离得到光学纯的配体 Xing-Phos 168（图 4-75）。

Štěpnička 等[161]以 2-二苯基膦基二茂铁甲酸与手性或非手性伯胺反应合成了配体 169。张俊良等[162]使用有机锂或镁进攻叔丁基亚磺酰亚胺高非对映选择性地合成了含有碳手性中心的配体 170 和 171（图 4-76）。

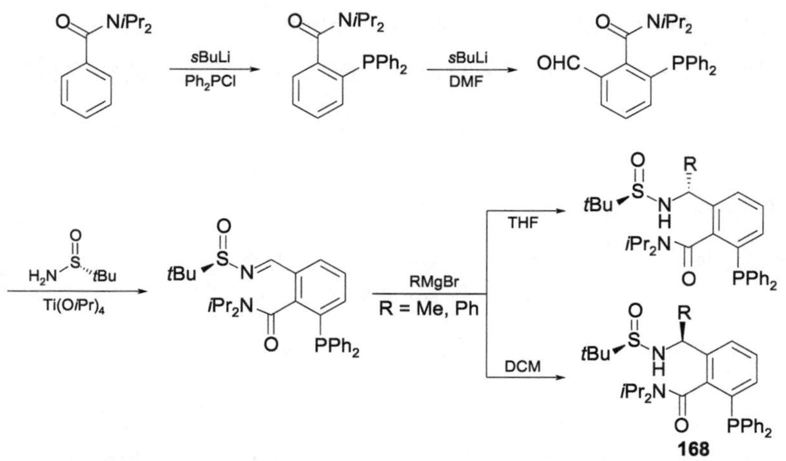

图 4-75 手性膦/酰胺配体 168(Xing-Phos) 的合成

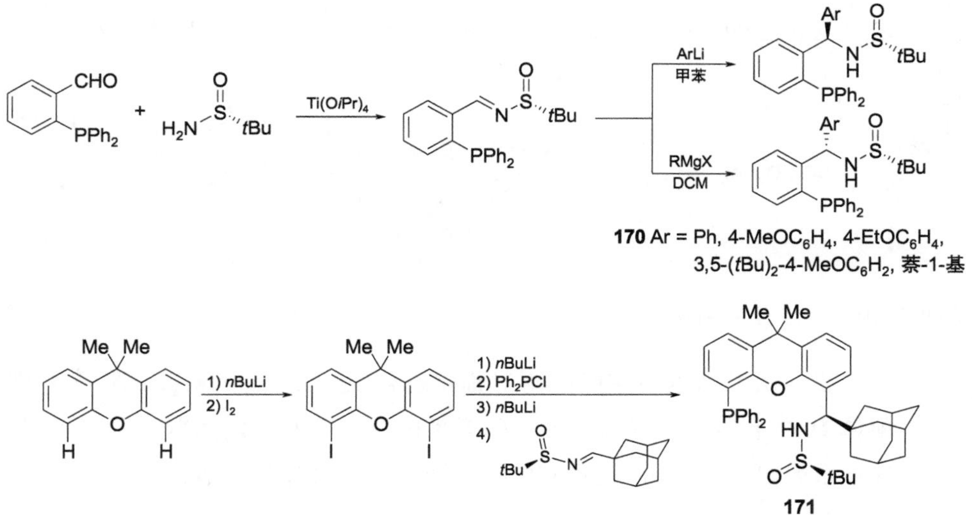

**170** Ar = Ph, 4-MeOC$_6$H$_4$, 4-EtOC$_6$H$_4$, 3,5-($t$Bu)$_2$-4-MeOC$_6$H$_2$, 萘-1-基

图 4-76 手性膦/亚磺酰胺配体 170 和 171 的合成

# 4.2 手性膦–杂原子双齿配体的应用

## 4.2.1 烯丙基化反应

在钯催化的烯丙基醋酸酯的不对称烷基化反应中，PHOX 类型配体

4 手性膦–杂原子双齿配体　207

大都表现出很好的活性和对映选择性。如 Pfaltz 等[1]发现，在 1,3-二苯基烯丙基醋酸酯与丙二酸二甲酯的反应中，配体 **1e** 具有优异的活性和选择性(图 4-77)。

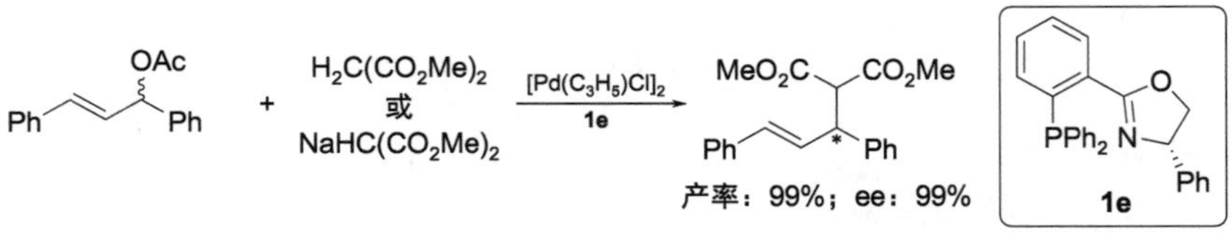

**图 4-77** 手性膦/噁唑啉配体 **1** 的应用

姜标等[86]发现，在上述 1,3-二苯基烯丙基醋酸酯与丙二酸二甲酯的反应中，同时具有平面手性和中心手性的膦/吡啶配体因取代基和构型不同对反应的活性和选择性有较大影响。如当使用配体 **90b** 时，反应在 5min 内可以高达 99% 的产率和 97% 的 ee 值得到相应的手性产物(图 4-78)。

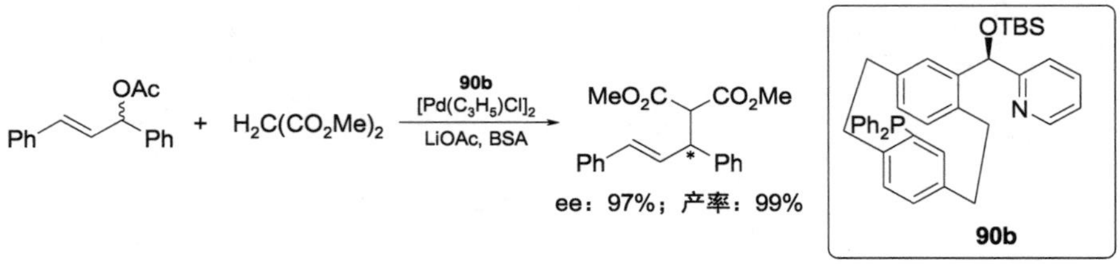

**图 4-78** 手性膦/吡啶配体 **90** 的应用

在环状烯丙基醋酸酯的不对称烷基化反应中，Helmchen 等[19]发现配体 **16** 中碳手性、平面手性、磷手性间存在手性匹配问题。研究表明，在该系列配体中，**16b** 在反应中有更好的活性和选择性(图 4-79)。

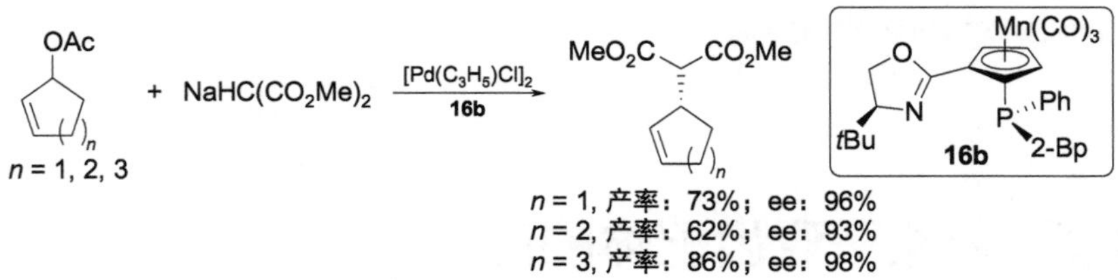

**图 4-79** 手性膦/噁唑啉配体 **16** 的应用

在环状烯丙基醋酸酯的不对称烷基化反应中，Gilbertson 等[29]发现配体 **25a** 具有更好的活性和选择性。研究发现，此类配体与其他膦/噁唑啉配体最明显的区别在于，产物的手性构型是由羟基脯氨酸的两个手性碳决定的，与噁唑啉环上的手性碳无关（图 4-80）。

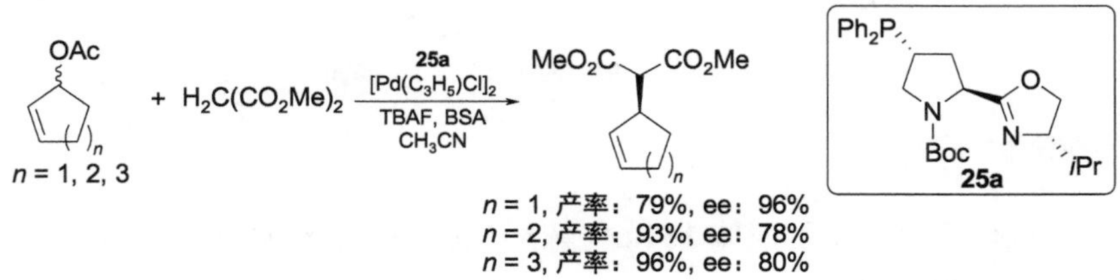

n = 1，产率：79%，ee：96%
n = 2，产率：93%，ee：78%
n = 3，产率：96%，ee：80%

图 4-80　手性膦/噁唑啉配体 25 的应用

侯雪龙、戴立信等[52]发现二茂铁骨架的亚膦酰胺酯/噁唑啉配体 **50d** 在单取代烯丙基醋酸酯的不对称烷基化反应中表现优异，能够以高活性和选择性得到手性产物（图 4-81）。

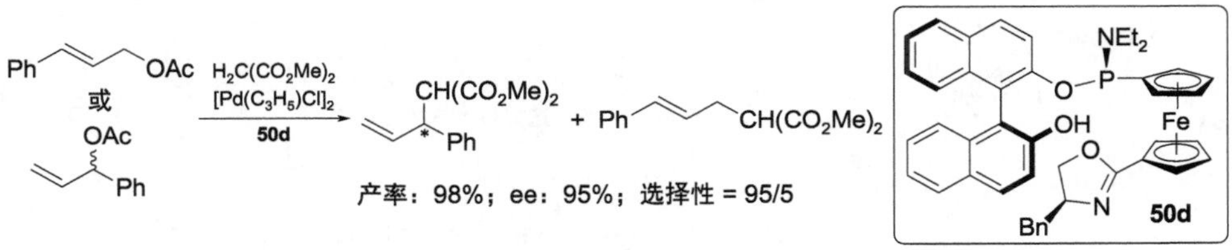

产率：98%；ee：95%；选择性 = 95/5

图 4-81　手性亚膦酰胺酯/噁唑啉配体 50 的应用

Pfaltz 等[41]发现亚磷酸酯/噁唑啉配体 38 与钯形成的络合物能够催化单取代烯丙基醋酸酯的不对称烷基化反应。当使用配体 **38b** 时，反应能够以 90% 的 ee 值得到相应的手性产物，支链/直链产物的比例为 76/24（图 4-82）。

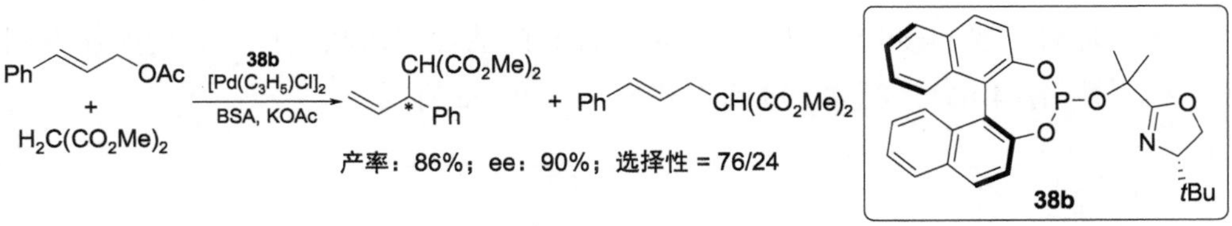

产率：86%；ee：90%；选择性 = 76/24

图 4-82　手性亚磷酸酯/噁唑啉配体 38 的应用

陈新滋等[125]发现同时含有平面手性和碳中心手性的配体 **135h** 在吲哚的不对称烯丙基烷基化反应中表现较好，各种取代的吲哚均能顺利生成目标产物，ee 值最高可达 96%（图 4-83）。

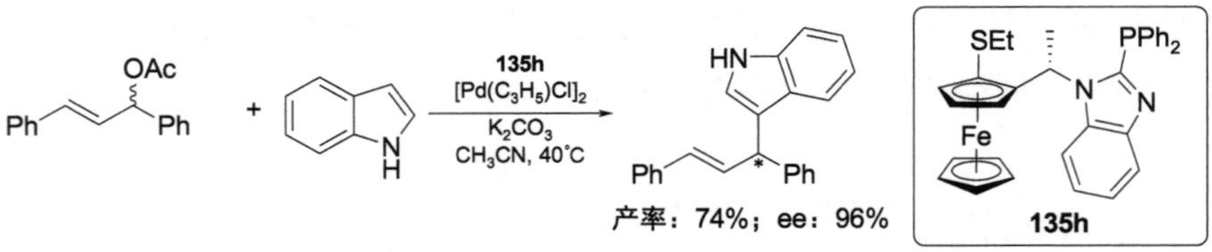

图 4-83　手性膦/硫醚配体 135 的应用

Takemoto 等[139]发现配体 **151a** 与铱形成的络合物催化剂在二苯亚甲胺酯的不对称烯丙基烷基化反应中具有优异的立体控制能力，目标产物 ee 值最高可达 97%。研究表明，该反应在使用相同的原料和催化剂条件下，只需改变其他反应参数（如溶剂和碱）就能选择性生成不同的非对映异构体化合物（图 4-84）。

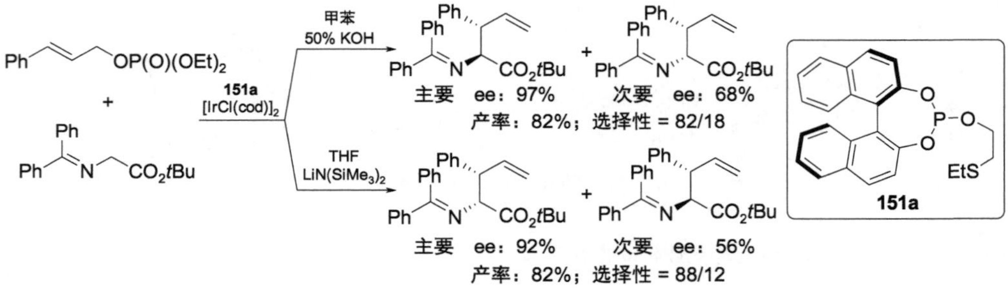

图 4-84　手性亚磷酸酯/硫醚配体 151 的应用

## 4.2.2　氢化反应

在铱催化的非环状 N-芳基亚胺的氢化反应中，周其林等[22]开发的螺环骨架膦/噁唑啉配体 **19** 展示了优异的活性和对映选择性。例如在下例中（图 4-85），配体 **19d** 可以获得 93% 的 ee 值。

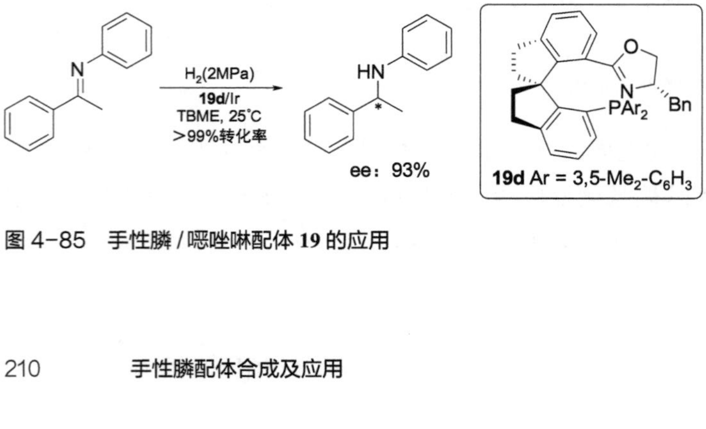

图 4-85　手性膦/噁唑啉配体 19 的应用

在 β-甲基肉桂酸甲酯的不对称氢化反应中，张绪穆等[35]开发的杂环骨架膦/噁唑啉配体 **32a** 与铱形成的络合物具有很好的对映选择性（图 4-86），还原产物的 ee 值高达 98%[34]。

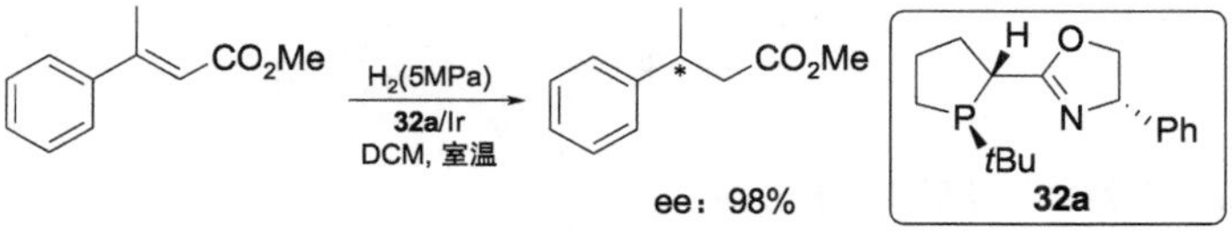

**图 4-86** 手性膦/噁唑啉配体 **32** 的应用

相对于 1,2-二芳基取代的烯烃，单芳基取代烯烃的不对称氢化反应更具挑战性。Pfaltz 等[38]发现亚次膦酸酯/噁唑啉配体 **35** 与铱形成的络合物催化剂在此类烯烃的不对称氢化反应中表现优异。如对于单芳基取代的烯烃，配体 **35c** 可以获得 96% 的 ee 值（图 4-87）。

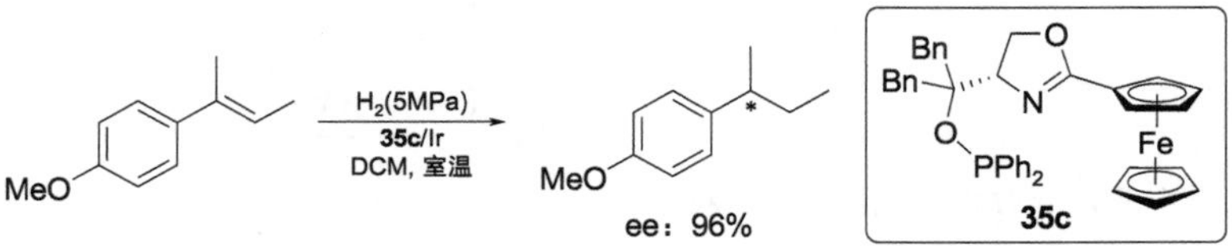

**图 4-87** 手性亚次膦酸酯/噁唑啉配体 **35** 的应用

Diéguez、Andersson 等[44]发现亚磷酸酯/噁唑啉配体 **41c** 在非官能化烯烃的不对称氢化反应中表现优异，能够以高活性和选择性得到手性还原产物。特别是构型难以控制的端烯烃的氢化反应，也能够以 99% 的对映选择性顺利得到相应的还原产物（图 4-88）。

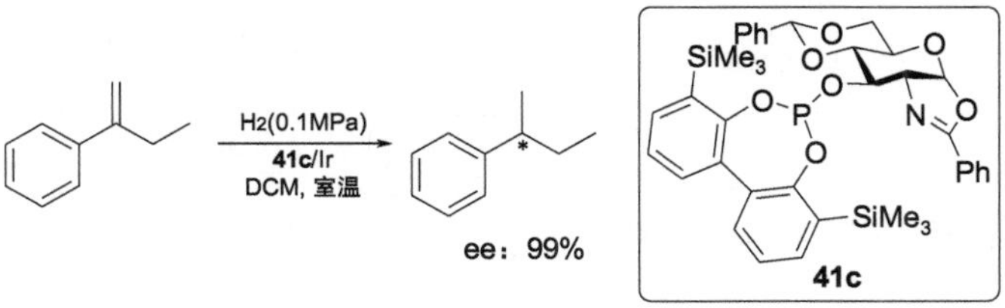

**图 4-88** 手性亚磷酸酯/噁唑啉配体 **41** 的应用

Andersson 等发现膦/噁唑配体 **56**[58] 及膦/噻唑配体 **57**[59] 的铱络合物在多种类型的烯烃不对称氢化反应中表现优异。研究表明，结构更为稳定的第二代配体 **57** 在顺式烯烃的氢化反应中有更好的活性和选择性

（图4-89）。

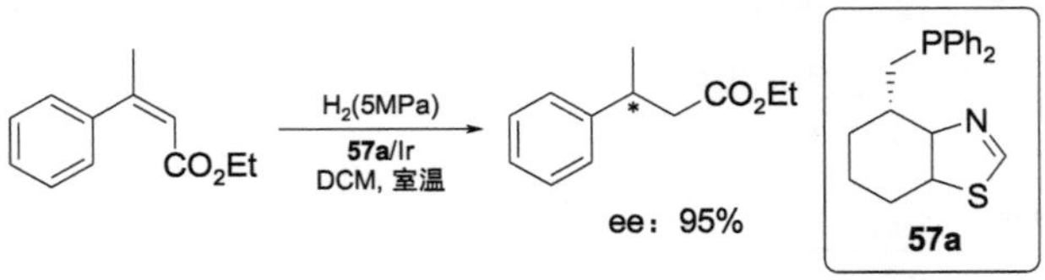

**图4-89 手性膦/噻唑配体57的应用**

Evans等发现手性膦/硫配体 **147**[135] 的铑络合物分别在烯烃不对称氢化及酮的不对称氢硅化反应中均表现出优异的选择性。如对于位阻较大的四取代烯烃的还原，产物的对映选择性高达95%（图4-90）。

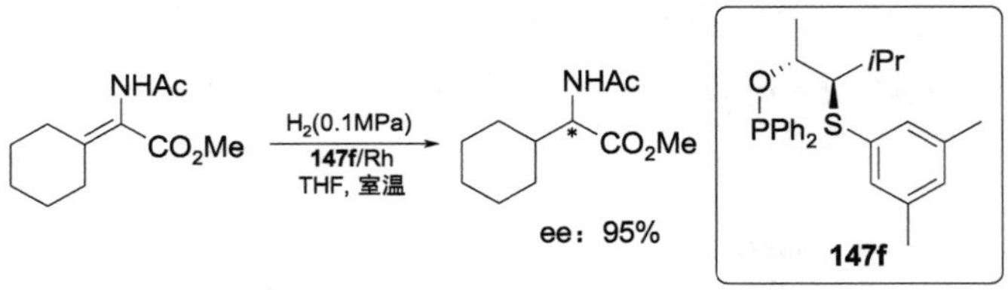

**图4-90 手性膦/硫配体147的应用**

Pfalts等发现手性膦/脲配体 **163d**[154] 的铱络合物在多种类型的烯烃不对称氢化反应中具有优异的活性和选择性。该类配体特别适用于 α,β-不饱和羧酸酯和酮的还原，所得产物的ee值高达99%，其不对称诱导能力甚至优于已报道最好的膦/氮、碳/氮配体（图4-91）。

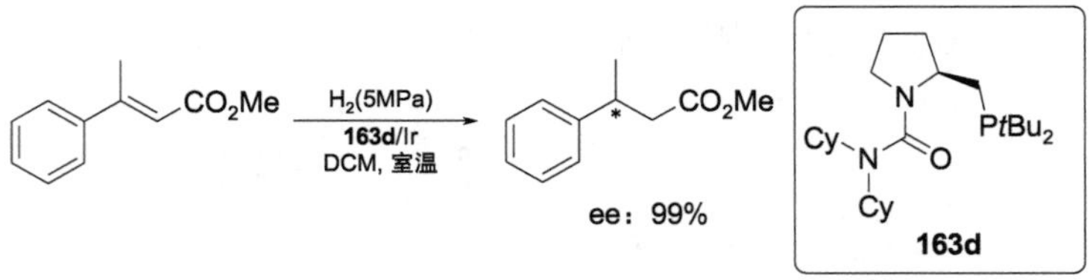

**图4-91 手性膦/脲配体163的应用**

## 4.2.3 共轭加成反应

有机金属试剂对 α,β-不饱和化合物的共轭加成反应是构建碳-碳键的常用方法之一。张绪穆等[84]发现手性NOBIN衍生的膦/吡啶配体 **88** 是该反应的有效催化剂。在非环状烯酮如查尔酮的加成反应中，可以取得

96% 的对映选择性，同时对于环己烯酮的不对称共轭加成产物也可以获得 92% 的 ee 值(图 4-92)。

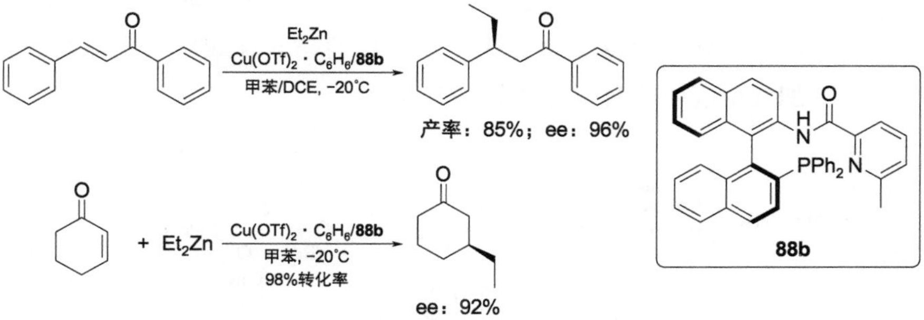

图 4-92　手性膦/吡啶配体 88 的应用

Hoveyda 等研究发现肽衍生的手性膦/亚胺配体 119[110] 与铜形成的络合物在有机锌对系列环状烯酮的不对称共轭加成反应中具有很高的对映选择性。比如，当使用二乙基锌时，可以 92%～98% 的产率及 98% 的对映选择性得到目标产物(图 4-93)。

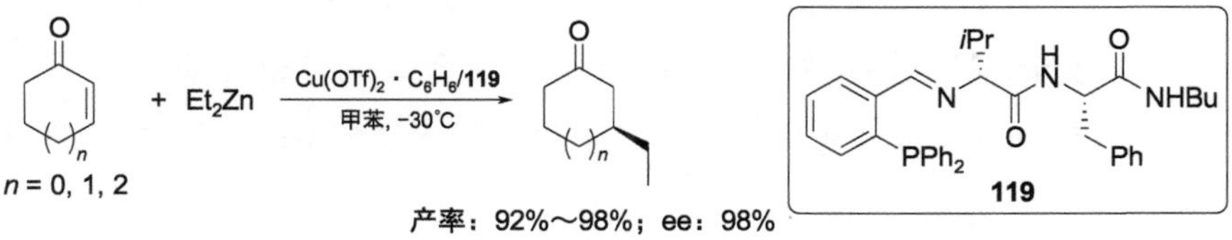

图 4-93　手性膦/亚胺配体 119 的应用

随后，Hoveyda 等[155]进一步研究了各种有机锌试剂对 $\alpha,\beta$-不饱和酰胺的不对称加成反应。研究发现，手性膦/三酰胺配体 164 与铜形成的络合物是极其有效的催化剂，可以高达 98% 的对映选择性得到相应的目标产物(图 4-94)。

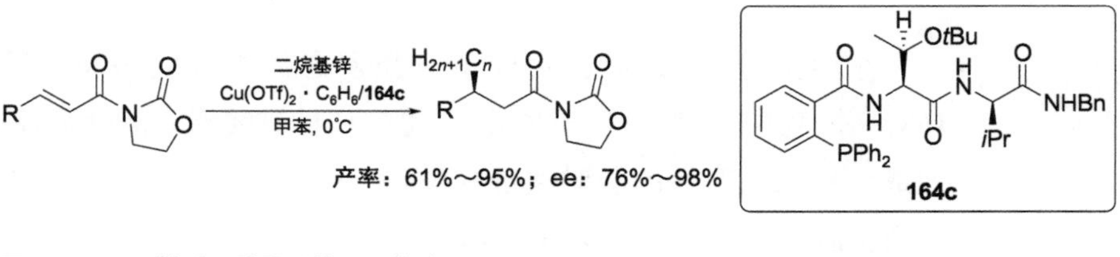

图 4-94　手性膦/酰胺配体 164 的应用

Tomioka 等[153]发现膦/酰胺配体 163 与铑配位后可催化苯硼酸与环烯酮的不对称共轭加成反应，形成 β-芳基取代的酮。在此络合物中，配体中磷原子与铑形成强配位键，而酰胺基中的氧原子对铑具有弱配位作用。配体的半易变特性决定了催化剂在苯硼酸与环戊烯酮的不对称共轭加成反应中的有效性（图 4-95）。

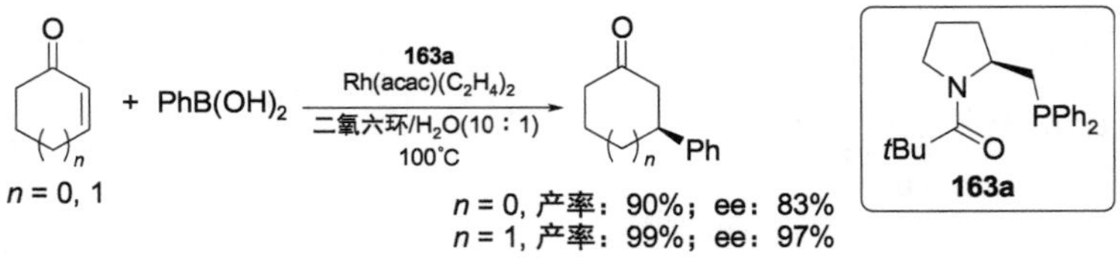

图 4-95 手性膦/酰胺配体 163 的应用

## 4.2.4 环加成反应

不对称催化环加成反应是合成手性杂环骨架化合物最高效和直接的方法之一。Carreira 等[83]基于手性辅基策略合成了 QUINAP 相似结构的配体 87。该配体在不同金属催化的反应中均表现出与 QUINAP 相当或更好的对映选择性。如在银催化的 N-亚烷基甘氨酸酯与丙烯酸酯的 [3+2] 环加成反应，可以极高的产率及对映选择性得到目标产物（图 4-96）。

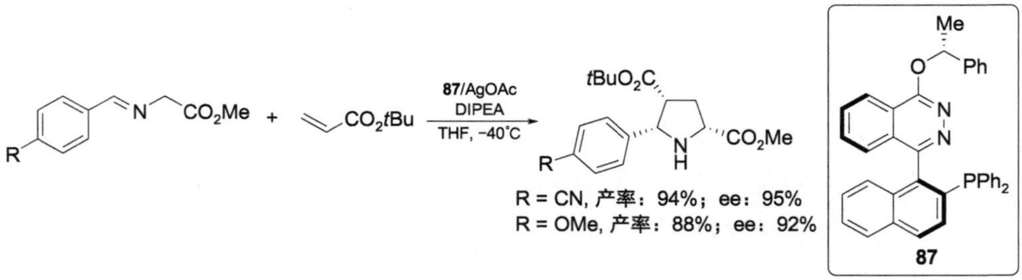

图 4-96 手性膦/酞嗪配体 87 的应用

徐利文等[160]设计、合成了一类多立体中心的酰胺轴手性膦配体 168，并将其与银结合用于催化 N-亚烷基甘氨酸酯与活性烯烃的 [3+2] 环加成反应，以高效的非对映选择性（> 99:1）以及对映选择性（高达 97%）得到相应的吡咯烷衍生物（图 4-97）。

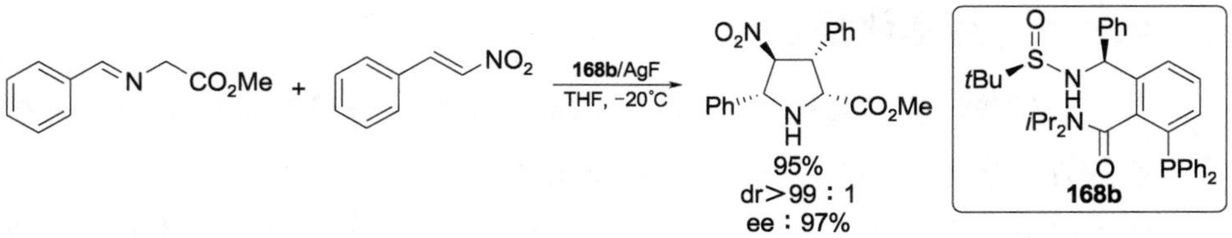

图 4-97 手性膦/酰胺配体 168 的应用

张俊良等[162]开发了一系列亚磺酰胺类手性膦配体 170，当与金络合时可用于催化炔基烯酮与硝酮的不对称环加成反应。值得注意的是，改变配体中碳手性中心的构型能够以良好的产率、优异的非对映选择性（＞95∶5）以及对映选择性（高达 94%）实现两种互为对映结构产物的合成（图 4-98）。

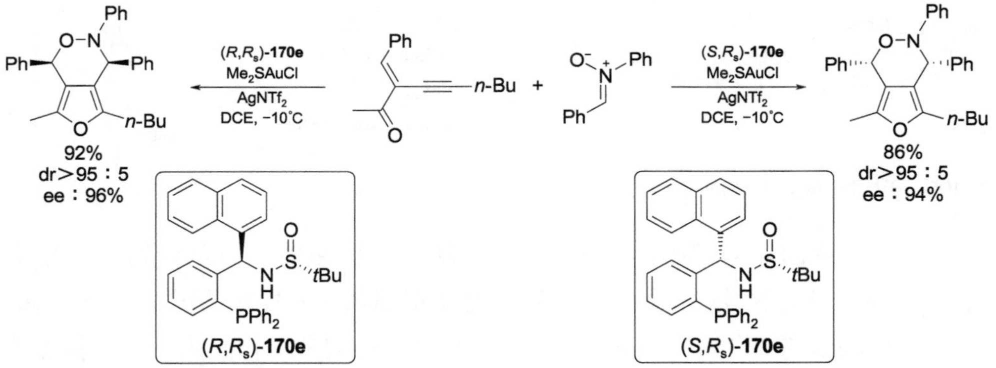

图 4-98 手性膦/亚磺酰胺配体 170 的应用

肖文精等[143]开发的配体 156 在钯催化的不对称脱羧 [4+2] 环加成反应中具有优异的立体控制能力。研究发现，配体中亚磷酰胺结构单元的存在可以显著提高催化反应活性，而 β-氨基硫醚骨架构型决定了反应的立体选择性。比如，当使用 156b 时，可以良好的产率、优异的非对映选择性以及对映选择性实现高度官能化四氢喹啉的合成（图 4-99）。

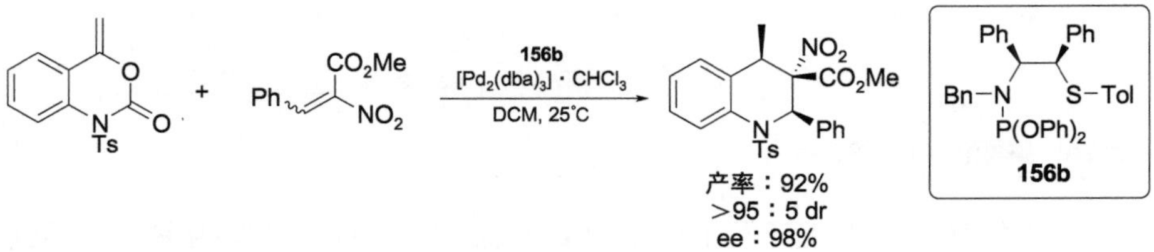

图 4-99 手性亚磷酰胺酯/硫醚配体 156 的应用

## 4.2.5 芳基化反应

卤代芳烃的芳基化如 Suzuki-Miyaura 偶联是合成联芳基化合物最有效的方法之一。不对称 Suzuki-Miyaura 偶联反应通常用来构筑邻位三取代的轴手性化合物。汤文军等[149]设计了含有叔醇作为氢键供体的手性膦配体 **160**，期望通过手性配体分别与两个偶联底物产生次级相互作用来调控反应的对映选择性。根据这一设想，作者成功发展了一个高效、普适的不对称 Suzuki-Miyaura 偶联反应，能够实现邻位四取代轴手性联芳基化合物的高效不对称合成（图 4-100）。

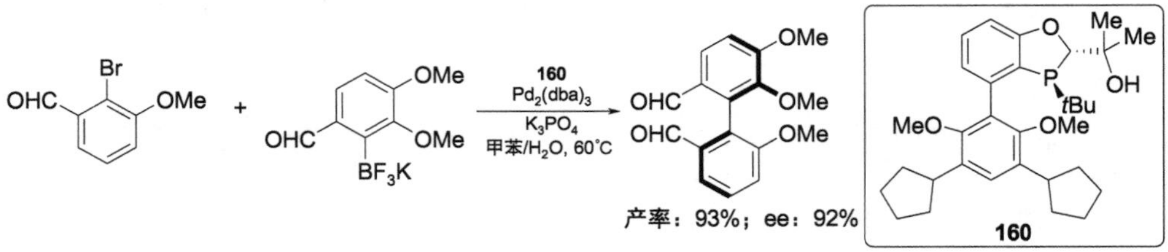

图 4-100　手性膦/醇配体 160 的应用

尽管不对称 Suzuki-Miyaura 偶联反应已在轴手性化合物合成中取得相当大的成就，但反应中不可避免用到预先官能化的试剂。Baudion、Cramer 等[163]通过钯催化的 C—H 对映选择性芳基化反应合成了相应的轴手性化合物。如在 1-甲基-4-苯基-1,2,3-三氮唑与 1-溴-2-甲氧基萘的偶联反应中，使用手性膦/膦氧配体 **161b** 与钯原位形成的络合物为催化剂，可以高达 93% 的产率和 90% 的对映选择性得到相应的偶联产物（图 4-101）。

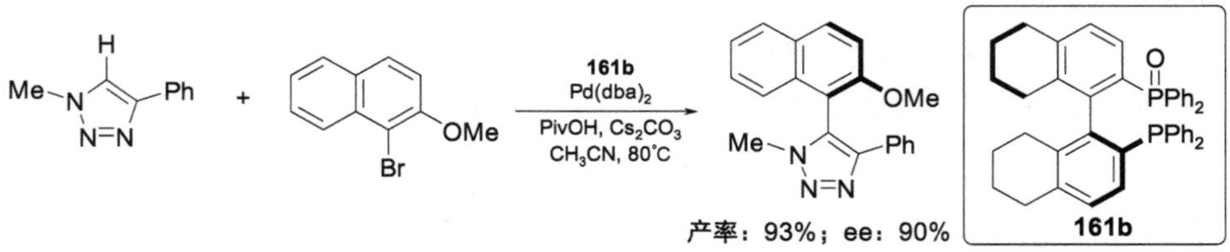

图 4-101　手性膦/膦氧配体 161 的应用

张俊良等[162a]还开发了一系列氧杂蒽骨架的叔丁基亚磺酰胺类手性膦配体 **171**，通过钯催化的对映选择性芳基化次磺酸阴离子合成手性亚砜。不仅实现了二芳基亚砜的高对映选择性合成，同时也解决了烷基次磺酸阴离子芳基化过程中对映选择性差的难题。该反应具有底物普适性广、对映选择性优异（ee 值高达 96%）等特点（图 4-102）。

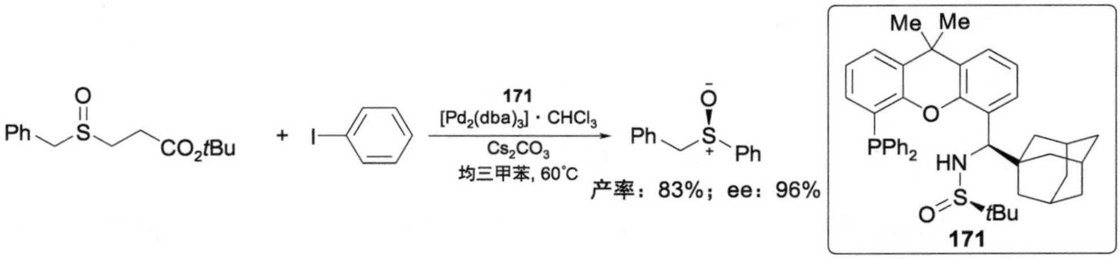

图 4-102　手性膦/亚磺酰胺配体 171 的应用

### 4.2.6　氢硅化反应

酮的不对称氢硅化反应可以用来合成具有重要价值的手性醇或者 $O$-保护产物。在 1974 年，Kumada 等[104]合成了具有平面手性和碳中心手性的膦胺配体 **114**，并将其用于铑催化的不对称氢硅化反应，能够以中等的对映选择性得到目标产物。随后，Hayashi 等[115]研究发现，对配体结构进一步修饰可以大幅提高反应效率和对映选择性。如当使用配体 **124** 时，可以高达 90% 的产率和 90% 的 ee 值得到相应的手性仲醇，反应在 10min 内完成（图 4-103）。

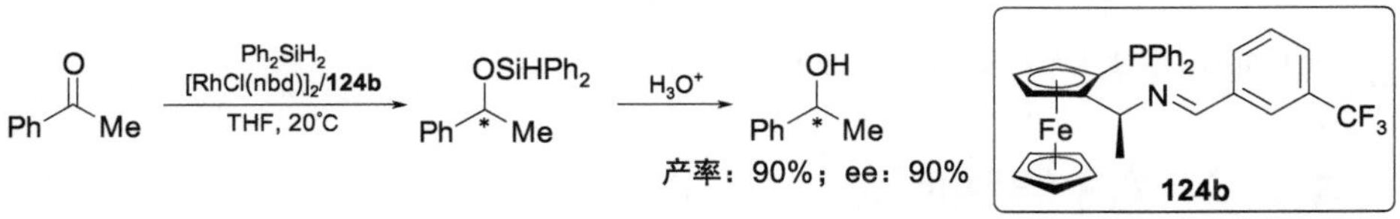

图 4-103　手性膦/亚胺配体 124 的应用

Fu 等[87]开发的一类平面手性膦/吡啶配体 **91** 与铑形成的络合物在酮的不对称氢硅化反应中表现优异。该催化体系不仅适用于芳基烷基酮的高对映选择性还原，同时对于挑战性的二烷基酮类底物也能解决硅氢化过程中手性识别困难的问题。当使用配体 **91** 时，苯乙酮和金刚烷基甲基酮分别以高达 98% 和 96% 的 ee 值被还原成相应的手性仲醇（图 4-104）。

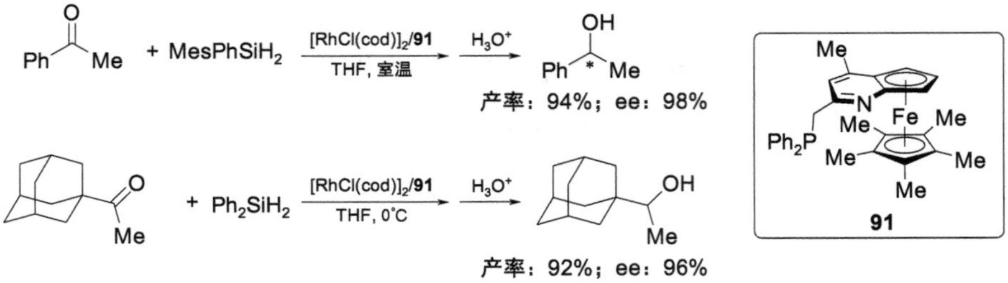

图 4-104　手性膦/吡啶配体 **91** 的应用

Evans 等[135]发现，手性膦/硫醚配体 **147** 与铑形成的络合物在前手性烯烃的氢化和酮的氢硅化反应中均具有优异的手性诱导能力。通过选择合适的铑金属前体及合适的氢硅烷作为还原剂，该催化体系具有底物普适性广、对映选择性优异（ee 值高达 99%）等特点。当使用配体 **147j** 时，苯乙酮和金刚烷基甲基酮分别以高达 95% 和 99% 的 ee 值被还原成相应的手性仲醇（图 4-105）。

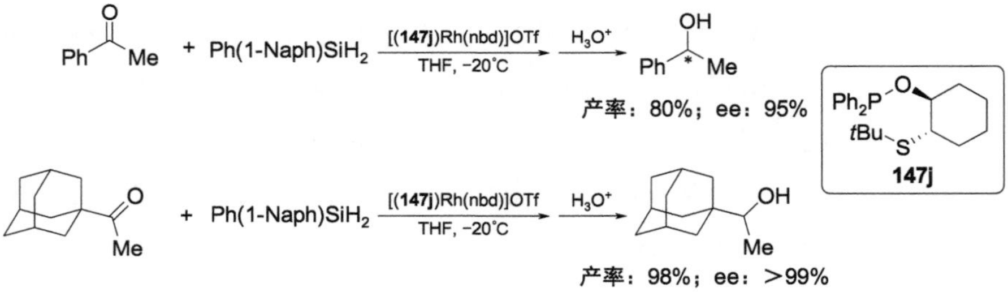

图 4-105　手性亚次膦酸酯/硫醚配体 **147** 的应用

## 4.2.7　A3 偶联反应

A3 偶联反应即醛、胺、炔的偶联反应，可以用于手性炔丙基胺的高效构建。Aponick 等[63]通过 π-堆积作用提高旋转能垒的方式合成了稳定的轴手性膦/氮配体 **65**。作者探索了其与铜形成的络合物在不对称 A3 偶联反应中的催化性能。该催化体系不仅适用于烷基醛的高对映选择性偶联反应，同时对于具有挑战性的芳基醛类底物也能解决偶联过程中手性识别困难的问题。当使用配体 **65** 时，环己醛、二苄胺、三甲基硅乙炔间的偶联反应以高达 97% 的对映选择性生成相应的手性胺产物（图 4-106）。

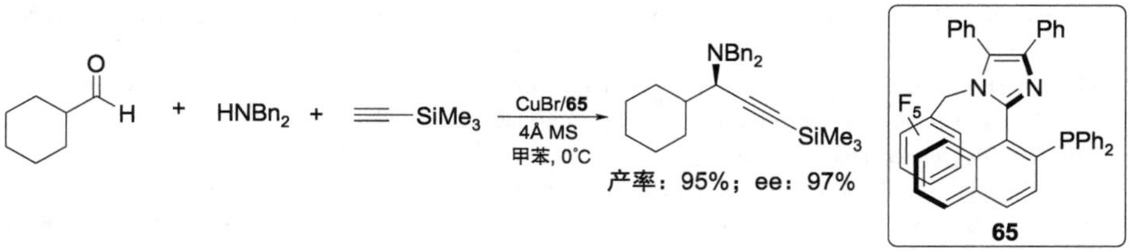

图 4-106　手性膦/咪唑配体 65 的应用

Guiry 等[57]为解决轴手性膦/氮配体拆分困难、拆分成本高等问题，在前期工作基础上，作者通过预装手性辅基、非对映异构体分步结晶策略合成了稳定的轴手性膦/咪唑啉配体 **55**。研究发现，配体分子中的五氟苯基与萘环存在平行 π-π 堆积作用。为研究其手性诱导性能，作者考察了其在不对称 A3 偶联反应中的催化性能。该催化体系对于烷基醛的偶联反应具有优异的活性和对映选择性（ee 值高达 98%），但对于具有挑战性的芳基醛类底物来说反应效果较差，产率和 ee 值均明显降低（图 4-107）。

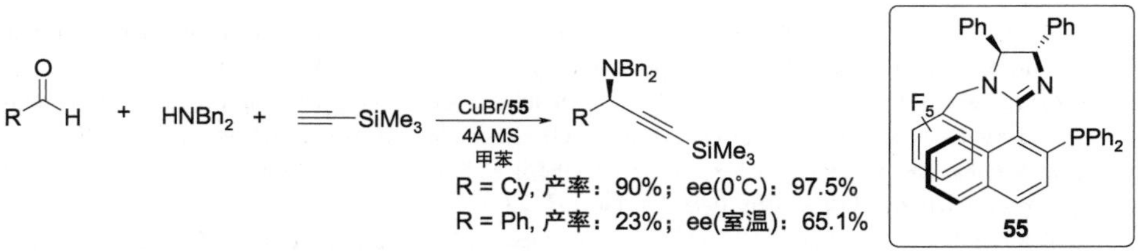

图 4-107　手性膦/咪唑啉配体 55 的应用

**参考文献**

[1] Von Matt P, Pfaltz A. Chiral Phosphinoaryldihydrooxazoles as ligands in asymmetric catalysis: Pd-catalyzed allylic substitution. Angewandte Chemie International Edition in Egnlish, 1993, 32(4): 566-568.
[2] Sprinz J, Helmchen G. Phosphinoaryl- and phosphinoalkyloxazolines as new chiral ligands for enantioselective catalysis: very high enantioselectivity in palladium catalyzed allylic substitutions. Tetrahedron Letters, 1993, 34(11): 1769-1772.
[3] Dawson G J, Frost C G, Williams J M J, et al. Asymmetric palladium catalyzed allylic substitution using phosphorus containing oxazoline ligands. Tetrahedron Letters, 1993, 34(19): 3149-3150.
[4] Krout M R, Mohr J T, Stoltz B M. Preparation of (S)-tert-butylPHOX. Organic Syntheses, 2009, 86: 181-193.
[5] Patti A, Lotz M, Knochel P. Synthesis of α,β-disubstituted ferrocenes via a ferrocenylepoxide intermediate. Preparation and catalytic activity of a new chiral ferrocenyloxazoline. Tetrahedron: Asymmetry, 2002, 12(24): 3375-3380.
[6] Froander A, Lutsenko S, Privalov T, et al. Conformational preferences and enantiodiscrimination of

phosphino-4-(1-hydroxyalkyl)oxazoline-metal-olefin complexes resulting from an OH-metal hydrogen bond. The Journal of Organic Chemistry, 2005, 70(24): 9882-9891.
[7] Stohler R, Wahl F, Pfaltz A. Enantio- and diastereoselective [3+2] cycloadditions of azomethine ylides with Ag(Ⅰ)-phosphinooxazoline catalysts. Synthesis, 2005, (9): 1431-1436.
[8] Wiese B, Helmchen G. Chiral phosphinooxazolines with a bi- or tricyclic oxazoline moiety-applications in Pd-catalyzed allylic alkylations. Tetrahedron Letters, 1998, 39(32): 5727-5730.
[9] Gläer B, Kunz H. Enantioselective allylic substitution using a novel (phosphino-α-D-glucopyranooxazoline)palladium catalyst. Synlett, 1998, (1): 53-54.
[10] Tian F T, Yao D M, Zhang Y J, et al. Phosphine-oxazoline ligands with an axial-unfixed biphenyl backbone: the effects of the substituent at oxazoline ring and P phenyl ring on Pd-catalyzed asymmetric allylic alkylation. Tetrahedron, 2009, 65(46): 9609-9615.
[11] Zhang W, Yoneda Y, Kida T, et al. Novel chiral P,N-ferrocene ligands in palladium-catalyzed asymmetric allylic alkylations. Tetrahedron: Asymmetry, 1998, 9(19):3371-3380.
[12] Tietze L F, Lohmann J K. Synthesis of novel chiral thiophene-, benzothiophene- and benzofuran-oxazoline ligands and their use in the enantioselective Pd-catalyzed allylation. Synlett, 2002, (12): 2083-2085.
[13] Kilroy T G, Cozzi P G, End N, et al. The application of HETPHOX ligands to the asymmetric intermolecular Heck reaction. Synlett, 2004, (1): 106-110.
[14] (a) Réthoré C, Fourmigué M, Avarvari N. Tetrathiafulvalene based phosphino-oxazolines: a new family of redox active chiral ligands. Chemical Communications, 2004, (12): 1384-1385; (b) Réthoré C, Suisse I, Agbossou-Niedercorn F, et al. Chiral tetrathiafulvalene based phosphine- and thiomethyl-oxazoline ligands. Evaluation in palladium catalyzed asymmetric allylic alkylation. Tetrahedron, 2006, 62(51): 11942-11947.
[15] Imai Y, Zhang W, Kida T, et al. Diphenylphosphinooxazoline ligands with a chiral binaphthyl backbone for Pd-catalyzed allylic alkylation. Tetrahedron Letters, 1998, 39(24): 4343-4346.
[16] Ogasawara M, Yoshida K, Kamei H, et al. Synthesis and application of novel chiral phosphino-oxazoline ligands with 1,1′-binaphthyl skeleton. Tetrahedron: Asymmetry, 1998, 9(10): 1779-1787.
[17] Nishibayashi Y, Uemura S. Asymmetric synthesis and highly diastereoselective *ortho*-lithiation of oxazolinylferrocenes. Synlett, 1995, (1): 79-81.
[18] Richards C J, Damalidis T, Hibbs D E, et al. Synthesis of 2-[2-(diphenylphosphino)ferrocenyl]oxazoline ligands. Synlett, 1995, (1): 74-76.
[19] Kudis S, Helmchen G. Enantioselective allylic substitution of cyclic substrates by catalysis with palladium complexes of P,N-chelate ligands with a cymantrene unit. Angewandte Chemie International Edition, 1998, 37(21): 3047-3050.
[20] Whelligan D K, Bolm C. Synthesis of pseudo-*geminal*-, pseudo-*ortho*-, and *ortho*-phosphinyl-oxazolinyl-[2.2]paracyclophanes for use as ligands in asymmetric catalysis. The Journal of Organic Chemistry, 2006, 71(12): 4609-4618.
[21] (a) Shintani R, Lo M M C, Fu G C. Synthesis and application of planar-chiral phosphaferrocene-oxazolines, a new class of P,N-ligands. Organic Letters, 2000, 2(23): 3695-3697; (b) Shintani R, Fu G C. Copper-catalyzed enantioselective conjugate addition of diethylzinc to acyclic enones in the presence of planar-chiral phosphaferrocene-oxazoline ligands. Organic Letters, 2002, 4(21): 3699-3702.
[22] Zhu S F, Xie J B, Zhang Y Z, et al. Well-defined chiral spiro iridium/phosphine-oxazoline cationic complexes for highly enantioselective hydrogenation of imines at ambient pressure. Journal of the American Chemical Society, 2006, 128(39): 12886-12891.
[23] Han Z B, Wang Z, Zhang X M, et al. Spiro[4,4]-1,6-nonadiene-based phosphine-oxazoline ligands for iridium-catalyzed enantioselective hydrogenation of ketimines. Angewandte Chemie International Edition, 2009, 48(29): 5345-5349.
[24] Trudeau S, Morken J P. StePHOX, a new family of optically active, tunable phosphine-oxazoline ligands: syntheses and applications. Tetrahedron, 2006, 62(49): 11470-11476.
[25] Wu X W, Yuan K, Sun W, et al. Novel planar chiral P,N-[2.2]paracyclophane ligands: synthesis and

application in palladium-catalyzed allylic alkylation. Tetrahedron: Asymmetry, 2003, 14(1): 107-112.
[26] Marchand A, Maxwell A, Mootoo B, et al. Oxazoline mediated routes to a unique amino-acid, 4-amino-13-carboxy[2.2]paracyclophane of planar chirality. Tetrahedron, 2000, 56(37): 7331-7338.
[27] Schrems M G, Pfaltz A. NeoPHOX-an easily accessible P,N-ligand for iridium-catalyzed asymmetric hydrogenation: preparation, scope and application in the synthesis of demethyl methoxycalamenene. Chemical Communications, 2009, (41): 6210-6212.
[28] Gilbertson S R, Chang C W T. A modular approach to ligands for asymmetric π-allyl palladium catalysed additions. Chemical Communications, 1997, (10): 975-976.
[29] Gilbertson S. R, Xie D. Proline-based P,N ligands in palladium-catalyzed asymmetric π-allyl additions. Angewandte Chemie International Edition, 1999, 38(18): 2750-2752.
[30] Porte A M, Reibenspies J, Burgess K. Design and optimization of new phosphine oxazoline ligands via high-throughput catalyst screening. Journal of the American Chemical Society, 1998, 120(36): 9180-9187.
[31] (a) Hou D R, Burgess K. JM-PHOS ligands: second-generation phosphine oxazolines for asymmetric catalysis. Organic Letters, 1999, 1(11): 1745-1747; (b) Hou D R, Reibenspies J H, Burgess K. New optically active phosphine oxazoline (JM-Phos) ligands: syntheses and applications in allylation reactions. The Journal of Organic Chemistry, 2001, 66(1): 206-215.
[32] Ezhova M B, Patrick B O, James B R, et al. New chiral N,P-oxazolines, and their Ir complexes in asymmetric hydrogenation of an imine. Journal of Molecular Catalysis A: Chemical, 2004, 224(1-2): 71-79.
[33] Gilbertson S R, Fu Z. Chiral P,N-ligands based on ketopinic acid in the asymmetric Heck reaction. Organic Letters, 2001, 3(2): 161-164.
[34] Gilbertson S R, Genov D G, Rheingold A L. Synthesis of new bicyclic P-N ligands and their application in asymmetric Pd-catalyzed π-allyl alkylation and Heck reaction. Organic Letters, 2000, 2(18): 2885-2888.
[35] Tang W J, Wang W M, Zhang X M. Phospholane-oxazoline ligands for Ir-catalyzed asymmetric hydrogenation. Angewandte Chemie International Edition, 2003, 42(8): 943-946.
[36] (a) Yonehara K, Hashizume T, Mori K, et al. Palladium-catalysed asymmetric allylic alkylation using new chiral phosphinite-nitrogen ligands derived from D-glucosamine. Chemical Communications, 1999, (5): 415-416; (b) Yonehara K, Hashizume T, Mori K, et al. Palladium-catalyzed asymmetric allylic substitution reactions using new chiral phosphinite-oxazoline ligands derived from D-glucosamine. The Journal of Organic Chemistry, 1999, 64(26): 9374-9380.
[37] Smidt S P, Menges F, Pfaltz A. SimplePHOX, a readily available chiral ligand system for Iridium-catalyzed asymmetric hydrogenation. Organic Letters, 2004, 6(12): 2023-2026.
[38] Blankenstein J, Pfaltz A. A new class of modular phosphinite-oxazoline ligands: Ir-catalyzed enantioselective hydrogenation of alkenes. Angewandte Chemie International Edition, 2001, 40(23): 4445-4447.
[39] Jones G, Richards C J. Simple phosphinite-oxazoline ligands for asymmetric catalysis. Tetrahedron Letters, 2001, 42(32): 5553-5555.
[40] Menges F, Pfaltz A. Threonine-derived phosphinite-oxazoline ligands for the Ir-catalyzedenantioselective hydrogenation. Advanced Synthesis & Catalysis, 2002, 344(1): 40-44.
[41] Prétôt R, Pfaltz A. New ligands for regio- and enantiocontrol in Pd-catalyzed allylic alkylations. Angewandte Chemie International Edition, 1998, 37(3): 323-325.
[42] Hilgraf R, Pfaltz A. Chiral bis(N-tosylamino)phosphine-and TADDOL-phosphite-oxazolines as ligands in asymmetric catalysis. Synlett, 1999, (11): 1814-1816.
[43] Pàmies O, Diéguez M, Claver C. New phosphite-oxazoline ligands for efficient Pd-catalyzed substitution reactions. Journal of the American Chemical Society, 2005, 127(11): 3646-3647.
[44] Diéguez M, Mazuela J, Pàmies O, et al. Chiral pyranoside phosphite-oxazolines: A new class of ligand for asymmetric catalytic hydrogenation of alkenes. Journal of the American Chemical Society, 2008, 130(23): 7208-7209.

[45] Xu G, Gilbertson S R. Facile synthesis of proline based phosphine_oxazoline ligands by formation of a P-N bond. Tetrahedron Letters, 2002, 43(15): 2811-2814.

[46] Blanc C, Hannedouche J, Agbossou-Niedercorn F. Two-fold amino acid-based chiral aminophosphine-oxazolines anduse in asymmetric allylic alkylation. Tetrahedron Letters, 2003, 44(34): 6469-6473.

[47] Cozzi P G, Zimmermann N, Hilgraf R, et al. Chiral Phosphinopyrrolyl-Oxazolines: A new class of easily prepared, modular P,N-ligands. Advanced Synthesis & Catalysis, 2001, 343(5): 450-454.

[48] Trifonova A, Diesen J S, Andersson P G. Asymmetric hydrogenation of imines and olefins using phosphine-oxazoline iridium complexes as catalysts. Chemistry-A European Journal, 2006, 12(8): 2318-2328.

[49] Bronger R P J, Guiry P J. Aminophosphine-oxazoline and phosphoramidite-oxazoline ligands and their application in asymmetric catalysis. Tetrahedron: Asymmetry, 2007, 18(9): 1094-1102.

[50] Hilgraf R, Pfaltz A. Chiral bis(N-sulfonylamino)phosphine- and TADDOL-phosphite-oxazoline ligands: synthesis and application in asymmetric catalysis. Advanced Synthesis & Catalysis, 2005, 347(1): 61-77.

[51] Gavrilov K N, Bondarev O G, Tsarev V N, et al. New oxazoline-containing phosphoramidite ligand for palladium-catalyzed asymmetric allylic sulfonylation, Russian Chemical Bulletin International Edition, 2003, 52(1): 122-125.

[52] (a) You S L, Zhu X Z, Luo Y M, et al. Highly regio- and enantioselective Pd-catalyzed allylic alkylation and amination of monosubstituted allylic acetates with novel ferrocene P,N-ligands, Journal of the American Chemical Society 2001, 123(30): 7471-7472; (b) Hou X L, Sun N. Construction of chiral quaternary carbon centers by Pd-catalyzed asymmetric allylic substitution with P,N-1,1'-ferrocene ligands. Organic Letters, 2004, 6(24): 4399-4401.

[53] Menges F, Neuburger M, Pfaltz A. Synthesis and application of chiral phosphino-imidazoline ligands: Ir-catalyzed enantioselective hydrogenation. Organic Letters, 2002, 4(26): 4713-4716.

[54] (a) Busacca C A, Grossbach D, So R C, et al. Probing electronic effects in the asymmetric heck reaction with the BIPIligands. Organic Letters, 2003, 5(4): 595-598; (b) Busacca C A, Grossbach D, Campbell S J, et al. Asymmetric hydrogenation of unsaturated ureas with the BIPI ligands. The Journal of Organic Chemistry, 2004, 69(16): 5187-5195.

[55] Guiuv E, Claver C, Benet-Buchholz J, et al. An efficient method for the synthesis of enantiopure phosphine-imidazoline ligands: application to the Ir-catalyzed hydrogenation of imines. Tetrahedron: Asymmetry, 2004, 15(21): 3365-3373.

[56] Menges F, Pfaltz A. (Solvias AG). European Patent EP1191030, 2005.

[57] Rokade B V, Guiry P J. Diastereofacial π-stacking as an approach to access an axially chiral P,N-ligand for asymmetric catalysis. ACS Catalysis, 2017, 7(4): 2334-2338.

[58] Kallstrom K, Hedberg C, Brandt P, et al. Rationally designed ligands for asymmetric iridium-catalyzed hydrogenation of olefins. Journal of the American Chemical Society, 2004, 126(44): 14308-14309.

[59] Hedberg C, Kallstrom K, Brandt P, et al. Asymmetric hydrogenation of trisubstituted olefins with iridium-phosphine thiazole complexes: a further investigation of the ligand structure. Journal of the American Chemical Society, 2006, 128(9): 2995-3001.

[60] Mazuela J, Paptchikhine A, Tolstoy P, et al. A new class of modular P,N-ligand library for asymmetric Pd-catalyzed allylic substitution reactions: a study of the key Pd-π-allyl intermediates. Chemistry-A European Journal, 2010, 16(2): 620-638.

[61] Kaukoranta P, Engman M, Hedberg C, et al. Iridium catalysts with chiral imidazole-phosphine ligands for asymmetric hydrogenation of vinyl fluorides and other olefins. Advanced Synthesis & Catalysis, 2008, 350(7-8): 1168-1176.

[62] Willms H, Frank W, Ganter C. Coordination chemistry and catalytic application of bidentate phosphaferrocene-pyrazole and -imidazole based P,N-ligands. Organometallics, 2009, 28(10): 3049-3058.

[63] Cardoso F S P, Abboud K A, Aponick A. Design, preparation, and implementation of an imidazole-based chiral biaryl P,N-ligand for asymmetric catalysis. Journal of the American Chemical Society, 2013,

135(39): 14548-14551.
[64] De Vaumas R, Marinetti A, Ricard L, et al. Use of prochiral phosphaalkene complexes in the synthesis of optically active phosphines. Journal of the American Chemical Society, 1992, 114(1): 261-266.
[65] He G, Loh S K, Vittal J J, et al. Palladium-complex-promoted asymmetric synthesis of stereoisomeric P-chiral pyridylphosphines via an unusual exo-endo stereochemically controlled asymmetric Diels-Alder reaction between 2-vinylpyridine and coordinated 3,4-dimethyl-1-phenylphosphole. Organometallics, 1998, 17(18): 3931-3936.
[66] Lyle M P A, Narine A A, Wilson D. A new class of chiral P,N-ligands and their application in palladium-catalyzed asymmetric allylic substitution reactions. The Journal of Organic Chemistry, 2004, 69(15): 5060-5064.
[67] Ito K, Kashiwagi R, Iwasaki K, Katsuki T. Asymmetric allylic alkylation using a palladium complex of chiral 2-(phosphinoaryl)pyridine ligands. Synlett, 1999, (10): 1563-1566.
[68] Chelucci G, Saba A, Soccolini F. Chiral 2-(2-diphenylphosphinophenyl)-5,6,7,8-tetrahydroquinolines: new P-N ligands for asymmetric catalysis. Tetrahedron, 2001, 57(50): 9989-9996.
[69] Malkov A V, Bella M, StaráI G, et al. Modular pyridine-type $P,N$-ligands derived from monoterpenes:application in asymmetric Heck addition. Tetrahedron Letters, 2001, 42(16): 3045-3048.
[70] Chelucci G, Orru G. Chiral 5-(diphenylphosphanyl)-1,2,3,4-tetrahydroacridines: new N,P-ligands for asymmetric catalysis. Tetrahedron Letters, 2005, 46(20): 3493-3496.
[71] Chelucci G, Cabras M A, Botteghi C, et al. Synthesis of homochiral pyridyl, bipyridyl and phosphino derivatives of 2,2-dimethyl-1,3-dioxolane: use in asymmetric catalysis. Tetrahedron: Asymmetry, 1996, 7(3): 885-895.
[72] Bunlaksananusorn T, Polborn K, Knochel P. New P,N ligands for asymmetric Ir-catalyzed reactions. Angewandte Chemie International Edition, 2003, 42(33): 3941-3943.
[73] Meng X Y, Li X S, Xu D C. Asymmetric hydrogenation and allylic substitution reaction with novel chiral pinene-derived N,P-ligands. Tetrahedron: Asymmetry, 2009, 20 (12), 1402-1406.
[74] Chen S, Ng J K P, Pullarkat S A,et al. Asymmetric synthesis of new diphosphines and pyridylphosphines via a kinetic resolution process promoted and controlled by a chiral palladacycle. Organometallics, 2010, 29(15): 3374-3386.
[75] Drury III W J, Zimmermann N, Keenan M, et al. Synthesis of versatile chiral N,P ligands derived from pyridine and quinoline. Angewandte Chemie International Edition, 2004, 43(1): 70-74.
[76] Alcock N W, Brown J M, Pearson M, et al. Synthesis and easy racemization of an atropisomerically chiral phosphinamine. Tetrahedron: Asymmetry, 1992, 3(1): 17-20.
[77] (a) Alcock N W, Brown J M, Hulmes D I. Synthesis and resolution of 1-(2-diphenylphosphino-1-naphthyl)isoquinoline; a phosphorus-nitrogen chelating ligand for asymmetric catalysis. Tetrahedron: Asymmetry, 1993, 4(4): 743-756; (b) Alcock N. W, Hulmes D I, Brown J M. Contrasting behavior of related palladium complex-derived resolving agents. 8-H conformational locking of the 1-naphthyl side-chain. Journal of the Chemical Society, Chemical Communications, 1995, 395-397.
[78] Valk J M, Claridge T D W, Brown J M, et al. Synthesis and chemistry of a new P-N chelating ligand; (R)- and (S)-6-(2'-diphenylphosphino-1'-naphthyl)phenanthridine. Tetrahedron: Asymmetry, 1995, 6(10): 2597-2610.
[79] Doucet H, Brown J M. Synthesis of 1'-(2-(diarylphosphino)-1-naphthyl)isoquinolines; variation of the aryl substituent. Tetrahedron: Asymmetry, 1997, 8(22): 3775-3784.
[80] McCarthy M, Goddard R, Guiry P J. The preparation and resolution of 2-phenyl-quinazolinap, a new atropisomeric phosphinamine ligand for asymmetric catalysis. Tetrahedron: Asymmetry, 1999, 10(14): 2797-2807.
[81] McCarthy M, Guiry P J. The preparation, resolution and chemistry of 1-(3,6-dimethylpyrazin-2-yl) (2-naphthyl)diphenylphosphine, an axially chiral phosphinamine. Tetrahedron, 1999, 55(10): 3061-3070.
[82] (a) Kwong F Y, Chan K S. A novel synthesis of atropisomeric P,N ligands by catalytic phosphination using triarylphosphines. Organometallics, 2001, 20(12): 2570-2578; (b) Kwong F Y, Chan K S. A new

atropisomeric P,N ligand for rhodium-catalyzed asymmetric hydroboration. The Journal of Organic Chemistry, 2002, 67(9): 2769-2777.

[83] Knofel T F, Aschwanden P, Ichikawa T, et al. Readily available biaryl P,N ligands for asymmetric catalysis. Angewandte Chemie International Edition, 2004, 43(44): 5971-5973.

[84] Hu X Q, Chen H L, Zhang X M. Development of new chiral P,N ligands and their application in the Cu-catalyzed enantioselective conjugate addition of diethylzinc to enones. Angewandte Chemie International Edition, 1999, 38(23): 3518-3521.

[85] Ruzziconi R, Santi C, Spizzichino S. Quinolinophane-derived alkyldiphenylphosphines: two homologous P,N-planar chiral ligands for palladium-catalysed allylic alkylation. Tetrahedron: Asymmetry, 2007, 18(14): 1742-1749.

[86] Jiang B, Lei Y, Zhao X L. [2.2]Paracyclophane-derived chiral P,N-Ligands: design, synthesis, and application in Palladium-catalyzed asymmetric allylic alkylation. The Journal of Organic Chemistry, 2008, 73(19):7833-7836.

[87] Tao B, Fu G C. Application of a new family of P,N Ligands to the highly enantioselective hydrosilylation of aryl alkyl and dialkyl ketones. Angewandte Chemie International Edition, 2002, 41(20): 3892-3894.

[88] Ganter C, Glinsökel C, Ganter B. New P,N-chelate ligands based on pyridyl-substituted phosphaferrocenes. European Journal of Inorganic Chemistry, 1998, 1998(8): 1163-1168.

[89] Cheemala M N, Knochel P. New P,N-Ferrocenyl ligands for the asymmetric Ir-catalyzed hydrogenation of imines. Organic Letters, 2007, 9(16): 3089-3092.

[90] Kloetzing R J, Knochel P. Ferrocenyl-QUINAP: a planar chiral P,N-ligand for palladium-catalyzed allylic substitution reactions. Tetrahedron: Asymmetry, 2006, 17(1): 116-123.

[91] Kloetzing R J, Lotz M, Knochel P. New P,N-ferrocenyl ligands for rhodium-catalyzed hydroboration and palladium-catalyzed allylic alkylation. Tetrahedron: Asymmetry, 2003, 14(2): 255-264.

[92] Arena C G, Nicolo F, Drommi D, et al. Enantioselective hydroformylation with the chiral bidentate P,N-ligand 2-[1-(1S,2S,5R)-(−)menthoxydiphenylphosphino]pyridine cationic rhodium(I) complexes. Journal of the Chemical Society, Chemical Communications, 1994, (19): 2251-2252.

[93] Goldfuss B, Lochmann T, Rominger F. Ligand bite governs enantioselectivity: electronic and steric control in Pd-catalyzed allylic alkylations by modular fenchyl phosphinites (FENOPs). Chemistry-A European Journal, 2004, 10(21): 5422-5431.

[94] Liu Q B, Zhou Y G. Synthesis of chiral cyclohexane-backbone P,N-ligands derivedfrom pyridine and their applications in asymmetric catalysis. Tetrahedron Letters, 2007, 48(12): 2101-2104.

[95] (a) Kaiser S, Smidt S P, Pfaltz A. Iridium catalysts with bicyclic pyridine-phosphinite ligands: asymmetric hydrogenation of olefins and furan derivatives. Angewandte Chemie International Edition, 2006, 45(31): 5194-5197; (b) Woodmansee D H, Müller M A, Neuburger M, et al. Chiral pyridyl phosphinites with large aryl substituents as efficient ligands for the asymmetric iridium-catalyzed hydrogenation of difficult substrates. Chemical Science, 2010, 1(1): 72-78.

[96] Liu, Q B, Yu C B, Zhou Y G. Synthesis of tunable phosphinite-pyridine ligands and their applications in asymmetric hydrogenation. Tetrahedron Letters, 2006, 47(27): 4733-4736.

[97] Meng X Y, GaoY G, Li X S, et al. Novel pyridine-phosphite ligands for Pd-catalyzed asymmetric allylic substitution reaction. Catalysis Communications, 2009, 10, (6): 950-954.

[98] (a) Brunel J M, Constanieux T, Labande A, et al. Enantioselective palladium catalyzed allylic substitution with new chiral pyridine-phosphine ligands. Tetrahedron Letters, 1997, 38(34): 5971-5974; (b) Constantieux T, Brunel J M, Labande A, et al. Enantioselective palladium catalyzed allylic amination using new chiralpyridine-phosphine ligands. Synlett, 1998, (1): 49-50; (c) Brunel J M, Constantieux T, Buono G. A practical method for the large-scalesynthesis of diastereomerically pure (2R,5S)- 3-Phenyl-2-(8-quinolinoxy)-1,3-diaza-2-phosphabicyclo-[3.3.0]-octane ligand (QUIPHOS). Synthesis and X-ray structure of its corresponding chiral π-allyl palladium complex. The Journal of Organic Chemistry, 1999, 64(24): 8940-8942; (d) Delapierre G, Brunel J M, Constantieux T, et al. Design of a new class of chiral quinoline-phosphine ligands. Synthesis and application in asymmetric catalysis. Tetrahedron: Asymmetry, 2001, 12(9): 1345-1352; (e) Delapierre G, Achard M, Buono G. Totally diastereoselective

synthesis of a new chiral quinolinediazaphospholidine ligand and its derivatives. Tetrahedron Letters, 2002, 43(22): 4025-4028.
[99] (a) Hayashi T, Fukushima M, Konishi M, et al. Chiral β-dimethylaminoalkylphosphines. Highly efficient ligands for a nickel complex catalyzed asymmetric Grignard cross-coupling reaction. Tetrahedron Letters, 1980, 21(1): 79-82; (b) Hayashi T, Konishi M, Fukushima M, et al. Chiral (β-Aminoalkyl) phosphines. Highly efficient phosphine ligands for catalytic asymmetric Grignard cross-coupling. The Journal of Organic Chemistry, 1983, 48(13): 2195-2202.
[100] Caiazzo A, Dalili S, Yudin A K. Design and development of cyclohexane-based P,N-ligands for transition metal catalysis. Organic Letters, 2002, 4(15):2597-2600.
[101] Ronholm P, Soergren M, Hilmersson G. Improved and efficient synthesis of chiral N,P-ligands via cyclic sulfamidates for asymmetric addition of butyllithium to benzaldehyde. Organic Letters, 2007, 9(19): 3781-3783.
[102] (a) Maj A M, Pietrusiewicz K M, Suisse I, et al. Chiral β-aminophosphine oxides as ligands for ruthenium assisted enantioselective transfer hydrogenation of ketones. Tetrahedron: Asymmetry, 1999, 10(5): 831-835; (b) A. Maj M, Pietrusiewicz K M, Suisse I, et al. P-chiral β-aminophosphine oxides vs. β-aminophosphines as auxiliaries for ruthenium catalysed enantioselective transfer hydrogenation of arylketones. Journal of Organometallic Chemistry, 2001, 626(1-2): 157-160.
[103] Cahill J P, Bohnen F, Goddard R, et al. The preparation of *trans*-2,5-dialkylpyrrolidinylbenzyldiphenylphosphines: new phosphinamine ligands for asymmetric catalysis. Tetrahedron: Asymmetry, 1998, 9(21): 3831-3839.
[104] Hayashi T, Yamamoto K, Kumada M. Asymmetric catalytic hydrosilylation of ketones preparation of chiral ferrocenylphosphines as chiral ligands. Tetrahedron Letters, 1974, (49-50): 4405-4408.
[105] Uemura M, Miyake R, Nishimura H, et al. New chiral phosphine ligands containing ($\eta^6$-arene) chromium and catalytic asymmetric cross-coupling reactions. Tetrahedron: Asymmetry, (1992), 3 (2) : 213-216.
[106] Vyskocil S, Smrcina M, Hanus V, et al. Derivatives of 2-amino-2′-diphenylphosphino-1,1′-binaphthyl(MAP) and their application in asymmetric palladium(0)-catalyzed allylic substitution. The Journal of Organic Chemistry, 1998, 63(22): 7738-7748.
[107] Stranne R, Vasse J L, Moberg C. Synthesis and application of chiral P,N-ligands with *pseudo-meso* and *pseudo-c*2 symmetry. Organic Letters, 2001, 3(16): 2525-2528.
[108] Brunner H, Rahman A F M M. Neue optisch aktive P,N-liganden und ihr einsatz in derRh-katalysierten asymmetrischen hydrierungund hydrosilylierung. Chemische Berichte, 1984, 117(2): 710-714.
[109] Kohara T, Hashimoto Y, Saigo K. Palladium-catalyzed allylic alkylation using a novel chiral iminophosphine ligand derived from 1-mesitylethylamine. Synlett, 2000, (4): 517-519.
[110] Degrado S J, Mizutani H, Hoveyda A H. Modular peptide-based phosphine ligands in asymmetric catalysis: efficient and enantioselective Cu-catalyzed conjugate additions to five-, six-, and seven-membered cyclic enones. Journal of the American Chemical Society, 2001, 123(4): 755-756.
[111] Park H J, Han J W, Seo H, et al. Preparation and application of polymer-supported π-allylpalladium complex as a chiral catalyst in the asymmetric allylic alkylation. Journal of Molecular Catalysis A: Chemical, 2001, 174(1-2): 151-157.
[112] Jang H Y, Seo H, Han J W, et al. Role of the planar chirality of imine-phosphine hybrid ligands bearing chromium tricarbonyl in the palladium-catalyzed asymmetric allylic alkylation. Tetrahedron Letters, 2000, 41(26): 5083-5087.
[113] Schenkel L B, Ellman J A. Novel sulfinyl imine ligands for asymmetric catalysis, Organic Letters, 2003, 5(4): 545-548.
[114] Lai H, Huang Z, Wu Q, et al. Synthesis of novel enantiopure biphenyl P,N-ligands and application in palladium-catalyzed asymmetric addition of arylboronic acids to *N*-benzylisatin. The Journal of Organic Chemistry, 2009, 74(1): 283-288.
[115] Hayashi T, Hayashi C, Uozumi Y. Catalytic asymmetric hydrosilylation of ketones with new chiral ferrocenylphosphine-imine ligands. Tetrahedron: Asymmetry, 1995, 6(10), 2503-2506.

[116] (a) Saitoh A, Morimoto T, Achiwa K. Asymmetric reactions catalyzed by chiral metal compounds. LXXX. A phosphorus-containing chiral amidine ligand for asymmetric reactions: enantioselective Pd-catalyzed allylic alkylation, Tetrahedron: Asymmetry, 1997, 8(21): 3567-3570; (b) Saitoh A, Achiwa K, Tanaka K, et al. Versatile chiral bidentate ligands derived from α-amino acids: synthetic applications and mechanistic considerations in the palladium-mediated asymmetric allylic substitutions. The Journal of Organic Chemistry, 2000, 65(14): 4227-4240; (c) Saitoh A, Achiwa K, Morimoto T. Asymmetric reactions catalyzed by chiral metal complexes, LXXXIII. Enantioselective allylic substitutions using ketene silyl acetals catalyzed by a palladium-chiral amidine complex. Tetrahedron: Asymmetry, 1998, 9(5): 741-744.

[117] Yuan Z L, Jiang J J, Shi M. The application of chiral phosphine-Schiff base type ligands in silver(Ⅰ)-catalyzed asymmetric vinylogous Mannich reaction of aldimines with trimethylsiloxyfuran. Tetrahedron, 2009, 65(31): 6001-6007.

[118] Gladiali S, Antonio D, Davide F. Novel heterobidentate ligands for asymmetric catalysis: synthesis and rhodium-catalyzed reactions of S-alkyl (R)-2-diphenylphosphino-1,1'-binaphthyl-2'-thiol. Tetrahedron: Asymmetry, 1994, 5(7): 1143-1146.

[119] (a) Zhang W, Xu Q, Shi M. Palladium-catalyzed asymmetric allylic substitutions by axially chiral P,S-, S,S-, and S,O-heterodonor ligands with a binaphthalene framework. Tetrahedron: Asymmetry, 2004, 15(19): 3161-3169; (b) ZhangW, Shi M. Axially chiral P,S-heterodonor ligands with a binaphthalene framework for palladium-catalyzed asymmetric allylic substitutions: experimental investigation on the reversal of enantioselectivity between different alkyl groups on sulfur atom. Tetrahedron: Asymmetry, 2004, 15(21): 3467-3476.

[120] Chapma, C J, Frost C G, Gill-Carey M P, et al. Diphosphine mono-sulfides: readily available chiral monophosphines. Tetrahedron: Asymmetry, 2003, 14(6): 705-710.

[121] Togni A, Hausel R. A new entry into sulfur-containing ferrocenylphosphine ligands for asymmetric catalysis. Synlett, 1990, (10): 633-635.

[122] Tu T, Zhou Y G, Hou X L, et al. Trans effect of different coordinated atoms of planar chiral ferrocene ligands with the same backbone in palladium-catalyzed allylic substitutions. Organometallics, 2003, 22(6): 1255-1265.

[123] Routaboul L, Vincendeau S, Daran J C, et al. New ferrocenyl P,S and S,S ligands for asymmetric catalysis. Tetrahedron: Asymmetry, 2005, 16(6): 2685-2690.

[124] Priego J, Mancheno O G, Cabrera S, et al. 1-Phosphino-2-sulfenylferrocenes: efficient ligands in enantioselective palladium-catalyzed allylic substitutions and ring opening of 7-oxabenzonorbornadienes. Chemical Communications, 2002, (21): 2512-2513.

[125] Cheung H Y, Yu W Y, Lam F L, et al. Enantioselective Pd-catalyzed allylic alkylation of indoles by a new class of chiral ferrocenyl P/S ligands. Organic Letters, 2007, 9(21): 4295-4298.

[126] Nakano H, Okuyama Y, Hongo H. New chiral phosphinooxathiane ligands for palladium-catalyzed asymmetric allylic substitution reactions. Tetrahedron Letters, 2000, 41(23): 4615-4618.

[127] Hiroi K, Suzuki Y, Abe I. Organosulfur functionality as an alternative enantiocontrollable coordinating element in chiral phosphine ligands for palladium-catalyzed asymmetric allylic alkylations. Chemistry Letters, 1999, (2): 149-150.

[128] (a) Yan Y Y, Rajanbabu T V. Ligand tuning in asymmetric catalysis: mono- and bis-phospholanes for a prototypical Pd-catalyzed asymmetric allylation reaction. Organic Letters, 2000, 2(2): 199-202 (b) Yan Y, Rajanbabu T V. Highly flexible synthetic routes to functionalized phospholanes from carbohydrates. The Journal of Organic Chemistry, 2000, 65(3): 900-906.

[129] (a) Verdaguer X, Moyano A, Pericàs M A, et al. A new chiral bidentate (P,S) ligand for the asymmetric intermolecular Pauson-Khand reaction. Journal of the American Chemical Society, 2000, 122(41): 10242-10243; (b) Verdaguer X, Pericàs M A, Riera A, et al. Design of new hemilabile (P,S) ligands for the highly diastereoselective coordination to alkyne dicobalt complexes: application to the asymmetric intermolecular Pauson-Khand reaction. Organometallics, 2003, 22(9): 1868-1877.

[130] Ferrer C, Riera A, Verdaguer X. Sulfinylmethyl phosphines as chiral ligands in the intermolecular

Pauson-Khand reaction, Organometallics, 2009, 28(15): 4571-4576.
[131] Hiroi K, Suzuki Y, Kawagishi R. Chiral $\beta$-phosphino sulfoxides as chiral ligands in palladium-catalyzed asymmetric allylic nucleophilic substitution reactions. Tetrahedron Letters, 1999, 40(4): 715-718.
[132] Nakamura S, Fukuzumi T, Toru T. Novel chiral sulfur-containing ferrocenyl ligands for palladium-catalyzed asymmetric allylic substitution. Chirality, 2004, 16(1): 10-12.
[133] Sugama H, Saito H, Danjo H, et al. P-chirogenic phosphine/sulfide hybrid ligands. Synthesis, 2001, (15): 2348-2353.
[134] Molander G A, Burke J P, Carroll P J. Synthesis and application of chiral cyclopropane-based ligands in palladium-catalyzed allylic alkylation. The Journal of Organic Chemistry, 2004, 69(23): 8062-8069.
[135] (a) Evans D A, Campos K R, Tedrow J S, et al. Chiral mixed phosphorus/sulfur ligands for palladium-catalyzed allylic alkylations and aminations. The Journal of Organic Chemistry, 1999, 64(9): 2994-2995; (b) Evans D A, Campos K R, Tedrow J S, et al. Application of chiral mixed phosphorus/sulfur ligands to palladium-catalyzed allylic substitutions. Journal of the American Chemical Society, 2000, 122(33): 7905-7920; (c) Evans D A, Michael F E, Tedrow J S, et al. Application of chiral mixed phosphorus/sulfur ligands to enantioselective rhodium-catalyzed dehydroamino acid hydrogenation and ketone hydrosilylation processes. Journal of the American Chemical Society, 2003, 125(12): 3534-3543.
[136] (a) Pàmies O, Diéguez M, Net G, et al. Synthesis and coordination chemistry of novel chiral P,S-ligands with a xylofuranose backbone: use in asymmetric hydroformylation and hydrogenation. Organometallics, 2000, 19(8): 1488-1496; (b) Pàmies O, Van Strijdonck G P F, Diéguez M, et al. Modular furanoside phosphite ligands for asymmetric Pd-catalyzed allylic substitution. The Journal of Organic Chemistry, 2001, 66(26): 8867-8871; (c) Guimet E, Diéguez M, Ruiz A, et al. Furanoside thioether-phosphinite ligands for Pd-catalyzed asymmetric allylic substitution reactions. Tetrahedron: Asymmetry, 2005, 16(5): 959-963; (d) Diéguez M, Pàmies O, Claver C. Furanoside thioether-phosphinite ligands for Pd-catalyzed asymmetric allylic substitution reactions: scope and limitations. Journal of Organometallic Chemistry, 2006, 691(10): 2257-2262.
[137] (a) Khiar N, Suárez B, Stiller M, et al. Mixed S/P ligands from carbohydrates: synthesis and utilization in asymmetric catalysis. Phosphorus, Sulfur, Silicon and the Related Elements, 2005, 180(5-6): 1253-1258; (b) Khiar N, Suárez B, Valdivia V, et al. Phosphinite thioglycosides derived from natural D-Sugars as useful P/S ligands for the synthesis of both enantiomers in palladium-catalyzed asymmetric substitution, Synlett, 2005,(19): 2963-2967.
[138] Selvakumar K, Valentini M, Pregosin P S, et al. Chiral phosphito-thioether complexes of palladium(0). Comments on the Pd, Rh, and Ir regio- and-enantioselective allylic alkylations of PhCH=CHCH(OAc) R, R = H, Me, Et. Organometallics, 1999, 18(22): 4591-4597.
[139] (a) Kanayama T, Yoshida K, Miyabe H, et al. Enantio- and diastereoselective Ir-catalyzed allylic substitutions for asymmetric synthesis of amino acid derivatives. Angewandte Chemie International Edition, 2003, 42(18): 2054-2056; (b) Kanayama T, Yoshida K, Miyabe H, et al. Synthesis of $\beta$-substituted $\alpha$-amino acids with use of iridium-catalyzed asymmetric allylic substitution. The Journal of Organic Chemistry, 2003, 68(16): 6197-6201.
[140] Solà J, Revés M, Riera A, et al. N-phosphino sulfinamide ligands: an efficient manner to combine sulfur chirality and phosphorus coordination behavior. Angewandte Chemie International Edition, 2007, 46(26): 5020-5023.
[141] Lam F L, Au-Yeung T T L, Cheung H Y, et al. Easily accessible ferrocenyl N-P/S type ligands and their applications in asymmetric allylic substitutions. Tetrahedron: Asymmetry, 2006, 17(4): 497-499.
[142] Boeda F, Rix D, Clavier H, et al. Design and synthesis of new bidentate phosphoramidite ligands for enantioselective copper-catalyzed conjugate addition of diethylzinc to enones. Tetrahedron: Asymmetry, 2006, 17(19): 2726-2729.
[143] Wei Y, Lu L Q, Li T R, et al. P,S Ligands for the asymmetric construction of quaternary stereocenters in palladium-catalyzed decarboxylative [4+2] cycloadditions. Angewandte Chemie International Edition,

2016, 55(6): 2200-2204.
[144] Hayashi T, Mise T, Fukushima M, et al. Asymmetric synthesis catalyzed by chiral ferrocenylphosphine-transition metal complexes. I. Preparation of chiral ferrocenylphosphines, Bulletin of the Chemical Society of Japan, 1980, 53(4): 1138-1151.
[145] Takahashi Y, Yamamoto Y, Katagiri K, et al. P-Chiral o-phosphinophenol as a P/O hybrid ligand: preparation and use in Cu-catalyzed asymmetric conjugate addition of diethylzinc to acyclic enones. The Journal of Organic Chemistry, 2005, 70(22): 9009-9012.
[146] Kurz L, Lee G, Morgans D Jr, et al. Stereospecific functionalization of (R)-(-)-1,1'-bi-2-naphthol triflate, Tetrahedron Letters, 1990, 31(44): 6321-6324.
[147] Uozumi Y, Tanahashi A, Lee S Y, et al. Synthesis of optically active 2-(diarylphosphino)-1,1'-binaphthyls, efficient chiral monodentate phosphine ligands. The Journal of Organic Chemistry 1993, 58(7): 1945-1948.
[148] Ito K, Eno S, Saito B, et al. Enantioselective conjugate addition of diethylzinc to acyclic enones using a copper phosphino-phenol complex as catalyst. Tetrahedron Letters, 2005, 46(23): 3981-3985.
[149] Yang H, Sun J W, Gu W, et al. Enantioselective cross-coupling for axially chiral tetra-ortho-substituted biaryls and asymmetric synthesis of gossypol. Journal of the American Chemical Society, 2020, 142(17): 8036-8043.
[150] Gladiali S, Pulacchini S, Fabbri D, et al. 2-diphenylphosphino-2'-diphenylphosphinyl-1,1'-binaphthalene(BINAPO), an axially chiral heterobidentate ligand for enantioselective catalysis. Tetrahedron: Asymmetry, 1998, 9(3): 391-395.
[151] Grushin V V. Catalysis for catalysis: synthesis of mixed phosphine-phosphine oxide ligands via highly selective, Pd-catalyzed monooxidation of bidentate phosphines. Journal of the American Chemical Society, 1999, 121(24): 5831-5832.
[152] Li C X, Chen T Y, Li B, et al. Efficient synthesis of sterically hindered arenes bearing acyclic secondary alkyl groups by suzuk-miyaura cross-couplings. Angewandte Chemie International Edition, 2015, 54(12): 3792-3796.
[153] Kanai M, Koga K, Tomioka K. Enatioselective conjugate addition of organocuprate using a chiral amidophosphine ligand. Tetrahedron Letters, 1992, 33(47): 7193-7196.
[154] Rageot D, Woodmansee D H, Pugin B, et al. Proline-based P,O ligand/iridium complexes as highly selective catalysts: asymmetric hydrogenation of trisubstituted alkenes. Angewandte Chemie International Edition, 2011, 50(41): 9598-9601.
[155] Hird A W, Hoveyda A H. Cu-catalyzed enantioselective conjugate additions of alkyl zinc reagents to unsaturated N-acyloxazolidinones promoted by a chiral triamide phosphane. Angewandte Chemie International Edition, 2003, 42(11): 1276-1279.
[156] Meeuwissen J, Detz R J, Sandee A J, et al. Rhodium-P,O-bidentate coordinated ureaphosphine ligands for asymmetric hydrogenation reactions. Dalton Trans., 2010, 39(8): 1929-1931.
[157] (a) Dai X, Wong A, Virgil S C. Synthesis and resolution of quinazolinone atropisomeric phosphine ligands. The Journal of Organic Chemistry, 1998, 63(8): 2597-2600; (b) Dai X, Virgil S. Asymmetric allylic alkylation catalyzed by palladium complexes with atropisomeric quinazolinone phosphine ligands. Tetrahedron Letters, 1999, 40(7): 1245-1248.
[158] Clayden J, Johnson P, Pink J H, et al. Atropisomeric amides as chiral ligands: using (−)-sparteine-directed enantioselective silylation to control the conformation of a stereogenic axis. The Journal of Organic Chemistry, 2000, 65(21): 7033-7040.
[159] Dai W M, Yeung K K Y, Liu J T, et al. A novel class of nonbiaryl atropisomeric P,O-ligands for palladium-catalyzed asymmetric allylic alkylation. Organic Letters, 2002, 4(9): 1615-1618.
[160] Bai X F, Song T, Xu Z, et al. Aromatic amide-derived non-biaryl atropisomers as highly efficient ligands in silver-catalyzed asymmetric cycloaddition reactions. Angewandte Chemie International Edition, 2015, 54(17): 5255-5259.
[161] Lamač M, Tauchman J, Císařová I, et al. Preparation of chiral phosphinoferrocene carboxamide ligands and their application to palladium-catalyzed asymmetric allylic alkylation. Organometallics, 2007,

26(20): 5042-5049.

[162] (a) Zhang Z M, Chen P, Li W B, et al. A new type of chiral sulfinamide monophosphine ligands: stereodivergent synthesis and application in enantioselective gold( I )-catalyzed cycloaddition reactions. Angewandte Chemie International Edition, 2014, 53(17): 4350-4354; (b) Wang L, Chen M, Zhang P, et al. Palladium/PC-Phos-Catalyzed Enantioselective Arylation of General Sulfenate Anions: Scope and Synthetic Applications. Journal of the American Chemical Society, 2018, 140(9): 3467-3473.

[163] Nguyen Q H, Guo S M, Royal T, et al. Intermolecular palladium(0)-catalyzed atropo-enantioselective C-H arylation of heteroarenes. Journal of the American Chemical Society, 2020, 142(5):2161-2167.

PHOSPHORUS 磷科学前沿与技术丛书

手性膦配体合成及应用

# 5 手性多官能化多齿型膦配体

5.1 多官能化单膦配体
5.2 多官能化双膦配体
5.3 P,P,P 型三齿配体

Synthesis and Application of Chiral Phosphine Ligands

多官能化手性膦配体是不对称催化领域中一类高附加值的含磷化合物，是有机合成中的重要组成部分。使用多官能化手性膦配体结合过渡金属离子的协同催化反应被认为是不对称催化合成中的有力工具，在诸多平常催化体系难以实现的催化反应中起着非常重要的作用，通过调控配体的各个区域可以实现对反应的活性和立体选择性进行有效控制。因此，设计和合成新型的、易于得到的多官能手性膦配体在快速优化催化体系、实现有价值的合成路线和更好地理解立体选择性控制的机理等方面起着举足轻重的作用。

非共价相互作用普遍存在于酶催化反应中，其通过多个活性位点的协作来实现酶催化中的精确控制，受此原理的启发，非共价相互作用辅助的过渡金属催化近年来已成为一种强大的工具，并引起了人们的浓厚兴趣。与传统的直接改变配体骨架的刚性或改变配位原子的电性的策略不同，将具有非共价相互作用的片段连接到刚性骨架上，其手性环境可以利用阴离子交换、分子间氢键和共轭作用等方式结合反应底物、金属离子、溶剂、添加剂等进行原位微调，以实现对手性配体手性微环境的控制。这类具有潜在的非等价配位点的新型的多功能手性膦配体不仅可直接作用于金属中心致使其电子配位能力不等，同时，第二个功能片段中的非共价相互作用位点（氢键，芳基-芳基相互作用等与第二种金属等配合）进一步使催化中心周围的空间不对称，从而实现高立体选择性催化反应。

本章节对近年来具有代表性的手性多官能化多齿膦配体进行了归纳、总结，并列举了这些配体在具体的不对称催化反应中的应用。根据其中磷原子的个数可分为：多官能化单膦配体、多官能化双膦配体、多膦配体及其他特殊结构配体。

# 5.1
## 多官能化单膦配体

通过设计在手性配体骨架中引入杂原子（一般包括杂原子 N、O、S）官能团，这些杂原子原子轨道最外层含有孤对电子，具有与 P 原子类似的路易斯碱性，故可以与金属中心原子进行配位。相对于传统单膦配体，多官能化单膦配体与金属形成的络合物结构更加稳定，不易异构化。同时，杂原子官能团中的 N、O、S 等原子可以通过氢键等非共价相互作用，影响反应的立体选择性。

## 5.1.1 多官能 P,N 配体

手性氨基酸广泛存在于自然界中，是一类天然的手性源，将氨基酸与叔膦片段结合制备的手性氨基酰胺化合物同样可作为手性膦配体应用于不对称催化反应中，其中氨基酸的结构多样性使得氨基酸衍生膦配体有着许多结构和功能的多样性。例如，Tomioka 课题组将 N-叔丁基羰基-L-缬氨酸与二苯基膦取代的手性四氢吡咯反应，以较高的收率得到了一种缬氨酸衍生酰胺膦配体 **L1**[1]（图 5-1）。

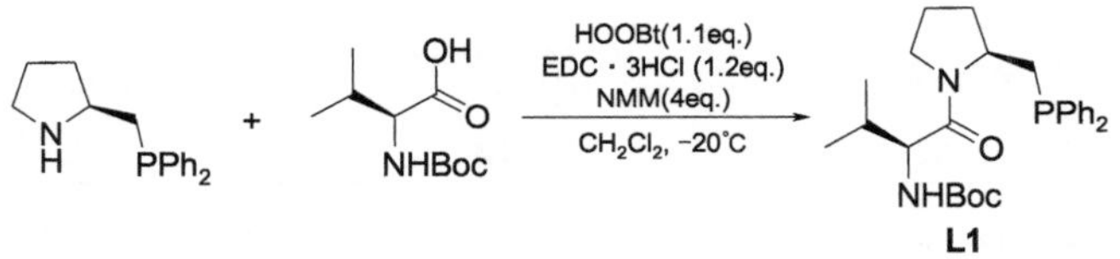

图 5-1 缬氨酸衍生酰胺膦配体的制备

2004 年，Tomioka 课题组研究了缬氨酸衍生酰胺膦配体 **L1** 与铑催化的亚胺与芳基硼酸的不对称芳基化反应，他们发现当使用 3%（摩尔分数）催化剂，反应以正丙醇作为溶剂在 60～100℃条件下，可以以较高的对映立体选择性制备一系列二芳基甲胺衍生物[1]（图 5-2）。

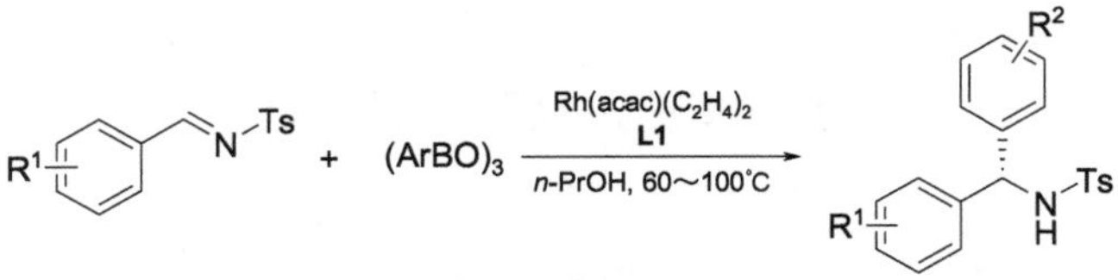

图 5-2 酰胺膦配体 **L1** 在铑催化亚胺不对称芳基化反应中的应用

天然产物金鸡纳生物碱具有结构稳定、廉价易得等特点，在不对称催化合成中发挥着非常重要的作用，其作为手性催化剂或者配体的优势骨架结构被广泛应用于有机合成中。2011 年，Dixon 课题组设计合成了一类新型金鸡纳生物碱衍生酰胺膦配体[2]。以商业易得的奎宁、辛克宁为原料，对其 9-位活性位点进行脱氧氨基化后，在 4-二甲氨基吡啶（DMAP）催化剂作用下，与 2-二苯基膦苯甲酸反应得到酰胺化目标产物。通过修饰两个片段的结构和取代基可以得到一系列不同基团修饰的金鸡纳生物碱衍生的酰胺膦配体(图 5-3)。

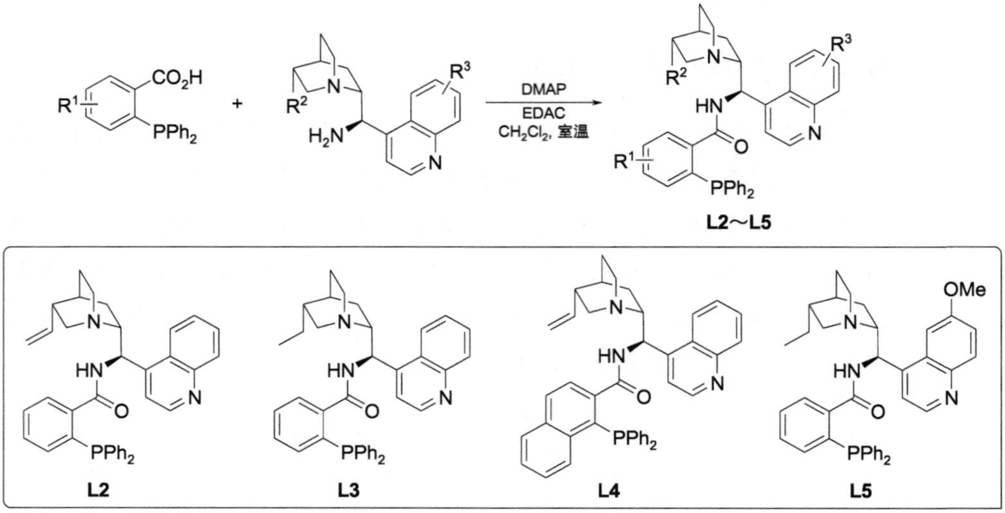

图 5-3 金鸡纳生物碱衍生酰胺膦配体的制备

Dixon 课题组研究了金鸡纳碱骨架衍生的酰胺膦配体 **L2** ～ **L5** 在不对称曼尼希(Mannich)反应中的反应活性[2]。使用 2.5%(摩尔分数)$Ag_2O$ 和 5%(摩尔分数)**L3**，在以甲基叔丁基醚(MTBE)作为溶剂的条件下，各种醛与甘氨酸衍生的异腈发生反应，以较高收率、非对映选择性和对映立体选择性制备了一系列手性噁唑啉化合物(图 5-4)。他们认为金鸡纳碱骨架衍生的双官能配体携带路易斯碱性的膦与银(一价)络合，而奎宁环的氮原子则作为布朗斯特碱加速了异腈的去质子化。

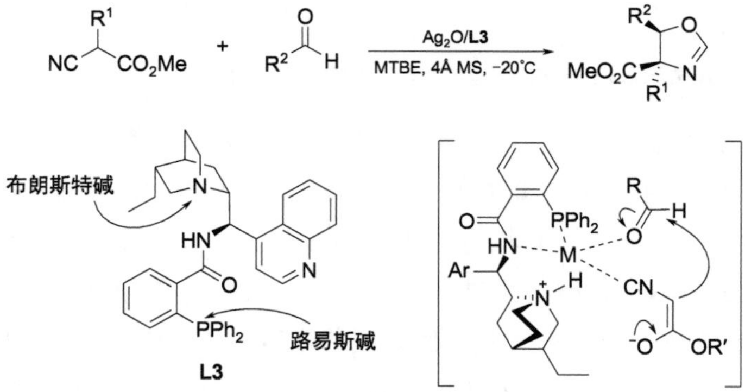

图 5-4 酰胺膦配体催化异腈与醛的不对称曼尼希反应

该课题组还研究了酮、亚胺在不对称曼尼希反应中的应用，使用上述方法他们合成了更具挑战性的化合物。比如，使用 N-双苯基氧膦保护的亚胺作为亲电试剂，合成了 β 位带有季碳手性中心的咪唑啉化合物[3]；使用 α-取代的异腈则可以得到具有两个相邻四取代立体中心的咪唑啉衍生物[4]。值得注意的是，近期研究中通过 DFT 计算揭示了产生主要对映异构体的优势反应过渡态，在两个最低能量过渡态上，强氢键接受体的膦氧通过氢键与银和配体的酰胺相互作用，这时酮亚胺采用 E 构型进行反应[5]（图 5-5）。

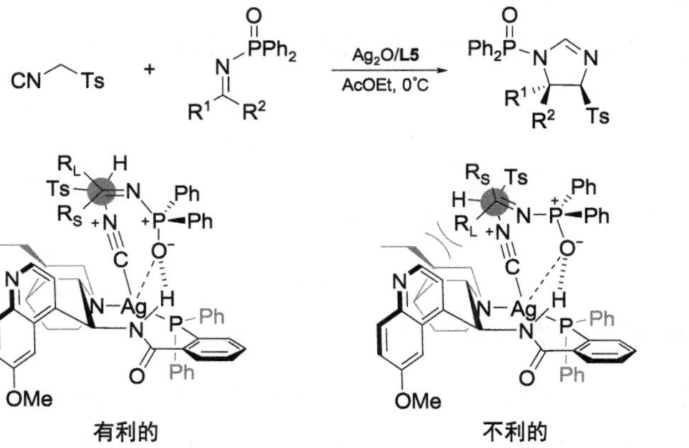

图 5-5 酰胺膦配体催化异腈与亚胺的不对称曼尼希反应

2019 年，刘心元课题组报道了一种外消旋卤代烷烃与末端炔烃的不对称铜催化 Sonogashira $C(sp^3)$-$C(sp)$ 交叉偶联反应[6]（图 5-6），其关键在于使用一种金鸡纳碱骨架衍生的三齿酰胺膦配体 L6 结合铜催化剂催化

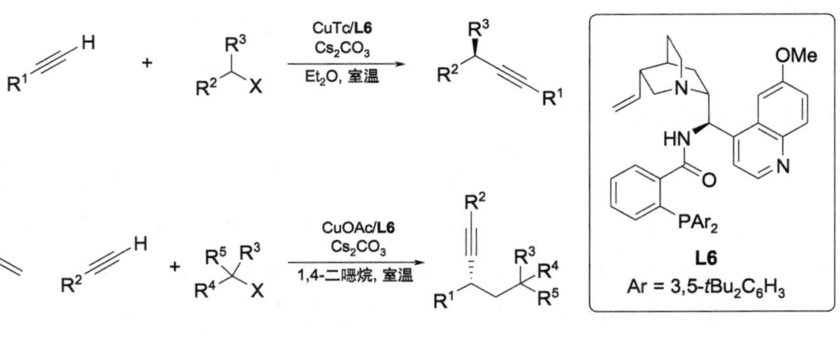

图 5-6 酰胺膦配体 L6 在不对称铜催化 Sonogashira 反应中的应用

5 手性多官能化多齿型膦配体   235

反应经历自由基炔基化过程。这种方法的潜在效用表现在能实现立体富集的生物活性以及功能分子衍生物、药用化合物和具有一系列手性$C(sp^3)$-$C(sp/sp^2/sp^3)$键天然产物的快速合成。基于这一催化体系，他们顺利开发出铜催化分子间三组分不对称烯烃1,2-双官能化反应，实现手性炔烃的高选择性制备[7]。

同年，刘心元课题组还报道了一例基于铜/手性金鸡纳生物碱衍生配体催化的氧化型Sonogashira反应[8]（图5-7），以高区域选择性、化学选择性和对映选择性实现了未活化的$C(sp^3)$-H键与末端炔烃的不对称$C(sp^3)$-C(sp)氧化交叉偶联。该反应使用N-氟酰胺作为氧化剂以选择性生成烷基自由基物种，同时也有效地避免了发生端位炔烃的Glaser自偶联。

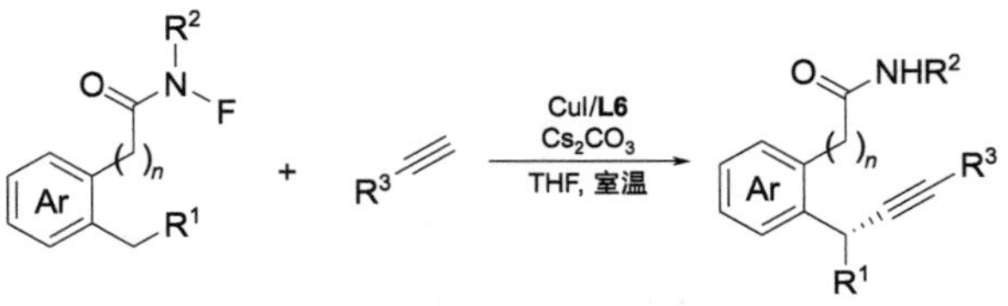

图5-7 不对称铜催化氧化型Sonogashira交叉偶联反应

2019年，祝介平课题组报道了一例通过金鸡纳骨架的双功能配体催化的不对称1,4-加成反应[9]，将α-异氰基乙酸酯与β-芳基-α,β-炔酮加成环化以合成具有轴手性的3-芳基吡咯化合物（图5-8）。

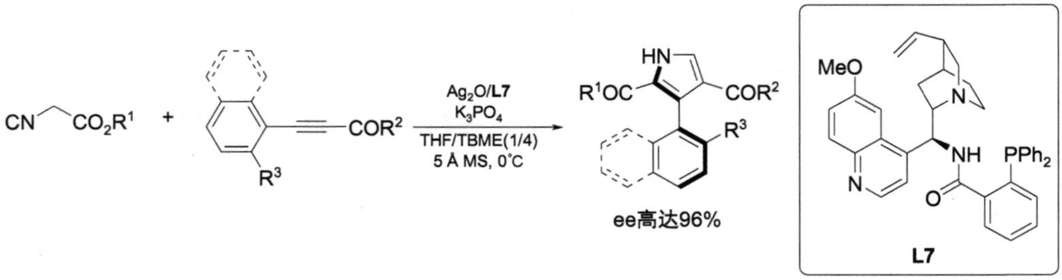

图5-8 不对称银催化1,4-加成反应

2016年，王海飞等人将α-氨基酸、9-氨基-(9-脱氧)辛克宁与2-二苯基膦苯甲酸片段三者结合制备了一种新型多功能膦/酰胺配体**L8**（图5-9）[10]。首先将金鸡纳碱与N-Boc-α-氨基酸反应，再在三氟甲磺酸作用下脱保护得到氨基酸衍生化辛克宁产物，最后再与2-二苯基膦苯甲酸反应得到目标产物。

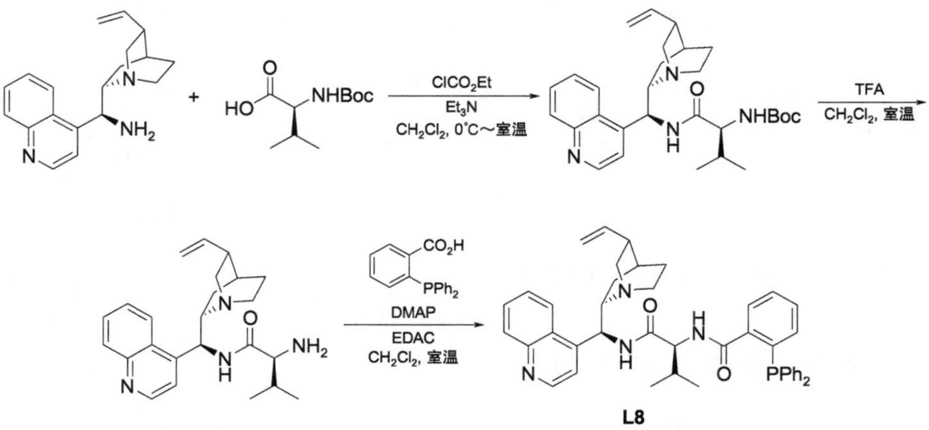

图 5-9 氨基酸-金鸡纳生物碱衍生氨基膦配体的制备及应用

这种金鸡纳生物碱-氨基酸衍生配体与金属络合形成一种新型的 Brønsted/Lewis 碱-Lewis 酸催化体系，可以同时实现对参与反应的亲核试剂和亲电试剂进行活化，共同控制反应中的过渡态，从而产生有效的手性催化环境（图 5-10）[10]。例如，该催化体系成功应用于偶氮甲碱叶立德与马来酸二酯的不对称 1,3-偶极环加成反应，以优异的产率和对映选择性制备了高度官能化的手性吡咯烷化合物。

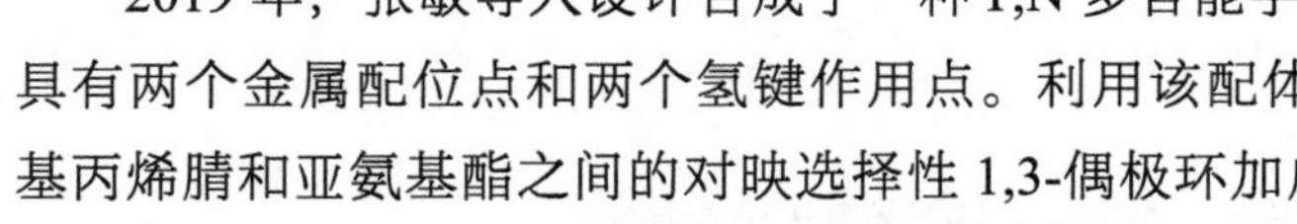

图 5-10 氨基酸-金鸡纳生物碱衍生氨基膦配体的制备及应用

2019 年，张敏等人设计合成了一种 P,N 多官能手性配体 L9，该配体具有两个金属配位点和两个氢键作用点。利用该配体他们成功解决了甲基丙烯腈和亚氨基酯之间的对映选择性 1,3-偶极环加成反应[11]（图 5-11）。在最优反应条件下，各种亚氨基酯和丙烯腈衍生物被转化为高度取代的手性吡咯烷产物。值得注意的是，DFT 量化计算表明尿素片段与甲基丙烯腈之间的氢键作用有助于控制非对映选择性和对映选择性，这也证明了合理利用配体和底物之间的非共价相互作用可以实现对反应的对映选择性控制。

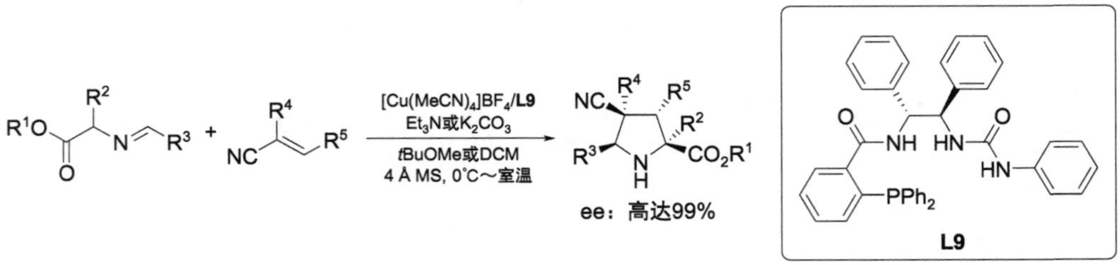

图 5-11 二苯基乙二胺-尿素衍生氨基膦配体的应用

## 5.1.2 多官能 P,O 配体

2013 年，周建荣课题组使用新型手性膦配体 **L10** 和 **L11** 开发了 $\gamma$-丁内酯衍生的甲硅烷基烯醇醚的 Heck 型不对称芳基化反应[12]（图 5-12）。甲硅烷基烯醇醚的芳基化和乙烯基化均获得了优异的产率和对映选择性。同时，该课题组还报道了环酮通过乙酰烯醇醚原位生成的烯醇锡的不对称芳基化[13]。**L10** 和 **L11** 在该反应中都得到了很好的结果。DFT 计算表明，配体和底物之间同样具有相似氢键相互作用，这对获得高对映选择性至关重要。

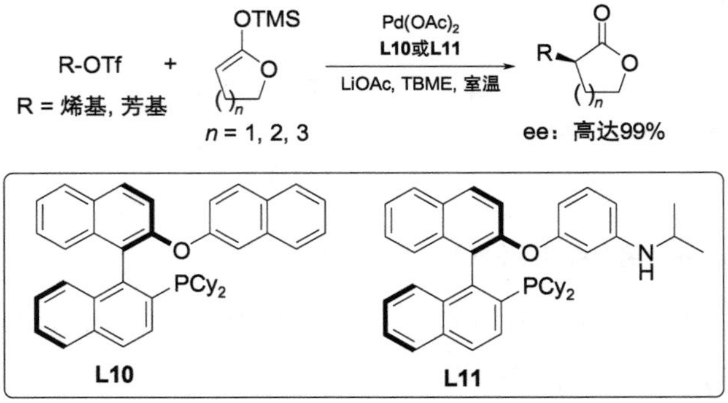

图 5-12 钯催化的烯醇化合物的芳基化和乙烯基化

2017 年，张立明、Zanoni 等人将碱性酰胺引入 BINOL 衍生的膦骨架的芳环上设计合成了膦配体 **L12**[14]，将其应用于 Au 催化高烯丙醇的分子内环化反应，以优秀的对映选择性和产率获得了手性杂环化合物（图 5-13）。在反应中，配体不仅可以通过氢键作用与游离底物结合调控其反应进攻方向，还能作为一种良好的氢受体，在随后的基元反应中作为碱增加醇的亲核性。

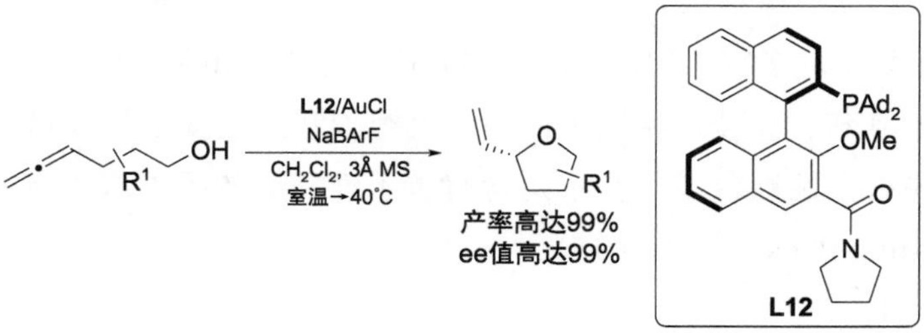

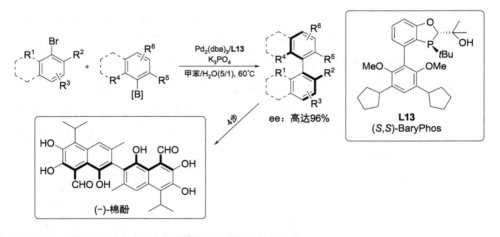

图 5-13　金催化的高烯丙醇化合物的分子内环化反应

2020 年，汤文军课题组经多步反应合成了含有磷手性中心的膦/醇配体 BaryPhos（**L13**），并成功地发展了一种不对称 Suzuki-Miyaura 交叉偶联反应[15,16]，构建了一系列四个邻位均有取代基的轴手性联芳基化合物（图 5-14）。在反应中，配体的羟基作为氢供体与其中一种底物发生氢键相互作用，在对映选择性还原消除步骤中起到了至关重要的作用。利用该催化体系，汤文军等人进一步成功地完成了棉酚的首次对映选择性合成，验证了该方法的实用性。

图 5-14　对映选择性合成具有轴手性的四取代联芳基化合物

## 5.1.3　多官能含 S 多齿单膦配体

具有手性硫基团的配体近年来受到了越来越多的关注，将手性亚砜或亚磺酰胺作为关键的手性导向基团与其他配位基团结合，已经开发出多种具有优异性能的新型配体。2014 年后，张俊良、肖文精和徐利文等中国有机化学家极大地发展了含有手性亚砜或亚磺酰胺结构单元的手性配体，为手性多官能化配体的发展和应用作出了重要贡献。

2014 年，张俊良课题组使用有机锂或镁进攻叔丁基亚磺酰亚胺，高非对映选择性地合成了一系列含有碳手性中心的手性叔丁基/亚磺酰胺配体 Ming-Phos[17,18]（图 5-15）。

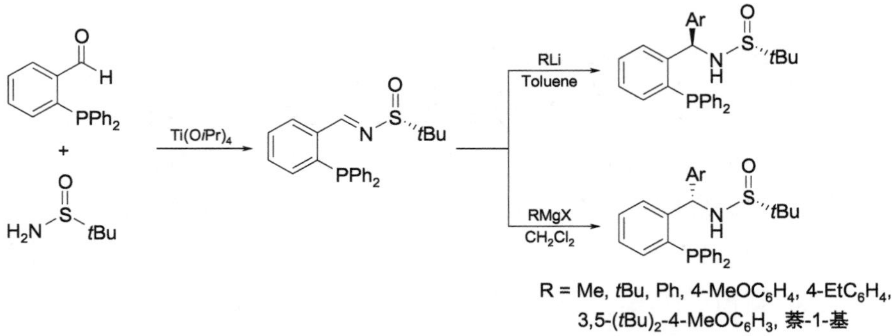

图 5-15　手性膦/亚磺酰胺配体 Ming-Phos 的合成

当碳手性中心的取代基为芳基时，手性膦/亚磺酰胺配体 Ming-Phos 可以成功应用于不对称金催化环化反应，合成一系列手性杂环化合物[17]。当碳手性中心的取代基为叔丁基时，这类配体在 Cu(Ⅰ) 催化的亚胺叶立德与 $\beta$-三氟甲基-$\beta,\beta$-二取代的烯酮的不对称 [3+2] 环加成反应中表现优异，以良好的产率、大于 20∶1 的非对映选择性和大于 98% 的对映选择性构建了一系列具有三氟甲基、季碳立体中心的高度取代的吡咯烷类化合物[18]（图 5-16）。

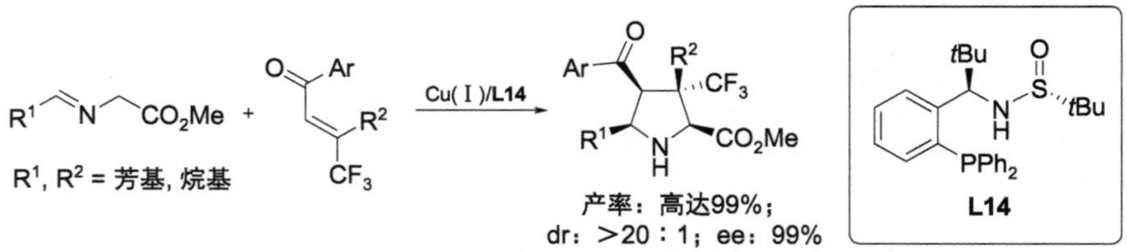

图 5-16　Ming-Phos 在铜催化不对称 [3+2] 环加成反应中的应用

基于前期的研究工作，张俊良课题组还开发了一种高效的聚合物型

Ming-Phos(**L15**)/金催化剂[19]，并成功将其应用于 Au(Ⅰ)催化的炔基取代烯酮与硝酮的不对称环加成反应。与均相催化剂相比，聚合物金催化剂表现出相似的催化活性，在没有对映选择性损失的情况下可实现多次循环使用（最高 8 次），适用于大规模合成反应（图 5-17）。

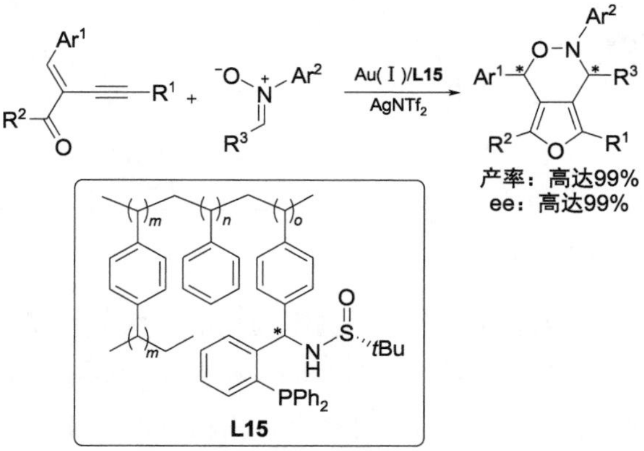

图 5-17 聚合型 Ming-Phos 在金催化不对称环加成反应中的应用

2016 年，张俊良课题组合成了具有更大空间位阻的二金刚烷基膦取代的手性膦/亚磺酰胺配体 Xiang-Phos(**L16**)[20]。其在 Au(Ⅰ)催化的 3-苯乙烯基吲哚和 N-烯二基噁唑烷酮的不对称 [2+2] 环加成反应中表现出优秀的催化效率，以很好的收率和对映选择性（ee 值最高 95%）得到了目标产物手性环丁烷（图 5-18）。

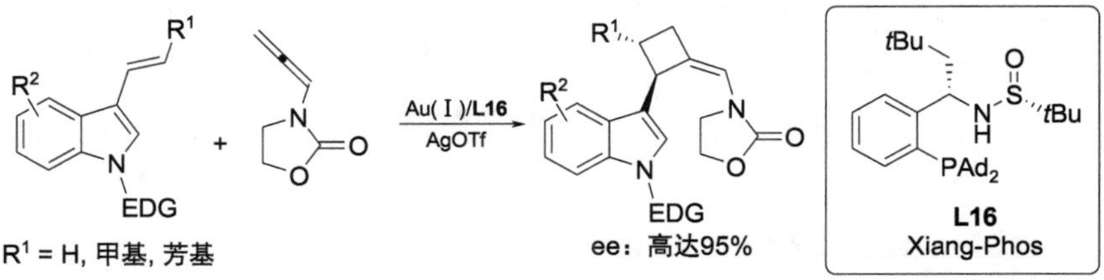

图 5-18 Au(Ⅰ)催化不对称 [2+2] 环加成反应

2017 年，张俊良课题组开发了一类双官能亚磺酰胺膦配体 SadPhos[21]，其中一种变体 PC-Phos(**L17**)可以很好地应用于不对称的 Au(Ⅰ)催化的联烯分子内环化反应。**L17** 是一个基于 XantPhos 骨架的单齿膦配体，配体骨架通过一个碳立体中心连接手性磺酰胺（图 5-19）。该反应中，亲核组分是取代的吲哚，亲电试剂是连接到 C-3 位置的侧链 N-联烯酰胺[22]。研究表明配体中磺酰胺的 NH 与底物的磺酰胺保护基团形成氢键，将底物联烯酰胺定向于 Re-face 面发生分子内进攻。

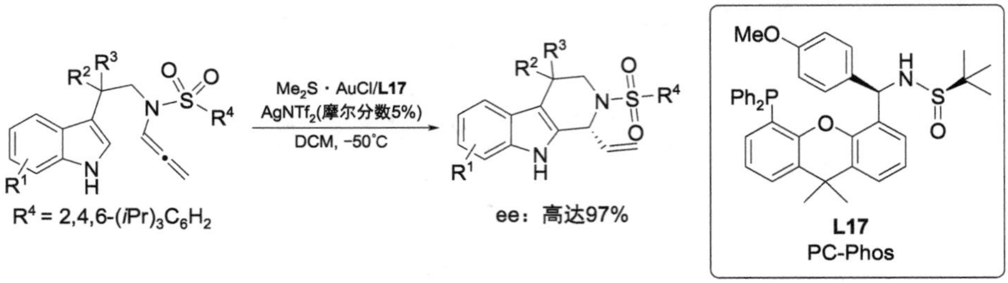

图 5-19　金催化对映选择性合成四氢咔啉化合物

2019 年，张俊良课题组报告了一例钯催化的二级氧化膦化合物的不对称芳基化反应[23]。有意思的是，PC-Phos 和 Ming-Phos 系列配体只得到外消旋产物，新配体 Xiao-Phos（**L18**）表现出极好的效果，以高达 97% 的对映选择性制备了一系列三取代手性膦氧化合物（图 5-20）。控制实验表明亚磺酰基的去除会阻止反应发生。通过 DFT 计算发现，在还原消去过渡态中钯与配体的膦和亚磺酰氨基两个配位点结合形成近似平面的钯配合物，同时氧化膦化合物的氧原子和亚磺酰胺的 NH 之间也存在氢键相互作用。

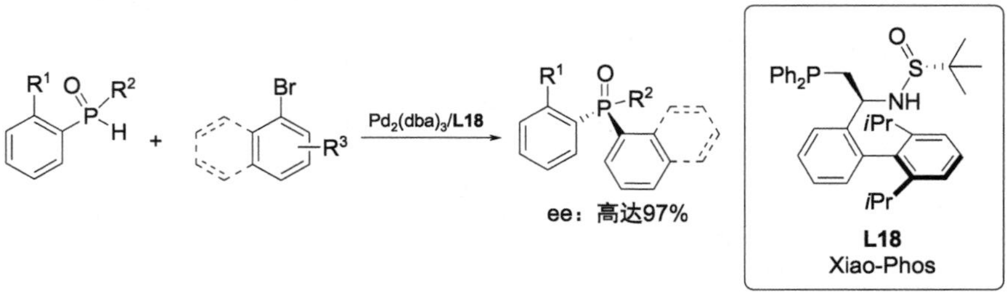

图 5-20　钯催化对映选择性合成手性膦氧化合物

肖文精课题组设计了新型含硫手性磷配体[24-28]，这些配体在钯催化不对称烯丙基取代反应中表现出优异的对映选择性控制性能（图 5-21）。2016 年，该课题组进一步开发了新型的 P,S 配体[27,28]，并实现了迈克尔加成/烯丙基取代串联环化，以良好的收率、优异的非对映和对映选择性得到目标产物手性四氢喹啉。

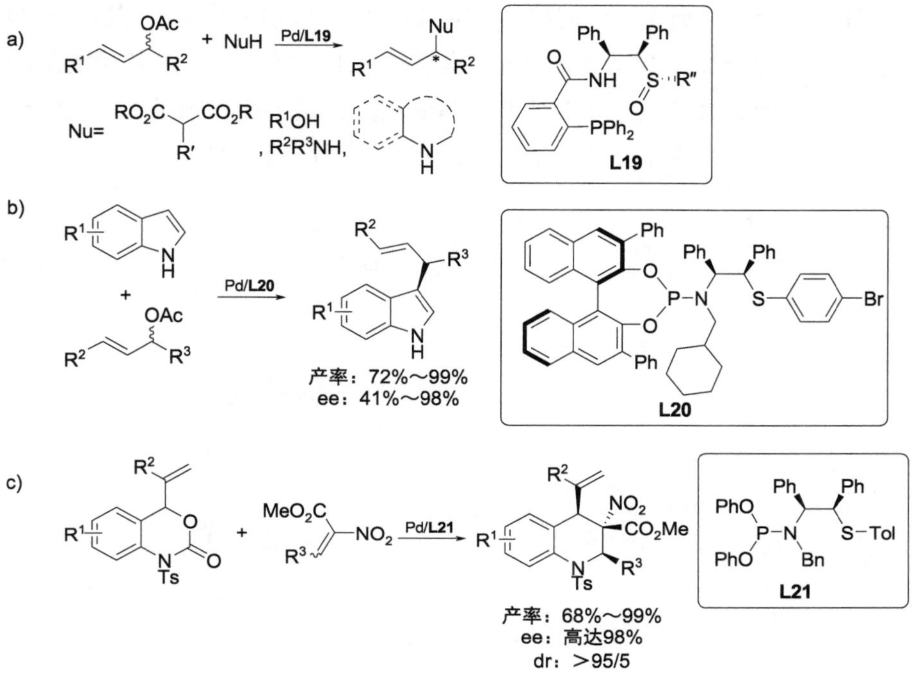

图 5-21 手性磷/硫配体在钯催化烯丙基取代反应中的应用

2021 年，钟为慧课题组利用二茂铁衍生的双功能膦配体作为催化剂，发展了一种改性 Morita-Baylis-Hillman（MBH）碳酸酯不对称烷基化/环化反应，以低负载量的催化剂，可以轻松获得一系列吡唑和喹啉衍生的手性杂环化合物，并且可以实现克级规模的制备[29]（图 5-22）。

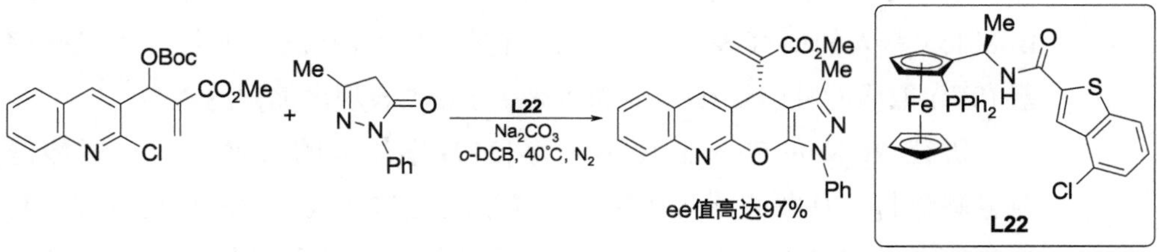

图 5-22 二茂铁骨架手性膦硫配体在钯催化不对称烷基化反应中的应用

2015 年，徐利文课题组通过格氏试剂对叔丁基亚磺酰亚胺进行不对称亲核反应首次合成了一种新型的非联芳基酰胺衍生的阻转异构体，通过柱色谱分离得到光学纯的配体 Xing-Phos[30]。区别于刚性联芳阻转异构膦配体，有一定柔性的 Xing-Phos 携带具有 N、O、S、P 的多功能基团和非等价的络合位点。通过调节芳基酰胺的芳基平面和轴手性以及手性 N-叔丁基亚磺酰胺片段可实现手性微环境的调控（图 5-23）。

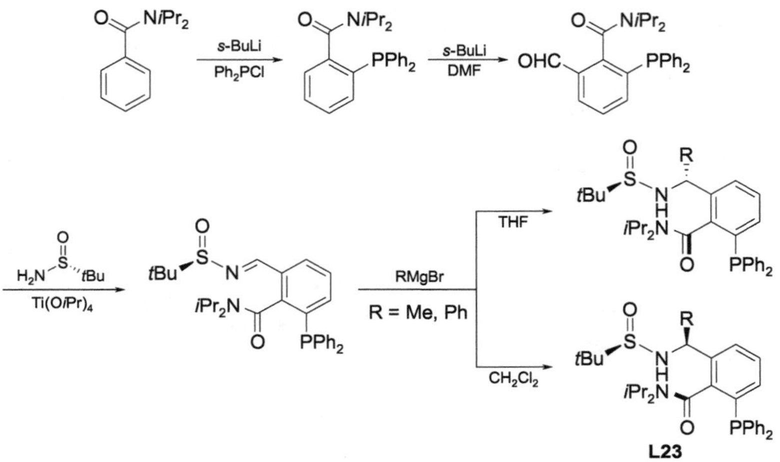

图 5-23　手性膦/酰胺配体 Xing-Phos 的合成

Xing-Phos（**L23**）被证明在银催化甘氨酸醛亚氨基酯与各种活化烯烃的不对称 [3+2] 环加成反应中非常有效[30-32]，硝基烯烃、马来酰亚胺、查耳酮和反应性较低的肉桂酸甲酯都能发生反应，并以良好的产率和非对映选择性（dr 最高 > 99∶1）以及优异的对映选择性（ee 值最高 99%）得到相应的富含多个立体中心的吡咯烷类化合物（图 5-24）。之后，Jubault、Bouillon 等人的研究表明 Xing-Phos 在催化合成高度官能化的五氟硫烷基（$SF_5$）-取代的吡咯烷（ee 值最高 98%）方面也表现出优异的性能[33]。

研究 Ag/Xing-Phos 催化不对称 [3+2] 环加成的机理发现甘氨酸醛亚氨基酯与查耳酮的催化环加成是通过迈克尔加成引发的环化过程，其中迈克尔加成物是关键的中间体。通过不同的处理过程，可以得到顺式-Δ(1)-吡咯啉和反式-Δ(1)-吡咯啉的立体发散性产物。有趣的是，通过 DBU 参与的差向异构化的后处理顺式异构体可以轻松转化为对应的反式异构体，同时保持良好非对映选择性和对映选择性（图 5-25）[34]。

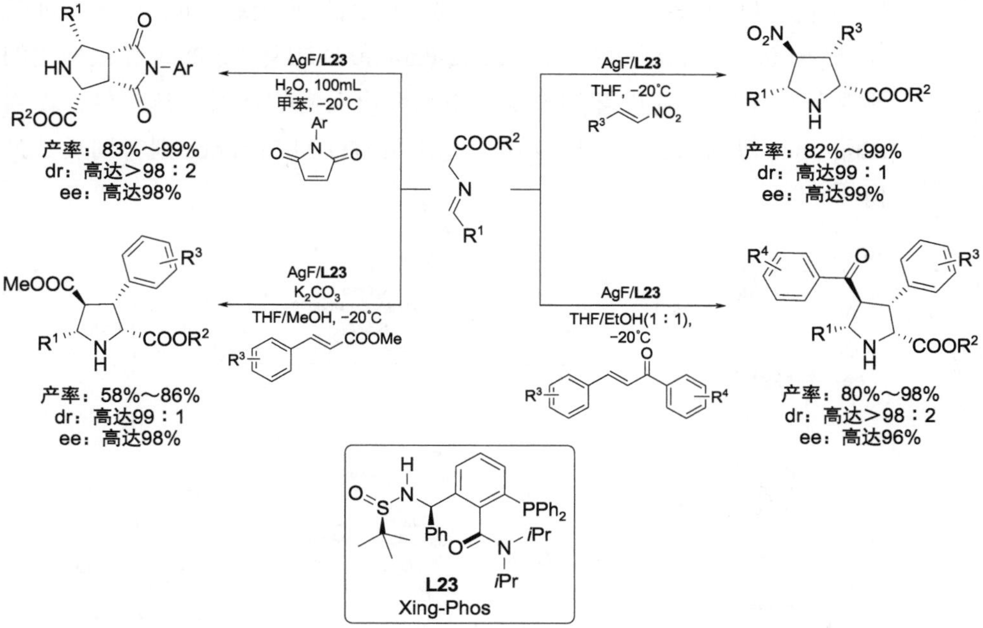

图 5-24 Xing-Phos（**L23**）在银催化不对称 [3+2] 环加成反应中的应用

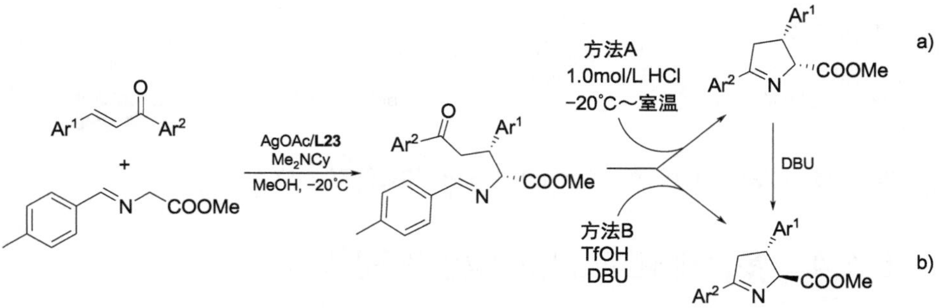

图 5-25 立体发散性合成顺式和反式的 Δ(1)-吡咯啉化合物

  徐利文课题组之后还探索了更多新型的 [3+2] 环加成反应，将甘氨酸醛亚氨基酯与各种亲电试剂反应，例如 MBH 型乙酸酯和甲硅烷基醚、1,1-二取代环丙烯和芳香族亚胺[35-37]（图 5-26）。使用 Ag/Xing-Phos（**L23**）作为催化剂，获得了各种具有高对映选择性(16 个底物实例，ee 值最高 99%)和非对映选择性(dr 最高 96∶4)的多立体中心的手性咪唑烷[38]。此外，甘氨酸醛亚氨基酯的聚合可以以高产率(91%～96%)获得手性聚咪唑烷，其中硅系聚咪唑烷可以得到非常窄的分子量分布（$M_n > 33000$，$M_w/M_n=1.08$）。值得注意的是，Xing-Phos 和 DTBM-SegPhos 在银催化的偶氮甲碱叶立德-亚胺环加成反应中表现出反向的非对映选择性，使用 Xing-Phos 得到顺式手性咪唑烷，而使用 DTBM-SegPhos 则得到反式产物[39]。

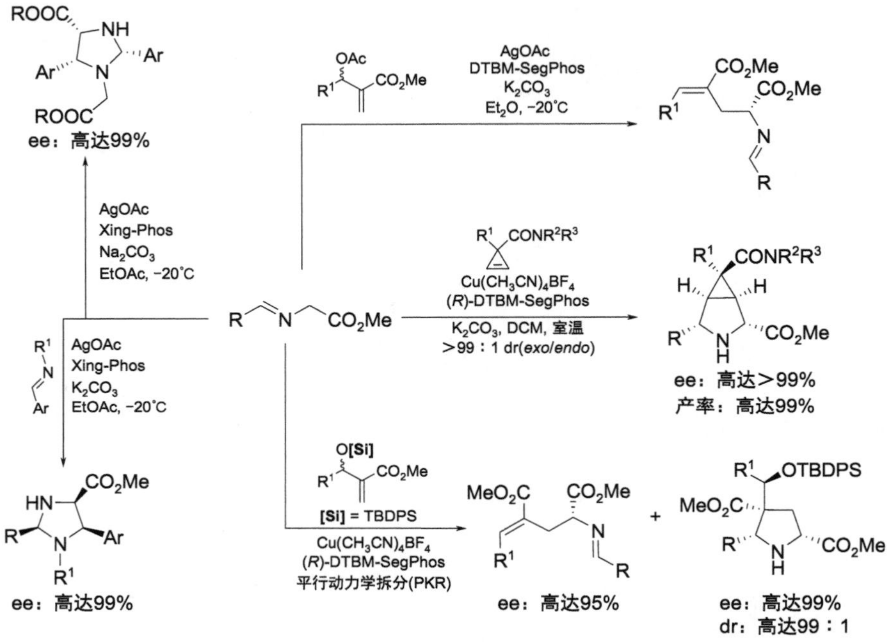

图 5-26 甘氨酸醛亚氨基酯与多种亲电试剂发生 [3+2] 环化反应

## 5.1.4 多官能化 P,N,N 型手性配体

Noyori 课题组报道的 Ru/ 双膦-二胺催化体系已被证明对酮的不对称直接氢化非常有效[40,41]。然而，这种催化系统也存在一定的局限性，比如底物范围局限：大位阻的酮、二烷基酮、杂环酮和亚胺还没有高效的氢化方法；二胺配体的潜在解离降低了催化剂的稳定性。基于 Noyori 的双膦-二胺配体，有机化学家们在手性二胺骨架上引入含膦基团合成了一系列新型的 P,N,N 型手性配体。例如，2004 年，Somanathan 等合成了基于环己二胺的膦配体 **L24**[42]，并将其用于 Ru 催化的酮类化合物的不对称氢化反应，尽管苯乙酮的反应具有高对映选择性，但产率仅

有15%～20%。随后，Clarke等开发了**L25**～**L29**[43-47]，相对于伯胺片段，仲胺片段的使用有效提升了反应活性[48]。2018年，Morris等开发了**L30**，并应用于锰催化的酮不对称转移氢化反应[49]，但是只有中等的对映选择性(图5-27)。

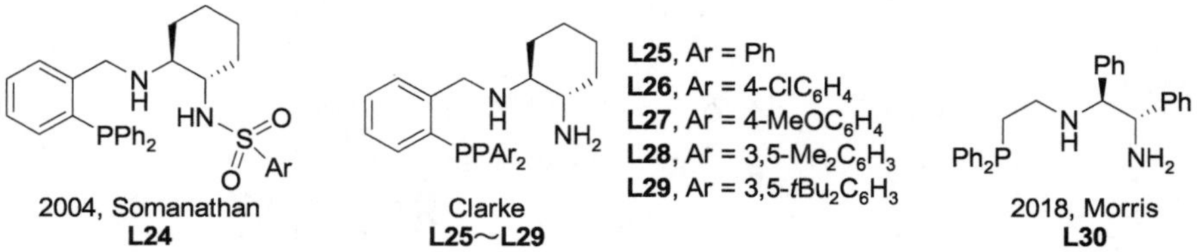

图5-27 基于手性二胺骨架制备的P,N,N型手性配体

一般来说，P,N,N配体可以被视为一个手性氨基膦(P,N部分)与一个二胺片段(N,N部分)的结合，这两个部分都是配体设计中非常有价值的模块，通过$S_N2$反应或还原胺化的模块化合成很容易得到P,N,N手性配体。例如，周其林课题组报道了基于螺环氨基膦骨架开发的三齿配体SpiroPAP (**L31**)[50,51]，其额外引入的吡啶末端可以防止二聚复合物的形成，从而达到提高催化效率的作用。SpiroPAP已经可以应用于Ir催化的酮不对称直接氢化的工业生产中，该催化体系对酮的不对称直接氢化反应具有广泛的底物普适性。2013年，Kitamura等基于BINOL骨架开发了**L33**[52]，其在叔烷基酮的不对称直接氢化中表现出出色的对映选择性控制。随后，张绪穆课题组在2018年开发了一个易合成的手性氧杂螺环骨架，在此基础上合成了可用于Ir催化的不对称直接氢化的新型P,N,N配体**L34**[53]。几乎同时，丁奎岭课题组报道了一个环己基融合的螺二茚烷骨架，在此基础上合成了**L35**[54]。2019年，周其林等人利用手性噁唑啉部分替换吡啶基片段来修饰SpiroPAP合成了新配体**L32**[55](图5-28)，其在Ir催化α-酮酰胺的不对称直接氢化中表现出优异的催化性能。

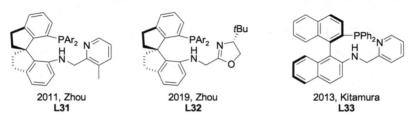

图5-28

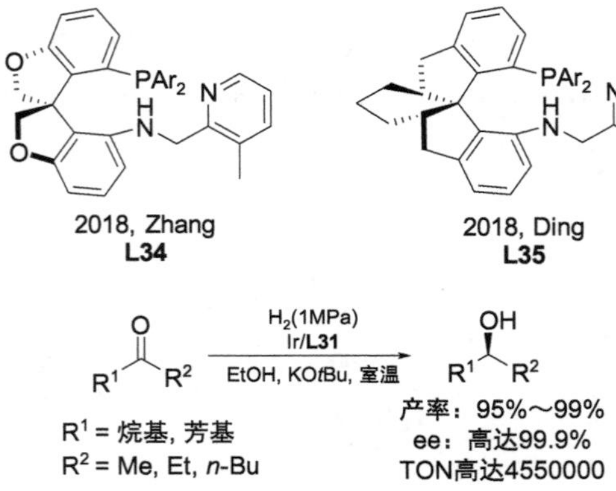

**图 5-28** 手性螺环吡啶氨基膦配体及其应用

手性二茂铁骨架配体同时具有中心手性和平面手性，是手性配体的一类优势骨架。近十年以来，化学家们基于手性二茂铁平面手性骨架开发了一系列手性配体[56-60]。2013 年，陈卫平和张生勇开发了第一个基于二茂铁的手性 P,N,N 配体（**L36**），在 Ir 催化酮的不对称氢化反应中表现出良好的对映选择性和高催化效率[61]。2017 年，Clarke 等人成功在 Mn 催化酮类不对称氢化中使用该配体[62]。2016 年，张绪穆课题组通过用膦基二茂铁基替换 Ambox 的一个噁唑侧臂，开发了二茂铁基手性 P,N,N 配体 f-Amphox（**L37**），其在 Ir 催化的苯乙酮不对称氢化反应中表现出高达 99.9% 的对映选择性，且具有广泛底物适用范围，并成功应用于百万吨级工业生产中[63]。后来，胡向平团队在 **L36** 的 N,N 部分加入手性基团合成了 **L38**，可以提高 Ir 催化酮类不对称氢化反应的对映选择性[64,65]。2018 年，张绪穆课题组将磺胺与 Ugi 胺结合合成了 f-Amphamide（**L39**）[66]。同年，钟为慧课题组报道了含磺酰胺的配体 **L40**[67]。这两种配体对于 Ir 催化的酮类不对称直接氢化都非常有效。该课题组还报道了掺入苯并[d]咪唑的 P,N,N 配体 **L41** 和 **L42** 并成功将其应用于 Mn 催化的酮类不对称直接氢化中[68,69]。2020 年，刘强及其同事报道了 **L43**，其在苯并[d]咪唑末端带有游离 NH 官能团[70]。其中，在喹啉的不对称直接氢化反应中发现配体与底物之间存在 π-π 堆积现象（图 5-29）。

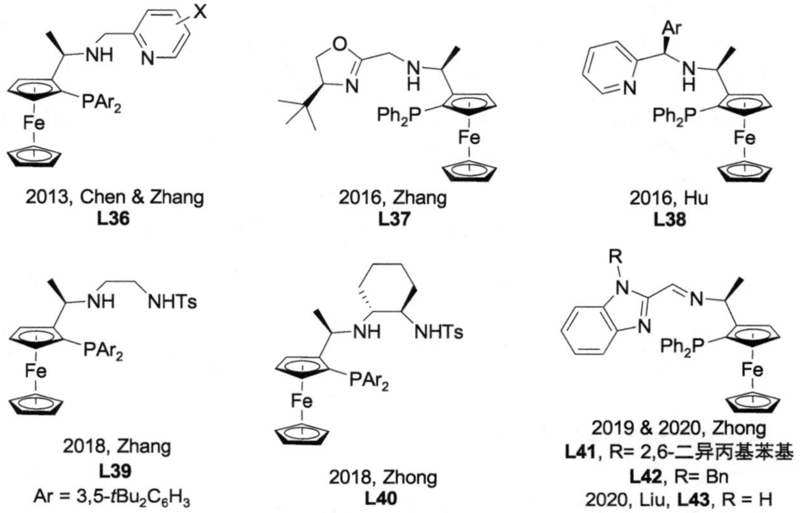

2013, Chen & Zhang
**L36**

2016, Zhang
**L37**

2016, Hu
**L38**

2018, Zhang
**L39**
Ar = 3,5-*t*Bu$_2$C$_6$H$_3$

2018, Zhong
**L40**

2019 & 2020, Zhong
**L41**, R= 2,6-二异丙基苯基
**L42**, R= Bn
2020, Liu, **L43**, R = H

图 5-29 手性二茂铁骨架氨基膦配体

除了手性二胺外，另一些手性 N,N 框架也被用于 P,N,N 配体的设计。2003 年，Brunner 课题组在 2-噁唑啉-2-基吡啶的基础上开发了 **L44**[71]。2018 年，黄正和刘桂霞开发了含氨基噁唑啉片段的配体 P,N,N-**L45**，并成功应用于 Ru 催化的亚胺不对称氢化，在反应中观察到了催化剂的二聚[72]。2019 年，胡向平课题组通过用磷酰苯取代磷酰二茂铁，开发了一种基于 f-Amphox 的新型 P,N,N 配体 **L46**，该配体失去了平面手性，在 Ir 催化的 ADH 反应中使用只能产生中等的对映选择性[73]。同年，黄正和刘桂霞报道了一种手性膦-噁唑啉配体（**L47**），其可以应用于钴催化乙烯基硅烷的不对称直接氢化[74]。2019 年，丁奎岭课题组报道了一种基于二甲基吡啶的手性配体 **L48**，它结合了 Milstein 发展的非手性 P,N,N 配体片段和手性磷烷基片段，该配体在 Mn 催化的酮类不对称直接氢化中具有高效反应活性和较广的底物范围[75]。2020 年，Farkas 等开发了 **L49** 和 **L50**，并探索了其在 Ir 催化烯酮加氢反应中金属环的环张力与化学选择性之间的关系[76]。2021 年，汤文军及其同事通过还原胺化引入手性氨基膦片段合成了 **L51**，该配体已应用于酮的共催化不对称氢化[77]（图 5-30）。

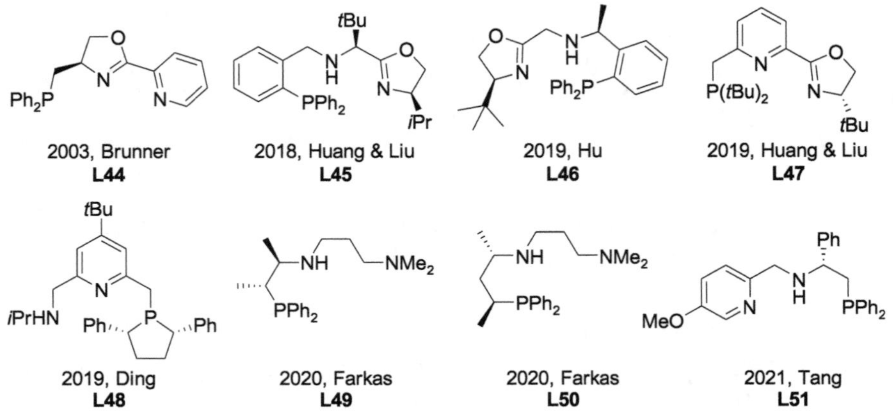

图 5-30　P,N,N 型手性配体

## 5.1.5　多官能化 N,P,N 型手性配体

1997 年，张绪穆课题组就设计合成了一类通过三价磷结合手性侧臂（如吡咯烷和噁唑啉）的 $C_2$-对称的手性 N,P,N 配体（**L52** 和 **L53**），这类配体被成功应用于 Ru 催化的不对称转移氢化反应[78,79]。2000 年，Braunstein 团队采用了类似的设计理念[80]，基于苯基亚膦酸酯制备了手性 N,P,N 配体 **L54**（图 5-31），该配体在苯乙烯与重氮乙酸乙酯的不对称环丙烷化和苯乙酮在丙-2-醇中的转移氢化反应中表现出有效的反应活性。

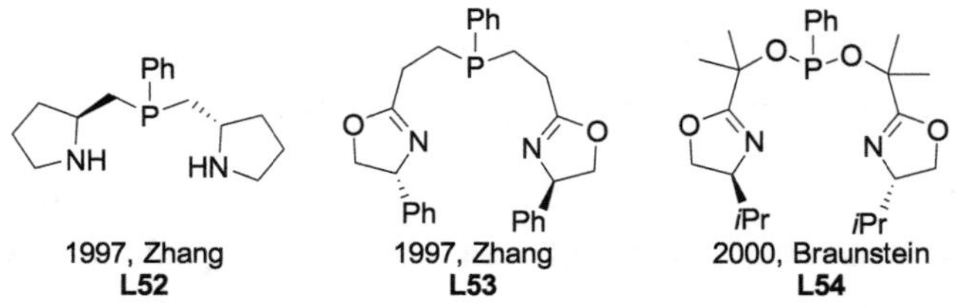

图 5-31　N,P,N 型多官能手性配体

## 5.1.6　多官能化 P,O,O 型手性配体

徐利文课题组以 BINOL 为原料开发了一种兼具 $C_2$-轴手性和 $sp^3$ 碳手性的双羟基化合物 Ar-BINMOLs，并应用于有机金属试剂与酮的不对称 1,2-加成和三甲基氰化硅烷与 $α,β$-不饱和羰基化合物的 1,4-共轭加成反

应[81-83]。基于非等价配位的多功能膦配体设计理念，通过在这类双羟基的优势骨架联萘环的 3 号位引入磷原子，成功开发了 Ar-BINMOLs 衍生的手性膦配体 Ar-BINMOL-Phos[84]（图 5-32）。

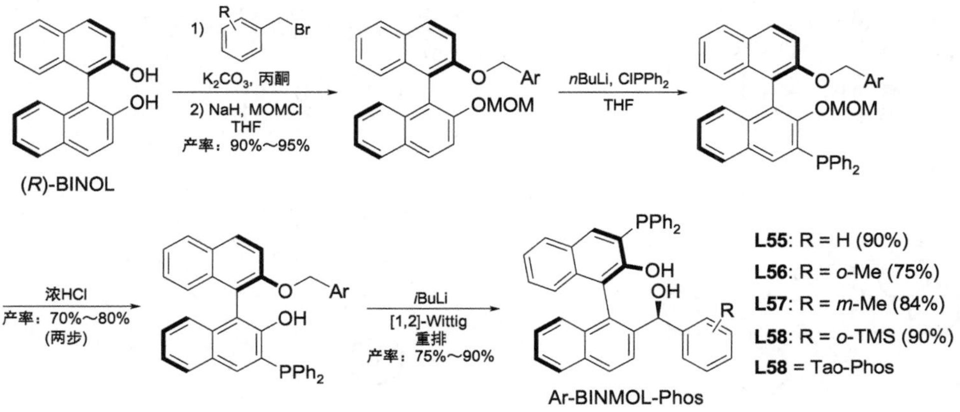

图 5-32 多官能手性配体 Ar-BINMOL-Phos 的合成

这类合成简便的 P,O,O 配体可应用于各种不对称金属催化反应中。例如，使用 (R,S)-Ar-BINMOL-Phos **L55** 作为配体，在 $Me_2Zn$ 存在下芳醛与末端炔烃的炔基化反应主要产生 S 型产物（ee 值为 44% ～ 70%，方法 A）[84]，而 $CaH_2$ 和 n-BuLi 与 **L55** 的组合以高产率生成了具有相反对映选择性的 R 型产物（ee 值为 78% ～ 99%，方法 B）。该反应能产生对映选择性逆转的原因可能是因为 n-BuLi/$CaH_2$ 与配体结合形成了双金属锂/钙基多功能金属络合物（模型 B，图 5-33）。此外，对照实验表明，Ar-BINMOL-Phos 携带的 $sp^3$-碳立体中心手性对反应至关重要，当使用没有 $sp^3$-碳立体中心手性的其他双羟基配体或 P,O 配体都会导致对映选择性和产率的降低[85]。

之后，徐利文课题组还探索了 Ar-BINMOL-Phos 在季碳立体中心的对映选择性构建中的应用。2015 年，他们研究了不对称铜催化的 Huisgen 环加成反应，Tao-Phos（**L58**）表现出优秀的催化能力，基于琥珀酰亚胺合成了含有全碳立体中心的三氮唑化合物[86]（图 5-34）。与其他手性膦配体相比，**L58** 在控制反应中的化学和对映选择性（ee 值＞99%）方面表现出优越的性能。有趣的是，对照实验和 ESI-MS 分析的机理研究

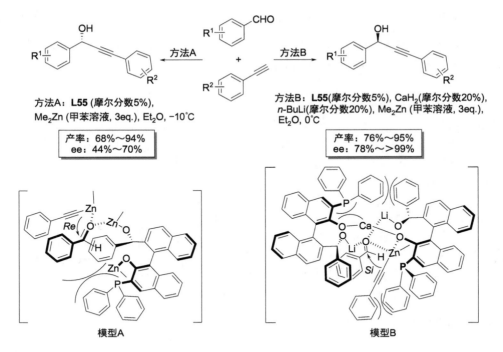

图 5-33 Ar-BINMOL-Phos 控制的对映选择性逆转在合成中的应用

表明，高对映选择性去对称化过程是基于 **L58** 的双核铜物种的形成。此外，**L58** 在纳米氧化亚铜催化的不对称 Huisgen 环加成反应中是唯一有效的膦配体，具有良好的产率，对映选择性最高可达 90%[87]。这是首例纳米 $Cu_2O$ 催化不对称叠氮化物-炔烃 Huisgen 环加成反应。机理研究进一步表明，**L58** 的双羟基部分对于在纳米氧化亚铜的铜表面形成活性催化剂物种至关重要。

手性噻咯的合成在合成化学和材料化学中非常重要，2020 年，徐利文团队报道了一种高对映选择性铑催化分子内硅氢加成反应，合成了一系列四取代的硅手性苯并噻咯化合物[88]（图 5-35）。值得一提的是，手性苯并噻咯产物表现出良好的聚集诱导发光（AIE）和循环偏振发光（CPL）活性。机理研究表明，Ar-BINMOL-Phos 的酚羟基和磷中心作为非等价配位配体的双齿供体，手性醇羟基可以有效调节 O-Rh 键的配位强度从而提高 Rh 中心的催化活性，同时还可以稳定手性 Rh 催化剂以增强对映选择性诱导。

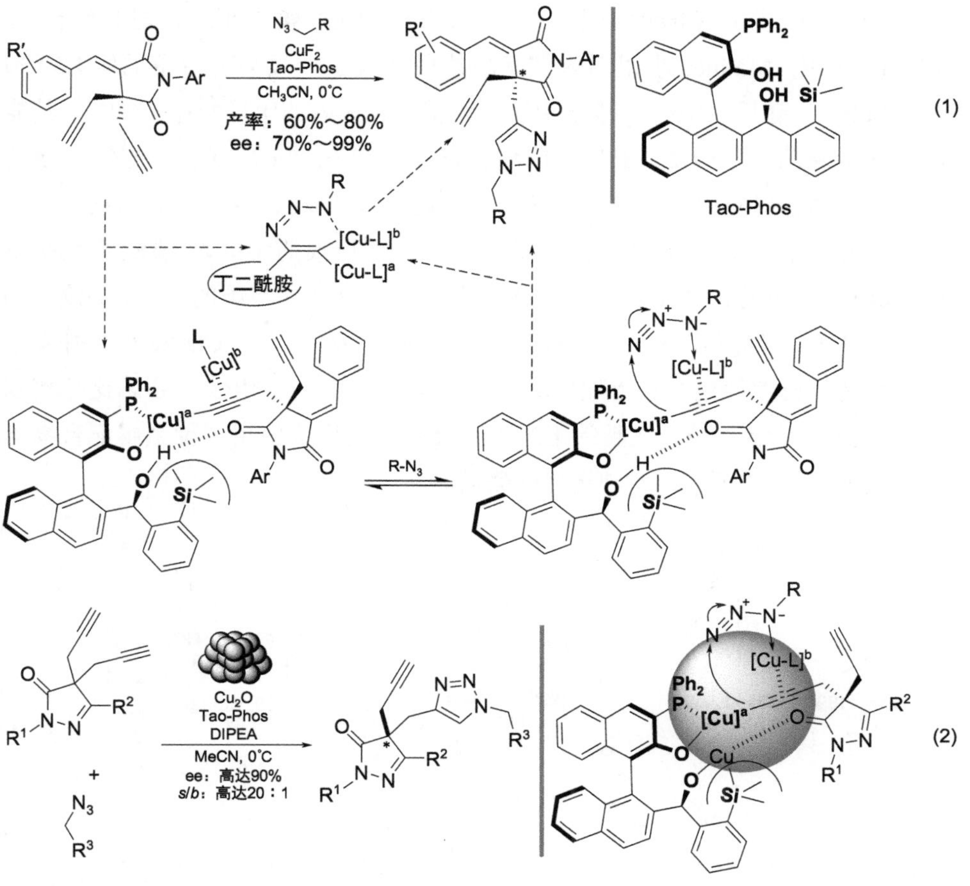

图 5-34 Tao-Phos 在不对称铜催化 Huisgen 环加成反应中的应用

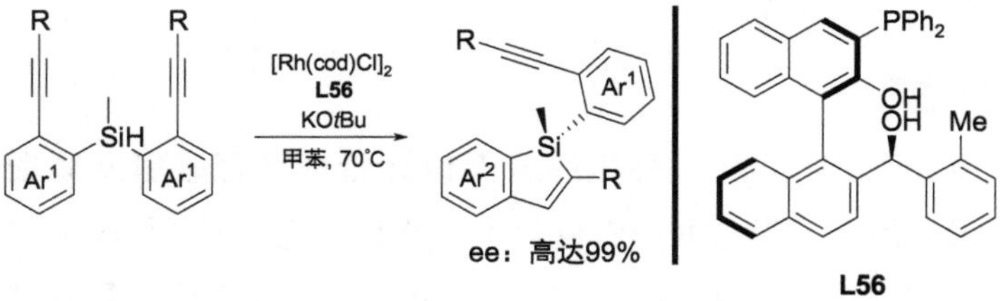

图 5-35 铑催化不对称合成硅立体中心苯并噻咯化合物

## 5.1.7 多官能化 P,N,O 型手性配体

与能为贵金属提供相对稳定配位能力的 P,N,N 配体不同，P,N,O 配体

(图 5-36)携带有一个强配位的磷原子和一个配位能力相对较弱的氧原子。1997 年，Mathieu 小组引入烷基醚片段合成了一类半稳定型配体 **L59** 和 **L60**[89]。随后，Kwong 等报道了一种基于希夫碱的手性配体 **L61**[90]，其能与金属阳离子以 2:1 配位形成配合物[91]。这三种 P,N,O 配体与钌形成的配合物都已应用于酮的不对称转移氢化中，并获得了中等的对映选择性。2003 年，郑卓及其同事探索了基于二茂铁骨架的希夫碱-膦配体 **L62** 及其在 Ru 催化的酮类不对称转移氢化中的应用[92]。有趣的是，将配体/Ru 比从 1:1 增加到 2:1 可显著提高对映选择性。2010 年，Clarke 小组开发了含有环己烷片段的三齿配体 **L63**，并成功地以 1:1 的配体/金属比分离到了金属配合物[93]。在酮的不对称直接氢化反应中使用这种金属配合物比其他 PNN 类似物的对映选择性略高。

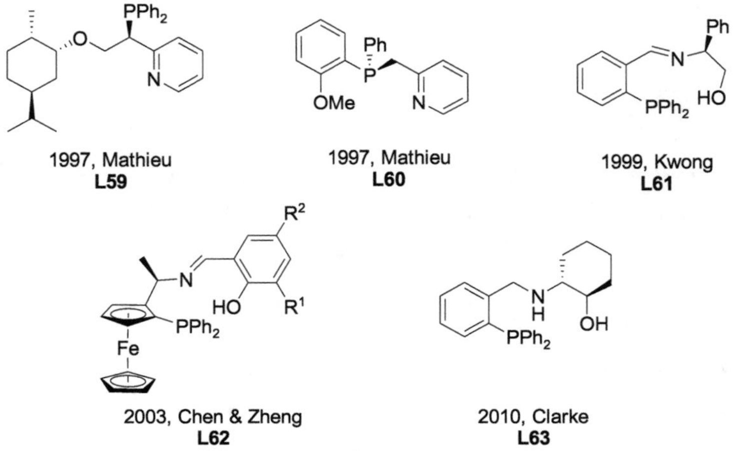

**图 5-36** PON 型手性配体

2013 年，Sawamura 团队基于脯氨醇骨架设计合成了羟基氨基膦配体 **L64**，并将其成功用于炔烃与醛的铜催化不对称加成反应[94](图 5-37)。DFT 计算表明，配体上的羟基和醛之间存在氢键作用，吡咯烷环和羰基氧之间还存在非经典的 $sp^3$-CH⋯O 氢键。这两处的氢键相互作用被认为是醛羰基定向的关键因素。

2017 年，Sawamura 等将这类配体应用于不对称铜催化的炔-硝酮偶联反应(Kinugasa reactions)，以高对映立体选择性合成了一系列手性 $\beta$-内酰胺，该反应具有很好的底物普适性[95]（图 5-38）。

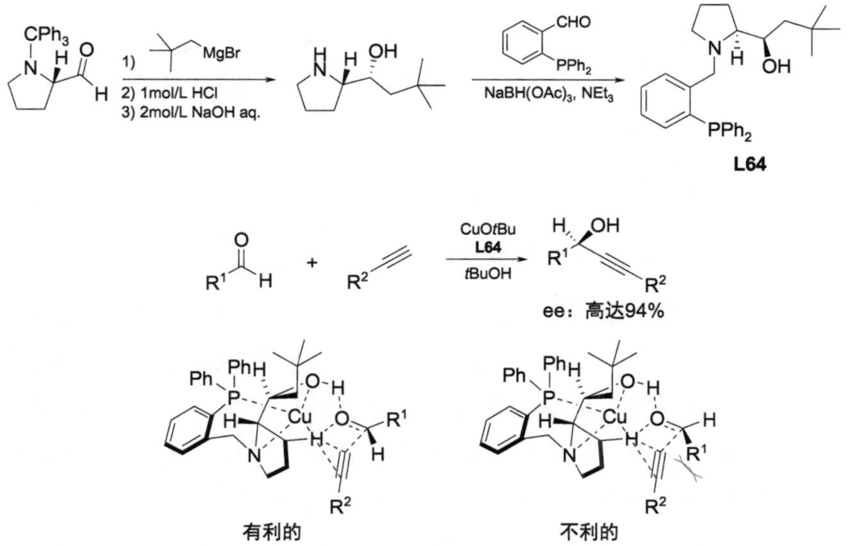

图 5-37 羟基氨基膦配体 L64 的制备及其应用

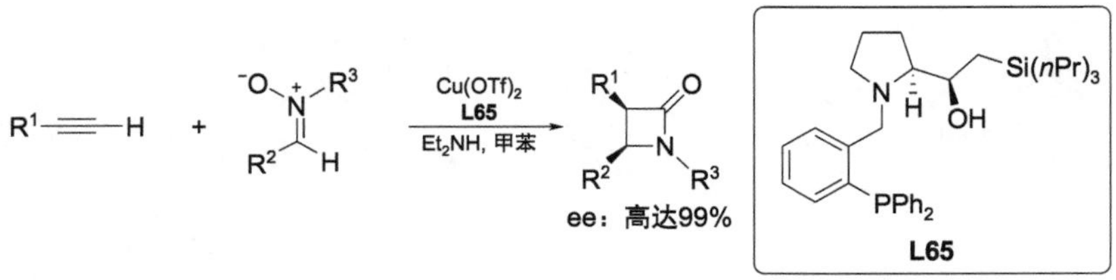

图 5-38 羟基氨基膦配体 L65 在 Kinugasa 反应中的应用

2017 年，张绪穆课题组报道了两种高效手性 P,N,O 配体 f-Ampha（**L66**）[96] 和 f-Amphol（**L67**）[97]（图 5-39），它们结合了优势手性二茂铁骨架与非手性氨基酸或手性氨基醇片段。在 Ir 催化的酮类不对称直接氢化中，这两种配体均表现出优秀的对映选择性和高效的转化效率。2018 年，Altan 课题组通过将手性氨基醇引入磷苯酰苯中，开发了一种新的配体 **L68**[98]。遗憾的是，该配体在 Ru 催化的不对称转移氢化反应中表现出较差的对映选择性。

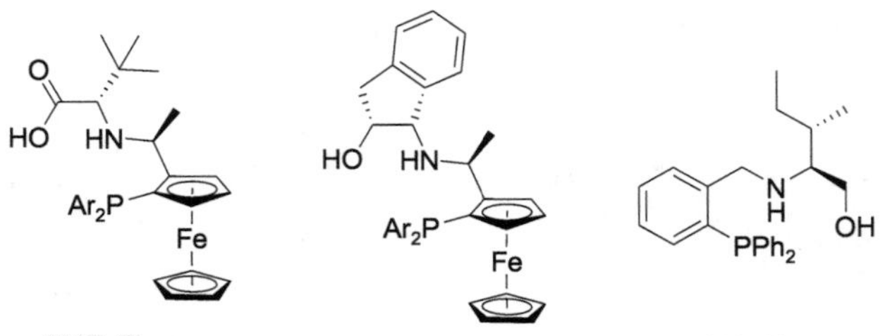

2017, Zhang
**L66**

2017, Zhang
**L67**

2018, Altan
**L68**

图 5-39　P,N,O 型手性配体

徐利文课题组在 BINOL 骨架上引入带有 $sp^3$-碳手性的亚胺二苯基膦片段开发了一种新型的 P,N,O 型手性配体 HZNU-Phos(**L69**)[99,100]。该配体在不对称催化反应中表现出优异的催化性能，一方面，在不对称 1,4-加成反应中，配体与金属铜和锌原位形成的双金属多核 Cu/Zn/**L69** 复合物被证明是催化该反应的优势催化剂，在这类反应中表现出迄今为止最高的催化效率（S/Cu=20000，TON=17600），且以 ee 值最高 99% 的对映选择性制备了目标产物[99]（图 5-40）。另一方面，在钯催化的 2-氰基乙酸

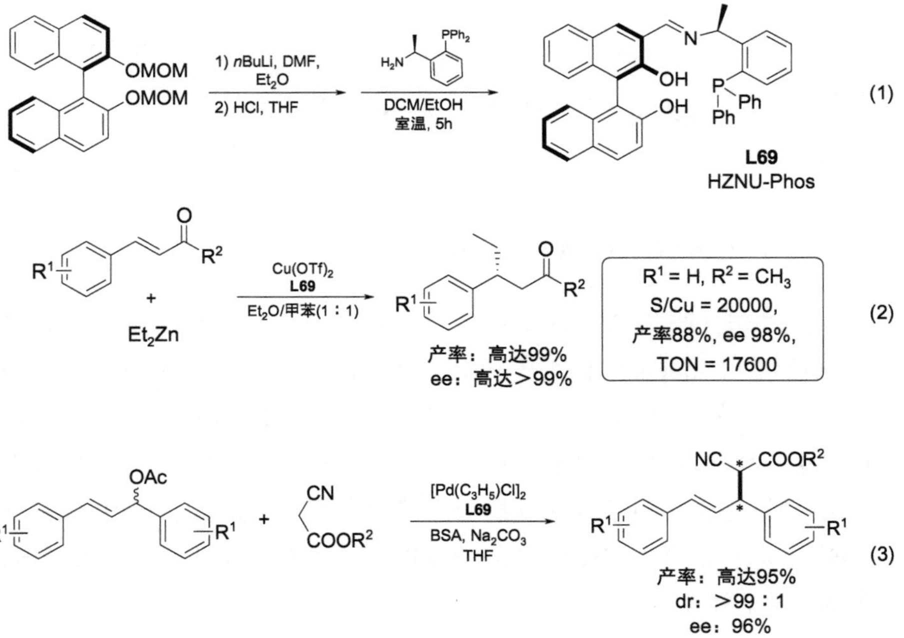

图 5-40　HZNU-Phos 制备及其在不对称催化中的应用

酯的烯丙基烷基化反应中，HZNU-Phos 也表现出优异的催化活性，以高非对映选择性和对应立体选择性构建了两个连续的碳手性中心[100]。

## 5.1.8 多官能化 P,N,S 型手性配体

手性亚磺酰基是常见的手性片段，被用于构建手性配体中，但相对较弱的配位作用难以形成稳定的催化剂。2012 年，邓金根小组制备了一类含手性亚磺酰基的三齿配体 **L70**[101]。作者选择了相对刚性的苯基作为连接基团，创建了较为稳定的配位环境。但该配体在 Ir 催化的酮不对称直接氢化中仅表现出中等的对映选择性。周其林课题组通过用二价硫原子取代 SpiroPAP 中的末端吡啶，开发了一系列高效的 P,N,S 配体 SpiroSAP(**L71** 和 **L72**)[102,103]。这两个配体(图 5-41)都被成功用于 *β*-烷基-*β*-酮酯的不对称直接氢化反应。

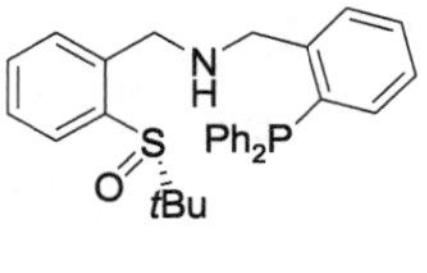


2012, Deng
**L70**

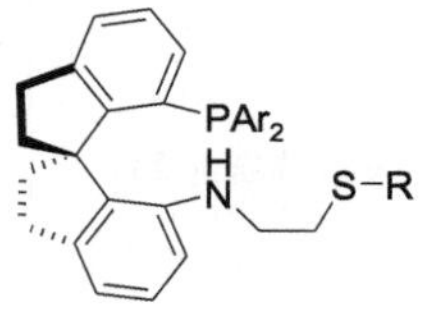

2015, Zhou
**L71**

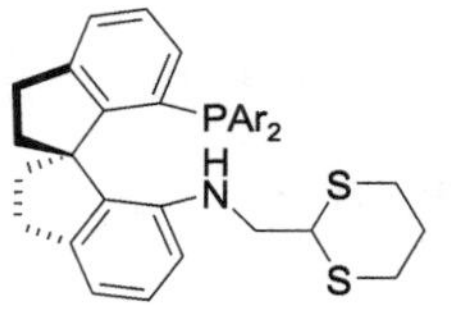

2017, Zhou
**L72**

图 5-41 多官能化 P,N,S 型手性配体

## 5.1.9 多官能化 N,P,O 型手性配体

2001 年，侯雪龙和戴立信课题组基于 PHOX 配体开发了同时具有碳手性、膦手性和轴手性的膦配体 SiocPhox[104]，其前体二茂铁二氨基膦在与 BINOL 反应过程中，在 P 原子上形成了一个新的手性中心(图 5-42)，

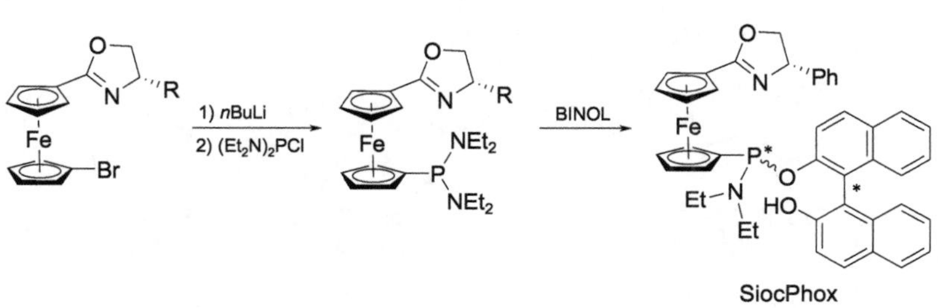


图 5-42 SiocPhox 的制备

所有非对映异构体很容易通过柱色谱分离。

SiocPhox 系列配体在烯丙基烷基化/胺化反应中均表现出良好区域选择性和对映选择性控制[105]，例如，配体 ($S,S_{phos},R_a$)-Bn-SiocPhox **L73** 在烷基化反应中得到最好的对映立体选择性，而 ($S,R_{phos},R_a$)-Ph-SiocPhox **L74** 在胺化反应中表现更好。使用其他亲核试剂的烯丙基取代反应[105-110]，也可

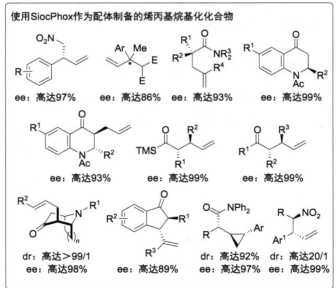

图 5-43 SiocPhox 在钯催化烯丙基取代反应中的应用

以以优异的区域选择性和良好对映选择性得到目标产物(图 5-43)。研究表明，在手性 P 原子上引入 BINOL 片段，庞大的位阻会将烯丙基底物中的取代基推向噁唑啉，从而影响产物的区域选择性[105]；膦手性中心的增加更有利于反应立体选择性的调控；BINOL 部分保留了游离 OH 官能团，这对于 Pd 催化的不对称烯丙基取代反应，尤其是胺化反应至关重要。

除了不对称烯丙基烷基化 / 胺化反应，SiocPhox 配体也已成功应用于不对称钯催化的乙烯基氮丙啶与烯酮的不对称 [3+2] 环加成反应[111]，以高对映选择性(ee值最高 99%)制备了一系列多取代的吡咯烷类产物(图 5-44)。

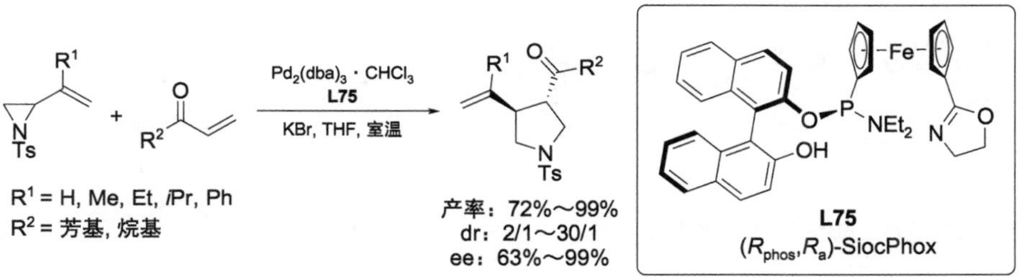

R1
N
Ts
+
O
R2
$Pd_2(dba)_3 \cdot CHCl_3$
L75
KBr, THF, 室温
R1
O
R2
N
Ts
R1 = H, Me, Et, iPr, Ph
R2 = 芳基, 烷基
产率：72%～99%
dr：2/1～30/1
ee：63%～99%
Fe
O
P
NEt2
OH
N
O
L75
($R_{phos}$,$R_a$)-SiocPhox


图 5-44　Pd/SiocPhox 催化 [3+2] 环加成反应

2010 年，戴立信课题组使用 Pd/($S_c$,$R_{phos}$,$S_a$)-*i*Pr-SiocPhox (**L76**) 作为催化剂，实现了单取代烯丙基底物与苯甲醇的不对称醚化，以高区域选择性和高对映选择性制备了烯丙基醚类化合物[112](图 5-45)。

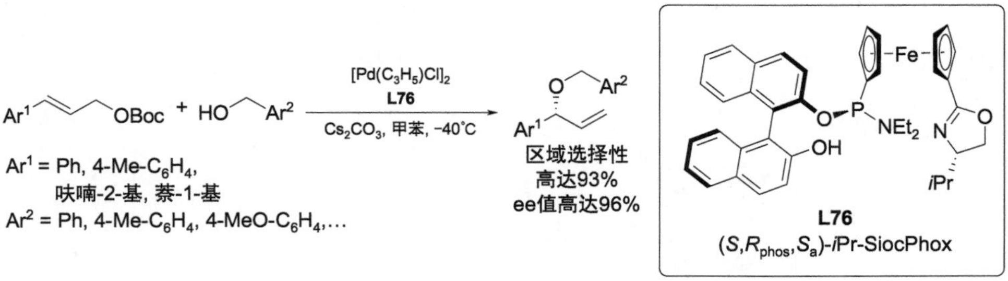

Ar1
OBoc
+ HO
Ar2
$[Pd(C_3H_5)Cl]_2$
L76
$Cs_2CO_3$, 甲苯, -40°C
O
Ar2
Ar1
Ar1 = Ph, 4-Me-$C_6H_4$,
呋喃-2-基, 萘-1-基
Ar2 = Ph, 4-Me-$C_6H_4$, 4-MeO-$C_6H_4$,…
区域选择性
高达93%
ee值高达96%
Fe
O
P
NEt2
OH
N
O
iPr
L76
(S,$R_{phos}$,$S_a$)-iPr-SiocPhox


图 5-45　SiocPhox 在钯催化烯丙基醚化反应中的应用

## 5.2 多官能化双膦配体

### 5.2.1 P,P,N 和 P,N,P 型手性配体

1980 年，Kumada 课题组利用 $N,N$-二甲基-1-二茂铁乙基胺作为起始原料，首先通过双负离子化引入二苯基膦，之后再与乙酸酐反应引入乙酰酯基，最后在甲醇溶液中与 $N,N,N'$-三甲基乙二胺反应制备得到了手性双叔氨基二茂铁双膦化合物 **L77**[113]（图 5-46）。

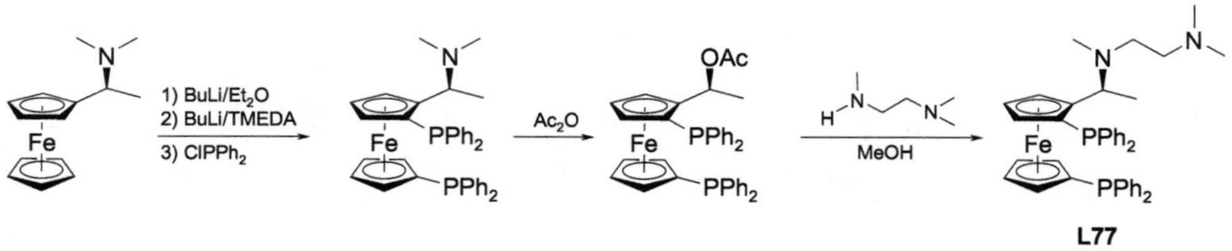

图 5-46　手性双叔氨基二茂铁双膦配体的制备

这类带有末端氨基和叔胺的柔性二茂铁基膦配体可以很好地应用在不对称氢化、烯丙基烷基化和胺化以及羟醛缩合反应中[114,115]。例如，Ito、Sawamura 和 Hayashi 研究了双官能手性双膦配体 **L77** 在不对称金催化 Aldol 反应中的应用，各种醛与甲基异氰乙酸酯之间发生缩合反应，以高收率、非对映选择性和对映选择性制备了一系列手性噁唑啉化合物（图 5-47）[116]。

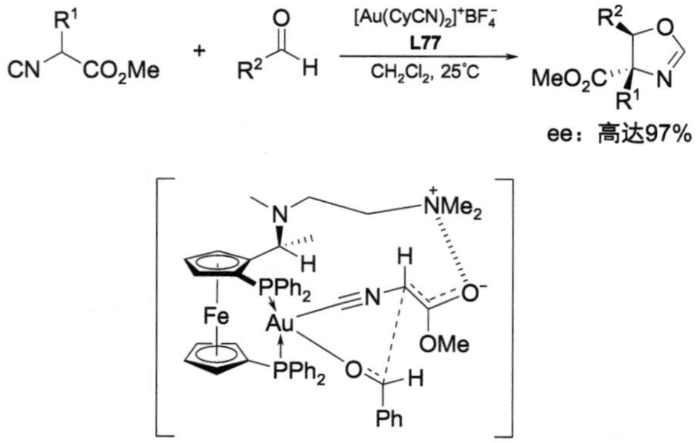

图 5-47　金催化不对称羟醛缩合反应

作者认为，配体中氨基链的长度和碱性基团对于实现高立体选择性都非常重要，两个氨基之间的链长增加或减少都会导致对映选择性的显著降低。这一例子表明多功能手性配体不仅可以连接和调节金属的反应活性，而且还能够同时与底物发生非共价相互作用调节对映立体选择性，为后来开发此类多官能新型手性配体及在新型不对称催化反应的应用奠定了基础。

2016 年，张绪穆、董秀琴等人报道了一类基于二茂铁骨架的新型手性双膦配体 WudaPhos（**L78**）。该配体具有一个碱性叔胺片段，该片段可以在氢化过程中去质子化并与不饱和酸形成离子对。研究结果表明具有三受阻象限的 WudaPhos 对 2-芳基和 2-烷基丙烯酸的对映选择性氢化反应具有很好的催化效果[117]，该催化体系对不饱和磺酸盐底物也同样适用[118]。类似地，该课题组最近开发了叔丁基取代的 *t*Bu-WudaPhos（**L79**），该配体也可以很好地催化 4-酮基丙烯酸[119]和乙烯基苯甲酸不对称氢化[120]（图 5-48）。

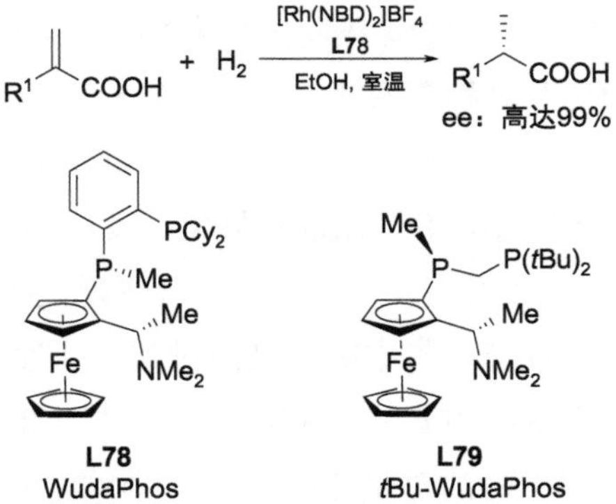

图 5-48　WudaPhos/Rh 催化的不饱和酸的对映选择性氢化

此后，该课题组用二级氧化膦（SPO）基团取代了 WudaPhos 中与二茂铁相连的膦基团合成了新型配体 SPO-Wudaphos **L80**，并成功应用于一系列不饱和羧酸或者膦酸的氢化（图 5-49）。对照实验和 DFT 计算表明，二级氧化膦基团可以连接金属并通过羟基部分的氢键与底物酮羰基相互作用，并且配体的质子化胺和底物的羧酸盐之间也存在离子对相互作用[121]。2017 年，该课题组利用 Rh/SPO-WudaPhos 催化体系，对 α-苯基取代的乙烯基膦酸进行氢化[122]，合成了一系列包括芳基、环己基和苯甲酰氧基 α-取代的手性乙基膦酸，ee 值最高达到 98%，TON（转化数）高达 2000。

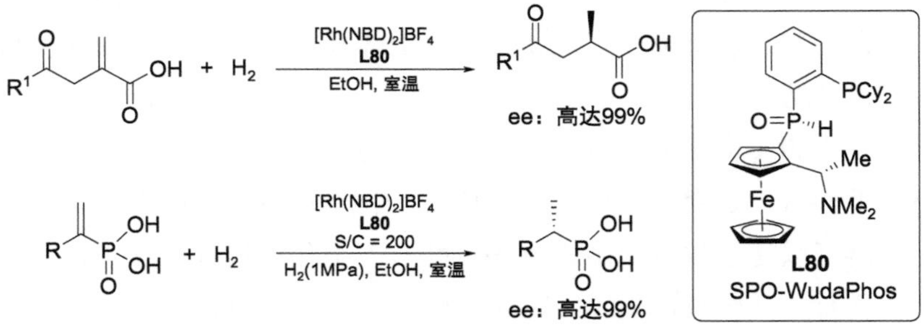

图 5-49　不饱和羧酸/膦酸的对映选择性氢化

1991 年，Burk 等报道了第一例具有两个手性含膦杂环戊烷的 P,N(H),P 配体 **L81**[123]。该配体与一系列过渡金属（如 Ir、Mn、Fe、Re 和 Ru）的络合物可应用于不对称氢化反应[124-126]，表现出中等至高度的对映选择性。此后，各课题组跟进了相关报道，陆续开发出多种类似结构的 P,N,P 配体（图 5-50），并广泛应用于不对称氢化反应中。1995 年，Bianchini 和 Graziani 等报道了手性 **L82**[127]，该配体在中心氮原子邻位上引入了手性片段，Ir/**L82** 配合物适用于烯丙基酮的不对称直接氢化，反应具有中等的对映选择性。1998 年，Bianchini 小组[128,129]在 2-甲基喹喔啉的不对称直接氢化中使用了 Ir/**L82** 配合物，反应具有中等至高的对映选择性。1996 年，张绪穆课题组报道了基于吡啶的 P,N,P 配体 **L83**，该配体在 Ru 催化的简单酮不对称直接氢化中表现出中等的对映选择性[130]。同年，Osborn 等报道了 **L84**[131]，发现这种基于吡啶的配体对亚胺的不对称直接氢化具有催化效果。2002 年，Large 课题组开发了一种衍生自 PPFA 的配体 **L85**，也称为 BoPhoz 配体，该配体被广泛应用于不对称氢化反应中[132]。2004 年，郑卓课题组开发了基于手性二茂铁的 **L86**[133]，该配体被证明在 Ru 催化的酮不对称直接氢化中是有效的。2005 年，Boaz 等人利用手性亚磷酰胺酯代替氨基膦部分[134]，在 BoPhoz 配体中引入额外的轴手性元素制备了 **L87**，相对于 BoPhoz，新配体在烯酰胺的氢化反应中表现出更好的对映选择性控制。2011 年，Kuriyama 及其同事报道了 P,N(H),P 配体 **L88**[135]，该配体被应用于 Ru 催化的不对称直接氢化反应。2014 年，Morris 小组报道了通过氨基膦与膦醛缩合制备的 P,N,P 配体 **L89**[136]，2017 年，该小组对该配体进行了修饰，通过还原胺化得到 **L90**[137]，这两种配体与 Fe(Ⅱ) 形成的配合物都可以应用于简单酮的不对称直接氢化，**L90** 在反应中表现出更好的对映选择性诱导能力。2016 年，Zirakzadeh 和 Kirchner 等开发了基于二茂铁的手性 P,N,P 配体 **L91**[138]，并成功将其用于 Fe(Ⅱ) 和 Mn(Ⅰ) 催化的不对称氢化反应[139]。2020 年，周其林团队开发了基于手性螺环骨架的 PNP 配体 **L92**[140]，其与铱的配合物形成的六元、八元双环构成一个非常窄的"手性口袋"，在二烷基酮的不对称直接氢化中表现出优异的对映选择性。同年，张绪穆课题组开发了二茂铁基手性配体 **L93**[141]，其与锰的配合物也可形成类似的五元、七元双环"手性口袋"，在各类酮的不对称直接氢化中表现出优秀的对映选择性。

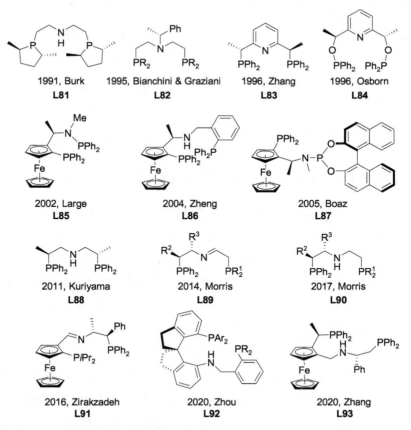

图 5-50　应用于不对称氢化的 P,N,P 配体

2020 年，殷亮课题组将 **L94** 应用于 α,β-不饱和酰胺的不对称共轭膦氢化反应[142]，构建了一系列带有酰胺片段的手性膦氧化合物，反应具有很好的收率和对映选择性(图 5-51)。值得一提的是目标产物可以与 Pd 金属形成钳状钯复合物，并且能以优秀的对映选择性控制实现查耳酮的不对称膦氢化反应(图 5-51)。

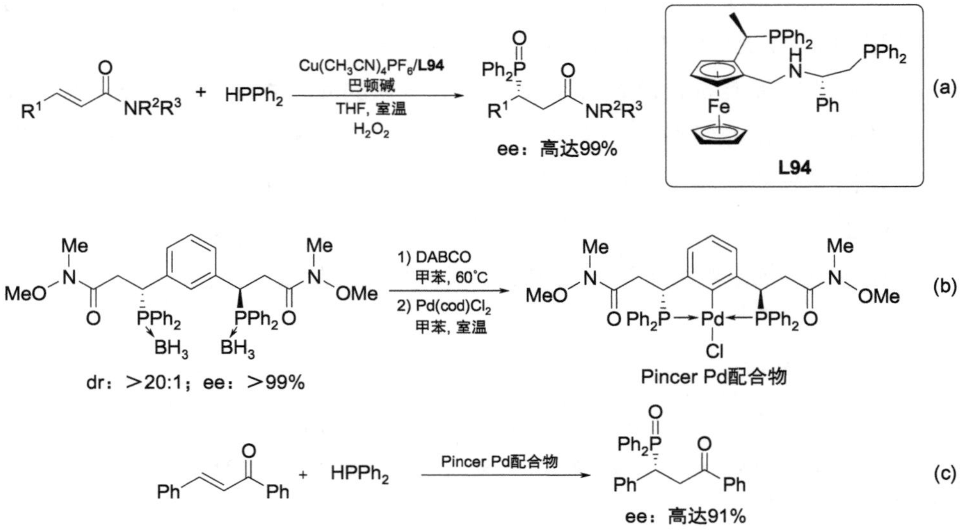

图 5-51　α,β-不饱和酰胺的不对称 1,4-共轭膦氢化及其衍生反应

## 5.2.2　P,P,N,N 型多齿配体

张绪穆课题组开发了一类新型二茂铁骨架 P,P,N,N 配体 ZhaoPhos(**L95**)，可用于铑催化的硝基烯烃不对称氢化(图 5-52)[143,144]。在进一步的研究中，使用 N-甲基化 ZhaoPhos(**L96**)可以很好地催化 β-氰基肉桂酸酯的氢化反应[145](图 5-52)。作者认为，配体的硫脲基团与底物之间存在氢键相互作用，各种带有氢键受体底物的不对称氢化反应活性和对映选择性都能得到提高[146-152]。

除此之外，ZhaoPhos 的硫脲官能团可以与亚胺盐卤化物的阴离子相互作用，充当与阴离子结合的位点[153]，NMR 研究也证实了配体中两个 NH 单元和底物氯离子之间存在双氢键作用。在 Rh(Ⅰ)/**L95** 催化体系下，酮亚胺类底物以优异的产率和对映选择性转化为铵盐产物(图 5-53)。此策略已成功应用于多种亚胺型底物的不对称氢化反应[154-157]。

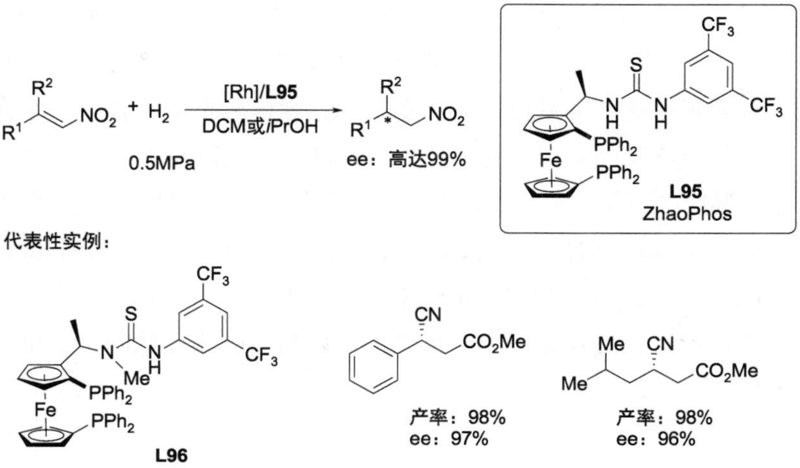

图 5-52 ZhaoPhos 催化的不对称氢化

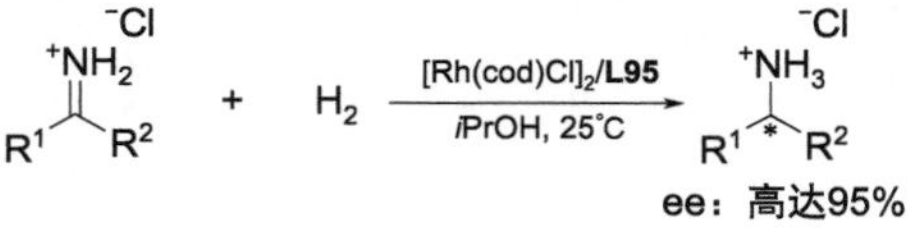

图 5-53 亚胺盐的不对称氢化反应

除了不对称氢化，ZhaoPhos 还被成功应用于 Pd 催化的 $\beta$-酮酯脱羧不对称烯丙基烷基化，对于在 $\alpha$ 位含有多种酯取代基的六元环烯丙醇碳酸酯，该反应能以高对映选择性得到目标产物[158]（图 5-54）。研究表明 ZhaoPhos 与底物之间的氢键作用是该反应成功的重要因素。

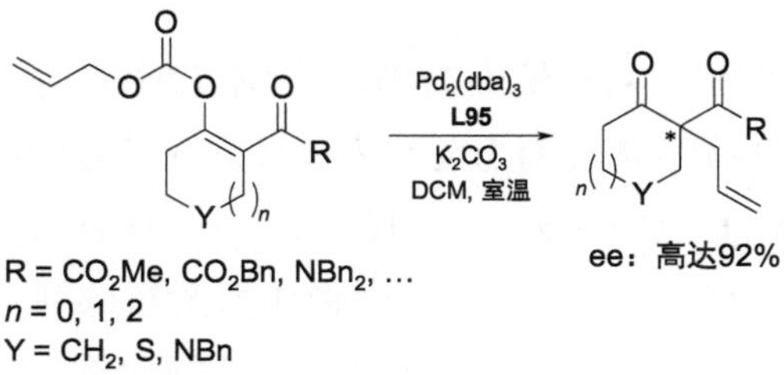

图 5-54 $\beta$-酮酯脱羧不对称烯丙基烷基化反应

### 5.2.3　P,N,N,P 型配体

Trost 等人开发了一类新型多官能双膦配体 (R,R)-DACH-naphthyl (**L97**)，并应用于不对称烯丙基取代反应[159-162]（图 5-55）。例如，他们利用钯催化乙烯基氮丙啶的动态动力学拆分反应构建了一系列烯丙基取代的吡咯和吲哚衍生物，反应具有高产率和高对映选择性[163]（ee 值最高 96%）。之后，该方法成功应用于药物和生物活性天然产物的合成，例如 Longamide B、Longamide B 甲酯、Hanishin、Agesamides A 以及环碘酮。2019 年，基于 Pd/**L98** 催化体系[164]，该课题组以 3-位具有内酰胺侧链取代的吲哚作为亲核试剂，以高对映选择性制备了单萜吲哚生物碱（图 5-55）。

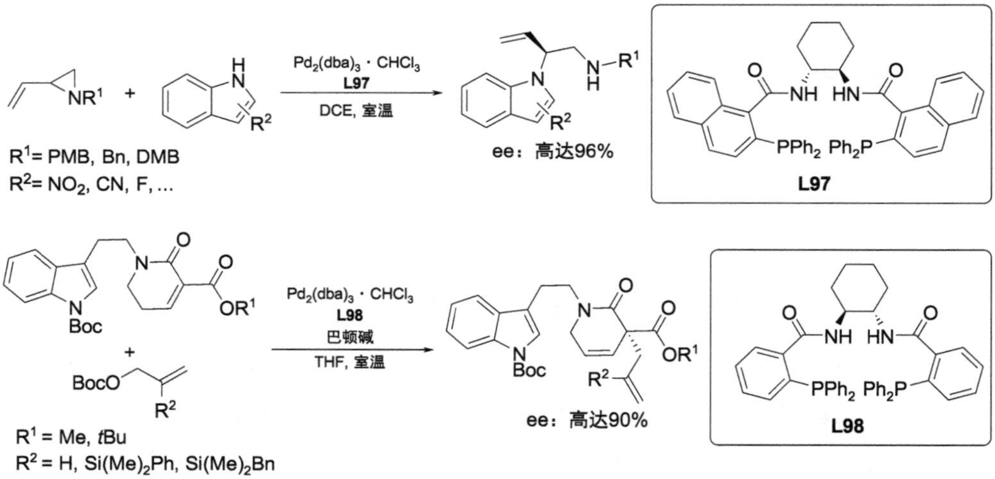

**图 5-55　钯催化不对称烯丙基取代反应**

2009 年，Trost 课题组成功将修饰的 Trost 配体 **L99** 应用于钯催化 4-甲氧基-N-(氨磺酰氧基)-苯磺酰胺对五元和六元环烯丙基碳酸酯的胺化反应[165]，产物 ee 值高达 96%（图 5-56）。2018 年，Trost 课题组使用乙烯基环丙烷作为底物，3-取代吲哚和色氨酸衍生物作为亲核试剂，成功合成了一系列高对映选择性（ee 值最高 98%）的 3,3-二取代的二氢吲哚[166]（图 5-56）。另外，钯/**L99** 催化体系可用于醛缩还原酶抑制剂雷尼司他的合成[167]。

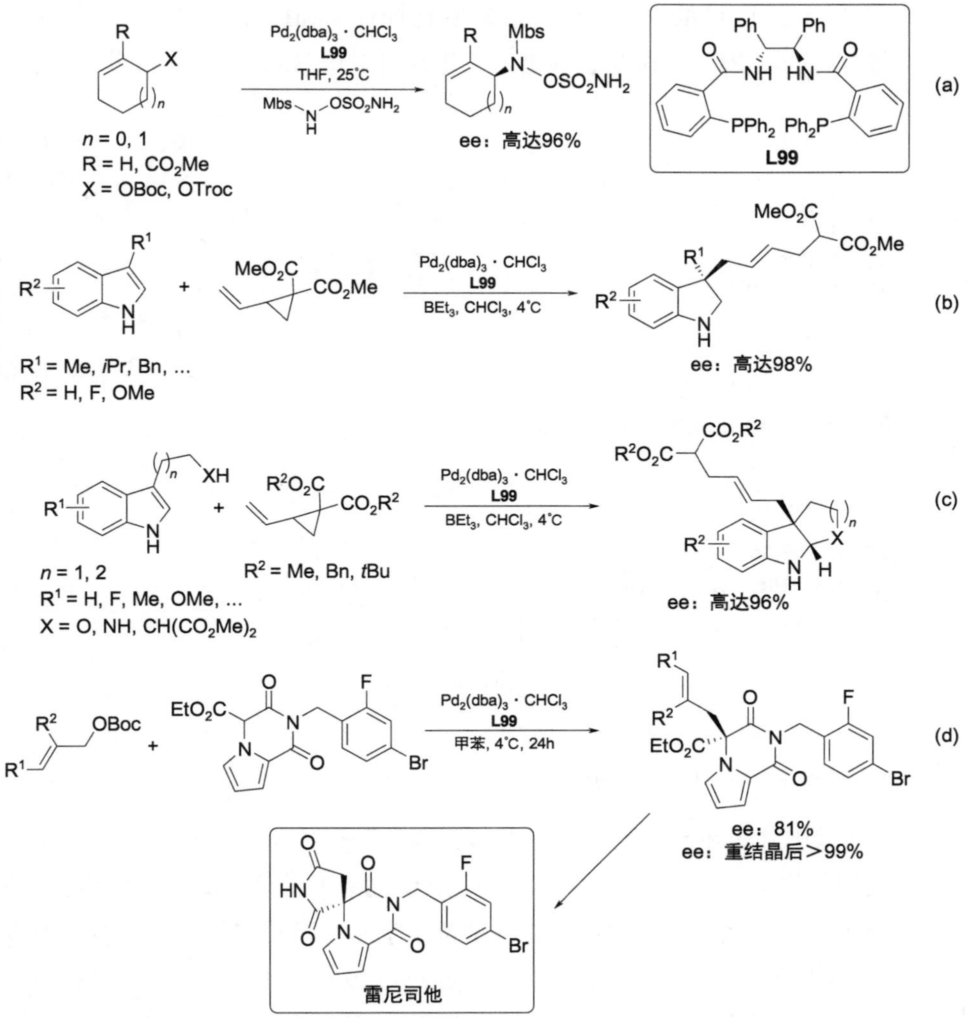

图 5-56　Pd/L99 催化的不对称烯丙基烷基化

近期，Trost 课题组以具有二氨基乙醇蒽骨架的 **L100** 作为配体，发展了一种高效的次膦酸去对称化反应。Pd/(*S*,*S*)-**L100** 催化体系能够有效区分两个对映体的氧原子，为具有高非对映选择性和对映选择性的手性次膦酸盐提供了一种新的合成途径[168]（图 5-57）。

2011 年，Ruffo 课题组以 β-1,2-D-葡糖二胺作为骨架合成了一种新型多官能化双膦配体 **L101**[169]（图 5-58）。Pd/**L101** 催化体系应用于内消旋环戊基-4-烯-1,3-二基双(甲苯磺酰基氨基甲酸酯)的去对称化反应，在 5min 内以定量收率和较高的对映选择性（ee 值最高 96%）合成了环戊烯并噁唑-2-酮。值得注意的是，催化剂可以用 BmpyBF$_4$ 溶剂进行回收。

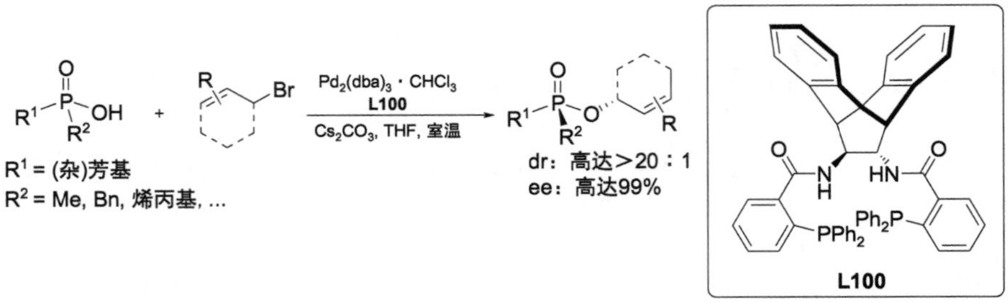

图 5-57　Pd/L100 催化构建次膦酸酯

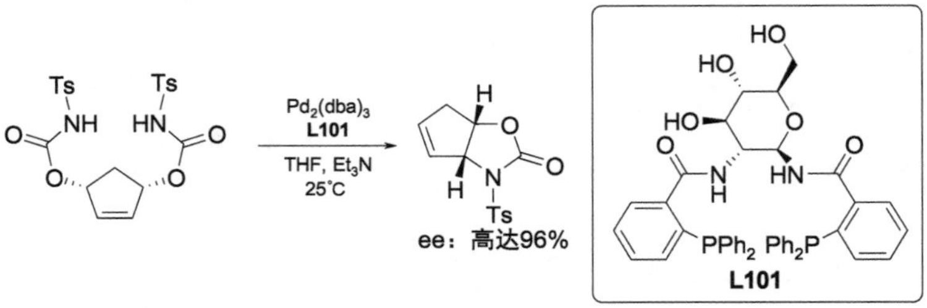

图 5-58　Pd/L101 催化去对称化反应

反式 1,2-环己二胺是一种理想且容易获得的手性原料，常被用于手性配体的构建[170,171]。2013 年，徐利文课题组通过手性反式 1,2-环己二胺衍生的席夫碱还原环化，设计并制备了空气中稳定的新型多官能化多齿 P,N,N,P 配体 Fei-Phos（**L102**）[172]。Fei-Phos 及其衍生膦配体在对映选择性钯催化烯丙基取代反应中得到很好的应用（图 5-59），醇、硅醇、芳香胺、吲哚、活化亚甲基化合物和儿茶酚及其类似物都能很好地适用于 Pd/**L102** 催化体系，以良好的收率和优秀的对映选择性（ee 值最高 99%）合成对应的烯丙基产物[172-174]。

值得一提的是，Fei-Phos 衍生配体在硅立体中心的构建方面也有一定的应用[175,176]。例如，铂/**L103** 催化炔醇与二氢硅烷的硅氢加成/环化反应中能以 32% 的 ee 值制备具有硅立体中心的手性硅醚化合物[175]（图 5-60）。

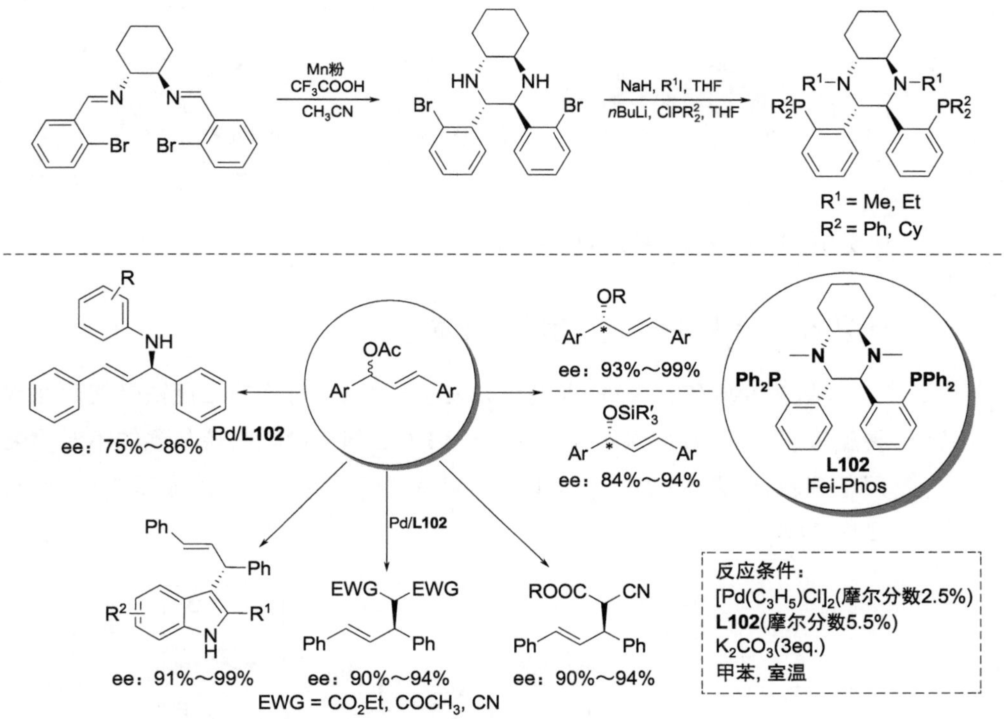

图 5-59　Pd/Fei-Phos 催化的不对称烯丙基取代反应

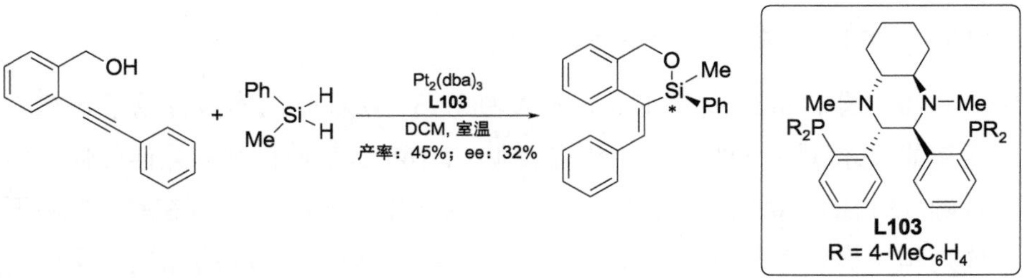

图 5-60　铂催化对映选择性催化构建硅立体中心

　　Morris 课题组发展了一种 Fe(Ⅱ) 催化酮的不对称转移氢化反应，以 PNNP 四齿双亚胺配体 **L104** 作为配体[177-179]（图 5-61）。NMR、X 射线衍射和对照实验表明，配体中的一个亚胺在反应中被氢化，底物酮羰基与配体的 N—H 之间存在氢键相互作用，诱导底物进入催化剂的手性环境，从而提升反应的对映选择性。

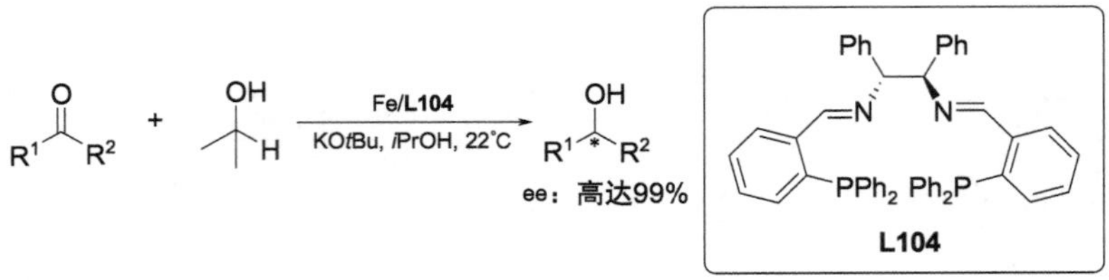

图 5-61　Fe/L104 催化酮的不对称转移氢化

### 5.2.4　N,P,P,N 型多齿配体

早在 1996 年，Ikeda 课题组开发了一类 $C_2$-对称的二茂铁骨架双氮双膦配体[180]。张万斌课题组研究了 Pd/L105 催化的吡咯烷与酮的不对称烯丙基烷基化（图 5-62），反应以高产率和对映选择性（ee 值最高 99%）得到目标产物[181-185]。

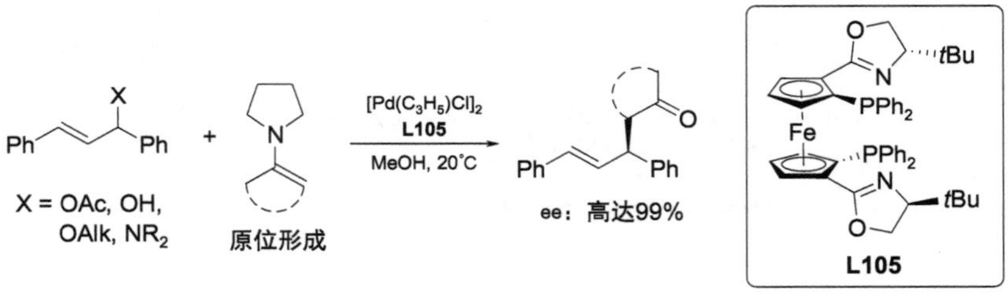

图 5-62　Pd/L105 催化的不对称烯丙基烷基化

2018 年，Shimuzu、Kanai 等人使用钯/硼双催化体系合成了 α-烯丙基羧酸（ee 值最高 99%）[186]。研究表明，L106 是钯催化剂的最佳配体，结合硼/(R)-SDP 催化体系，两种不同手性催化剂的组合可以在 C—C 键形成过程中微调手性环境，从而构建了一系列具有高对映选择性的季碳手性化合物（图 5-63）。

2007 年，张万斌等人开发了一系列基于二茂钌骨架的手性双膦配体 RuPHOX(L107)[187]。相对于二茂铁骨架，二茂钌的两个茂环距离更长，因此与金属的螯合角也更大。此外，茂金属的不同电子性质也会对配体产生相应的影响。该类配体目前已在 Ru 催化酮和 α,β-不饱和羧酸的不对称氢化中展现出优异的催化活性和对映选择性诱导能力[188-190]（图 5-64）。

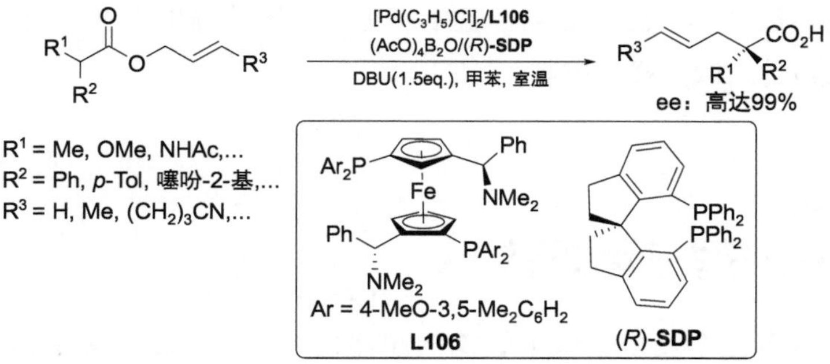

图 5-63 钯/硼双催化合成 α-烯丙基羧酸

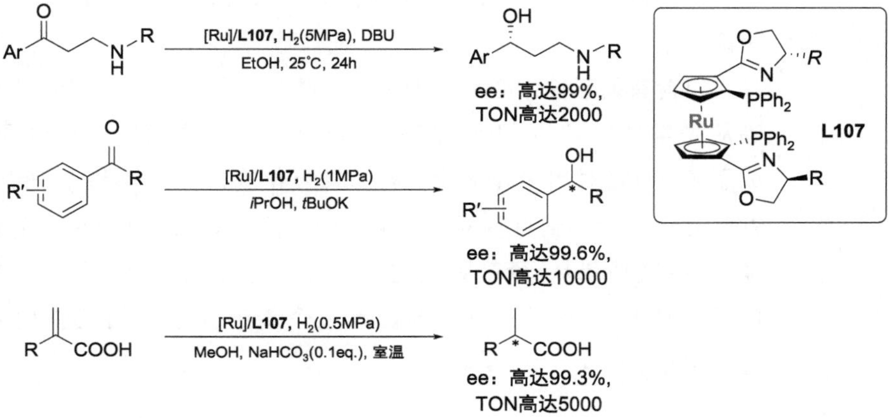

图 5-64 钌催化酮的不对称氢化

同年，Chen、McCormack 等人报道了一种包含三个二茂铁单元的 $C_2$-对称双膦双氮配体 **TriFer**(**L108**)[191]。该配体被应用于 α-取代肉桂酸的对映选择性氢化反应（图 5-65）。作者推测，该配体中的二甲基氨基使

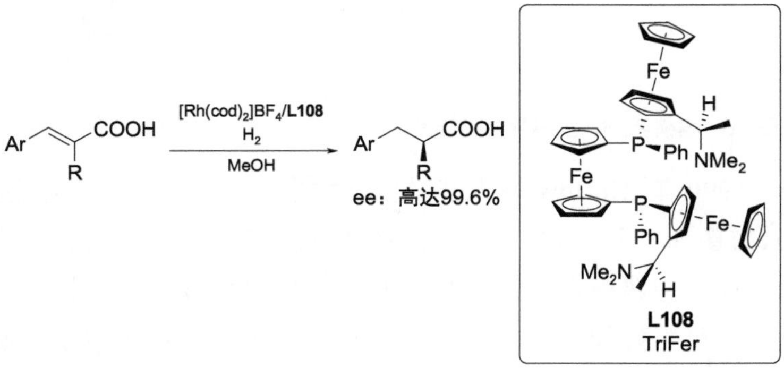

图 5-65 Rh/**L108** 催化 α-取代肉桂酸的不对称氢化

5 手性多官能化多齿型膦配体　271

底物羧酸去质子化,加氢步骤中产生了关键的静电相互作用,从而做到了反应的对映选择性控制。

# 5.3
## P,P,P型三齿配体

含有三个及以上叔膦结构的配体相对于单膦和双膦配体较为少见,统称为多膦配体。本章节只讨论为数较多的三膦配体,四个及以上磷原子的多膦配体大部分以大分子或高分子聚合物的形式存在,不在本章节讨论。

包含有磷手性中心片段的手性多膦配体研究较早,1987年,Imamoto等从光学纯甲基(对甲氧基苯基)苯基膦氧化物出发,经过两步转化合成了乙烯基(对甲氧基苯基)苯基膦氧化物,然后通过与$PhPH_2$加成和$SiHCl_3$还原后得到具有两个磷手性中心的三膦配体 **L109**。另外,将乙烯基(对甲氧基苯基)苯基膦氧化物与另一种α-酯基手性膦氧化物加成后,再经脱酯基和还原反应得到具有三个磷手性中心的三膦配体 **L110**[192](图 5-66)。

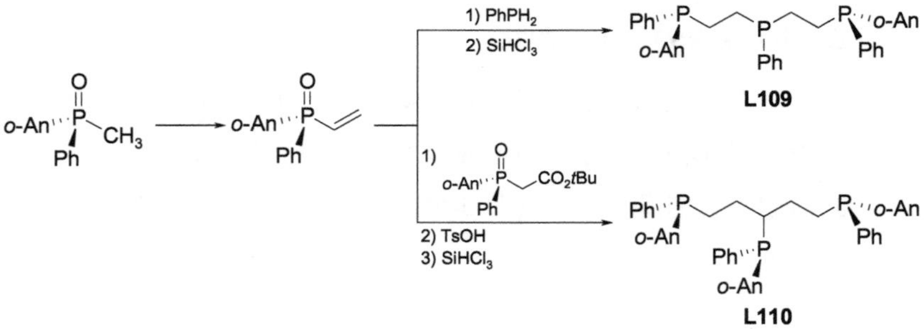

图 5-66 含磷手性中心三膦配体的制备

1990年,Harlow 等人[193-196]开发了 $C_2$ 对称性的三膦配体 **L111** 和 **L112**(图 5-67),与具有 $C_2$ 对称性的双齿手性配体相比,这类配体的八面体过渡金属配合物中减少了非等价残留配位点的数量,因而在催化反应中可更好地控制对映选择性。这两种配体都可以很容易地从 2,5-二甲基-1-苯基膦烷开始制备,目前已知这类配体可与 Fe、Rh 和 Ru 形成络合物，并在官能化烯烃的不对称氢化反应中有较好的活性和选择性。

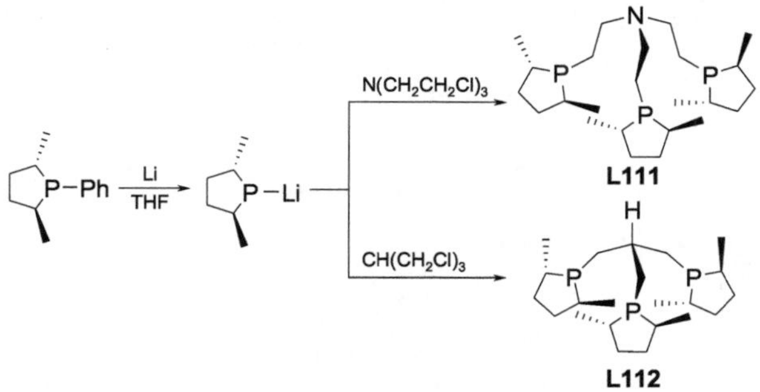


图 5-67　携带 2,5-二甲基磷杂环戊基三膦配体的制备

1995 年，Togni 课题组开发了手性 P,P,P 配体 Pigiphos（**L113**）（图 5-68）[197]。将手性二苯基膦二茂铁部分引入两侧臂，形成伪 $C_2$ 对称性。该配体在与 Ru(Ⅱ) 配位时会形成两种不同空间构象的异构体，其中面异构体在苯乙酮的不对称转移氢化中表现出更高的对映选择性控制(ee 值 72%)。

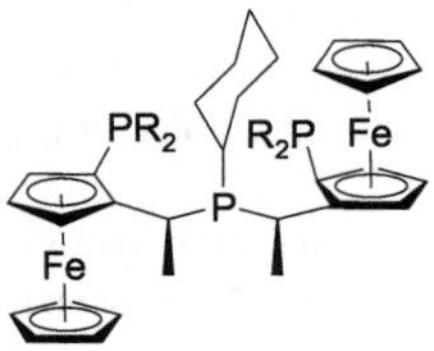


1995, Togni
**L113**
(*S*,*R*)-Pigiphos

图 5-68　手性三膦配体 Pigiphos

2003 年，Togni 课题组基于二茂铁骨架又开发了一种新型的三膦配体(图 5-69)，该配体由一个二茂铁骨架和一个柔性的侧臂组成。他们使用二苯基乙烯基膦氧化物为原料，经过与 $PhPH_2$ 加成，再与二茂铁骨架的手性氨基膦片段反应得到一对非对映异构体膦氧配体混合物。通过进一步柱色谱分离和还原反应，最终分别得到了非对映异构纯含有一个磷手性中心的二茂铁骨架的三叔膦配体 **L114** 和 **L115**。遗憾的是，该配体在 Ru 催化的酮不对称转移氢化中表现出较差的对映选择性[198]。

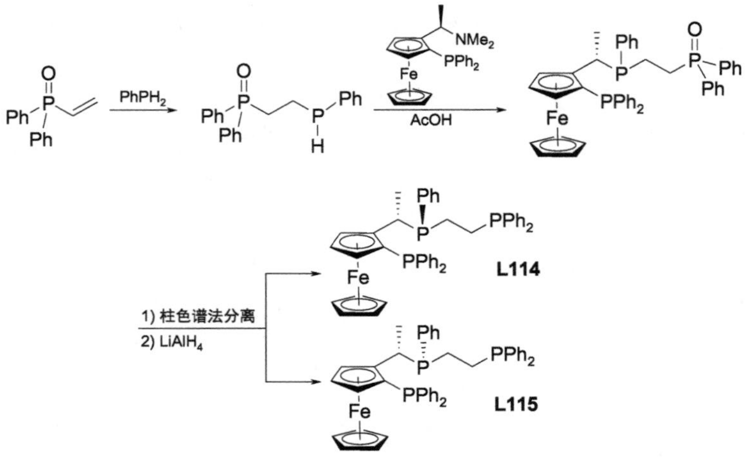

图 5-69　二茂铁骨架手性三叔膦配体

## 参考文献

[1] Kuriyama M, Soeta T, Hao X, et al. *N*-Boc-L-Valine-connected amidomonophosphane rhodium(Ⅰ) catalyst for asymmetric arylation of *N*-tosylarylimines with arylboroxines. Journal of the American Chemical Society, 2004, 126(26): 8128-8129.

[2] Sladojevich F, Trabocchi A, Guarna A, et al. A new family of cinchona-derived amino phosphine precatalysts: application to the highly enantio- and diastereoselective silver-catalyzed isocyanoacetate aldol reaction. Journal of the American Chemical Society, 2011, 122(6): 1710-1713.

[3] Ortín I, Dixon D J. Direct catalytic enantio- and diastereoselective Mannich reaction of isocyanoacetates and ketimines. Angewandte Chemie International Edition, 2014, 53(13), 3462-3465.

[4] Campa R, Yamagata A D G, Ortín I, et al. Catalytic enantio- and diastereoselective Mannich reaction of α-substituted isocyanoacetates and ketimines. Chemical Communications, 2016, 52, 10632-10635.

[5] Franchino A, Chapman J, Funes-Ardoiz I, et al. Catalytic enantio- and diastereoselective Mannich addition of TosMIC to ketimines. Chemistry A European Journal, 2018, 24(67): 17660-17664.

[6] Dong X Y, Zhang Y F, Ma C L, et al. A general asymmetric copper-catalysed Sonogashira C($sp^3$)-C(sp) coupling [J]. Nature Chemistry, 2019, 11, 1158-1166.

[7] Dong X Y, Cheng J T, Zhang Y F, et al. Copper-catalyzed asymmetric radical 1,2-carboalkynylation of alkenes with alkyl halides and terminal alkynes [J]. Journal of the American Chemical Society, 2020, 142(20): 9501-9509.

[8] Zhang Z H, Dong X Y, Du X Y, et al. Copper-catalyzed enantioselective Sonogashira-type oxidative cross-coupling of unactivated C($sp^3$)-H bonds with alkynes [J]. Nature Communications, 2019, 10, 5689.

[9] Zheng S C, Wang Q, Zhu J P. Catalytic atropenantioselective heteroannulation between isocyanoacetates and alkynyl ketones: synthesis of enantioenriched axially chiral 3-arylpyrroles [J]. Angewandte Chemie International Edition, 2019, 58, 1494-1498.

[10] Wang H F, Deng Q F, Zhou Z P, et al. $Ag_2CO_3$/CA-AA-AmidPhos multifunctional catalysis in the enantioselective 1,3-dipolar cycloaddition of azomethine ylides [J]. Organic Letters, 2016, 18(3): 404-407.

[11] Xiong Y, Du Z Z, Chen H H, et al. Well-designed Phosphine-Urea ligand for highly diastereo- and enantioselective 1,3-Dipolar cycloaddition of methacrylonitrile: a combined experimental and theoretical study [J]. Journal of the American Chemical Society, 2019, 141(2): 961-971.

[12] Huang Z Y, Chen Z L, Lim L H, et al. Weak arene C-H···O hydrogen bonding in palladium-catalyzed arylation and vinylation of lactones [J]. Angew. Chem., Int. Ed. 2013, 52: 5807-5812.

[13] Huang Z Y, Lim L H, Chen Z L, et al. Arene CH-O hydrogen bonding: a stereo- controlling tool in palladium-catalyzed arylation and vinylation of ketones [J]. Angewandte Chemie International Edition, 2013, 52: 4906-4911.

[14] Wang Z X, Nicolini C, Hervieu C, et al. Remote cooperative group strategy enables ligands for accelerative asymmetric gold catalysis [J]. Journal of the American Chemical Society, 2017, 139: 16064-16067.

[15] Yang H, Sun J W, Gu W, et al. Enantioselective cross-coupling for axially chiral tetra-ortho-substituted biaryls and asymmetric synthesis of gossypol [J]. Journal of the American Chemical Society, 2020, 142: 8036-8043.

[16] Xu G Q, Senanayake C H, Tang W J. P-chiral phosphorus ligands based on a 2,3-dihydrobenzo[d][1,3]oxaphosphole motif for asymmetric catalysis [J]. Accounts of Chemical Research, 2019, 52: 1101-1112.

[17] Zhang Z M, Chen P, Li W B, et al. A new type of chiral sulfonamide monophos-phine ligands: stereodivergent synthesis and application in enantioselective gold(I)-catalyzed cycloaddition reactions [J]. Angewandte Chemie International Edition, 2014; 53(17): 4350-4354.

[18] Zhang Z M, Xu B, Xu S, et al. Diastereo- and enantioselective copper(I)-catalyzed intermolecular [3+2] cycloaddition of azomethine ylides with β-trifluoromethyl β,β-disubstituted enones [J]. Angewandte Chemie International Edition, 2016, 55(18): 6324-6328.

[19] Zhang Z M, Chen P, Li W B, et al. Polymer-bound chiral gold-based complexes as efficient heterogeneous catalysts for enantioselectivity tunable cycloaddition [J]. ACS Catalysis, 2015, 5(12): 7488-7492.

[20] Hu H X, Wang Y D, Qian D Y, et al. Enantioselective gold-catalyzed intermolecular [2+2]-cycloadditions of 3-styrylindoles with $N$-allenyl oxazolidinone [J]. Organic Chemistry Frontiers, 2016, 3: 759-763.

[21] Wang Y D, Zhang P C, Di X Y, et al. Gold-catalyzed asymmetric intramolecular cyclization of N-allenamides for the synthesis of chiral tetrahydrocarbolines. Angewandte Chemie International Edition, 2017, 56:15905-15909.

[22] Zhang Z M, Chen P, Li W B, et al. A new type of chiral sulfinamide monophosphine ligands: stereodivergent synthesis and application in enantioselective gold-(I)-catalyzed cycloaddition reactions. Angewandte Chemie International Edition, 2014, 53:4350-4354.

[23] Dai Q, Li W B, Li Z M, et al. P-chiral phosphines enabled by palladium/Xiao-Phos-catalyzed asymmetric P-C cross-coupling of secondary phosphine oxides and aryl bromides. Journal of the American Chemical Society, 2019, 141: 20556-20564.

[24] Cheng H G, Feng B, Chen L Y, et al. Rational design of sulfoxide-phosphine ligands for Pd-catalyzed enantioselective allylic alkylation reactions. Chemical Communications, 2014, 50: 2873-2875.

[25] Feng B, Cheng H G, Chen J R, et al. Palladium/sulfoxide-phosphine-catalyzed highly enantioselective allylic etherification and amination. Chemical Communications, 2014, 50: 9550-9553.

[26] Chen L Y, Yu X Y, Chen J R, et al. Enantioselective direct functionalization of indoles by Pd/sulfoxide-phosphine-catalyzed N-allylic alkylation. Organic Letters, 2015, 17: 1381-1384.

[27] Wei Y, Liu L Q, Li T R, et al. P,S ligands for the asymmetric construction of quaternary stereocenters in palladium-catalyzed decarboxylative [4+2] cycloadditions. Angewandte Chemie International Edition, 2016, 55: 2200-2204.

[28] Feng B, Pu X Y, Liu Z C, et al. Highly enantioselective Pd-catalyzed indole allylic alkylation using

binaphthyl-based phosphoramidite-thioether ligands. Organic Chemistry Frontiers, 2016, 3: 1246-1249.

[29] Xiao X, Shao B X, Li J Y, et al. Enantioselective synthesis of functionalized 1,4-dihydropyrazolo-[4′,3′:5,6]pyrano[2,3-b]quinolines through ferrocenyl-phosphine-catalyzed annulation of modified MBH carbonates and pyrazolones. Chemical Communications, 2021, 57:4690-4693.

[30] Bai X F, Song T, Xu Z, et al. Aromatic amide-derived non-biaryl atropisomers as highly efficient ligands in silver-catalyzed asymmetric cycloaddition reactions. Angewandte Chemie International Edition, 2015, 54:5255-5259.

[31] Bai X F, Xu Z, Xia C G, et al. Aromatic-amide-derived non-biaryl atropisomer as highly efficient ligand for asymmetric silver-catalyzed [3+2] cycloaddition. ACS Catalysis, 2015, 5, 6016-6020.

[32] Bai X F, Zhang J, Xia C G, et al. N-tert-butanesulfinyl imine and aromatic-amide -derived non-biaryl atropisomers as chiral ligands for silver-catalyzed *endo*-selective [3+2] cycloaddition of azomethine ylides with maleimides. Tetrahedron, 2016, 72: 2690-2699.

[33] Zhao Q, Vuong T M H, Bai X F, et al. Catalytic enantioselective synthesis of highly functionalized pentafluorosulfanylated pyrrolidines. Chemistry A European Journal, 2018, 24: 5644-5651.

[34] Bai X F, Li L, Xu Z, et al. Asymmetric michael addition of aldimino esters with chalcones catalyzed by silver/Xing-Phos: mechanism-oriented divergent synthesis of chiral pyrrolines. Chemistry A European Journal, 2016, 22: 10399-10404.

[35] Yuan Y, Yu B, Bai X F, et al. Asymmetric synthesis of glutamic acid derivatives by silver-catalyzed conjugate addition-elimination reactions. Organic Letters, 2017, 19: 4896-4899.

[36] Yuan Y, Zheng Z J, Ye F, et al. Highly efficient desymmetrization of cyclopropenes to azabicyclo[3.1.0]hexanes with five continuous stereogenic centers by copper-catalyzed [3+2] cycloadditions. Organic Chemistry Frontiers, 2018, 5: 2759-2764.

[37] Yuan Y, Zheng Z J, Li L, et al. Silicon-based bulky group tuned parallel kinetic resolution in copper-catalyzed 1,3-dipolar additions. Advanced Synthesis & Catalysis, 2018, 360: 3002-3008.

[38] Yu B, Bai X F, Lv J Y, et al. Enantioselective synthesis of chiral imidazolidine derivatives by asymmetric Ag/Xing-Phos-catalyzed homo-1,3-dipolar [3+2] cycloaddition of azomethine ylides. Advanced Synthesis & Catalysis, 2017, 359:3577-3584.

[39] Yu B, Yang K F, Bai X F, et al. Ligand-controlled inversion of diastereo- and enantioselectivity in silver-catalyzed azomethine ylide-imine cyclo-addition of glycine aldimino esters with imines. Organic Letters, 2018, 20: 2551-2554.

[40] Ohkuma T, Ooka H, Hashiguchi S, et al. Practical enantioselective hydrogenation of aromatic ketones. Journal of the American Chemical Society, 1995, 117: 2675-2676.

[41] Doucet H, Ohkuma T, Murata K, et al. *Trans*-[RuCl$_2$(phosphane)$_2$(1,2-diamine)] and chiral *trans*-[RuCl$_2$(diphosphane)(1,2-diamine)]: shelf-stable precatalysts for the rapid, productive, and stereoselective hydrogenation of ketones. Angewandte Chemie International Edition, 1998, 37:1703-1707.

[42] Flores-López C Z, Flores-López L A Z, Aguirre G, et al. Ruthenium(II)-assisted asymmetric hydrogen transfer reduction of acetophenone using chiral tridentate phosphorus-containing ligands derived from (1*R*, 2*R*)-1,2-diaminocyclohexane. Journal of Molecular Catalysis A: Chemical, 2004, 215: 73-79.

[43] Clarke M L, Diaz-Valenzuela M B, Slawin A M Z. Hydrogenation of aldehydes, esters, imines, and ketones catalyzed by a ruthenium complex of a chiral tridentate ligand. Organometallics, 2007, 26: 16-19.

[44] Diaz-Valenzuela M B, Phillips S D, France M B, et al. Enantioselective hydrogenation and transfer hydrogenation of bulky ketones catalysed by a ruthenium complex of a chiral tridentate ligand. Chemistry A European Journal, 2009, 15: 1227-1232.

[45] Phillips S D, Andersson K H O, Kann N, et al. Exploring the role of phosphorus substituents on the enantioselectivity of Ru-catalysed ketone hydrogenation using tridentate phosphine-diamine ligands. Catalysis Science & Technology, 2011, 1: 1336-1339.

[46] Fuentes J A, Carpenter I, Kann N, et al. Highly enantioselective hydrogenation and transfer hydrogenation of cyclo-alkyl and heterocyclic ketones catalysed by an iridium complex of a tridentate phosphine-diamine ligand. Chemical Communications, 2013, 49: 10245-10247.

[47] Clarke M. Catalytic hydrogenation of low-reactivity carbonyl groups using bifunc-tional chiral tridentate ligands. Synlett, 2014, 25: 1371-1380.

[48] Phillips S D, Fuentes J A, Clarke M L. On the NH effect in ruthenium-catalysed hydrogenation of ketones: rational design of phosphine-amino-alcohol ligands for asymmetric hydrogenation of ketones. Chemistry A European Journal, 2010, 16: 8002-8005.

[49] Demmans K Z, Olson M E, Morris R H. Asymmetric transfer hydrogenation of ketones with well-defined manganese(I) PNN and PNNP complexes. Organometallics, 2018, 37: 4608-4618.

[50] Xie J H, Liu X Y, Xie J B, et al. An additional coordination group leads to extremely efficient chiral iridium catalysts for asymmetric hydrogenation of ketones. Angewandte Chemie International Edition, 2011, 50: 7329-7332.

[51] Xie J H, Liu X Y, Yang X H, et al. Chiral iridium catalysts bearing spiro pyridine-aminophosphine ligands enable highly efficient asymmetric hydrogenation of beta-aryl beta-ketoesters. Angewandte Chemie International Edition, 2012, 51: 201-203.

[52] Yamamura T, Nakatsuka H, Tanaka S, et al. Asymmetric hydrogenation of tert-alkyl ketones: DMSO effect in unification of stereoisomeric ruthenium complexes. Angewandte Chemie International Edition, 2013, 52: 9313-9315.

[53] Chen G Q, Lin B J, Huang J M, et al. Design and synthesis of chiral oxa-spirocyclic ligands for Ir-catalyzed direct asymmetric reduction of Bringmann's lactones with molecular $H_2$. Journal of the American Chemical Society, 2018, 140: 8064-8068.

[54] Zheng Z Y, Cao Y X, Chong Q L, et al. Chiral cyclohexyl-fused spirobiindanes: practical synthesis, ligand development, and asymmetric catalysis. Journal of the American Chemical Society, 2018, 140: 10374-10381.

[55] Zhang F H, Wang C, Xie J H, et al. Synthesis of tridentate chiral spiro aminophosphine-oxazoline ligands and application to asymmetric hydrogenation of α-keto amides. Advanced Synthesis & Catalysis, 2019, 361: 2832-2835.

[56] Gomez Arrayas R, Adrio J, Carretero J C. Recent applications of chiral ferrocene ligands in asymmetric catalysis. Angewandte Chemie International Edition, 2006, 45: 7674-7715.

[57] Stepnicka P. Ferrocenes: ligands, materials and biomolecules; Wiley, 2008.

[58] Dai L X, Hou X L. Chiral ferrocenes in asymmetric catalysis: synthesis and applications; John Wiley & Sons, 2010.

[59] Hayashi T. Asymmetric catalytic hydrosilylation of ketones. preparation of chiral ferrocenylphosphines as chiral ligands. Tetrahedron Letters, 1974, 15: 4405-4408.

[60] Hayashi T, Mise T, Fukushima M, et al. Asymmetric synthesis catalyzed by chiral ferrocenylphosphine-transition metal complexes. I. Preparation of chiral ferrocenylphosphines. Bulletin of the Chemical Society of Japan, 1980, 53: 1138-1151.

[61] Nie H F, Zhou G, Wang Q, et al. Asymmetric hydrogenation of aromatic ketones using an iridium(Ⅰ) catalyst containing ferrocene-based P-N-N tridentate ligands. Tetrahedron: Asymmetry, 2013, 24: 1567-1571.

[62] Widegren M B, Harkness G J, Slawin A M Z, et al. A highly active manganese catalyst for enantioselective ketone and ester hydrogenation. Angewandte Chemie International Edition, 2017, 56: 5825-5828.

[63] Wu W L, Liu S D, Duan M, et al. Iridium catalysts with f-amphox ligands: asymmetric hydrogenation of simple ketones. Organic Letters, 2016, 18: 2938-2941.

[64] Hou C J, Hu X P. Sterically hindered chiral ferrocenyl P,N,N-ligands for highly diastereo-/enantioselective Ir-catalyzed hydrogenation of α-alkyl-β-ketoesters via dynamic kinetic resolution. Organic Letters, 2016, 18: 5592-5595.

[65] Qin C, Hou C J, Liu H, et al. Ir-catalyzed asymmetric hydrogenation of simple ketones with chiral

ferrocenyl P,N,N-ligands. Tetrahedron Letters, 2018, 59: 719-722.
[66] Liang Z Q, Yang T D, Gu G X, et al. Scope and mechanism on iridium-f-amphamide catalyzed asymmetric hydrogenation of ketones. Chinese Journal of Chemistry, 2018, 36: 851-856.
[67] Ling F, Nian S F, Chen J C, et al. Development of ferrocene-based diamine-phosphine-sulfonamide ligands for iridium-catalyzed asymmetric hydrogenation of ketones. The Journal of Organic Chemistry, 2018, 83: 10749-10761.
[68] Ling F, Hou H C, Chen J C, et al. Highly enantioselective synthesis of chiral benzhydrols via manganese catalyzed asymmetric hydrogenation of unsymmetrical Benzophenones using an imidazole-based chiral PNN tridentate ligand. Organic Letters, 2019, 21: 3937-3941.
[69] Ling F, Chen J C, Nian S F, et al. Manganese-catalyzed enantioselective hydrogenation of simple ketones using an imidazole-based chiral PNN tridentate ligand. Synlett, 2020, 31: 285-289.
[70] Liu C, Wang M, Liu S, et al. Manganese-catalyzed asymmetric hydrogenation of quinolines enabled by π-π interaction. Angewandte Chemie International Edition, 2021, 60: 5108-5113.
[71] Brunner H, Zettler C, Zabel M. Asymmetric catalysis. Part 149 [1]. Synthesis of new chiral tridentate ligands for enantioselective catalysis. Monatshefte für Chemie / Chemical Monthly, 2003, 134: 1253-1269.
[72] Ma X, Qiao L, Liu G, et al. A new phosphine-amine-oxazoline ligand for Ru-catalyzed asymmetric hydrogenation of N-phosphinylimines. Chin. J. Chem., 2018, 36:1151-1155.
[73] Wei D Q, Chen X S, Hou C J, et al. Iridium-catalyzed asymmetric hydrogenation of β-keto esters with new phenethylaminederived tridentate P,N,N-ligands. Synthetic Communications, 2019, 49: 237-243.
[74] Zuo Z Q, Xu S G, Zhang L, et al. Cobalt-catalyzed asymmetric hydrogenation of vinylsilanes with a phosphine-pyridine-oxazoline ligand: synthesis of optically active organosilanes and silacycles. Organometallics, 2019, 38: 3906-3911.
[75] Zhang L L, Tang Y T, Han Z B, et al. Lutidine-based chiral pincer manganese catalysts for enantioselective hydrogenation of ketones. Angewandte Chemie International Edition, 2019, 58: 4973-4977.
[76] Császár Z, Szabó E Z, Bényei A C, et al. Chelate ring size effects of Ir(P,N,N) complexes: Chemoselectivity switch in the asymmetric hydrogenation of α,β-unsaturated ketones. Catalysis Communications, 2020, 146: 106128.
[77] Du T, Wang B, Wang C, et al. Cobalt-catalyzed asymmetric hydrogenation of ketones: a remarkable additive effect on enantioselectivity. Chinese Chemical Letters, 2021, 32: 1241-1244.
[78] Jiang Y T, Jiang Q Z, Zhu G X, et al. New chiral ligands for catalytic transfer hydrogenation of ketones. Tetrahedron Letters, 1997, 38: 6565-6568.
[79] Jiang Y T, Jiang Q Z, Zhu G X, et al. Highly effective NPN-type tridentate ligands for asymmetric transfer hydrogenation of ketones. Tetrahedron Letters, 1997, 38: 215-218.
[80] Braunstein P, Naud F, Pfaltz A, et al. Ruthenium complexes with novel tridentate N, P, N ligands containing a phosphonite bridge between two chiral oxazolines. Catalytic activity in cyclopropanation of olefins and transfer hydrogenation of acetophenone. Organometallics, 2000, 19: 2676-2683.
[81] Gao G, Gu F L, Jiang J X, et al. Neighboring lithium-assisted [1,2]-Wittig rearrangement: practical access to diarylmethanol-based 1,4-diols and optically active BINOL derivatives with axial and $sp^3$-central chirality. Chemistry A European Journal, 2011, 17: 2698-2703.
[82] Gao G, Bai X F, Yang H M, et al. Ar-BINMOLs with axial and $sp^3$ central chirality-characterization, chiroptical properties, and application in asymmetric catalysis. Eur. J. Org. Chem., 2011, 2011(26):5039-5046.
[83] Dong C, Song T, Bai X F, et al. Enantioselective conjugate addition of cyanide to chalcones catalyzed by magnesium-Py-BINMOL Complex. Catalysis Science & Technology, 2015, 5: 4755-4759.
[84] Song T, Zheng L S, Ye F, et al. Modular synthesis of Ar-BINMOL-Phos for catalytic asymmetric

alkynylation of aromatic aldehydes with unexpected reversal of enantioselectivity. Advanced Synthesis & Catalysis, 2014, 356: 1708-1718.

[85] Beletskaya I P, Nájera C, Yus M. Stereodivergent catalysis. Chemical Review, 2018, 118: 5080-5200.

[86] Song T, Li L, Zhou W, et al. Enantioselective Copper-catalyzed azide-alkyne click cyclo-addition to desymmetrization of maleimide-based bis-alkynes. Chemistry A European Journal, 2015, 21: 554-558.

[87] Chen M Y, Xu Z, Chen L, et al. Catalytic asymmetric Huisgen alkyne-azide cycloaddition of bisalkynes by copper(I) nanoparticles. ChemCatChem, 2018, 10: 280-286.

[88] Tang R H, Xu Z, Nie Y X, et al. Catalytic asymmetric *trans*-selective hydrosilylation of bisalkynes to access AIE and CPL-active silicon-stereogenic benzosiloles. iScience, 2020, 23: 101268.

[89] Yang H, AlvarezGressier M, Lugan N, et al. Ruthenium(Ⅱ) complexes containing optically active hemilabile P,N,O-tridentate ligands. Synthesis and evaluation in catalytic asymmetric transfer hydrogenation of acetophenone by propan-2-ol. Organometallics, 1997, 16: 1401-1409.

[90] Kwong H L, Lee W S, Lai T S, et al. Ruthenium catalyzed asymmetric transfer hydrogenation based on chiral P,N,O Schiff base ligands and crystal structure of a ruthenium(Ⅱ) complexbearing chiral P,N,O Schiff base ligands. Inorganic Chemistry Communications, 1999, 2: 66-69.

[91] Alvarez M, Lugan N, Mathieu R. Synthesis and evaluation of the bonding properties of a potentially tridentate ligand: 1-(diphenylphosphino)-2-ethoxy-1-(2-pyridyl) ethane. Journal of the Chemical Society, Dalton Transactions, 1994, 2755-2760.

[92] Dai H C, Hu X P, Chen H L, et al. New efficient P,N,O-tridentate ligands for Ru-catalyzed asymmetric transfer hydrogenation. Tetrahedron: Asymmetry, 2003, 14: 1467-1472.

[93] Fuentes J A, Phillips S D, Clarke M L. New phosphine-diamine and phosphine-amino-alcohol tridentate ligands for ruthenium catalysed enantioselective hydrogenation of ketones and a concise lactone synthesis enabled by asymmetric reduction of cyano-ketones. Chemistry Central Journal, 2012, 6: 151-156.

[94] Ishii T, Watanabe R, Moriya T, et al. Cooperative catalysis of metal and O-H···O/$sp^3$-C-H···O two-point hydrogen bonds in alcoholic solvents: Cu-catalyzed enantioselective direct alkynylation of aldehydes with terminal alkynes. Chemistry A European Journal, 2013, 19: 13547-13553.

[95] Takayama Y, Ishii T, Ohmiya H, et al. Asymmetric synthesis of *β*-lactams through copper-catalyzed alkyne-nitrone coupling with a prolinol-phosphine chiral ligand. Chemistry A European Journal, 2017, 23: 8400-8404.

[96] Yu J F, Long J, Yang Y H, et al. Iridium-catalyzed asymmetric hydrogenation of ketones with accessible and modular ferrocene-based amino-phosphine acid (f-Ampha) ligands. Organic Letters, 2017, 19: 690-693.

[97] Yu J F, Duan M, Wu W L, et al. Readily accessible and highly efficient ferrocene-based amino-phosphine-alcohol (f-Amphol) ligands for iridium-catalyzed asymmetric hydrogenation of simple ketones. Chemistry A European Journal, 2017, 23: 970-975.

[98] Altan O, Yılmaz M K. New phosphine-amino-alcohol tridentate ligands for ruthenium catalyzed asymmetric transfer hydrogenation of ketones. Journal of Organometallic Chemistry, 2018, 861: 252-262.

[99] Ye F, Zheng Z J, Deng W H, et al. Modulation of multifunctional N,O,P-ligands for enantioselective copper-catalyzed Conjugate addition of diethylzinc and trapping of the zinc enolate. Chemistry An Asian Journal, 2013, 8: 2242-2253.

[100] Deng W H, Ye F, Bai X F, et al. Multistereogenic phosphine ligand-promoted palladium-catalyzed allylic alkylation of cyanoesters. ChemCatChem, 2015: 7, 75-79.

[101] Tang L, Wang Q W, Wang J J, et al. A new chiral sulfinyl-*NH*-pyridine ligand for Ir-catalyzed asymmetric transfer hydrogenation reaction. Tetrahedron Letters, 2012, 53: 3839-3842.

[102] Bao D H, Wu H L, Liu C L, et al. Development of chiral spiro P-N-S ligands for iridium-catalyzed asymmetric hydrogenation of beta-alkyl-beta-ketoesters. Angewandte Chemie International Edition,

2015, 54: 8791-8794.

[103] Bao D H, Gu X S, Xie J H, et al. Iridium-catalyzed asymmetric hydrogenation of racemic beta-keto lactams via dynamic kinetic resolution. Organic Letters, 2017, 19: 118-121.

[104] You S L, Zhu X Z, Luo Y M, et al. Highly regio- and enantioselective Pd-catalyzed allylic alkylation and amination of monosubstituted allylic acetates with novel ferrocene P,N-ligands. Journal of the American Chemical Society, 2001, 123(30): 7471-7472.

[105] Dai L X, Tu T, You S L, et al. Asymmetric catalysis with chiral ferrocene ligands. Accounts of Chemical Research, 2003; 36: 659-67.

[106] Yu Y, Yang X F, Xu C F, et al. Desymmetrization of bicyclo[3.$n$.1]-3-one derivatives by palladium-catalyzed asymmetric allylic alkylation. Organic Letters, 2013, 15: 3880-3883.

[107] Li X H, Zheng B H, Ding C H, et al. Enantioselective synthesis of 2,3-disubstituted indanones via Pd-catalyzed intramolecular asymmetric allylic alkylation of ketones. Organic Letters, 2013, 15: 6086-6089.

[108] Liu W, Chen D, Zhu X Z, et al. Highly diastereo- and enantioselective Pd-catalyzed cyclopropanation of acyclic amides with substituted allyl carbonates. Journal of the American Chemical Society, 2009; 131: 8734-8735.

[109] Yang X F, Ding C H, Li X H, et al. Regio- and enantioselective palladium-catalyzed allylic alkylation of nitromethane with monosubstituted allyl substrates: synthesis of ($R$)-Rolipram and ($R$)-Baclofen. The Journal of Organic Chemistry, 2012, 77: 8980-8985.

[110] Yang X F, Yu W H, Ding C H, et al. Palladium-catalyzed regio-, diastereo-, and enantioselective allylation of nitroalkanes with monosubstituted allylic substrates. The Journal of Organic Chemistry, 2013, 78: 6503-6509.

[111] Xu C F, Zheng B H, Suo J J, et al. Highly diastereo- and enantioselective palladium-catalyzed [3+2] cycloaddition of vinyl aziridines and $α,β$-unsaturated ketones. Angewandte Chemie International Edition, 2015; 54: 1604-1607.

[112] Fang P, Ding C H, Hou X L, et al. Palladium-catalyzed regio- and enantio-selective allylic substitution reaction of monosubstituted allyl substrates with benzyl alcohols. Tetrahedron: Asymmetry, 2010, 21: 1176-1178.

[113] Hayashi T, Mise T, Fukushima M, et al. Asymmetric synthesis catalyzed by chiral ferrocenylphosphine-transition metal complexes. I. preparation of chiral ferrocenylphosphines. Bulletin of the Chemical Society of Japan, 1980, 53(4): 1138-1151.

[114] Ito Y, Sawamura M, Hayashi T. Catalytic asymmetric aldol reaction: reaction of aldehydes with isocyanoacetate catalyzed by a chiral ferrocenylphosphine-gold(I) complex. Journal of the American Chemical Society, 1986, 108: 6405-6406.

[115] Sawamura M, Ito Y. Catalytic asymmetric synthesis by means of secondary interaction between chiral ligands and substrates. Chemical Review, 1992, 92: 857-871.

[116] Ito Y, Sawamura M, Hayashi T. Asymmetric aldol reaction of an isocyanoacetate with aldehydes bychiral ferrocenylphosphine-gold(Ⅰ) complexes: design and preparation of new efficient ferrocenylphosphine ligands. Tetrahedron Letters, 1987, 28: 6215-6218.

[117] Chen C Y, Wang H, Zhang Z F, et al. Ferrocenyl chiral bisphosphorus ligands for highly enantioselective asymmetric hydrogenation via noncovalent ion pair interaction. Chemical Science, 2016, 7: 6669-6673.

[118] Yin X G, Chen C Y, Dong X Q, et al. Rh/Wudaphos-catalyzed asymmetric hydrogenation of sodium $α$-arylethenylsulfonates: a method to access chiral $α$-arylethylsulfonic acids. Organic Letters, 2017, 19: 2678-2681.

[119] Chen C Y, Wen S W, Geng M Y, et al. A new ferrocenyl bisphosphorus ligand for the asymmetric hydrogenation of $α$-methylene-$γ$-keto-carboxylic acids. Chemical Communications, 2017, 53: 9785-9788.

[120] Wen S W, Chen C Y, Du S C, et al. Highly enantioselective asymmetric hydrogenation of carboxy-directed $α,α$-disubstituted terminal olefins via the ion pair noncovalent interaction. Organic Letters,

2017, 19: 6474-6477.

[121] Chen C Y, Zhang Z F, Jin S C, et al. Enzyme-inspired chiral secondary-phosphine-oxide ligand with dual noncovalent interactions for asymmetric hydrogenation. Angewandte Chemie International Edition, 2017, 56: 6808-6812.

[122] Yin X G, Chen C Y, Li X, et al. Rh/SPO-WudaPhos-catalyzed asymmetric hydrogenation of α-substituted ethenylphosphonic acids via noncovalent ion-pair interaction. Organic Letters, 2017, 19: 4375-4378.

[123] Burk M J, Feaster J E, Harlow R L. New chiral phospholanes-synthesis, characterization, and use in asymmetric hydrogenation reactions. Tetrahedron: Asymmetry, 1991, 2: 569-592.

[124] Abdur-Rashid K. Transfer hydrogenation processes and catalysts. US Patent: US 20050107638 A1.

[125] Garbe M, Junge K, Walker S, et al. Manganese(Ⅰ)-catalyzed enantioselective hydrogenation of ketones using a defined chiral PNP pincer ligand. Angewandte Chemie International Edition, 2017, 56: 11237-11241.

[126] Garbe M, Wei Z, Tannert B, et al. Enantioselective hydrogenation of ketones using different metal complexes with a chiral PNP pincer ligand. Advanced Synthesis & Catalysis, 2019, 361: 1913-1920.

[127] Bianchini C, Farnetti E, Glendenning L, et al. Synthesis of the new chiral aminodiphosphine ligands (R)-(alpha-methylbenzyl)bis(2-(diphenylphosphino)-ethyl)amine and (S)-(alpha-methylbenzyl)bis(2-(diphenylphosphino)ethyl)amine and their use in the enantioselective reduction of alpha, beta-unsaturated ketones to allylic alcohols by iridium catalysis. Organometallics, 1995, 14: 1489-1502.

[128] Bianchini C, Barbaro P, Scapacci G, et al. Enantioselective hydrogenation of 2-methylquinoxaline to (-)-(2S)-2-methyl-1,2,3,4-tetrahydroquinoxaline by iridium catalysis. Organometallics, 1998, 17: 3308-3310.

[129] Bianchini C, Glendenning L, Zanobini F, et al. Asymmetric hydrogen-transfer reduction of prochiral and α,β-unsaturated ketones by iridium complexes containing optically pure aminodiphosphine ligands. Journal of Molecular Catalysis A: Chemical, 1998, 132: 13-19.

[130] Jiang Q Z, VanPlew D, Murtuza S, et al. Synthesis of (1R,1R′)-2,6-bis[1-(diphenylphosphino)ethyl] pyridine and its application in asymmetric transfer hydrogenation. Tetrahedron Letters, 1996, 37: 797-800.

[131] Sablong R, Osborn J A. The asymmetric hydrogenation of imines using tridentate C-2 diphosphine complexes of iridium(Ⅰ) and rhodium(Ⅰ). Tetrahedron Letters, 1996, 37: 4937-4940.

[132] Boaz N W, Debenham S D, Mackenzie E B, et al. Phosphinoferrocenylaminophosphines as novel and practical ligands for asymmetric catalysis. Organic Letters, 2002, 4, 2421.

[133] Dai H C, Hu X P, Chen H L, et al. New chiral ferrocenyldiphosphine ligand for catalytic asymmetric transfer hydrogenation. Journal of Molecular Catalysis A: Chemical, 2004, 209: 19-22.

[134] Boaz N W, Jr. Ponasik J A, Large S E A versatile synthesis of phosphine-aminophosphine ligands for asymmetric catalysis. Tetrahedron: Asymmetry, 2005, 16: 2063.

[135] Kuriyama W, Matsumoto T, Ino Y, et al. Novel ruthenium carbonyl complex having a tridentate ligand and manufacturing method and usage therefor. Patent, WO 2011048727A1.

[136] Lagaditis P O, Sues P E, Sonnenberg J F, et al. Iron(Ⅱ) complexes containing unsymmetrical P-N-P′ pincer ligands for the catalytic asymmetric hydrogenation of ketones and imines. Journal of the American Chemical Society, 2014, 136: 1367-1380.

[137] Smith S A M, Lagaditis P O, Lupke A, et al. Unsymmetrical iron P-NH-P′ catalysts for the asymmetric pressure hydrogenation of aryl ketones. Chemistry - A European Journal, 2017, 23: 7212-7216.

[138] Zirakzadeh A, Kirchner K, Roller A, et al. Iron(Ⅱ) complexes containing chiral unsymmetrical PNP′ pincer ligands: synthesis and application in asymmetric hydrogenations. Organometallics, 2016, 35: 3781-3787.

[139] Zirakzadeh A, de Aguiar S R M M, Stöger B, et al. Enantioselective transfer hydrogenation of

ketones catalyzed by a manganese complex containing an unsymmetrical chiral PNP′ tridentate ligand. ChemCatChem, 2017, 9: 1744-1748.

[140] Zhang F H, Zhang F J, Li M L, et al. Enantioselective hydrogenation of dialkyl ketones. Nature Catalysis, 2020, 3: 621-627.

[141] Zeng L, Yang H, Zhao M, et al. C1-Symmetric PNP ligands for manganese-catalyzed enantioselective hydrogenation of ketones: reaction scope and enantioinduction model. ACS Catalysis, 2020, 10: 13794-13799.

[142] Li Y B, Tian H, Yin L. Copper(Ⅰ)-catalyzed asymmetric 1,4-conjugate hydrophosphination of $α,β$-unsaturated amides. Journal of the American Chemical Society, 2020, 142: 20098-20106.

[143] Zhao Q Y, Li S K, Huang K X, et al. A Novel chiral bisphosphine-thiourea ligand for asymmetric hydrogenation of $β,β$-disubstituted nitroalkenes. Organic Letters, 2013, 15: 4014-4017.

[144] Li S K, Huang K X, Cao B N, et al. Highly enantioselective hydrogenation of $β,β$-disubstituted nitro-alkenes. Angewandte Chemie International Edition, 2012, 51: 8573-8576.

[145] Li X X, You C, Yang Y S, et al. Rhodium-catalyzed asymmetric hydrogenation of $β$-cyanocinnamic esters with the assistance of a single hydrogen bond in a precise position. Chemical Science, 2018, 9: 1919-1924.

[146] Han Z Y, Li P, Zhang Z P, et al. Highly enantioselective synthesis of chiral succinimides via Rh/bisphosphine-thiourea-catalyzed asymmetric hydrogenation. ACS Catalysis. 2016, 6: 6214-6218.

[147] Yin X G, Huang Y, Chen Z Y, et al. Enantioselective access to chiral 2-substituted 2,3-dihydrobenzo[1,4]dioxane derivatives through Rh-catalyzed asymmetric hydrogenation. Organic Letters, 2018, 20: 4173-4177.

[148] Liu G, Li A Q, Qin X Y, et al. Efficient access to chiral $β$-borylated carboxylic esters via Rh-catalyzed hydrogenation. Advanced Synthesis & Catalysis, 2019, 361: 2844-2848.

[149] Han Z Y, Liu G, Wang R, et al. Highly efficient Ir-catalyzed asymmetric hydrogenation of benzoxazinones and derivatives with a Brønsted acid cocatalyst. Chemical Science, 2019, 10: 4328-4333.

[150] Sun Y J, Jiang J, Guo X C, et al. Asymmetric hydrogenation of $α,β$-unsaturated sulfones by a rhodium/thiourea-bisphosphine complex. Organic Chemistry Frontiers, 2019, 6: 1438-1441.

[151] Yang J X, Li X X, You C, et al. Rhodium-catalyzed asymmetric hydrogenation of exocyclic $α,β$-unsaturated carbonyl compounds. Organic & Biomolecular Chemistry, 2020, 18: 856-859.

[152] Yin C C, Yang T, Pan Y M, et al. Rh-catalyzed asymmetric hydrogenation of unsaturated medium-ring NH lactams: highly enantioselective synthesis of N-unprotected 2,3-dihydro-1,5-benzothiazepinones. Organic Letters, 2020, 22: 920-923.

[153] Zhao Q Y, Wen J L, Tan R C, et al. Rhodium-catalyzed asymmetric hydrogenation of unprotected NH imines assisted by a thiourea. Angewandte Chemie International Edition, 2014, 53: 8467-8470.

[154] Wen J L, Tan R C, Liu S D, et al. Strong Brønsted acid promoted asymmetric hydrogenation of isoquinolines and quinolines catalyzed by a Rh-thiourea chiral phosphine complex via anion binding. Chemical Science, 2016, 7: 3047-3051.

[155] Li P, Huang Y, Hu X Q, et al. Access to chiral seven-member cyclic amines via Rh-catalyzed asymmetric hydrogenation. Organic Letters, 2017, 19: 3855-3858.

[156] Wen J L, Fan X R, Tan R C, et al. Brønsted-acid-promoted Rh-catalyzed asymmetric hydrogenation of N-unprotected indoles: a cocatalysis of transition metal and anion binding. Organic Letters, 2018, 20: 2143-2147.

[157] Yang T L, Sun Y J, Wang H, et al. Iridium-catalyzed enantioselective hydrogenation of oxocarbenium ions: a case of ionic hydrogenation. Angewandte Chemie International Edition, 2020, 59: 6108-6114.

[158] Qian H, Gu G, Zhou Q, et al. Enantioselective palladium-catalyzed decarboxylative allylation of $β$-keto esters assisted by a thiourea. Synlett, 2018, 29: 51-56.

[159] Trost B M, Crawley M L. Asymmetric transition-metal-catalyzed allylic alkylations: applications in total synthesis. Chemical Review, 2003, 103: 2921-2944.

[160] Trost B M, Zhang T, Sieber J D. Catalytic asymmetric allylic alkylation employing heteroatom nucleophiles: a powerful method for C-X bond formation. Chemical Science, 2010, 1: 427-440.

[161] Trost B M. Pd- and Mo-catalyzed asymmetric allylic alkylation. Organic Process Research & Development, 2012, 16: 185-194.

[162] Trost B M. Metal catalyzed allylic alkylation: its development in the Trost laboratories. Tetrahedron, 2015, 71: 5708-5733.

[163] Trost B M, Osipov M, Dong G. Palladium-catalyzed dynamic kinetic asymmetric transformations of vinyl aziridines with nitrogen heterocycles: rapid access to biologically active pyrroles and indoles. Journal of the American Chemical Society, 2010, 132: 15800-15807.

[164] Trost B M, Bai Y, Bai W J, et al. Enantioselective divergent synthesis of C19-oxo eburnane alkaloids via palladium-catalyzed asymmetric allylic alkylation of an $N$-alkyl-$\alpha,\beta$-unsaturated lactam. Journal of the American Chemical Society, 2019, 141: 4811-4814.

[165] Trost B M, Malhotra S, Olson D E, et al. Asymmetric synthesis of diamine derivatives via sequential palladium and rhodium catalysis. Journal of the American Chemical Society, 2009, 131: 4190-4191.

[166] Trost B M, Bai W J, Hohn C, et al. Palladium-catalyzed asymmetric allylic alkylation of 3-substituted 1H-indoles and tryptophan derivatives with vinylcyclopropanes. Journal of the American Chemical Society, 2018, 140: 6710-6717.

[167] Trost B M, Osipov M, Dong G A concise enantioselective synthesis of (−)-ranirestat. Organic Letters, 2010, 12: 1276-1279.

[168] Trost B M, Spohr S M, Rolka A B, et al. Desymmetrization of phosphinic acids via Pd-catalyzed asymmetric allylic alkylation: rapid access to P-chiral phosphinates. Journal of the American Chemical Society, 2019, 141: 14098-14103.

[169] Benessere V, De Roma A, Del Litto R, et al. Naplephos through the looking-glass: chiral bis(phosphanylamides) based on $\beta$-1,2-d-glucodiamine and their application in enantiose-lective allylic substitutions. European Journal of Organic Chemical, 2011, 2011, 5779-5782.

[170] Zheng L S, Li L, Yang K F, et al. New silver(I)-monophosphine complex derived from chiral Ar-BINMOL: synthesis and catalytic activity in asymmetric vinylogous Mannich reaction. Tetrahedron, 2013, 69, 8777-8784.

[171] Zhao B, Wang Z, Ding K. Practical by ligand design: a new class of monodentate phosphoramidite ligands for rhodium-catalyzed enantioselective hydrogenations. Advanced Synthesis & Catalysis, 2006, 348, 1049-1057.

[172] Ye F, Zheng Z J, Li L, et al. Development of a novel multifunctional N,P-ligand for highly enantioselective palladium-catalyzed asymmetric allylic etherification of alcohols and silanols. Chemistry A European Journal, 2013, 19, 15452-15457.

[173] Xu J X, Ye F, Bai X F, et al. Fei-Phos ligand-controlled asymmetric palladium-catalyzed allylic substitutions with structurally diverse nucleophiles: scope and limitation. RSC Advances, 2016, 6, 45495-45502.

[174] Xu J X, Ye F, Bai X F, et al. A mechanistic study on multifunctional Fei-Phos ligand-controlled asymmetric palladium-catalyzed allylic substitutions. RSC Advances, 2016, 6, 70624-70631.

[175] Long P W, Bai X F, Ye F, et al. Construction of six-membered silacyclic skeletons via platinum-catalyzed tandem hydrosilylation/cyclization with dihydrosilanes. Advanced Synthesis & Catalysis, 2018, 360, 2825-2830.

[176] Long P W, Xie J L, Yang J J, et al. Stereo-andregio-selective synthesis of silicon containing diboryl alkenes via platinum-catalyzed mono-lateral diboration of dialkynylsilanes. Chemical Communications, 2020, 56, 4188-4191.

[177] Mikhailine A, Lough A J, Morris R H. Efficient asymmetric transfer hydrogenation of ketones catalyzed by an iron complex containing a P-N-N-P tetradentate ligand formed by template synthesis. Journal of the American Chemical Society, 2009, 131: 1394-1395.

[178] Mikhailine A A, Maishan M I, Lough A J, et al. The mechanism of efficient asymmetric

transfer hydrogenation of acetophenone using an iron( II ) complex containing an $(S,S)$-Ph$_2$PCH$_2$CH=NCHPhCHPhN=CHCH$_2$PPh$_2$ ligand: partial ligand reduction is the key. Journal of the American Chemical Society, 2012, 134: 12266-12280.

[179] Morris R H. Exploiting metal-ligand bifunctional reactions in the design of iron asymmetric hydrogenation catalysts. Accounts of Chemical Research, 2015, 48: 1494-1502.

[180] Zhang W, Adachi Y, Hirao T, et al. Highly diastereoselective *ortho*-lithiation of 1,1′-bis-(oxazolinyl) ferrocene directed to $C_2$-symmetric chiral ligands. Tetrahedron: Asymmetry, 1996, 7: 451-460.

[181] Zhao X H, Liu D L, Guo H, et al. C-N Bond cleavage of allylic amines via hydrogen bond activation with alcohol solvents in Pd-catalyzed allylic alkylation of carbonyl compounds. Journal of the American Chemical Society, 2011, 133: 19354-19357.

[182] Zhao X H, Liu D L, Xie F, et al. Efficient palladium-catalyzed asymmetric allylic alkylation of ketones and aldehydes. Organic & Biomolecular Chemistry, 2011, 9: 1871-1875.

[183] Huo X H, Yang G Q, Liu D L, et al. Palladium-catalyzed allylic alkylation of simple ketones with allylic alcohols and its mechanistic study. Angewandte Chemie International Edition, 2014, 53: 6776-6780.

[184] Huo X H, Quan M, Yang G Q, et al. Hydrogen-bond-activated palladium-catalyzed allylic alkylation via allylic alkyl ethers: challenging leaving groups. Organic Letters, 2014, 16: 1570-1573.

[185] Zhao X H, Liu D L, Xie F, et al. Enamines: efficient nucleophiles for the palladium-catalyzed asymmetric allylic alkylation. Tetrahedron, 2009, 65: 512-517.

[186] Fujita T, Yamamoto T, Morita Y, et al. Chemo- and enantioselective Pd/B hybrid catalysis for the construction of acyclic quaternary carbons: Migratory allylation of *O*-allyl esters to α-C-allyl carboxylic acids. Journal of the American Chemical Society, 2018, 140: 5899-5903.

[187] Liu D L, Xie F, Zhang W B. The synthesis of novel $C_2$-symmetric P,N-chelation ruthenocene ligands and their application in palladium-catalyzed asymmetric allylic substitution. Tetrahedron Lett, 2007; 48: 585-588.

[188] Guo, H, Liu, D. L.; Butt, N. et al. Efficient Ru( II )-catalyzed asymmetric hydrogenation of simple ketones with $C_2$-symmetric planar chiral metallocenyl phosphinooxazoline ligands. Tetrahedron, 2012; 68: 3295-3299.

[189] Wang J X, Wang Y Z, Liu D L, et al. Asymmetric hydrogenation of β-secondary amino ketones catalyzed by a ruthenocenyl phosphino-oxazoline-ruthenium complex (RuPHOX-Ru): the synthesis of γ-secondary amino alcohols. Advanced Synthesis & Catalysis, 2015; 357: 3262-3272.

[190] Li J, Shen J F, Xia C, et al. Asymmetric hydrogenation of α-substituted acrylic acids catalyzed by a ruthenocenyl phosphino-oxazoline-ruthenium complex. Organic Letters, 2016; 18: 2122-2125.

[191] Chen W, McCormack P J, Mohammed K, et al. Stereoselective synthesis of ferrocene-based $C_2$-symmetric diphosphine ligands: application to the highly enantioselective hydrogenation of α-substituted cinnamic acids. Angewandte Chemie International Edition, 2007, 46: 4141-4144.

[192] Johnson C R, Imamoto T. Synthesis of polydentate ligands with homochiral phosphine centers. The Journal of Organic Chemistry, 1987, 52(11): 2170-2174.

[193] Burk M J, Harlow R L. New chiral $C_3$-symmetric tripodal phosphanes. Angewandte Chemie International Edition, 1990, 29: 1462-1464.

[194] Heidel H, Scherer J, Asam A, et al. Fe(NCMe)$_2^{3+}$-komplexe von chiralen tripodliganden mit dreiverschiedenen donorgruppen: dynamik und struktur. Chemische. Berichte. 1995, 128: 293-301.

[195] Jia G, Lee H M, Williams I D. Synthesis of the chiral triphosphine $(S, S)$-PhP(CH$_2$CHMeCH$_2$PPh$_2$)$_2$ and its metal complexes. Organometallics, 1996, 15: 4235-4239.

[196] Lee H M, Bianchini C, Jia G, et al. Styrene cyclopropanation and ethyl diazoacetate dimerization catalyzed by Ruthenium complexes containing chiral tridentate phosphine ligands. Organometallics, 1999, 18: 1961-1966.

[197] Barbaro P, Bianchini C, Togni A. Synthesis and characterization of ruthenium(II) complexes containing chiral bis(ferrocenyl)-$P_3$ or-$P_2S$ ligands. Asymmetric transfer hydrogenation of acetophenone. Organometallics, 1997, 16: 3004-3014.

[198] Barbaro P, Bianchini C, Giambastiani G, et al. Ruthenium(II) complexes with triphosphane ligands combining planar, phosphorus, and carbon chirality: application to asymmetric reduction of trifluoroacetophenone. European Journal of Inorganic Chemistry, 2003, 2003: 4166-4172.

PHOSPHORUS 磷科学前沿与技术丛书

手性膦配体合成及应用

# 6

# 手性膦配体的合成及应用实例

6.1 手性单膦配体的代表性合成及应用实例

6.2 手性双膦配体的代表性合成及应用实例

6.3 手性膦-杂原子双齿配体的代表性合成及应用实例

6.4 手性多官能化多齿型膦配体的代表性合成及应用实例

Synthesis and Application of Chiral Phosphine Ligands

# 6.1 手性单膦配体的代表性合成及应用实例

## 6.1.1 Ph-BINEPINE 配体的合成及应用实例

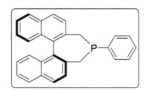

### (a) 配体的合成反应过程

**步骤1:** 化合物 2 的合成

将 (S)-联萘酚 (2.86g, 10mmol) 溶于二氯甲烷 (80mL) 中, 然后在室温下缓慢加入吡啶 (3.0mL, 37.5mmol), 并在 0℃下缓慢滴加三氟甲磺酸酐 (3.5mL, 21mmol) 到上述溶液体系中。在室温下继续搅拌 5h, 利用薄层色谱（TLC）监测反应进程，反应完成后，反应体系在 0℃下用 1mol/L 盐酸溶液淬灭，乙酸乙酯萃取，饱和碳酸氢钠水溶液和盐水洗涤有机层，无水硫酸镁干燥得粗产物。通过硅胶快速色谱法(正己烷：乙酸乙酯 = 10：1)纯化得到白色固体化合物 **2**(5.55g，收率 99%)。

**步骤 2：化合物 3 的合成**

将化合物 **2**(5.55g，10mmol)、$NiCl_2(dppp)$ (379.4mg，0.7mmol)溶于乙醚(80mL)中，在 0℃下将甲基溴化镁(13.3mL)逐滴滴加到反应体系中，并在 0℃下搅拌 30min。然后将反应体系回流 20h，TLC 监测反应进行程度，反应完成后，在室温下用水淬灭反应，并在 0℃下用 1mol/L 盐酸溶液稀释，用正己烷萃取得有机层，用饱和碳酸氢钠水溶液和盐水洗涤有机层，无水硫酸镁干燥。剩余物质通过柱色谱法(硅胶，己烷：乙酸乙酯 =50：1)纯化得到白色固体 2.76g，收率 98%。

**步骤 3：化合物 4 的合成**

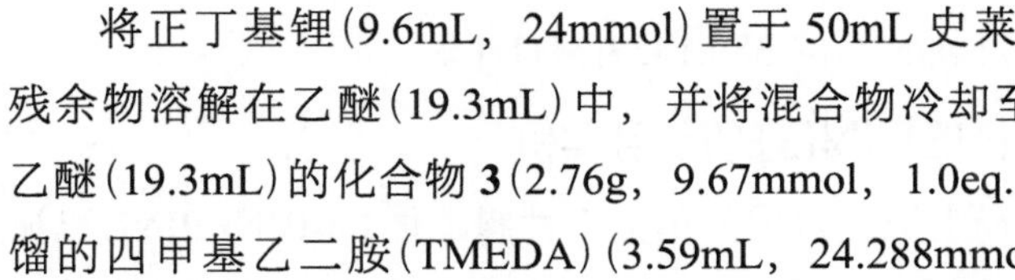

将正丁基锂(9.6mL，24mmol)置于 50mL 史莱克(Schlenk)管中，将残余物溶解在乙醚(19.3mL)中，并将混合物冷却至 0℃。逐滴添加溶于乙醚(19.3mL)的化合物 **3**(2.76g，9.67mmol，1.0eq.)溶液，然后加入新蒸馏的四甲基乙二胺(TMEDA)(3.59mL，24.288mmol)。将所得悬浮液在室温下继续搅拌 20h。反应完成后，在氮气气氛下过滤得到混合物，并用干燥的正己烷(50mL)洗涤所得固体。在真空下干燥得到红色固体二锂化合物 **4**(2.54g，收率 51%)。

### 步骤4：Ph-BINEPINE 的合成

将二锂化合物 4(2.54g，4.572mmol) 溶于正己烷(25mL)中得悬浮液，冷却至 0℃。在此温度下，加入使用正己烷(12mL)稀释的苯基二氯化膦(0.69mL，5.08mmol) 溶液，继续加热回流 2h。反应完成后，用水：甲苯(1:1，50mL)淬灭，萃取，减压下除去有机相的溶剂。快速硅胶柱色谱法(环己烷：甲苯 = 5:1)分离得到白色固体形式的磷化合物 1.33g，收率 75%。

### (b) Ph-BINEPINE 的结构表征信息 [1]

$^1$H NMR (700 MHz，CD$_2$Cl$_2$)$\delta$：2.81～2.82(m，2H)，2.85(dd，$J$ = 11.6Hz，$J$ = 3.0Hz，1H)，3.05(dd，$J$ = 16.9Hz，$J$ = 11.6Hz，1H)，6.93(d，$J$ = 8.3Hz，1H)，7.18～7.31(m，9H)，7.40(ddd，$J$ = 7.9Hz，$J$ = 5.6Hz，$J$ = 2.2Hz，1H)，7.44(m，1H)，7.69(d，$J$ = 8.4Hz，1H)，7.71(d，$J$ = 8.4Hz，1H)，7.88(d，$J$ = 8.2Hz，1H)，7.95(d，$J$ = 7.4Hz，1H)，7.97(d，$J$ = 8.4Hz，1H)。

$^{13}$C NMR (126 MHz，CD$_2$Cl$_2$)$\delta$：30.8，32.3，125.2，125.5，126.2，126.3，126.9，126.9，127.7，127.8，128.3，128.6，128.6，128，128.7，128.8，131.7，132.3，132.6，132.8，133.1，133.3，134.1，134.1，134.9，138.5。

$^{31}$P NMR (202 MHz，C$_6$D$_6$)$\delta$：6.2。

**HRMS**(ESI)：$m/z$(%) = 389.1453[M+H]$^+$。

### (c) Ph-BINEPINE 的应用实例

由联萘酚为原料合成的单齿手性膦配体 Ph-BINEPINE 已被广泛应用，在不对称催化反应中具有良好的手性诱导性能，可用于铑催化的各种化合物的不对称氢化反应。如基于 $\beta$-去氢氨基甲酸衍生物的不对称氢化来制备具有光学活性的 $\beta$-氨基酸 [2]。在不对称铑催化反应中，对映选择性在很大程度上取决于磷原子上取代基的性质和底物的结构。通过改变 P 原子上的取代基，可以很容易地对它们进行结构优化。在铑和钌催化的不对称氢化反应中可高对映选择性地获得预期产物。

Ph-BINEPINE 除了作手性配体外，本身也是很有效的有机膦催化剂。2017 年，Kalek 和 Fu 课题组[3]研究了 4-苄基-2-叔丁基噁唑啉-5-酮在 Ph-BINEPINE 的催化作用下，对外消旋烯的对映选择性加成反应，以 92% 的产率和 91% 的 ee 值得到含有季碳手性中心的产物。

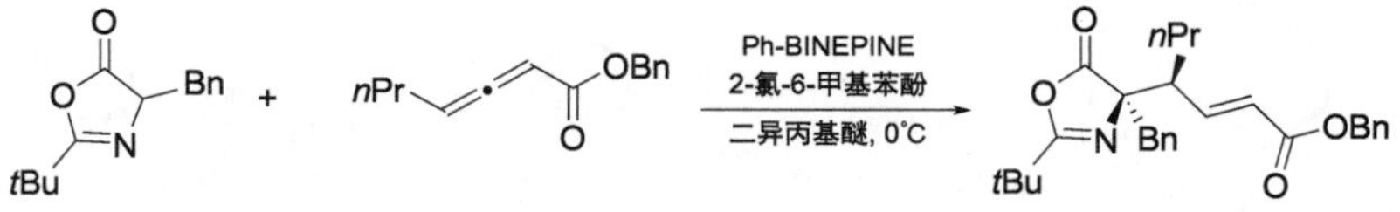

## 具体反应示例[3]：

在充氮手套箱中，将适量手性 Ph-BINEPINE 的无水二异丙基醚溶液 (0.0125mol/L) 置于干燥小瓶中，然后添加 (±)4-苄基-2-叔丁基噁唑啉-5-酮的无水二异丙基醚溶液 (0.50mol/L，100μL，0.050mmol) 和 2-氯-6-甲基苯酚的无水二异丙基醚溶液 (0.10mol/L，50μL，0.0050mmol) 的混合物，并冷却至 0℃。添加预冷至 0℃的苄基 (±)-庚-2,3-二烯酸酯的无水二异丙基醚溶液 (0.40mol/L，150μL，0.060mmol) 储备溶液，并在 0℃下搅拌反应混合物 24h。然后，加入叔丁基过氧化氢的癸烷溶液 (5.0～6.0mol/L，5μL) 以淬灭反应。将混合物在 0℃下再搅拌 10min，然后将小瓶从手套箱中取出并缓慢升温至室温。添加二苄基醚 (内标物，9.5μL，9.9mg，0.050mmol)，然后在减压下浓缩反应混合物。通过 $^1$H NMR 分析确定所有反应的转化率 > 90%。产物通过 TLC [二氧化硅；己烷:乙酸乙酯 (9:1) 和乙酸乙酯] 纯化。

## 参考文献

[1] Junge K, Oehme G, Monsees A, et al. *Tetrahedron Lett.*, **2002**, *43*: 4977-4980.
[2] Ehtbaler S, Erre G, Junge K, et al. *Org. Proc. Res. Dev.*, **2007**, *11*: 569-577.
[3] Kalek M, Fu G C. *J. Am. Chem. Soc.*, **2017**, *139*: 4225-4229.

## 6.1.2 Monophosphine-olefin 配体的合成及应用实例

### (a) 配体的合成反应过程

**步骤1：** 化合物1的合成

氩气保护下，将(S)-联萘酚(2.0g，7mmol，1.0eq.)加入100mL圆底烧瓶中。在氩气气氛下真空干燥15min后，加入20mL无水二氯甲烷形成悬浮液。将三乙胺(3mL，21mmol，3.0eq.)加入混合溶液中，然后将三氟甲磺酸酐(3mL，17mmol，2.4eq.)滴加到通过冰浴冷却的溶液中。在添加过程中，溶液变成深棕色。TLC监测反应，反应液在减压下通过旋转蒸发浓缩，并在硅胶柱上进行柱色谱纯化，用石油醚和乙酸乙酯作为洗脱剂，得到所需产物。经上述步骤得到的中间体在氩气气氛下真空干燥 15min 后直接进行下一步反应。在手套箱中称取 1,3-双(二苯基膦)丙烷二氯化镍(0.05eq.)，然后用 30mL 无水四氢呋喃稀释。将甲基溴化镁的乙醚溶液(6.0eq., 3mol/L)滴加到通过冰浴冷却的溶液中。添加后，将反应体系移至室温，加热回流 24h。反应结束后，反应液用 1mol/L 盐酸水溶液淬灭。用乙醚(30mL×4)萃取水层，然后合并有机层，无水硫酸钠干燥，过滤，减压除去溶剂，并在硅胶柱上纯化，收率 95%。

**步骤 2**：化合物 2 的合成

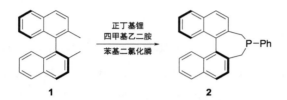

(S)-构型的 2,2′-二甲基联萘（化合物 **1**）与正丁基锂/四甲基乙二胺的金属化反应可以 70%～80% 的产率得到二锂化中间体。然后在 0℃下，往 35mL 己烷的手性 2,2′-二甲基联萘基二锂盐溶液(12mmol)中，加入 13.6mmol 苯基二氯化膦(溶于 15mL 己烷)。回流 2h 后，将反应混合物用水/甲苯淬灭。萃取分离有机层，并用无水硫酸镁干燥，过滤，减压除去溶剂。所得粗品在甲苯/甲醇中结晶纯化，产率为 83%。

**步骤 3**：化合物 3 的合成

将化合物 **2**(100mmol)充分溶解于 50mmol 的四甲基乙二胺，并加入水(35mL)和二氯甲烷(200mL)，水层用二氯甲烷(25mL×2)萃取。用水和盐水洗涤合并的有机层，并用无水硫酸镁干燥。过滤，减压除去溶剂，将粗产物溶解在四氢呋喃(60mL)中，冷却至 0℃。向该溶液中逐滴加入硼烷的四氢呋喃溶液(1mol/L, 40mL, 40mmol)，并将反应缓慢升温至室温。搅拌 2h 后，小心地加入水(4mL)。然后将混合物浓缩并将剩余物溶于二氯甲烷(200mL)。有机层用水(25mL×2)和盐水洗涤，并用无水硫酸钠干燥，过滤，减压除去溶剂，剩余物质通过柱色谱法(硅胶，石油醚：二氯甲烷＝7：3)纯化，得到白色粉末状固体 5.5g，收率 64%。

**步骤 4：** 化合物 4 的合成

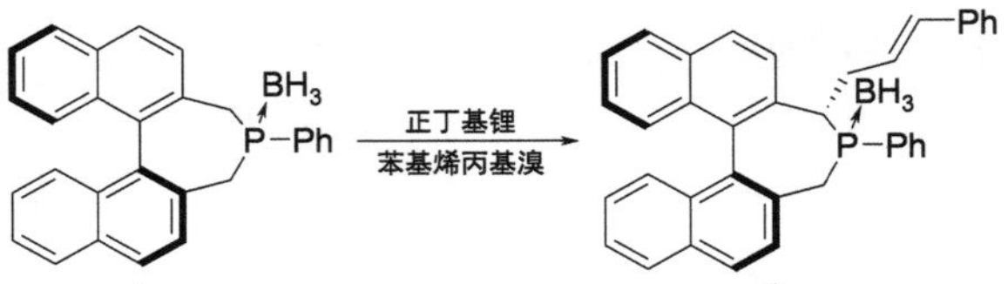

氮气保护下，在冷却时(-78℃)向 **3**(600mg，1.5mmol)的四氢呋喃(15mL)溶液中逐滴加入正丁基锂(1.3mL，2.08mmol，1.6mol/L 的己烷溶液)。使暗红色溶液在 2h 内逐渐升温至-40℃，然后再次冷却至-78℃。将苯基烯丙基溴(500mg，2.5mmol)在 15min 内溶于无水四氢呋喃(5mL)，并继续搅拌 1h。加入几滴水淬灭后，浓缩反应液，并将剩余物溶于二氯甲烷(50mL)中。有机相依次用水和盐水洗涤，并用硫酸镁干燥，过滤，减压除去溶剂，残留物用丙酮(20mL)处理并超声处理 10min。过滤悬浮液，并将固体用丙酮(10mL)洗涤，得到纯的白色固体硼烷络合物 563mg，收率 71%。

**步骤 5：** 化合物 5 的合成

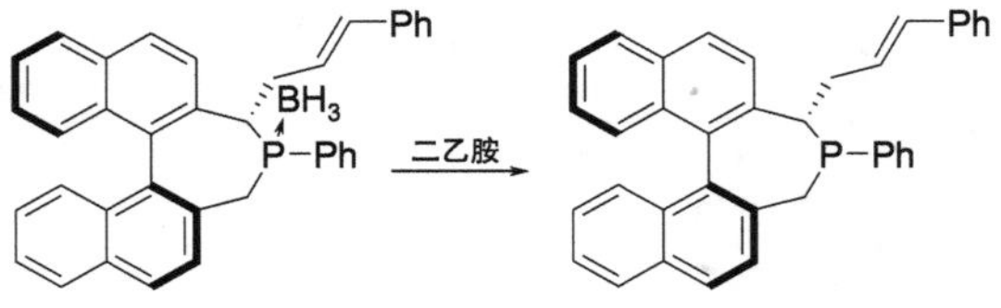

将硼烷络合物 **4**(540mg，1.04mmol)在 50℃下加入二乙胺(5mL)和四氢呋喃(5mL)的溶液中，搅拌 12h，冷却后，减压除去溶剂，而后再加入甲醇(15mL)，并将悬浮液超声 10min。分离出固体并进行重结晶纯化：通过甲苯/戊烷结晶得到白色固体 461mg，收率 86%。

### (b) Monophosphine-olefin 的结构表征信息 [1]

$^1$H NMR δ：7.91(d，J＝8.6Hz，1H)；7.89(d，J＝8.6Hz，1H)；7.83

(d, $J$ = 8.1Hz, 1H); 7.69(d, $J$ = 8.3Hz, 1H); 7.63(d, $J$ = 8.6Hz, 1H); 7.40~6.96(m, 16H); 6.90(d, $J$ = 8.6Hz, 1H); 5.84~5.80(m, 1H); 5.60(d, $J$ = 16.0Hz, 1H); 3.03~2.90(m, 3H); 2.11~2.10(m, 1H), 1.53~1.48(m, 1H)。

**$^{13}$C NMR** $\delta$：138.86(C); 137.46; 137.36(d, $J$ = 2.7Hz, C); 134.39(d, $J$ = 5.5Hz, C); 132.98(d, $J$ = 31.5Hz, C); 132.79(C); 132.78(C); 132.76(C); 132.75(C); 132.58(C); 131.75(d, $J$ = 18.2Hz); 130.77; 130.31; 129.44; 129.28; 129.02; 128.71(d, $J$ = 1.7Hz); 128.39; 128.35; 128.29; 128.26; 128.04; 127.82; 127.10(d, $J$ = 2.5Hz); 126.76; 126.60; 125.99; 125.90; 125.81; 125.20(d, $J$ = 4.6Hz); 48.68(d, $J$ = 20.8Hz); 38.60(d, $J$ = 31.4Hz, $CH_2$); 31.86(d, $J$ = 16.8Hz, $CH_2$)。

**$^{31}$P NMR** $\delta$：16.43(s)。

**HRMS**(EI)：$m/z$(%) = 504.1996[M+H]$^+$。

**$[\alpha]_D^{30}$**：+286($c$ = 0.5，氯仿)。

**mp**：178~180℃。

### (c) Monophosphine-olefin 的应用实例

该配体可用于 Rh 催化环戊烯酮或环类 α, β-不饱和内酯与芳基硼酸 Ar-B(OH)$_2$ 的不对称共轭加成反应，能够以中等到较好的收率得到预期芳基化的产物(64%~88%)，同时具有较高的对映选择性(ee值为88%~98%)。

**参考文献**

[1] Kasak P, Arion V B, Widhalm M. *Tetrahedron: Asymmetry*, **2006**, *17*: 3084-3090.

## 6.1.3 Ding's secondary phosphine oxide (SPO) 配体的合成及应用实例

## (a) 配体的合成反应过程

**步骤1：** 化合物 2 的合成

氮气保护下，在 1L 的配备磁搅拌棒和回流冷凝器圆底烧瓶中加入甲醇(300mL)、2-甲氧基苯甲醛(27.2g，0.2mol)、铝粉(10.8g，0.4mol)后搅拌，分三次加入氢氧化钾(67g，1.2mol)，再搅拌反应 0.5h。冷却到室温后，加入 300mL 水。用二氯甲烷(300mL×3)提取混合物，有机相用盐水洗涤，无水硫酸镁干燥，过滤，减压除去溶剂，残余物转移到500mL圆底烧瓶中，加入250mL的二甲基亚砜和40mL的氢溴酸(40%)。将混合物加热到12℃。搅拌反应3h后，加入碳酸钠，调节pH=7。将混合物倒入1L水中搅拌过夜后冷却至室温，通过过滤收集混合物中的固体，用水洗涤4次，并在减压下干燥，获得棕色固体。正己烷/乙酸乙酯重结晶纯化，得到产物17.9g，收率65%。

**步骤2：** 化合物3的合成

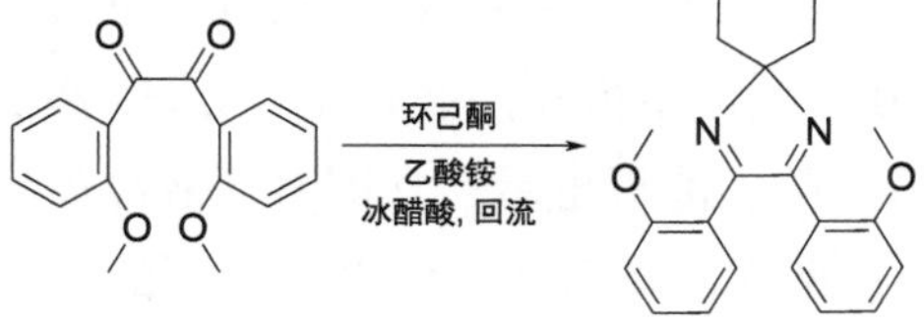

氮气保护下，在配备磁搅拌棒和回流冷凝器的0.5L圆底烧瓶中加入冰醋酸(150mL)、化合物**2**(17.9g，1.3mol)、乙酸铵(38g)和环己酮(8.5mL)。混合液搅拌回流2h。趁热将混合物倒入1L水中搅拌过夜后冷却至室温，固体经过滤收集，用水洗涤两次，无水硫酸钠干燥，过滤，减压除去溶剂，得到黄绿色固体，通过水/甲醇重结晶纯化，得到产物21.5g，收率93.5%。

**步骤3：** 化合物4的合成及其化合物5、6的手性拆分

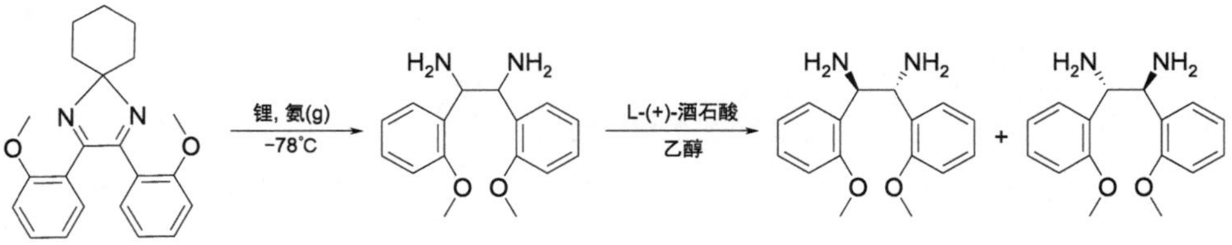

氩气保护下，在装有机械搅拌器、温度计和干冰冷凝器的1L四颈圆底烧瓶中加入化合物**3**(28g，0.073mol)。烧瓶置换氩气，加入150mL的无水四氢呋喃。将混合物冷却到-78℃，用气态氨处理，直到液体体积为150mL。在氩气状态下，缓慢地加入锂(2.8g，0.46mol)。混合液搅拌30min，加入乙醇(15mL)。混合液搅拌20min，加入氯化铵(12.4g)。取

出冷却槽，使混合物加热至 0℃，然后加 150mL 的水，并分离相。用乙醚（200mL×3）提取混合物。合并有机相用盐水洗涤，用硫酸镁干燥，过滤，减压除去溶剂，冷却到 0℃，再用 100mL 的 2mol/L 盐酸水洗涤。将混合物搅拌反应 1h，然后加入 200mL 的水，分离两相混合物。用水（30mL×3）洗涤有机相，用二氯甲烷 100mL 洗涤水相。用 2mol/L 氢氧化钠溶液 100mL 处理水溶液，用二氯甲烷 100mL 提取混合液 4 次，结合有机相用盐水洗涤，硫酸镁干燥，过滤，减压除去溶剂。得到淡黄色固体化合物 **4**，收率 94%（19g）。

氮气保护下，将消旋的二胺混合物 **4**（13.5g, 0.05mol）溶于乙醇（60mL）溶液中，并置于 70 ℃油浴中搅拌，然后加入 L-(+)-酒石酸（7.5g, 0.05mol）进行反应。酒石酸盐沉淀析出后，立即将溶液冷却到室温。固体经过滤收集，用乙醇洗涤两次，减压除去溶剂。得到的酒石酸盐用乙醇/水重结晶两次，得到 5g 无色晶体。将晶体转移到 250mL 的圆底烧瓶中，冷却至 0～5℃后，在搅拌下，滴加 50% 氢氧化钠水溶液（6mL），然后加入 50mL 二氯甲烷，继续搅拌反应 1h，用二氯甲烷（50mL×3）萃取水相。有机相用盐水洗涤，无水硫酸镁干燥，过滤，减压除去溶剂。得到的化合物 **5**（2.6g）为白色固体，收率 40%。过滤固体后的母液同上处理，经碱中和、萃取和浓缩得到的化合物 **6**（3g）为淡黄色固体，收率 45%。化合物 **5** 和 **6** 是构型相反的对映体。

### 步骤 4：化合物 7 的合成

将化合物 **5**（5.4g, 20mmol）分批加入二碳酸二叔丁酯（12.5g, 60mmol）、二甲基氨基吡啶（0.25g, 2mmol）的乙腈溶液中，在室温下搅拌半小时后，用水淬灭（10mL），并用乙醚萃取 3 次（每次 50mL），合并有机相，用盐水洗涤后用无水硫酸镁干燥。浓缩后，残余物溶于二氯甲烷（100mL），加入三氟乙酸（20mL），然后在室温下搅拌半小时，用水淬灭(8mL)。用二氯甲烷萃取三次(每次 50mL)后,用饱和碳酸氢钠水溶液和盐水依次洗多次,用无水硫酸钠干燥,浓缩后得到的固体,用重结晶的方法进行纯化(正己烷/乙酸乙酯),得到白色固体 5.72g,产率 96%。

**步骤 5:** 化合物 8 的合成

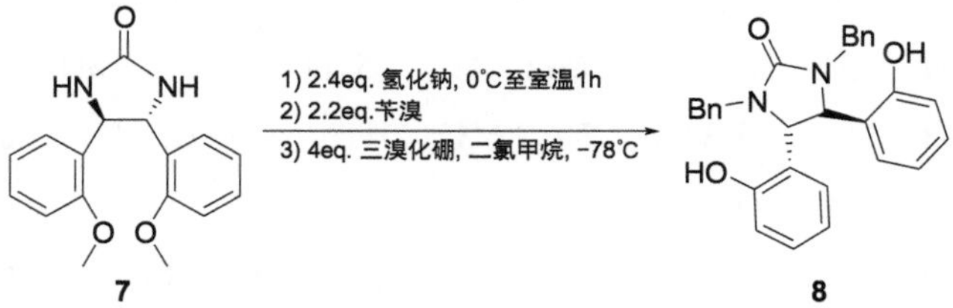

将化合物 **7**(715mg,2.4mmol)的四氢呋喃溶液(2mL)于 0℃下滴加到氢化钠(含量为 60%的矿物油悬浮液)的四氢呋喃(5mL)溶液中,而后在室温下搅拌 0.5h。在 0℃下加入苄溴(5mmol),升温至室温后继续搅拌反应 5~10h 直到反应完全。用水(5mL)淬灭反应,用乙醚萃取(10mL×3),合并有机相并用盐水洗,无水硫酸镁干燥,浓缩后的残余物用硅胶柱色谱法分离(洗脱剂为乙酸乙酯/正己烷 = 1:10 ~ 1:5),得到相应油状产物,产率 95%。而后取该化合物(0.5mmol)溶于干燥的二氯甲烷(3mL)中,冷却到-78℃,加入三溴化硼(2mmol),在-78℃下搅拌半小时后撤去冷却浴,继续搅拌 24h。反应结束后,用水淬灭(3mL),蒸去二氯甲烷,水相用氢氧化钠水溶液中和,而后用乙酸乙酯萃取(10mL×3),有机相用无水硫酸钠干燥。浓缩后得到的残余物用硅胶柱色谱法分离(洗脱剂为乙酸乙酯/正己烷 = 1:3 ~ 1:1),得到化合物 **8**。

**步骤 6:** Ding's SPO 的合成

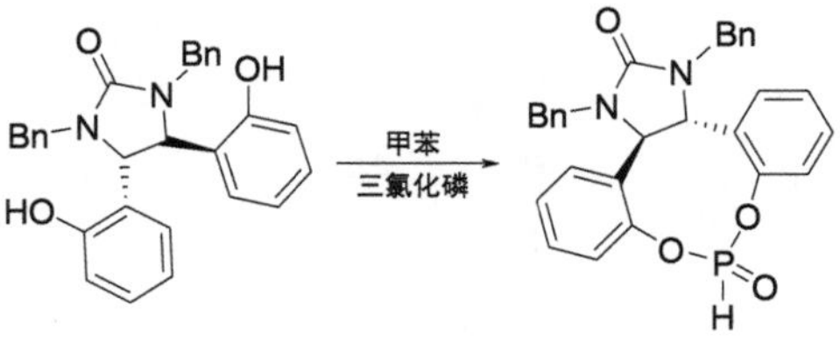

在氩气保护下,干燥的史莱克管中加入化合物 **8**(0.6mmol)、三氯化磷(0.3mL,3.2mmol)和甲苯(3mL)。将混合物加热回流反应 3h,在减压

下除去溶剂和过量的三氯化磷。残余物溶于甲苯(3mL)中，0℃下，加水(15μL，0.8mmol)。将混合物在室温下搅拌0.5h，减压除去溶剂，得到相应的纯的二次膦氧化物，为白色固体，收率91%。

## (b) SPO 的结构表征信息 [1-3]

$^1$H NMR(400 MHz，CDCl$_3$)δ：7.40～7.02(m，18H)，7.08(d，$J$ = 736.0Hz，1H)，5.26(d，$J$ = 15.2Hz，1H)，5.09(d，$J$ = 15.2Hz，1H)，4.49(d，$J$ = 3.6Hz，1H)，3.34(d，$J$ = 3.6Hz，1H)，3.74(d，$J$ = 15.2Hz，1H)，3.70(d，$J$ = 15.2Hz，1H)。

$^{13}$C NMR(100 MHz，CDCl$_3$)δ：159.6，145.6(d，$J$ = 8.6Hz)，145.0(d，$J$ = 11.1Hz)，136.4，136.0，132.6(d，$J$ = 3.0Hz)，131.4(d，$J$ = 3.7Hz)，129.6，129.2，128.6，128.5，127.8，127.6，127.5，127.2，126.7，126.5，125.3，123.4，122.0(d，$J$ = 4.5Hz)，58.2，57.8，45.7，45.4。

$^{31}$P NMR(161 MHz，CDCl$_3$)δ：6.4。

**FTIR**(KBr)υ(cm$^{-1}$)：3062，3031，2921，1698，1584，1493，1452，1289，1214，1094，923，700。

**MS**(ESI)：$m/z$= 513.1 [M + OH$^-$]。

**HRMS**(MALDI/DHB)：$m/z$(%)= 497.1636 [M + H$^+$]。

$[α]_D^{20}$：+257.4°($c$ = 1.0，二氯甲烷)。

**mp**：138～140℃。

## (c) Ding's SPO 的应用实例

2012年，丁奎岭课题组[2]使用他们创制的SPO配体解决了Rh(Ⅰ)催化的α-苯乙烯基膦酸加氢反应，获得了90%～95%的收率和ee值＞99%的对映选择性，为合成光学活性膦酸提供了一条便捷的途径。

R$^1$R$^2$C=C(P(OH)$_2$(O)) → [Rh(cod)$_2$]OTf/SPO配体(1∶2) (摩尔分数0.01%～1%), CH$_2$Cl$_2$, Et$_3$N(1eq.), H$_2$(1MPa), 12h, 室温 → R$^1$R$^2$CH-CH(P(OH)$_2$(O))

R$^1$ = H, Et
R$^2$ = 芳基, 环己基, 异丙基
ee：＞99%

**具体反应示例**[3]：

在氩气下，将手性单齿氧膦配体(SPO)(1.9mg，0.005mmol)和

[Rh(μ-Cl)(cod)]₂(0.62mg，0.00125mmol)溶于甲苯(1.0mL)，混合液在室温下搅拌 10min 后得到催化剂配合物溶液。向装有底物(0.25mmol)和甲苯(1.5mL)的史莱克反应管里加入三乙胺(17μL，0.125mmol)，室温下搅拌 10min 后，在手套箱里将底物溶液转移到含有催化剂的套管并放入不锈钢反应釜，封好后取出，用氢气置换 3 次后加氢气压力到 5MPa。反应液在 0℃下搅拌 24h。反应结束后放出氢气（通风橱中进行处理），用 NMR 分析转化率，用 HPLC 检测产物的 ee 值。

### 参考文献

[1] Liu Y, Ding K L. *J. Am. Chem. Soc.*, 2005, *127*: 10488-10489.
[2] Dong KW, Wang Z, Ding K L. *J. Am. Chem. Soc.*, 2012, *134*: 12474-12477.
[3] Dong KW, Li Y, Wang Z, et al. *Org. Chem. Front.*, 2014, *1*: 155-160.

## 6.1.4　Taddol 类亚磷酰胺配体的合成及应用实例

### (a) 配体的合成反应过程

### 步骤 1：化合物 2 的合成

取 250mL 史莱克瓶，加入(+)-酒石酸二甲酯(30mmol，5.34g)、对甲苯磺酸(1.5mmol，127mg)。在氮气氛围下，加入 3-戊酮(420mmol，44mL)、原甲酸三甲酯(75mmol，8.2mL)。在 60℃下反应 24h，反应完全后，用饱和碳酸氢钠溶液淬灭，乙酸乙酯萃取，饱和食盐水洗 3 次，将上层有机溶剂用无水硫酸钠干燥，减压浓缩，最后用柱色谱法（石油醚 / 乙酸乙酯 = 10 : 1）分离得到无色液体(R,R)-2(6.6g，收率 89%)。

### 步骤 2：化合物 3 的合成

取 250mL 三口瓶，加入 1.1g 镁屑。在氮气氛围下，加入碘粒。将反应接通回流冷凝装置，加入 10mL 经干燥处理的四氢呋喃，将 3,5-二甲基溴苯(6.1mL，45.0mmol)溶于 45mL 四氢呋喃，通过恒压滴液漏斗缓慢滴加。随着反应的进行溶液中碘的颜色褪去，镁屑消失，3h 后反应结束，格氏试剂逐渐冷却至室温，将产物 2(2.46g，10.0mmol)溶于 10mL 四氢呋喃通过恒压滴液漏斗缓慢滴加，该反应为放热反应，室温下滴加完全后，将该反应在室温下反应 12h。反应结束后用氯化铵的饱和溶液淬灭，用乙醚和水洗涤 3 次，合并有机相，减压旋蒸除去溶剂。最后用柱色谱法（石油醚 / 乙酸乙酯 = 50 : 1）分离得到白色固体(R,R)-3(3.65g，收率 60%)。

### 步骤 3：(R,R)-Taddol 类亚磷酰胺配体 4 的合成

取 250mL 史莱克反应瓶，加入 (R,R)-**3** (2.4g, 4.0mmol)，在氮气氛围下加入 80mL 四氢呋喃，待溶液变为无色澄清时加入三乙胺 (2.2mL, 16.0mmol, 4.0eq.)，反应溶液呈淡黄色。在 0℃下于 2min 内缓慢逐滴加入三氯化磷 (0.37mL, 4.2mmol, 1.05eq.)，溶液呈白色浑浊态。将反应体系移至室温下搅拌 1h 后将反应体系重新置于 0℃冰水浴后缓慢加入吲哚啉 (12.0mmol, 3.0eq.)，将反应体系移至室温下搅拌 12h。用硅藻土过滤，并用四氢呋喃冲洗滤饼数次，减压旋蒸除去溶剂后用柱色谱法 (石油醚/乙酸乙酯 = 90∶1) 分离得到泡沫状固体，即 (R,R)-Taddol 类亚磷酰胺配体 **4** (2.21g, 产率 73%)。

### (b) (R, R)-Taddol 类亚磷酰胺配体 4 的结构表征信息 [1]

**mp**：191～193℃。

**$^1$H NMR** (400 MHz, CDCl$_3$) $\delta$：7.19 (d, $J$ = 4.8Hz, 4H), 7.10 (s, 2H), 7.05 (d, $J$ = 7.2Hz, 1H), 7.00～6.91 (m, 5H), 6.83 (s, 3H), 6.77～6.68 (m, 1H), 5.11 (d, $J$ = 8.4Hz, 1H), 5.00 (d, $J$ = 8.8Hz, 1H), 4.10～4.04 (m, 1H), 3.45～3.35 (m, 1H), 3.10～2.90 (m, 1H), 2.83～2.66 (m, 1H), 2.30 (s, 6H), 2.25 (s, 6H), 2.24 (s, 6H), 2.15 (s, 6H), 1.42～1.28 (m, 1H), 1.25～1.15 (m, 1H), 0.92～0.84 (m, 1H), 0.60～0.52 (m, 4H), 0.33 (t, $J$ = 7.2Hz, 3H)。

**$^{13}$C NMR** (100 MHz, CDCl$_3$) $\delta$：148.3 (d, $J$ = 24.6Hz), 146.3, 145.9, 142.1 (d, $J$ = 3.1Hz), 142.0 (d, $J$ = 2.0Hz), 137.6, 137.1, 136.4, 136.2, 131.2 (d, $J$ = 4.8Hz), 129.6, 128.8, 128.7, 127.1, 126.8, 125.4, 125.1, 124.9, 119.6, 116.3, 110.1 (d, $J$ = 22.0Hz), 84.5 (d, $J$ = 6.0Hz), 82.8 (d, $J$ = 7.5Hz), 81.9 (d, $J$ = 15.0Hz), 81.7 (d, $J$ = 3.5Hz), 46.0 (d, $J$ = 4.7Hz), 29.8, 29.1, 29.0, 21.8, 21.7, 21.6, 21.5, 8.17, 8.15。

**HRMS** (ESI)：$m/z$ (%) = 754.4037 [M + H]$^+$。

### (c) (R, R)-Taddol 类亚磷酰胺配体的应用实例

2014 年，Morken 课题组[1]以钯/手性 Taddol 衍生的亚磷酰胺配体的络合物作为催化剂，实现了不对称催化 Suzuki 偶联反应，基于去对称化的策略构建了高对映选择性的有机硼酸酯。

2019 年，徐利文课题组也利用该类配体实现了多类具有挑战性的不对称钯催化碳-碳键活化转化反应，以环丁酮为原料合成了结构多样性的

手性茚酮衍生物，绝大部分产物 ee 值都在 90% 以上 [2]。

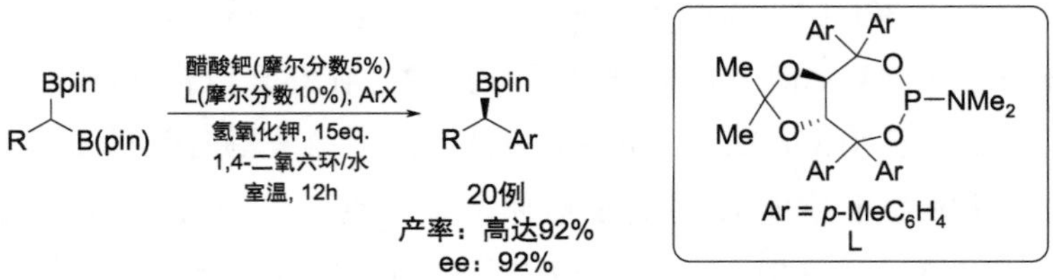

2010 年，Suginome 课题组 [3] 将 (R,R)-Taddol 类亚磷酰胺配体应用于不对称镍催化的炔基化反应，能够得到较高的产率和优异的对映选择性。

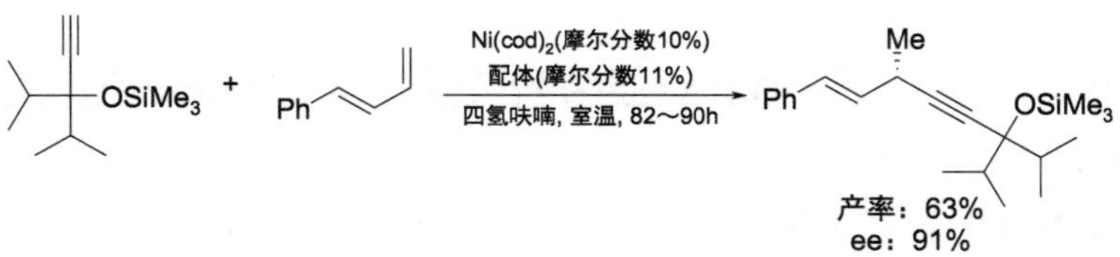

## 具体反应示例[3]：

1-苯基-1,3-丁二烯（反式结构 > 98%，0.30mmol）、Ni(cod)$_2$（8.3mg，0.30mmol）和 (R,R)-Taddol 类亚磷酰胺配体（25.6mg，0.033mmol）溶于 0.10mL 四氢呋喃中，在室温下用注射泵缓慢加入乙炔取代物（95.6mg，0.450mmol），反应大约 82h。然后，将反应混合物通过一个硅胶短柱。有机相浓缩后，残留物进行快速柱色谱法纯化（己烷：乙酸乙酯 = 400：1），得到 64.6mg 纯产物（63% 产率）。

**参考文献**

[1] Sun C R, Potter B, Morken J P. *J. Am. Chem. Soc.*, **2014**, *136*: 6534-6537.
[2] Cao J, Chen L, Sun F N, Sun, et al. *Angew. Chem. Int. Ed.*, **2019**, *58*: 897-901.
[3] Shirakura M, Suginome M. *Angew. Chem. Int. Ed.*, **2010**, *49*: 3827-3829.

# 6.1.5　Me-BI-DIME 配体的合成及应用实例

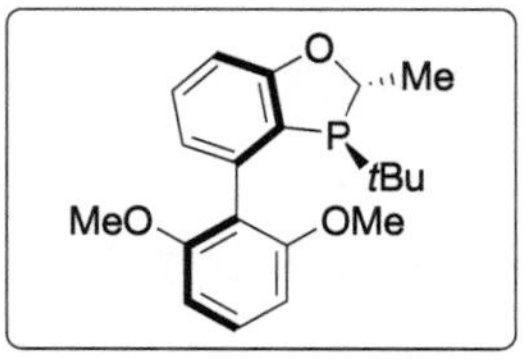

### (a) 配体的合成反应过程

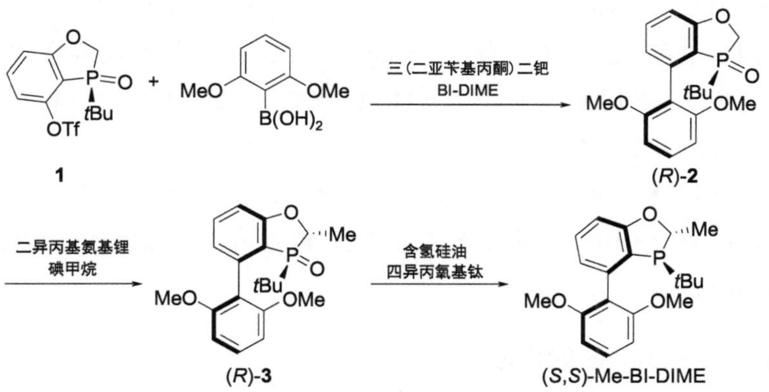

**步骤 1**：化合物 2 的合成

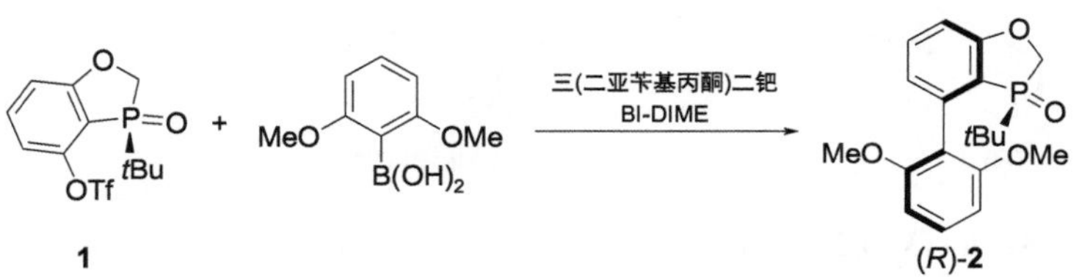

将 **1**(215g，600mmol)、2,6-二甲氧基苯硼酸(163.8g，900mmol，1.5eq.)、三(二亚苄基丙酮)二钯(6.9g，7.5mmol，0.013eq.)、BI-DIME (5.9g，18mmol，0.03eq.)和氟化钾(138.5g，2.4mol，4.0eq.)在脱气的 1,4-二噁烷(2L)中混合。将混合物在氮气保护下 100℃下搅拌 3h，然后冷却至室温。向混合物中加入 2mol/L 氢氧化钠溶液(0.5L)，并将所得混合物在室温下搅拌 15min，然后减压蒸馏除去大多数二噁烷，向残余物中加入二氯甲烷(1L)。所得混合物经硅藻土过滤。分离二氯甲烷层，依次用 2mol/L 氢氧化钠溶液(0.5L)和 5％氯化钠溶液(0.5L)洗涤，并浓缩。将残余物用甲基环己烷(0.4L)处理，并将得到的浆液过滤得到(R)-**2** 为白色结晶固体，产率 91％。

**步骤 2**：化合物 3 的合成

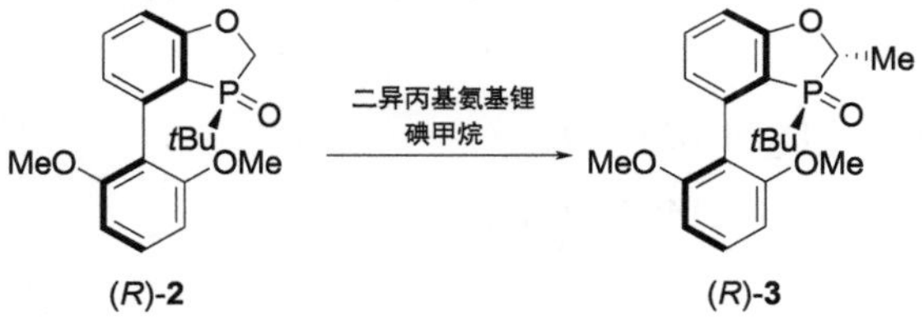

将 (R)-2(0.51g, 1.48mmol, 1eq.) 溶解在 5mL 四氢呋喃溶液中，在 -78℃下滴加二异丙基氨基锂溶液(1.1eq.)(1.0mL，1.8mol/L 的正庚烷/四氢呋喃/氯化苯溶液，1.77mmol，1.2eq.)。将反应混合物在 -78℃下搅拌 1h，然后加入碘甲烷(0.11mL，1.77mmol，1.2eq.)，在 -78℃下搅拌 1h，随后在 1h 内升温至室温。反应完成后，加入氯化铵水溶液进行淬灭，用乙酸乙酯萃取。用硫酸钠干燥，过滤并减压浓缩。粗产物通过色谱柱(5%～10% 甲醇溶于乙酸乙酯)得到白色结晶固体 (R)-3(0.32g，收率 61%)。

### 步骤 3：Me-BI-DIME 的合成

将 (R)-3(0.32g，0.90mmol) 加入 4mL 四氢呋喃溶液中，然后加入含氢硅油(0.26g) 和四异丙氧基钛(0.31mL，1.08mmol，1.2eq.)，混合物回流 12h，旋蒸除去四氢呋喃溶液。然后加入 4mL 30% 质量分数的氢氧化钠溶液，并于 60℃反应 0.5h，加入 4mL 甲基叔丁基醚，分离甲基叔丁基醚层，用甲基叔丁基醚洗水层，干燥，旋干，通过纯化得到白色固体产物(0.28g，收率 90%)。

### (b) Me-BI-DIME 的结构表征信息 [1]

$^{1}$H NMR(400 MHz，CD$_2$Cl$_2$)δ：7.35～7.20(m，2H)，6.85～6.75(m，2H)，6.64(d，J=8.4Hz，1H)，6.61(d，J=8.4Hz，1H)，4.97(q，J=7.0Hz，1H)，3.73(s，3H)，3.71(s，3H)，1.40(dd，J=16.4，7.0Hz，3H)，0.71(d，J=12.0Hz，9H)。

$^{13}$C NMR(100 MHz，CD$_2$Cl$_2$)δ：163.1，158.2，157.4，139.5，130.7，129.4，125.78，125.0，124.0，120.0，109.9，104.6，104.0，79.2，56.0，31.1，25.6，21.6。

$^{31}$P NMR(162 MHz，CD$_2$Cl$_2$)δ：9.9。

HRMS(ESI)：m/z(%)= 345.1605[M+H]$^{+}$。

### (c) Me-BI-DIME 的应用实例

含有 2,3-二氢苯并噁唑骨架的联芳基单膦配体对于钯催化的 Suzuki-Miyaura 交叉偶联反应非常有效，表现出独特的空间和电子性质，与 Buchwald 等人报道的联芳基配体相比，即使是消旋的 BI-DIME 也可解决对映选择性难题，说明此类骨架特别适用于交叉偶联反应[2]，如汤文军课题组首次实现了 2,4,6-三异丙基苯基硼酸与邻位取代的芳基溴化物的不对称催化偶联反应。与 Me-BI-DIME 类似的膦手性中心配体在很多反应中都表现出非常好的手性诱导性，更为详细的信息可参阅汤文军教授课题组的专题综述论文[3]。

**参考文献**

[1] Tang W J, Patel N D, Xu G Q, et al. *Org. Lett.*, **2012**, *14*: 2258-2261.
[2] Tang W J, Capacci A G, Wei X D, et al. *Angew. Chem. Int. Ed.*, **2010**, *49*: 5879-5883.
[3] Yang H, Tang W J. *J. Chem. Rec.*, **2020**, *20*: 23-40.

## 6.1.6　*N*-芳基亚磷酰胺配体的合成及应用实例

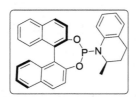

### (a) 配体的合成反应过程

$N_2$ 保护下，向干燥的 250mL 两口烧瓶中加入甲苯(50mL)和三氯化磷(0.67mL，7.7mmol)，将该混合物冷却至 0℃。在另一个干燥的 25mL 烧瓶中将(*R*)-**1**、甲基四氢喹啉(1.13g，7.7mmol)溶于甲苯(8mL)和三乙胺(1.8mL，12.9mmol)。将该混合物在 0℃下逐滴滴加到上述三氯化磷混合溶液中。滴加完毕后,将反应体系在80℃下加热6h,然后缓慢冷却至-78℃。在-78℃下,向该烧瓶中缓慢加入(R, S)-联萘酚(2.0g,7.0mmol)[溶于三乙胺(3.5mL,25.2mmol)、甲苯(30mL)和四氢呋喃(6mL)混合溶液中]。将上述反应体系在室温下搅拌过夜,TLC监测反应。反应完成后,将反应体系通过硅藻土过滤,用乙醚洗涤,将所得有机相真空浓缩。所得产物通过柱色谱法纯化(石油醚:乙酸乙酯:三乙胺 = 10:1:0.01),得到2.97g白色粉末状的固体,产率92%。

### (b) N-芳基亚磷酰胺的结构表征信息 [1]

**$^1$H NMR** (300 MHz, CDCl$_3$) $\delta$ : 8.01~7.86(m, 4 H), 7.57(d, $J$= 8.7Hz, 1 H), 7.46~7.37(m, 4 H), 7.32~7.23(m, 3 H), 7.13(d, $J$= 6.9Hz, 1 H), 6.94~6.79(m, 3 H), 3.73(m, 1 H), 3.12(dd, $J$= 8.4, 15.6Hz, 1 H), 2.38(d, $J$= 15.6Hz, 1 H), 1.06(d, $J$= 6.0Hz, 3 H)。

**$^{13}$C NMR** (75 MHz, CDCl$_3$) $\delta$ : 149.76, 149.70, 149.4, 145.2, 145.0, 132.77, 132.75, 132.60, 132.59, 131.5, 130.9, 130.64, 130.60, 130.5, 130.1, 128.3, 127.05, 127.01, 126.91, 126.22, 126.17, 125.5, 125.0, 124.7, 124.17, 124.10, 122.69, 122.67, 121.78, 121.76, 121.5, 121.06, 121.03, 112.7, 112.5, 55.84(d, $J$ = 4.6Hz), 37.5, 23.23(d, $J$ = 1.7Hz)。

**$^{31}$P NMR** (121 MHz, CDCl$_3$) $\delta$ : 147.0。

**IR** (KBr) $\gamma$ (cm$^{-1}$) : 3054, 2966, 2924, 2849, 1620, 1590, 1508, 1479, 1464, 1457, 1432, 1369, 1253, 1234, 1222, 1204, 1160, 1106, 1071, 1025, 986, 950, 939, 924, 823, 807, 748。

**HRMS** (ESI) : $m/z$(%) = 447.1392[M+H]$^+$。

**mp** : 201~204℃。

$[\alpha]_D^{20}$ : -315.4°($c$=0.5,氯仿)。

### (c) N-芳基亚磷酰胺的应用实例

2012年,游书力课题组[1]研究了不对称催化丙二酸二甲酯钠盐的烯丙基烷基化反应,发现N-芳基亚磷酰胺配体与铱形成的配合物可作为高效的催化剂,产率高达99%,主要的支链产物的对映体过量值高达99%,支链产物与直链产物的产率比值达到>99:1。

2021 年 Eelco Ruijter 课题组将 N-芳基亚磷酰胺配体成功应用到不对称催化 [4+3] 环缩合反应，实现了 98% 的产率和 60% 的 ee 值。

**具体反应示例**[2]：

在惰性气氛下，在干燥的小瓶中依次加入取代环丙烷(0.826mmol，3.5eq.)、二氯化钯的溶液(4.2mg，0.024mmol，0.1eq.)、N-芳基亚磷酰胺配体(47mg，0.094mmol，0.4eq.)、氯化锂(566μL，0.283mmol，0.5mol/L 四氢呋喃溶液)、水(26μL，1.416mmol，6.0eq.)和四氢呋喃(614μL)，在室温下搅拌 1h 后，加入水杨醛(0.236mmol，1.0eq.)和叔丁醇钾(53mg，0.472mmol，2.0eq.)，混合物搅拌过夜。反应结束后，过滤掉固体，真空除去溶剂，粗品直接通过柱色谱法纯化得到产物。

**参考文献**

[1] Liu W B, Zheng C, Zhuo C X, et al. *J. Am. Chem. Soc.*, **2012**, *134*: 4812-4821.
[2] Faltracco M, van de Vrande K N A, Dijkstra M, et al. *Angew. Chem. Int. Ed.*, **2021**, *60*: 14410-14414.

## 6.1.7　Feringaphos 配体的合成及应用实例

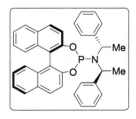

## (a) 配体的合成反应过程

## 步骤1：化合物2的合成

向烧瓶中加入(R)-(+)-1,1′-联-2-萘酚 1 (1eq.)与三氯化磷(9.6eq.)及1-甲基-2-吡咯烷酮(0.008eq.)，置于92℃的油浴中，固体在5min内完全溶解，并保持回流10min。反应过程中氯化氢(盐酸)气体迅速放出，溶液变成黄色。然后将该反应冷却至室温。待反应结束后，减压7mmHg (1mmHg=133.3Pa)除去残留的三氯化磷和氯化氢，并用氮气将烧瓶排气后，加入适量乙醚以溶解固体。再次减压7mmHg除去溶剂和三氯化磷，得到浅黄色固体。将烧瓶连接至高真空(0.1mmHg)系统12h，所得到的产物为灰白色粉末，用于后续反应。

## 步骤2：化合物Feringaphos的合成

将装有磁力搅拌子的史莱克瓶经过火焰干燥并用氮气吹扫。向反应瓶内注入无水四氢呋喃和(S,S)-双(1-苯基乙基)胺 **3**,并移至-78℃环境中。在 15min 内逐滴滴加正丁基锂(1.0eq.),滴加完成时,反应液呈浅灰粉色。将反应液在约 30min 内加热至-30℃,并立即冷却至-78℃搅拌 1h,溶液变成深粉红色。然后将(R)-(-)-1,1'-联萘-2,2'-亚磷酰氯 **2**(1.1eq.)用适量无水四氢呋喃稀释,通过注射器缓慢滴加,并将该反应液于-78℃下再保持 2h,然后将其加热至环境温度,并搅拌 12h。通过旋转蒸发除去溶剂,得到浅黄色油与白色固体混合物。将残余物溶于微量二氯甲烷中,然后通过硅胶柱(戊烷/二氯甲烷=4∶1)进行纯化,得到白色泡沫状物,产率为 41%。

### (b) Feringaphos 的结构表征信息 [1]

$^1$H NMR (400MHz, CDCl$_3$) $\delta$: 7.93 (d, $J$ = 8.8Hz, 2H), 7.88 (dd, $J$ = 7.8Hz, $J$ = 4.2Hz, 2H), 7.58 (d, $J$ = 8.8Hz, 1H), 7.43 (d, $J$ = 8.8Hz, 1H), 7.40~7.34 (m, 3H), 7.28 (d, $J$ = 8.0Hz, 1H), 7.25~7.16 (m, 2H), 7.15~7.06 (m, 10H), 4.49 (m, 2H), 1.72 (d, $J$ = 6.8Hz, 6H)。

$^{13}$C NMR (100.6MHz, CDCl$_3$) $\delta$: 150.1, 150.0, 149.5, 142.8, 132.8, 132.7, 131.4, 130.4, 130.2, 129.4, 128.3, 128.1, 127.9, 127.7, 127.1, 127.07, 126.6, 126.0, 125.9, 124.7, 124.1, 124.0, 122.4, 122.3, 121.73, 121.71, 52.3, 52.2, 21.9。

$^{31}$P NMR (101.3MHz, CDCl$_3$) $\delta$: 145.6。

IR (薄膜) $\gamma$ (cm$^{-1}$): 3069, 2971, 1617, 1590, 1506, 1495, 1375, 1326, 1231, 1203, 1134, 1070。

### (c) Feringaphos 的应用实例

1997 年,Feringa 等[1]在联芳基二酚的亚磷酰胺配体中引入手性基部分,合成了手性二(1-苯乙基)氨基取代的配体(S,R,R)-**L4**,并应用于铜催化的二乙基锌对环状烯酮的加成反应中,首次获得了几乎单一构型的产物。

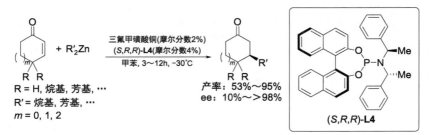

2020年，陆平课题组[2]以二烷基锌为亲核试剂，二苯基磷酰氯为亲电试剂，三氟甲磺酸铜作为铜源，对环丁烯酮进行1,4-加成/捕获反应从而获得手性环丁烯基磷酸酯。在该反应中，手性亚磷酰胺配体($R,R,R$)-**L4**呈现出色的催化活性和立体选择性，其对映选择性高达97%。

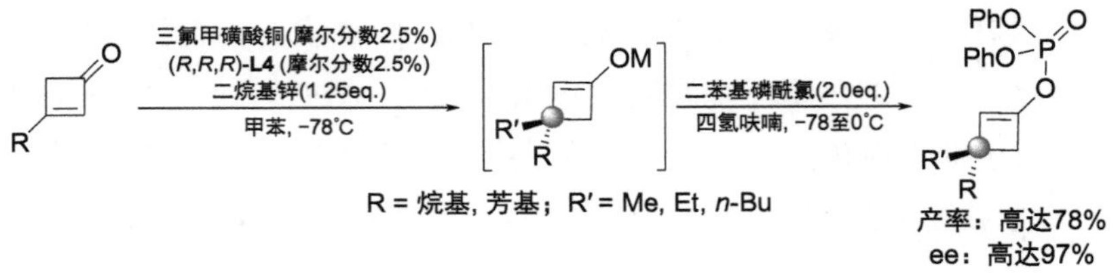

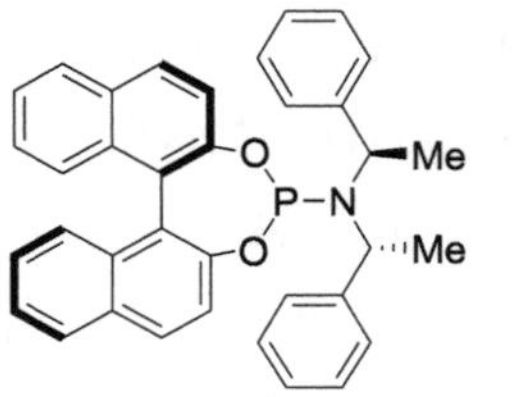

## 具体反应示例[3]：

将乙烯基苯并噁嗪酮(2.0mmol，0.6580g)、亚烷基吡唑啉酮(2.0mmol，0.5240g)、四(三苯基膦)钯(0.1mmol，0.1156g)和Feringaphos(0.4mmol，0.210g)混溶于10mL四氢呋喃中，并将混合物在室温下搅拌12h。所得反应物经过减压浓缩后，粗产物以快速硅胶色谱柱分离纯化(石油醚：乙酸乙酯＝5∶1～8∶1)，可得到产物(1.0721g，产率98%，dr＞99∶1，ee值为99%)。

## 参考文献

[1] Feringa B L, Pineschi M, Arnold L A, et al. *Angew. Chem., Int. Ed.*, **1997**, *36*: 2620-2623.
[2] Zhong C X, Huang Y C, Zhang H C, et al. *Angew. Chem. Int. Ed.*, **2020**, *59*: 2750-2754.
[3] Guo J M, Fan X Z, Wu H H, et al. *J. Org. Chem.*, **2021**, *86*: 1712-1720.

# 6.2 手性双膦配体的代表性合成及应用实例

## 6.2.1 BINAP 配体的合成及应用实例

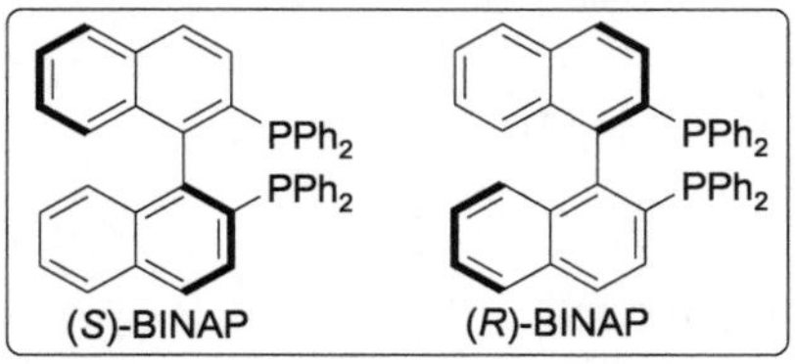


### (a) 配体的合成反应过程

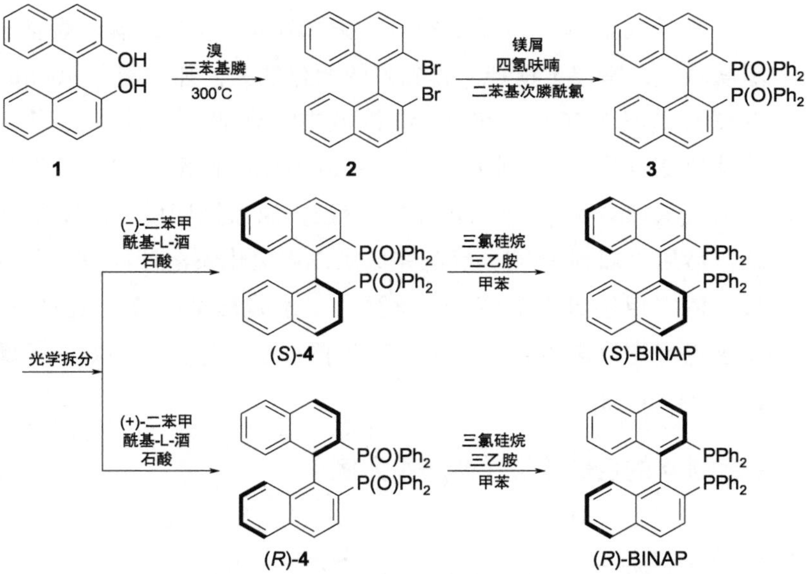


**步骤 1：** 化合物 2 的合成

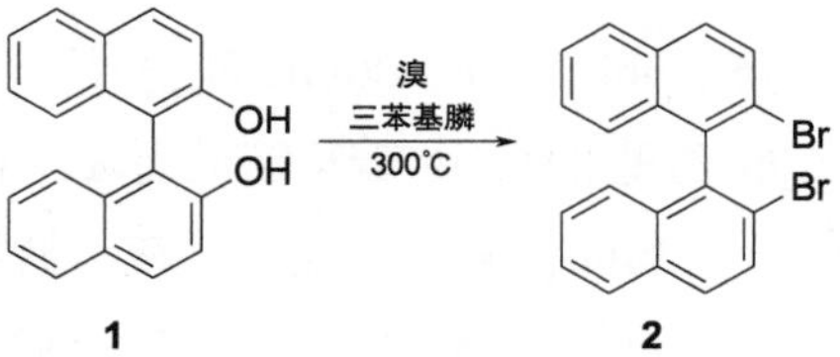


1　　2

将三苯基膦(240g, 0.915mol)溶于500mL乙腈中，0℃下加入溴(155g, 55mL, 0.969mol)，1h内滴加完毕。室温下，加入1,1′-联-2-萘酚(120g, 0.42mol)，60℃下反应30min，然后逐渐升温，300℃下回流。反应完成后，将反应体系冷却到70℃，用500mL(苯：正己烷=1：1)萃取，浓缩有机层成黄色油，将黄色油溶于200mL乙醇中，置于冰箱中放置两天，经过滤，乙醇重结晶得纯二溴化合物78.0g，收率45%。

**步骤2**：化合物3的合成

氮气气氛下，镁屑(2.84g)、碘(0.05g)、1,2-二溴乙烷溶于50mL四氢呋喃中，室温下搅拌，直到碘的颜色褪去，加入溶于400mL四氢呋喃的联萘二溴**2**(20.0g, 50mmol)溶液中，滴加3h，滴加完毕后，在75℃下反应2h然后冷却到10℃。在20min内滴加溶于200mL甲苯的二苯基次膦酰氯(28.4g, 120mmol)混合溶液，滴加完毕后，继续在60℃下反应2h，然后冷却到15℃。反应完成后，用氯化铵溶液淬灭，分离得到有机层，依次用氢氧化钠溶液、水洗，无水硫酸钠干燥，浓缩得黄色固体。将该粗产物溶于甲苯中，室温下放置过夜，过滤，70℃下干燥得淡黄色固体24.5g，收率75%。

**步骤3**：化合物4的合成 [以(*S*)-4为例]

外消旋化合物**3**(10.5g, 16mmol)溶于700mL氯仿中，然后加热回流，并加入溶于460mL乙酸乙酯的(−)-二苯甲酰基-L-酒石酸(6g, 16mmol)。将混合物继续回流2～3min，室温下反应过夜。滤液可回收，得到的固体产物在室温下干燥6h，可得(*S*)-**4** 7.2g，收率89%。

**步骤 4:** BINAP 的合成 [ 以 ($S$)-BINAP 为例 ]

氮气气氛下，将 ($S$)-**4** (4.5g，6.9mmol) 溶于 100mL 干燥的甲苯中，然后逐滴加入 4.2mL 三乙胺 (3.1g，30mmol) 和 3.0mL (4.0g，29mmol) 三氯硅烷。该反应体系在 100℃下搅拌并加热 1h，在 120℃下加热 1h，最后在回流温度下加热 6h。反应完成后，将反应体系冷却至室温后，小心加入 70mL 30% 氢氧化钠水溶液。然后在 60℃下搅拌该反应体系，直到有机层和水层变得清楚，并转移到 300mL 的分离漏斗中。将有机层分离，用两份 50mL 的热甲苯提取水层。组合后的有机层用 70mL 的 30% 氢氧化钠溶液和 3 份 100mL 的水洗涤，然后用无水硫酸钠干燥。将该有机层在减压下浓缩至约 15mL 的体积，并向其添加 15mL 的脱气甲醇。沉淀收集在烧结玻璃漏斗上，用 15mL 甲醇洗涤，80℃干燥 6h，得到 ($S$)-BINAP 无色固体 4.2g，收率 97%。

## (b) BINAP 的结构表征信息 [1]

$^1$**H NMR** (300MHz，CDCl$_3$) $\delta$：7.93 (2H，d，$J$ = 8.5Hz,)，7.88 (2H，d，$J$ = 8.1Hz)，7.51 (2H，d，$J$ = 8.5Hz)，7.39 (2H，t，$J$ = 7.9Hz)，7.24~7.10 (20H，m)，6.95 (2H，dd，$J$ = 8.1Hz)，6.89 (2H，d，$J$ = 8.5Hz)。

$^{31}$**P NMR** (81 MHz，CDCl$_3$) $\delta$：-14.3。

**mp**：236~238℃。

$[\alpha]_D^{25}$：-223 ($c$ = 0.502，苯)。

## (c) BINAP 的应用实例

BINAP 是使用非常广泛的手性膦配体，如 Ru-BINAP 配合物就可以广泛应用于一系列不饱和底物的不对称氢化反应，很早就发现潜手性烯烃底物包括 $\alpha$-酰基丙烯酸、烯丙醇、3-不饱和羧酸和烯酰胺，可以获得 ee 值为 90%~99% 的氢化产物 [2]。据不完全统计，BINAP 已被应用于上万种手性化合物的合成，是一类优势配体。

2020 年，Bisai 课题组将 BINAP 应用于不对称 1,4-共轭加成反应，得到 94% 的产率和 96% 的 ee 值。

**具体反应示例**[3]：

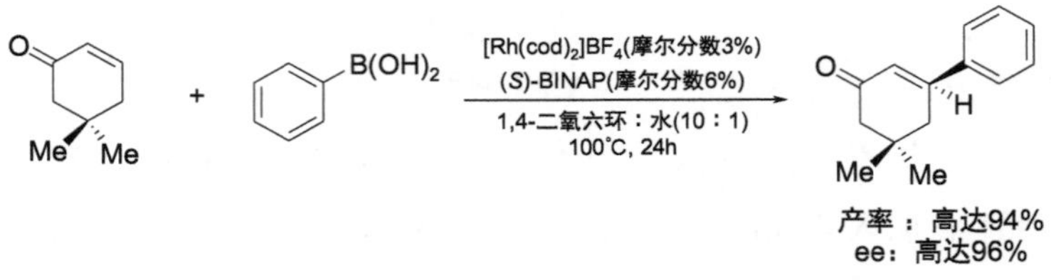

在干燥的史莱克管（含搅拌棒）里加入 [Rh(cod)$_2$]BF$_4$（摩尔分数 3%，15mg，0.01mmol，0.03eq.）和手性配体 (S)-BINAP（摩尔分数 6%，10mg，0.022mmol，0.06eq.），溶于干燥的 1,4-二氧六环和水的混合溶剂（体积比为 10∶1，10mL）中，在 N$_2$ 氛围下脱气 30min。所得溶液为浑浊状，可在史莱克管底部观察到暗黄色沉淀物，加少量水后会澄清。然后在反应混合物中加入烯酮（50mg，0.4mmol，1.0eq.）和芳基硼酸（101mg，0.84mmol，1.4eq.），在预热的油浴中加热至 100℃，继续反应 24h，经 TLC 监测反应完成后，用乙酸乙酯（10mL）萃取，有机相用 2mL 水洗。有机层经无水硫酸钠干燥后，浓缩（旋蒸去溶剂），粗产物通过色谱柱分离（正己烷／乙酸乙酯＝9∶1）得到目标产物（无色油状物）。

**参考文献**

[1] Miyashita A, Yasuda A, Takaya H, et al. *J. Am. Chem. Soc.*, 1980, *102*: 7932-7934.
[2] Akutagawa S. *Appl. Catal. A: General*, 1995, *128*: 171-207.
[3] Khatua A, Pal S, Bisai V. *Eur. J. Org. Chem.*, 2020, *2020*: 2435-2438.

## 6.2.2　SegPhos 配体的合成及应用实例

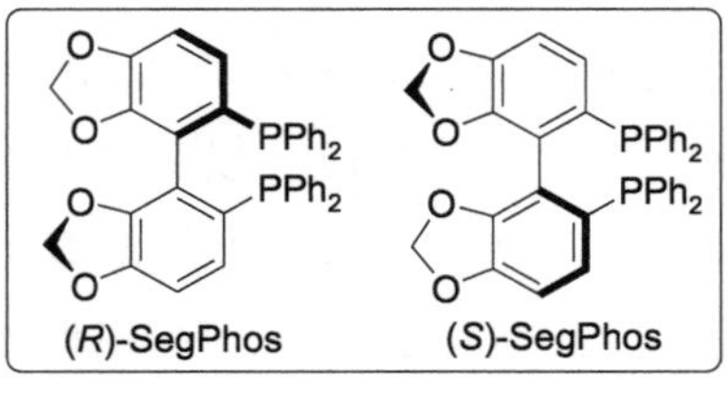

## (a) 配体的合成反应过程 [以(R)-SegPhos的合成为例]

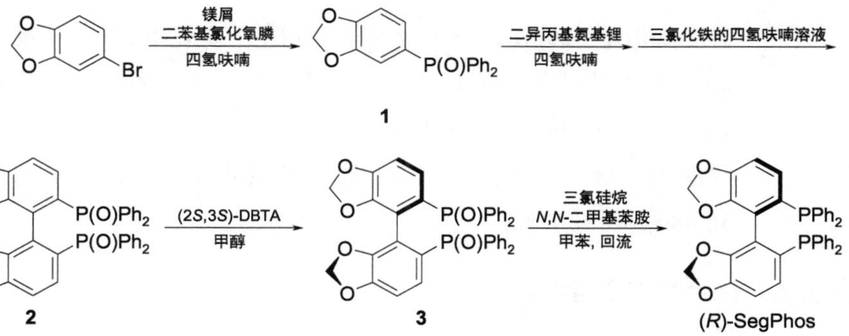

**步骤1：** 化合物1的合成

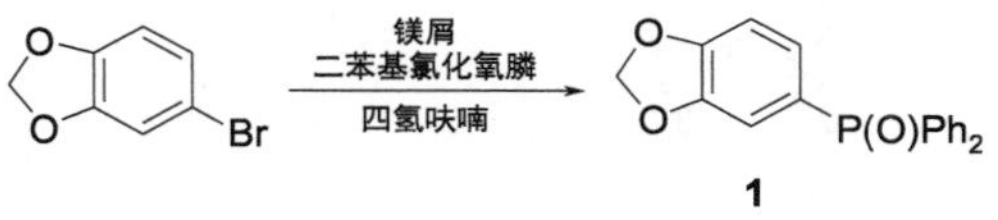

在室温下，向反应瓶中加入镁屑(132g, 5.5mol, 1.1eq.)并加入500mL 无水四氢呋喃，使其液面没过镁屑。于1h内滴加溶于300mL 四氢呋喃的 5-溴苯并 [d]1,3-二氧戊环(1000g, 5mol, 1eq.)混合溶液于反应瓶中。将得到的悬浮液在室温下再搅拌2h。然后，在20℃条件下滴加二苯基氯化氧膦(1112g, 5.5mol, 1.1eq.)，将溶液在20℃下搅拌，反应结束后，用水和1mol/L 的盐酸水溶液进行淬灭，并用乙酸乙酯萃取。分离有机层，依次用 1mol/L 盐酸水溶液、碳酸氢钾饱和水溶液和纯净水洗涤。有机层经硫酸钠干燥，过滤并减压浓缩，得到的深棕色油在硅胶柱上纯化，用乙酸乙酯/庚烷(50/50 至 100/0)洗脱，得到棕色油状物 106.0g，收率 66%。

**步骤2：** 化合物2的合成

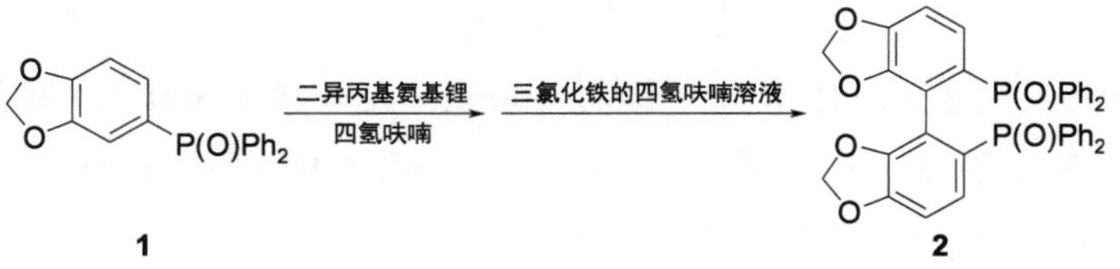

氮气氛下，将化合物 **1**(75.22g，233mmol) 溶于 300mL 四氢呋喃中，将反应瓶置于-15℃冷却浴中，向该反应液中滴加二异丙基氨基锂 (0.7mol/L，400mL，280mmol)，随后在 0℃下加入三氯化铁(45.79g，282mmol) 的四氢呋喃(300mL)溶液。反应结束后浓缩四氢呋喃得到产物粗品，加入 500mL 二氯甲烷溶解，用 10% 盐酸和水洗，有机相用无水硫酸钠干燥，减压浓缩，残留物用热的乙酸乙酯洗涤得到白色固体产物 56.08g，收率 75%。

## 步骤 3：化合物 3 的合成

将 (2S,3S)-DBTA (11.68g，32.6mmol) 溶于 30mL 甲醇中加入化合物 **2** (20.73g，32.3mmol) 的甲醇(60mL)溶液中，回流反应 5min，沉淀物用甲醇洗涤得到无色晶体。用乙酸乙酯萃取，分离有机层。有机层经硫酸钠干燥，过滤并减压浓缩，得到白色目标产物(9.12g，产率 44%)。

## 步骤 4：(R)-SegPhos 的合成

氮气氛下，将化合物 **3**(1.50g，2.34mmol)、N,N-二甲基苯胺(3.11g，25.6mmol) 和三氯硅烷(3.22g，23.3mmol) 加入甲苯(25mL) 中，在 110℃下反应 5h，随后将反应混合物冷却至 5℃，加入 15% 氢氧化钠溶液(30mL)，混合物在室温下搅拌 30min，用甲苯(15mL×2) 萃取，用水、1mol/L 盐酸(30mL×2) 洗有机层。旋干，粗品经硅胶色谱柱纯化得到白色固体(R)-SegPhos 1.35g，收率 95%。

### (b)(R)-SegPhos 的结构表征信息 [1]

**¹H NMR**(CDCl₃)δ：5.03(2H, d, $J$ = 1.6Hz), 5.66(2H, d, $J$ = 1.6Hz), 6.51(2H, dd, $J$ = 7.9Hz, 3.1Hz), 6.66(2H, d, $J$ = 8.1Hz), 7.11~7.21(20H, m)。

**³¹P NMR**(CDCl₃)δ：-12.6。

**mp**：215~217℃。

$[\alpha]_D^{24}$：+133.5($c$= 0.502，氯仿)。

### (c) SegPhos 的应用实例

2020 年，Dixon 课题组[2]发现在(R)-DM-SegPhos 存在下，三氟甲磺酸银(I)能够实现重氮乙酸酯与 N-Boc 醛亚胺的高对映选择性曼尼希反应(ee 值高达 98%)。关于(杂)芳族醛衍生的亚胺，反应的范围很广，可以跟结构多样性的重氮乙酸酯衍生物发生反应。产率和对映选择性高，并且反应可以以克级进行，催化剂负载量可低至 1%(摩尔分数)。

**参考文献**

[1] Saito T, Yokozawa T, Ishizaki T, et al. *Adv. Synth. Catal.*, **2001**, *343*: 264-267.
[2] Robertson G P, Farley A J M, Dixon D J *J. Org. Chem.*, **2020**, *85*: 2785-2792.

## 6.2.3 Spiro Diphosphines（SDP）配体的合成及应用实例

## (a) 配体的合成反应过程

### 步骤1：化合物2的合成

氮气保护下，将(S)-1(5.0g，19.8mmol)溶解在100mL二氯甲烷中，加入吡啶(7.0mL，86.7mmol)，然后在0℃下滴加三氟甲磺酸酐(8.2mL，43.7mmol)。反应体系在室温下搅拌过夜。减压除去溶剂后，将残余物用乙酸乙酯(80mL)稀释，然后用5%盐酸水溶液、饱和碳酸氢钠和盐水依次洗涤。有机层经无水硫酸钠干燥，过滤，减压除去溶剂，剩余粗品通过硅胶色谱柱(洗脱剂：二氯甲烷)纯化，得到白色固体9.9g，收率97%。

### 步骤2：化合物3的合成

氮气保护下，向化合物 **2**(4.0g，7.75mmol)、二苯基氧膦(3.13g，15.5mmol)和醋酸钯催化剂的混合物中依次加入25mL脱气的二甲基亚砜、二异丙基乙胺(4.1g，32mmol)和1,4-双(二苯基膦基)丁烷(dppb，166mg，0.39mmol)，随后置反应瓶于100℃油浴中加热反应6h。冷却至室温后，将反应混合物用乙酸乙酯稀释，随后用5％盐酸水溶液和饱和碳酸氢钠洗涤。有机相经无水硫酸钠干燥，过滤，减压除去溶剂，得到的粗品通过硅胶柱色谱法(洗脱剂为石油醚：乙酸乙酯=3：1)纯化，得到白色固体4.0g，收率90%。

**步骤 3：化合物 4 的合成**

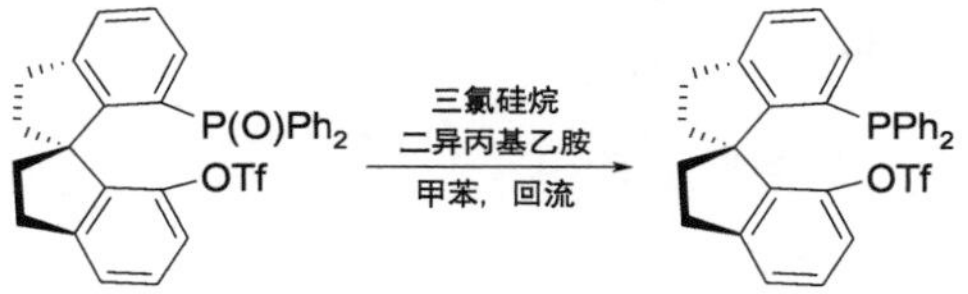

氮气保护下，将 **3**(1.4g，2.5mmol)和二异丙基乙胺(13.2g，102mmol)加入甲苯(25mL)溶液中，在 0℃下加入三氯硅烷(4.0mL，39mmol)。将反应混合物在110℃下搅拌反应5d。反应结束后，用乙醚稀释并用少量饱和碳酸氢钠淬灭。悬浮液通过硅藻土过滤，并用乙醚多次洗涤固体滤渣。合并有机层并经无水硫酸镁干燥，过滤，减压除去溶剂，剩余物质通过柱色谱法(硅胶，乙醚：乙酸乙酯=30：1)纯化，得到白色固体1.26g，收率91%。

**步骤 4：化合物 5 的合成**

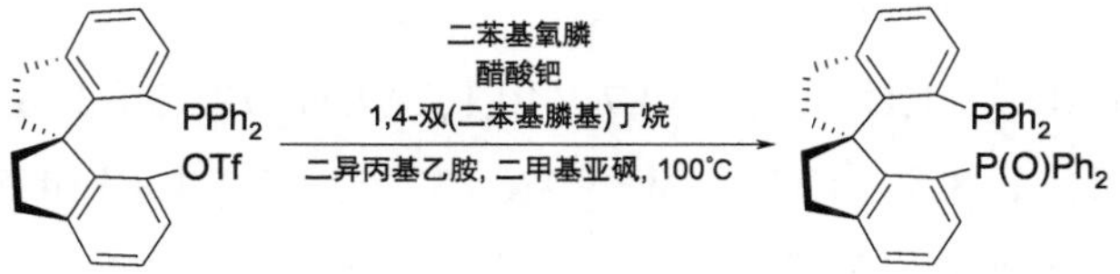

氮气保护下，将 **4**(1.2g，2.17mmol)、二苯基氧膦(0.87g，4.3mmol)、醋酸钯(22.4mg，0.1mmol)和1,4-双(二苯基膦基)丁烷(dppb，42.6mg，

0.1mmol)称入反应瓶中,置换氮气后依次加入 15mL 脱气的二甲基亚砜和二异丙基乙胺(1.0g,8.7mmol),并将混合物在 100℃搅拌加热 10h。反应结束后,反应混合物用乙酸乙酯稀释,用 5% 盐酸水溶液和饱和碳酸氢钠洗涤。有机相经无水硫酸钠干燥,过滤,减压除去溶剂,剩余粗品通过硅胶柱色谱法(洗脱剂为石油醚:乙酸乙酯 =3:1)纯化,得到白色固体 1.2g,收率 92%。

**步骤 5:** SDP 的合成

氮气保护下,0℃下,将(S)-**5**(0.36g,0.6mmol)、二异丙基乙胺(3.1g,24mmol)加入甲苯(10mL)溶液中,缓慢加入三氯硅烷(0.9mL,9mmol)。将反应混合物在 110℃下搅拌 3d。反应结束后,反应混合物用乙醚稀释,并用少量饱和碳酸氢钠淬灭。将得到的悬浮液通过硅藻土过滤,并将过滤得到的固体用乙醚多次洗涤。合并的有机层经硫酸镁干燥,过滤,减压除去溶剂,剩余粗品通过硅胶柱色谱法(洗脱剂为乙醚:乙酸乙酯 =30:1)纯化,得到白色固体 0.3g,收率 86%。

## (b) SDP 配体的结构表征信息

$^1$**H NMR**(300 MHz,CDCl$_3$)$\delta$:1.80~2.00(m,4H,CH$_2$),2.62~2.80(m,2H,CH$_2$),2.80~2.96(m,2H,CH$_2$),6.96(m,2H,Ar-H),7.08~7.30(m,14H,Ar-H)。

$^{13}$**C NMR**(75 MHz,CDCl$_3$)$\delta$:30.7,40.1,64.1,125.3,126.9,127.5,127.9,128.2,128.5,133.3,133.4,133.5,134.4,134.5,134.6,138.4,144.5,144.6,154.5,154.7,154.8。

$^{31}$**P NMR**(121 MHz,CDCl$_3$)$\delta$:-17.26(s)。

**MS(ESI)**:$m/z$(%)=588[M+H]$^+$;

$[\alpha]_D$=-81($c$ = 0.5,二氯甲烷)。

### (c) SDP 的应用实例

周其林等人[1]报道的 SDP 配体对于 Rh 催化官能化烯烃的不对称氢化反应是一类优势配体,表现出极为卓越的手性诱导能力。2017 年施敏课题组[2]通过使用银、铑双金属与 SDP 协同催化实现了酮-亚乙烯基环丙烷(VDCPs)与末端炔烃的不对称环异构化/交叉偶联反应,获得了良好的收率和较高的 ee 值,形成了结构多样化的四氢吡啶类烯炔化合物,显示 SDP 在很多反应中都具有值得开发的应用潜力。

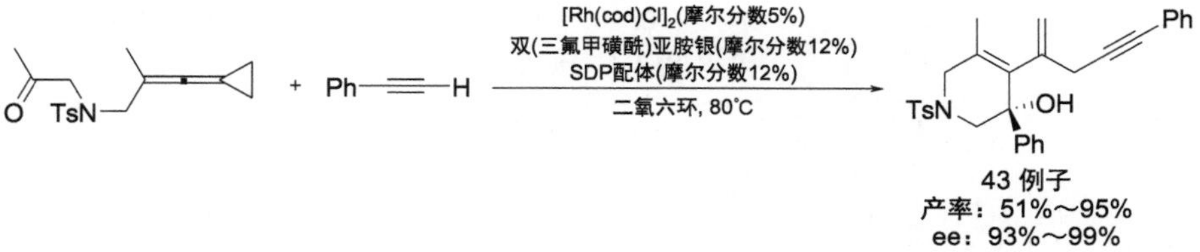

### 具体反应示例[2]:

在氮气氛下,将酮-亚乙烯基环丙烷(VDCPs)(39.6mg,0.10mmol)、末端炔烃(15.3mg, 0.15mmol)、双(三氟甲磺酰)亚胺银(3.88mg, 0.01mmol)、[Rh(cod)Cl]$_2$(0.005mmol, 2.5mg)、(R)-SDP(0.01mmol, 5.9mg)和 1.5mL 二氧六环加入史莱克管中。反应混合物在温度为 80℃下进行反应,用 TLC 监测,直至完全消耗底物。最后通过硅胶柱色谱法纯化,得到所需产物。

**参考文献**

[1] Xie J H, Wang L X, Fu Y, et al. *J. Am. Chem. Soc.*, **2003**, *125*: 4404-4405.
[2] Yang S, Rui K H, Tang X Y, et al. *J. Am. Chem. Soc.*, **2017**, *139*: 5957-5964.

## 6.2.4 Spiroketal-based diphosphine(SKP)配体的合成及应用实例

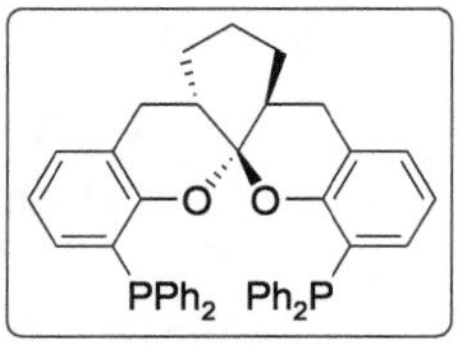

## (a) 配体的合成反应过程

### 步骤1：化合物1的合成

氮气保护下，将3-溴水杨醛(10.25g，51.0mmol)和环己酮(2.5mL，25.0mmol)溶于乙醇(20.0mL)中，随后加入氢氧化钠水溶液[20%(质量分数)，15mL]，室温搅拌24h，反应结束后，向反应体系中加入100mL蒸馏水，用6mol/L的盐酸水溶液调节pH=5，过滤，滤饼用蒸馏水洗涤后浓缩干燥，剩余粗品通过重结晶(丙酮-石油醚)，得黄色固体4.6g，收率60%。

### 步骤2：化合物3的合成

将化合物1(46.4mg，0.1mmol)、催化剂2(1.6mmol)、无水二氯甲

烷(2mL)加入氢化反应瓶中，在手套箱中转移到高压反应釜。置换氢气3次后，充入氢气50个大气压强，室温反应24h。放空氢气后，打开反应釜，减压除去溶剂，由核磁粗谱确定产物的顺反比，剩余粗品通过硅胶柱色谱法分离得到化合物(S,S,S)-**3**的收率为91%，ee值大于99%。

### 步骤3：SKP 的合成

氮气保护下，将(S,S,S)-**3**(350mg，0.77 mmol)加入四氢呋喃(6.0mL)溶液中，冷却至-78℃，随后缓慢滴加正丁基锂(0.8mL，2.5mol/L己烷溶液，1.9mmol)，反应30min后，逐滴加入二苯基氯化膦(0.36mL，1.9mmol)，加完后将反应体系移至室温下继续搅拌10h。反应体系用15mL蒸馏水淬灭，二氯甲烷萃取(20mL×3)，有机相用无水硫酸钠干燥，过滤，减压除去溶剂，剩余物质通过硅胶柱色谱法分离，得目标产物375mg，收率72%。

## (b) SKP 配体的结构表征信息 [1]

$^1$**H NMR**(400 MHz，CDCl$_3$)$\delta$：7.30～7.26(m，20H)，6.89(d，$J$ = 7.2Hz，2H)，6.74(t，$J$ = 7.2Hz，2H)，6.53～6.50(m，2H)，2.34～2.30(m，4H)，1.95～1.92(m，2H)，1.30～1.29(m，2H)，1.17～1.15(m，4H)。

$^{13}$**C NMR**(100 MHz，CDCzl$_3$)$\delta$：153.1(d，$J_{(P,C)}$ = 14.2Hz)，137.1(d，$J_{(P,C)}$ = 11.8Hz)，136.7(d，$J_{(P,C)}$ = 10.9Hz)，134.2(d，$J_{(P,C)}$ = 21.9Hz)，133.9(d，$J_{(P,C)}$ = 20.2Hz)，130.9(d，$J_{(P,C)}$ = 3.2Hz)，129.9(s)，128.5(s)，128.2～128.1(m)，124.9(d，$J_{(P,C)}$ = 14.1Hz)，120.4～120.3(m)，101.3，33.5，27.6，26.7，19.4。

$^{31}$**P**(162 MHz，CDCl$_3$)$\delta$：-15.8(s)。

**IR** $\nu$(cm$^{-1}$)：3050，2922，2858，1573，1423，1243，995，961，770，741，695。

**MS** (ESI)：$m/z$(%) = 661.4 [M+H]$^+$。

**HRMS**(MALDI)：$m/z$(%) = 661.2433[M+H]$^+$。

### (c) SKP 的应用实例

通过使用芳香族螺酮基二膦(SKP)作为手性配体和Cu(Ⅱ)盐作为氧化剂，可以良好的收率得到结构多样的 $\alpha$-亚甲基-$\beta$-氨基芳基酸酯(36个成功实施例)，具有优异的对映选择性(高达96% ee)和高区域选择性(支链B/线型L > 98∶2)[2]。

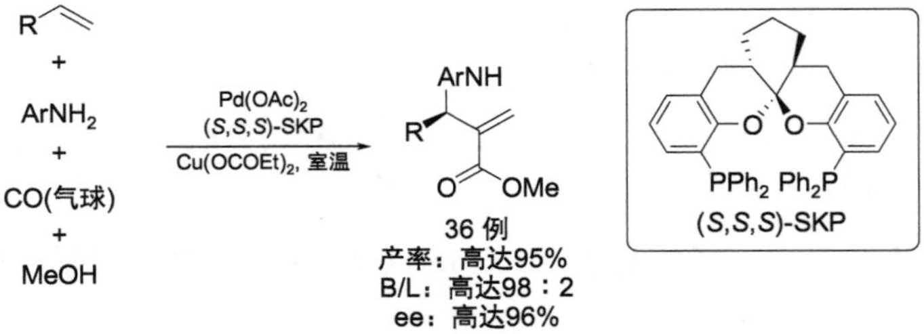

2019年，丁奎岭课题组利用手性配体($S,S,S$)-SKP 实现了铜催化 $\alpha,\alpha$-二氟酮的对映选择性炔丙基化，可得到产率为99%、ee值为97%的预期产物。

**具体反应示例**[3]：

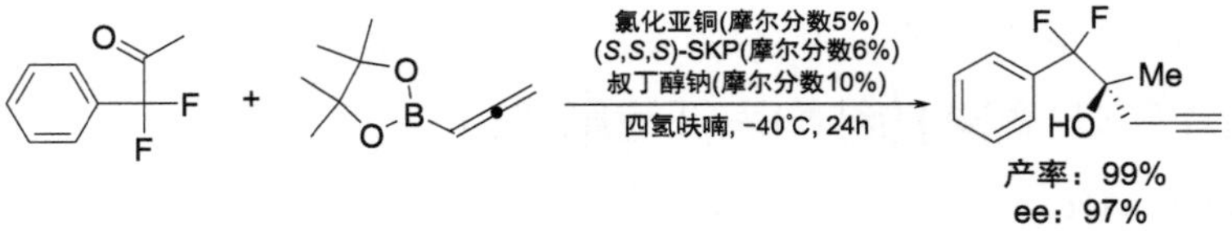

在充满氮气的手套箱中，向装有磁子搅拌棒的25mL史莱克管中依次加入($S,S,S$)-SKP(19.8mg，0.03mmol，摩尔分数6%)、氯化亚铜(2.5mg，0.025mmol，摩尔分数5%)、叔丁醇钠(4.8mg，0.05mmol，摩尔分数10%)和四氢呋喃(1.5mL)。将混合物在25℃下搅拌5～10min，然后冷却至-40℃。通过注射器依次向混合物中加入(频哪醇)烯基硼(133μL，0.75mmol，1.5eq.)和 $\alpha,\alpha$-二氟苯丙酮(74μL，0.5mmol，1.0eq.)。所得混合物在-40℃搅拌24h(用薄层色谱法监测)。反应结束后加入饱和的氯化铵(5mL)淬灭，在室温下用二氯甲烷萃取(15mL×4)，用盐水(25mL)洗涤，合并的有机层用硫酸钠干燥，过滤，浓缩。以石油醚/乙酸乙酯(体积比5∶1)为洗脱剂，在硅胶上通过快速色谱法纯化残余物，得到预期产物。

### 参考文献

[1] Wang X M, Meng F Y, Wang Y, et al. *Angew. Chem. Int. Ed.*, **2012**, *51*: 9276-9282.
[2] Liu J W, Han Z B, Wang X M, et al. *J. Am. Chem. Soc.*, **2015**, *137*: 15346-15349.
[3] Guo P H, Wang X M, Ding K L, et al. *Chem. Eur. J.*, **2019**, *25*: 16425-16434.

## 6.2.5 C$_3$-TunePhos 配体的合成及应用实例

### (a) 配体的合成反应过程

**步骤 1：** 化合物 1 的合成

氮气保护下，将(R,R)-戊烷-2,4-二醇(5.2g，50mmol)、N-(3-羟苯基)、新戊酰胺(20g，104mmol)和三苯基膦(27.5g，105mmol)溶于无水四氢呋喃(200mL)中。于 0℃下将偶氮二甲酸二异丙酯(21.2g，105mmol)滴加到反应体系中。反应体系室温搅拌过夜(约 10～12h)。减压除去溶剂，加入乙醚和石油醚的混合溶剂(150mL/150mL)。沉淀三苯基氧膦，过滤除去杂质。将残余物固体用乙醚/石油醚混合物（150mL/150mL）洗涤，合并滤液，减压除去溶剂，剩余物质通过硅胶柱（洗脱剂为石油醚：乙酸乙酯 = 10：1）纯化，得到白色固体18.41g，收率81%。

**步骤2：化合物2的合成**

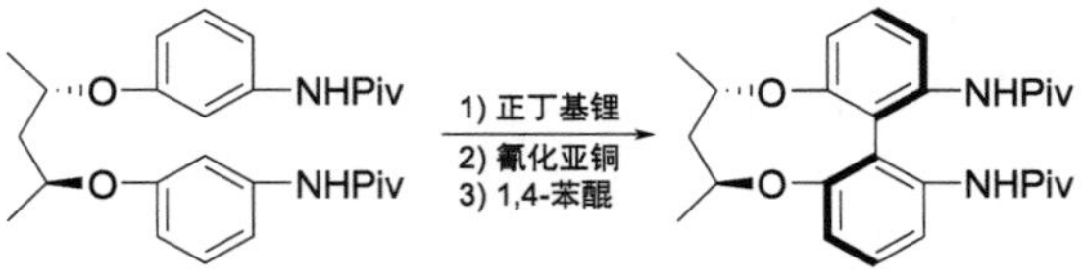

氮气保护下，−20℃下，30min 内将正丁基锂的己烷溶液（28.3mL，2.4mol/L，68mmol）滴加到化合物 **1**（6.18g，13.6mmol）的四氢呋喃（100mL）溶液中。将所得混合物在−20℃下搅拌反应4h，然后缓慢升温至0℃，并搅拌2h。将反应体系移至−78℃，加入氰化亚铜粉末（2.45g，27.2mmol）。5min 后，将反应体系移至室温并搅拌40min，将反应体系移至−78℃，加入 1,4-苯醌（4.4g，1mmol）。将反应体系移至室温，搅拌过夜。反应完成后，加入2mL水淬灭反应，硅藻土过滤。滤饼用二氯甲烷重复洗涤（500mL×8）。合并滤液，减压除去溶剂，剩余粗品通过硅胶柱（洗脱剂为石油醚：乙酸乙酯 = 15：1）纯化，得到白色固体3.4g，收率55%。

**步骤3：化合物3的合成**

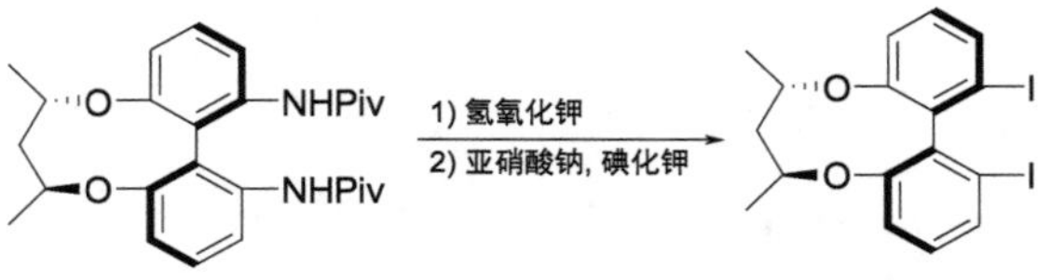

氮气保护下，在密封管中，用乙醇（45mL）和水（9mL）的混合溶剂将化合物 **2**（4.5g）、氢氧化钾（10g）溶解。然后将反应体系移至140℃油浴中反应过夜。冷却至室温后，将反应液缓慢加入冰冷的盐酸溶液中（150mL，2mol/L）。反应结束后，有机相用乙醚（100mL×2）萃取，水相用饱和碳酸氢钠水溶液中和后再用乙酸乙酯（300mL×3）萃取。合并有机相，减压除去溶剂，得到白色固体，将以上获得的二胺产物（0.57g，

2.0mmol)溶解于盐酸(25mL，4mol/L)和乙腈(25mL)的混合溶剂中。-20℃下，将亚硝酸钠(1.6g，22mmol)的水溶液(10mL)和碘化钾的水溶液(8g，50mmol)在20min内加到溶剂中。并将所得混合物加热至80℃反应3h。冷却后，反应体系用二氯甲烷(50mL)萃取，合并有机相，亚硫酸氢钠水溶液(100mL)洗涤，无水硫酸钠干燥。过滤，减压除去溶剂，剩余物质通过硅胶柱(洗脱剂为正己烷：乙酸乙酯 = 20：1)纯化，得到白色固体 2.0g，收率 40%。

**步骤 4：** $C_3$-TunePhos 的合成

氮气保护下，在-78℃下，将化合物 3(0.253g，0.5mmol)溶解在乙醚(8mL)溶液中，缓慢滴加正丁基锂(0.5mL，2.4mol/L，1.2mmol)。反应20min后，将相应的二芳基氯化膦(1.2mmol)(室温下将二芳基氧化膦和三氯化磷在甲苯中混合搅拌 1h，然后在高真空下除去溶剂)逐滴加入，并将所得混合物移至室温，并继续搅拌过夜。反应体系用饱和氯化铵溶液淬灭，乙酸乙酯(5mL×3)萃取，无水硫酸钠干燥。过滤，减压除去溶剂，剩余粗品通过柱色谱法(硅胶，正己烷：乙酸乙酯 = 100：1)纯化，得到白色粉末状固体，收率 61%。

## (b) $C_3$-TunePhos 配体的结构表征信息 [1]

$^1$**H NMR**(500 MHz，CDCl$_3$)$\delta$：7.63-7.56(m，4H)，7.40～7.34(m，6H)，7.21～7.12(m，8H)，7.11～7.05(m，4H)，6.85(d，$J$ = 8.0Hz，2H)，6.71(dd，$J$ = 7.6Hz，0.9Hz，2H)，4.48～4.35(m，2H)，1.74(t，$J$ = 4.0Hz，2H)，1.25(d，$J$ = 6.4Hz，6H)。

$^{31}$**P NMR** {1H}(202 MHz，CDCl$_3$)$\delta$：-11.0。

## (c) $C_3$-TunePhos 的应用实例

2018 年张绪穆课题组利用 TunePhos 配体及钌催化剂实现了铵盐和氢气参与的烷基芳基酮的直接还原胺化反应，可顺利得到手性伯胺，该反应具有广泛的底物范围、良好的官能团耐受性和出色的对映体控制性（最高 ee 值为 98%）[2]，相应的产物可以容易地衍生为更有价值的产物。

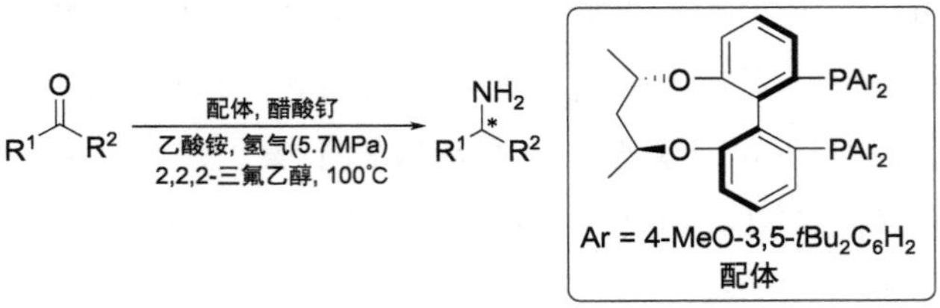

2021 年，Zhang 课题组[3] 将原位生成的氢化铜与 (R)-DTBM-$C_3^*$-TunePhos 联用，在温和条件下实现了 α,β-不饱和酮的 1,2-硅氢还原反应，得到 ee 值较高的手性环烯丙醇。特别是对 α-溴代环烯酮的对映选择性可高达 98%，产率最高可达到 99%。

### 具体反应示例[3]：

在氩气保护的手套箱内，在干燥的含有磁力搅拌棒的 10mL 史莱克管中加入醋酸铜 (0.5mg，0.003mmol)、(R)-DTBM-$C_3^*$-TunePhos (4.0mg，0.0033mol) 和三苯基膦 (0.8mg，0.003mol)。从手套箱中取出史莱克管，加入脱气无水的四氢呋喃 (0.5mL)，搅拌 30min。再向反应瓶中加入苯基硅烷 (0.12mmol)，搅拌 30min 后 (生成氢化铜) 冷却至 −25℃，在氩气下加入溶于 0.5mL 四氢呋喃的底物 (0.1mmol，于 −25℃注射器转移至生成的溶液中)。搅拌 2h 后，反应转移至室温。加入含有甲醇 (0.5mL) 饱和的氟化铵淬灭反应 (搅拌 10min)。产生的混合物经由短的硅藻土柱子过滤，除去沉淀物。真空浓缩滤液，在硅胶上进行快速柱色谱分离 [洗脱液为石油醚:乙酸乙酯=10:1 (体积比)]，得到预期产物。

**参考文献**

[1] Zhang Z G, Qian H, Longmire J, et al. *J. Org. Chem.*, 2000, 65: 6223-6226.
[2] Tan X F, Gao S, Zeng W J, et al. *J. Am. Chem. Soc.*, 2018, 140: 2024-2027.
[3] Shi Y J, Wang J X, Yin Q, et al. *Org. Lett.*, 2021, 23: 5658-5663.

## 6.2.6 DuanPhos 配体的合成及应用实例

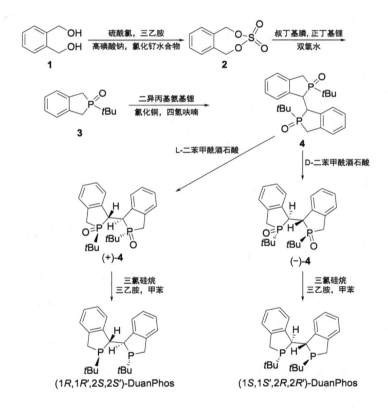

### (a) 配体的合成反应过程

**步骤1：** 化合物 2 的合成

氮气保护下，在 50mL 的二氯甲烷中，加入 1,2-苯二甲醇(2.00g,

14.5mmol)和三乙胺(8.08mL，58.0mmol)，在 0℃时滴加硫酰氯(1.59mL，21.7mmol)到上述溶液里。将得到的深棕色溶液在 0℃下搅拌 1h，然后加入 30mL 的氯化钠溶液。分离有机层，用二氯甲烷(20mL)萃取水层一次。合并有机相，用无水硫酸钠干燥，过滤，减压除去溶剂，得到环亚硫酸盐粗品(2.48g，13.5mmol)。将环亚硫酸盐粗品溶于乙腈、氯仿和水(分别为 20mL、20mL、30mL)的混合液中。然后在 0℃下加入高碘酸钠(4.33g，20.3mmol)和氯化钌水合物(60mg)。将反应混合物继续在 0℃下剧烈搅拌 1h，加入 30mL 盐水，然后加入 60mL 的乙醚。用乙醚(30mL×2)萃取水层。合并有机相，用无水硫酸钠干燥，过滤，减压除去溶剂，剩余物质通过硅胶柱色谱法(洗脱剂为二氯甲烷)纯化，得到白色固体 2.55g，收率 88%。

**步骤 2**：化合物 3 的合成

氮气保护下，将叔丁基膦(1.40mL，11.5mmol)加入四氢呋喃(40mL)溶液中，-78℃下滴加正丁基锂(11.5mmol)得到黄色溶液。将反应体系移至室温下继续搅拌 1h。同时在四氢呋喃(20mL)中加入 **2**(2.29g，11.5mmol)的溶液，在-78℃时通过套管加入叔丁基膦和正丁基锂的溶液中。将混合的反应液逐渐加热到室温，并在室温下搅拌 5h。再次冷却到-78℃后，滴加正丁基锂(11.5mmol)。将反应体系移至室温，搅拌反应过夜。反应结束后，在 0℃下，加入 2.5mL 双氧水(30% aq.)，并搅拌 1h，反应结束后，用饱和亚硫酸钠溶液(10mL)在 0℃淬灭反应。通过重结晶(二氯甲烷和正己烷)纯化，得到白色晶体 2.26g，收率 94%。

**步骤 3**：化合物 4 的合成

氮气保护下，将 **3**(2.95g，23.8mmol)溶于四氢呋喃(160mL)中，而

后在-78℃下滴加二异丙基氨基锂(26.2mmol)，得到红色溶液。将反应体系移至室温下继续搅拌反应 2h，然后再冷却到-78℃后，加入氯化铜。产生的棕色悬浮液，在剧烈搅拌下逐渐加热到室温。反应结束后，加入 50mL 的浓氨水，混匀 10min 后，用二氯甲烷(30mL×3)分离和提取深蓝色水层。合并有机相，无水硫酸钠干燥，过滤，减压除去溶剂，硅胶柱纯化，得到淡黄色固体 2.45g。

**步骤 4：化合物（+）-4 和（-）-4 的合成**

氮气保护下，在 25mL 的甲醇中加入 4(2.45g)，并缓慢滴加 L-二苯甲酰酒石酸溶液(2.34g)。所得悬浮液在回流下加热 4h，冷却至室温后，用少量甲醇过滤和洗涤白色沉淀。然后用 2mol/L 氢氧化钠溶液(10mL)洗涤沉淀，水相用二氯甲烷萃取，无水硫酸钠干燥，过滤，减压除去溶剂，得到纯(+)-4(ee 值＞99%)白色固体 0.73g，收率 30%。

用 2mol/L 氢氧化钠溶液洗涤后，用二氯甲烷萃取浓缩，用 D-二苯甲酰酒石酸溶液以类似的方式处理残渣，得到其他对映体(-)-4(ee 值＞99%)白色固体 0.81g，收率 34%。

**步骤 5：DuanPhos 的合成**

氮气保护下，在干燥脱气甲苯(8mL)中加入(+)-4(207mg，0.5mmol)，加入三乙胺(0.694mL，5.0mmol)和三氯硅烷(0.303mL，3.0mmol)。所得混合物在 70℃下加热反应 16h。0℃下，加入 5mL 的脱气氢氧化钠溶液(30% 水溶液)，然后在 60℃下加热反应 1h。用脱气乙醚(10mL×3)萃取水层。合并有机相，无水硫酸钠干燥，过滤，减压除去溶剂。剩余物质通过柱色谱法(硅胶,正己烷:乙醚 = 9:1)纯化得到(1$R$,1$R'$,2$S$,2$S'$)-DuanPhos 白色固体 170mg,收率 89%。另一种对映体(1$S$,1$S'$,2$R$,2$R'$)-DuanPhos 从(−)-**4** 获得,收率 89%。

## (b) DuanPhos 的结构表征信息 [1]

**$^1$H NMR**(CDCl$_3$,300MHz)$\delta$:7.35~7.32(m,2H),7.19~7.05(m,6H),3.71~3.62(m,2H),3.31~3.18(m,2H),2.73(d,$J$=17.6Hz,2H),0.59(t,$J$=6.0Hz,18H)。

$^{13}$**C NMR**(CDCl$_3$,75 MHz)$\delta$:146.0,144.2,126.7,126.6,126.2,126.1,52.7(t,$J$=4.8Hz),29.4(t,$J$=6.1Hz),27.9(t,$J$=5.4Hz),26.9(t,$J$=7.3Hz)。

$^{31}$**P NMR**(CDCl$_3$,145Hz)$\delta$:4.70。

**HRMS**(ESI):$m/z$(%)=383.20528[M+H]$^+$。

## (c) DuanPhos 的应用实例

张绪穆课题组合成的 DuanPhos 可用于氨基酮的不对称催化加氢反应和环加成反应。在铑催化功能化的烯烃时,能够以非常高的对映选择性(ee 值>99%)和转化率得到预期产物,TON 可达到 10000[2]。

最近 Kong 课题组[3]将 DuanPhos 应用于不对称催化氰化反应,实现了 64% 的产率和 73% 的 ee 值。

**具体反应示例**[3]:

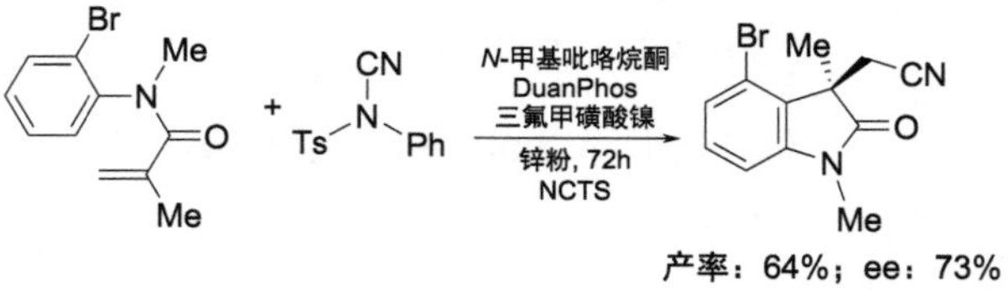

在惰性气体保护下,取 1mL $N$-甲基吡咯烷酮装入有聚四氟乙烯涂层搅拌棒的干燥密封管中,再分别装入 $\alpha$,$\beta$-不饱和酰胺取代的芳基溴化物

(0.1mmol)、三氟甲磺酸镍（摩尔分数 10%）、DuanPhos（摩尔分数 15%）、锌粉（3eq.）和 NCTS（1.0eq.）。反应混合物在铝珠浴中加热到 100℃，继续反应 72h 直到反应完成（用薄层色谱法监测）。所得到的混合物用饱和氯化铵溶液（5mL）淬灭，再用水（10mL）稀释。用乙酸乙酯（15mL×3）萃取水层，有机层用盐水（20mL×2）洗涤，硫酸镁干燥，过滤，减压浓缩。残余物用硅胶色谱法纯化（洗脱剂为乙酸乙酯/石油醚 1∶25～1∶10），分离得到氰基取代的手性羟吲哚产物。

### 参考文献

[1] Liu D, Zhang X M. *Eur. J. Org. Chem.*, 2005, 4: 646-649.
[2] Zhong Y, Zhao X Y, Gan L, et al. *Org. Lett.*, 2018, 20: 4250-4254.
[3] Li H X, Chen J C, Dong J Q, et al. *Org. Lett.*, 2021, 23: 6466-6470.

## 6.2.7　QuinoxP 配体的合成及应用实例

### (a) 配体的合成反应过程

**步骤 1：化合物 2 的合成**

氮气氛下，向化合物 **1**(2.22g，15.0mmol)中加入吡啶(10mL)，而后在 0℃下滴加苯甲酰氯(2.1mL，18mmol)，混合液升至室温，1h 后，反应混合物用水稀释，乙醚萃取 3 次。有机层依次用 1mol/L 盐酸溶液、碳酸氢钠溶液、盐水洗，硫酸钠干燥，浓缩，粗品通过硅胶色谱法(正己烷/乙酸乙酯 = 3/1)纯化，后再由正己烷和乙酸乙酯的混合溶剂重结晶两次得到无色固体 **2**(2.34g，收率 62%)。

**步骤 2：化合物 3 的合成**

氮气氛下，将化合物 **2**(6.05g，24.0mmol)溶于 25mL 乙醇中，逐滴加氢氧化钾(4.0g，72mmol)水溶液(15mL)，水解完成后(1h)加水稀释，乙醚萃取 3 次。用盐水洗涤有机层，无水硫酸钠干燥，用旋转蒸发器去除溶剂，用硅胶柱色谱法纯化残渣(正己烷/乙酸乙酯 = 3/1)，得到(R)-叔丁基(羟甲基)甲基膦硼烷。将此化合物溶解在丙酮中，冷却至 0℃，逐滴加入氢氧化钾(13.5g，240mmol)、过硫酸钾(19.4g，72.0mmol)和三氯化钌三水合物(624mg，2.4mmol)的水溶液(150mL)，并剧烈搅拌。反应 2h 后，用 3mol/L 盐酸中和反应混合物，用乙醚萃取三次，盐水洗涤有机相，无水硫酸钠干燥。在室温下旋转蒸发器去除溶剂，在硅胶上进行色谱纯化(戊烷/醚 = 8/1)，得到化合物 **3**(2.27g，收率 80%)。

**步骤 3：QuinoxP 的合成**

氮气氛下，将化合物 **3**(236mg，2.0mmol)溶于 4mL 四氢呋喃中，置于 -78℃中，逐滴加入正丁基锂(1.25mL，1.60mol/L，2.0mmol)，反应 15min 后，将 2,3-二氯喹啉(133mg，0.67mmol)溶于四氢呋喃(4mL)中加

入上述体系，用力搅拌，并在 1h 内将混合物加热到室温。再搅拌 3h 后加入四甲基乙二胺(1mL)继续搅拌 2h。反应用 1mol/L 盐酸淬灭，混合物用己烷萃取。有机相用 1mol/L 盐酸和盐水洗涤，无水硫酸钠干燥。浓缩，残留物通过硅胶色谱法分离(正己烷/乙酸乙酯 = 30/1)得橙色固体179mg，收率 80%。

## (b) QuinoxP 的结构表征信息 [1]

$^1$**H NMR**(395.75 MHz，CDCl$_3$)$\delta$：1.00～1.03(m，18H)，1.42～1.44(m，6H)，7.70～7.74(m，2H)，8.08～8.12(m，2H)。

$^{13}$**C NMR**(99.45 MHz，CDCl$_3$)$\delta$：4.77(t，$J$ = 4.1Hz)，27.59(t，$J$ = 7.4Hz)，31.90(t，$J$ = 7.4Hz)，129.50，129.60，141.63，165.12(dd，$J$ = 5.7，2.4Hz)。

$^{31}$**P NMR**(202.35 MHz，CDCl$_3$)$\delta$：-17.7(s)。

**IR**(KBr) $\gamma$(cm$^{-1}$)：2950，1470，780。

**HRMS**(FAB)：$m/z$(%)=335.1826[M+H]$^+$；

**mp**：102～103℃；

$[\alpha]_D^{22}$：-54.3($c$= 1.00，氯仿)。

## (c) QuinoxP 的应用实例

2020 年，祝介平课题组 [2] 成功构建了钯催化的对映选择性 Cacchi 反应，实现了轴手性 2,3-二取代吲哚的不对称合成。在醋酸钯/配体的催化作用下，N-芳基(烷基)磺酰基-2-炔基苯胺与芳基硼酸在氧气气氛下反应，得到高收率和对映选择性的 2,3-二取代吲哚。这是钯催化用于合成带有手性 C$_2$ 芳基轴的吲哚的对映选择性 Cacchi 反应的第一个实例，显示出 QuinoxP 的独特性。

2021 年，樊保敏课题组开发了铑/铜共催化体系，并成功应用于以

水为氢源的芳香族 α-脱氢氨基酸酯的不对称还原反应。在三氟甲磺酸铜、(R,R)-QuinoxP* 的催化作用下，芳香族 α-脱氢氨基酸酯苯环上对位、间位和邻位上含有不同取代基的底物均能够有效地反应，得到具有优异产率和对映选择性的相应手性氨基酸酯，底物普适性好。

**具体反应示例**[3]：

![反应示例]

在氩气气氛下，将水合氯化铑(5.4mg，0.01mmol)、(R,R)-QuinoxP*(4.0mg，0.012mmol)和 1mL 1,2-二氯乙烷加入史莱克管中。将所得溶液在室温下搅拌 30min，然后加入三氟甲磺酸铜(7.3mg，0.02mmol)再搅拌 10min。将锌粉(39mg，0.6mmol)加入上述混合物中，而后继续加入底物 α-脱氢氨基酸酯(0.2mmol)的 1,2-二氯乙烷(1.0mL)溶液。加入水(36μL，2mmol)后，在氩气气氛下于 70℃油浴下继续反应，TLC 监测至反应结束。残余物通过硅胶柱色谱法纯化，得到所需产物。

**参考文献**

[1] Imamoto T, Sugita K, Yoshida K. *J. Am. Chem. Soc.*, 2005, *127*: 11934-11935.
[2] He Y P, Wu H, Wang Q, et al. *Angew. Chem. Int. Ed.*, 2020, *59*: 2105-2109.
[3] Dai Y Z, Chen J C, Wang Z T, et al. *J. Org. Chem.*, 2021, *86*: 7141-7147.

## 6.2.8　DuPhos 配体的合成及应用实例

## (a) 配体的合成反应过程

## 步骤 1：化合物 1 的合成

取干燥后的 250mL 圆底烧瓶，氮气氛下中加入 1,2-双(膦基)苯(6.0g，4.27mmol)和四氢呋喃(100mL)，然后将正丁基锂(1.6mol/L，5.4mL，8.6mmol)逐滴加入上述体系中，搅拌 1h。向所得混合物中加入溶于四氢呋喃(10mL)的环己二醇硫酸酯(1.54g，8.55mmol)。将该白色浑浊的混合物继续搅拌 2h。随后通过注射器滴加正丁基锂(1.6mol/L，5.9mL，9.4mmol)，得到不透明的红色溶液。反应搅拌过夜，然后逐滴加入硼烷·四氢呋喃(1.0mol/L，12.8mL，12.8mmol)，将该灰色混合物搅拌 2h。加入 1eq. 盐酸水溶液(50mL)，反应用乙醚(50mL×3)萃取，水(50mL)和盐水(50mL)洗涤，无水硫酸镁干燥，蒸发溶剂，残留物通过硅胶柱色谱法纯化(洗脱剂：正己烷/乙醚=40∶1)，分别得到白色固体 meso-Me-DuPhos(高 $R_f$，442mg，收率 32%)和 rac-Me-DuPhos 单硼烷络合物(低 $R_f$，477mg，收率 35%)。

**步骤 2:** DuPhos 的合成

氮气保护下，向 50mL 的史莱克管中加入化合物 **1**（320mg，1.0mmol）、1,4-二氮杂双环[2.2.2]辛烷（123mg，1.1mmol）、10mL 甲苯，50℃下搅拌 15h。冷却至室温后，通过氧化铝过滤并将滤饼用 150mL 脱气后的正己烷与乙醚混合物溶液洗脱，旋干滤液后将粗品放置于真空干燥箱中 12h 后得白色固体 276mg，收率 90%。

### (b) DuPhos 的结构表征信息

$^1$H NMR（400MHz，$C_6D_6$）$\delta$：7.33～7.28（m，2H），7.14～7.10（m，2H），2.59～2.46（m，4H），2.05～1.94（m，2H），1.92～1.84（m，2H），1.38～1.21（m，10H），1.03～0.99（m，6H）。

$^{13}$C NMR（100 MHz，$C_6D_6$）$\delta$：144.4（d，$J$ = 2.9Hz），131.5（t，$J$ = 2.4Hz），127.9，36.4，35.8（t，$J$ = 2.4Hz），34.3（t，$J$ = 6.7Hz），32.8，20.7（t，$J$ = 18.1Hz），18.7（t，$J$ = 2.9Hz）。

$^{31}$P NMR（160 MHz，$C_6D_6$）$\delta$：3.3。

**HRMS**（ESI）：$m/z$(%) = 307.1727[M+H]$^+$。

### (c) DuPhos 的应用实例

2014 年，Tsuji 课题组[1]报道了一种在 $CO_2$ 气氛下，铜作为催化剂，实现了 PhMe$_2$Si-B(pin) 与联烯类化合物的区域选择性羰化硅烷反应，DuPhos 配体可以很好地控制区域选择性。最近，Shinada 课题组[2]利用 DuPhos 制备了一种特殊的氨基酸，显示出铑/DuPhos 催化体系能够适用于高度官能化的氨基酸合成。

**具体反应示例**[2]：

将[(S,S)-Et-DuPhos-Rh](OTf)(4.2mg, 5.89μmol)和反应底物(295mg, 0.589mmol)溶于甲醇(2mL)，并在氩气下将反应物置于高压氢气反应釜中，多次真空置换氢气后，在0.8MPa氢气压力和室温下，反应48h。反应结束后，小心放出氢气，浓缩反应液，残留物通过快速柱色谱法(正己烷/乙酸乙酯=1∶1)分离得到无色固体纯品，产率88%。

**参考文献**

[1] Tani Y, Fujihara T, Terao J, et al. *J. Am. Chem. Soc.*, **2014**, *136*: 17706-17709.
[2] Okamura H, Yasuno N, Taikawa H, et al. *Eur. J. Org. Chem.*, **2021**, *9*: 1396-1401.

## 6.2.9　DIOP配体的合成及应用实例

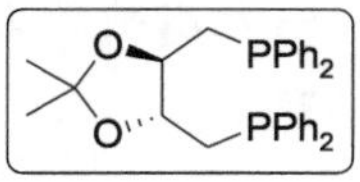

### (a) 配体的合成反应过程

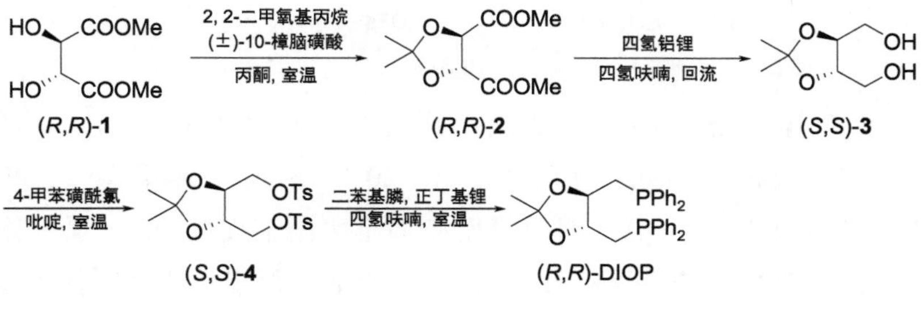

**步骤1：** 化合物2的合成

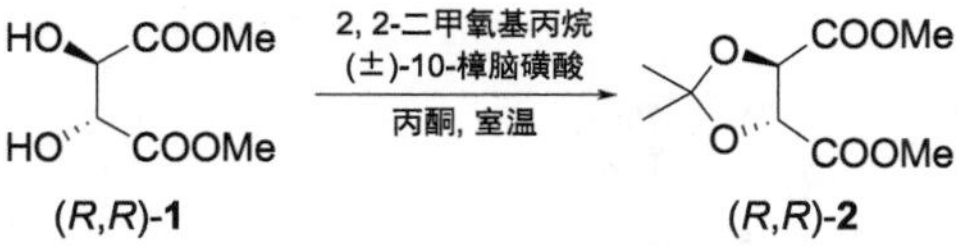

取干燥后的单口烧瓶，氮气氛下向反应瓶中加入(R,R)-酒石酸二甲酯[(R,R)-1](20.4g, 115mmol, 1.0eq.)、2,2-二甲氧基丙烷(31.2mL, 26.3g, 253mmol, 2.2eq.)、(±)-10-樟脑磺酸(13.6g, 57.5mmol, 0.5eq.)和140mL

丙酮，在室温下反应 20h。用饱和碳酸氢钠溶液(50mL)进行中和，加入 50mL 水，乙酸乙酯(100mL×3)萃取水相，合并有机相，无水硫酸钠干燥后真空除去溶剂。减压蒸馏粗产物，得到无色液体($R,R$)-**2**(19.0g，收率 67%)。

**步骤 2**：化合物 3 的合成

($R,R$)-**2** →[四氢铝锂 / 四氢呋喃, 回流]→ ($S,S$)-**3**

取干燥后的单口烧瓶，氮气氛下向反应瓶中加入四氢铝锂(7.32g, 193mmol, 2.5eq.)和超干四氢呋喃(200mL)，在 0℃下缓慢滴加($R,R$)-**2** (19.0g, 77.2mmol, 1.0eq.)的四氢呋喃溶液，滴加完毕后升温至回流，反应 3h。反应冷却至 0℃，加入 15mL 水和 15mL 6mol/L 的氢氧化钠溶液进行后处理，过滤得到的悬浮液用乙醚洗涤并用无水硫酸钠干燥，真空除去溶剂，得到无色液体($S,S$)-**3**(9.03g，收率 72%)。

**步骤 3**：化合物 4 的合成

($S,S$)-**3** →[4-甲苯磺酰氯 / 吡啶, 室温]→ ($S,S$)-**4**

取干燥后的单口烧瓶，氮气氛下向反应瓶中加入($S,S$)-**3**(9.03g, 55.7mmol, 1.0eq.)和吡啶，在-20℃下加入 4-甲苯磺酰氯(32.8g, 172mmol, 3.1eq.)，随后将反应移至室温下搅拌 20h。反应体系中加入 600mL 水，并将反应瓶移至 0℃数小时，滤出，所得固体用水和乙醇洗涤，在真空下干燥，得到无色固体($S,S$)-**4**(15.1g，收率 58%)。

**步骤 4**：DIOP 的合成

($S,S$)-**4** →[二苯基膦, 正丁基锂 / 四氢呋喃, 室温]→ ($R,R$)-DIOP

取干燥后的单口烧瓶，氮气氛下向反应瓶中加入二苯基膦(10.0g, 53.7mmol, 2.4eq.)和 72mL 四氢呋喃，在-78℃下缓慢滴加正丁基锂(2.5mol/L, 26.9mL, 67.1mmol, 3.0eq.)。滴加完毕后，将反应瓶移至室温搅拌 2h，反应溶液颜色由黄色变为橙色，最后变为深红色。之后向反应瓶中缓慢滴加($S,S$)-**4**(10.5g，22.4mmol，1.0eq.)的四氢呋喃(70mL)溶液，反应在室温下搅拌 16h。后处理：用脱气后的甲醇(1.5mL)淬灭反应，通过减压除去溶剂，加入脱气后的正戊烷/二氯甲烷(1:1，150mL)和脱气后的水(50mL)，用脱气后的正戊烷/二氯甲烷(1:1，50mL×3)萃取有机相，合并有机相，无水硫酸钠干燥，并在真空下除去溶剂。用乙醇重结晶，得到无色固体($R,R$)-DIOP(8.97g，收率 80%)。

## (b) DIOP 的结构表征信息 [1]

$^1$**H NMR**(400.13 MHz，CDCl$_3$)$\delta$：1.35(s，6H)，2.31～2.45(m，4H)，3.92(m$_c$，2H)，7.29～7.34(m，12H)，7.37～7.47(m，8H)。

$^{13}$**C NMR**(100.61 MHz，CDCl$_3$)$\delta$：27.4，32.5(dd，$J$ = 15.7，3.4Hz)，79.8(m$_c$)，109.6，128.5，128.6，128.6，128.7，128.9，133.0(dd，$J$ = 34.3，19.4Hz)，138.6(dd，$J$ = 28.7，13.0Hz)。

**HRMS**(pos. APCI)：$m/z$(%) = 498.1880[M+H]$^+$。

## (c) DIOP 的应用实例

2011 年，Breit 小组[2]采用 DIOP 配体与铑催化体系首次实现了末端联烯的不对称催化羧酸加成反应，合成了一系列具有应用价值的支链烯丙基酯。2015 年，该课题组进一步使用该催化体系，完成了带有联烯和羧酸基团分子的二聚反应，成功合成了 Clavosolide(一种细胞毒性分子)。

2021 年清华大学李必杰课题组[3]报道了将(+)-DIOP 配体与铑催化体系应用于 α,β-不饱和酰胺的不对称硼氢化反应，成功构建了含季碳手性中心的叔硼酯，ee 值可达 95%。

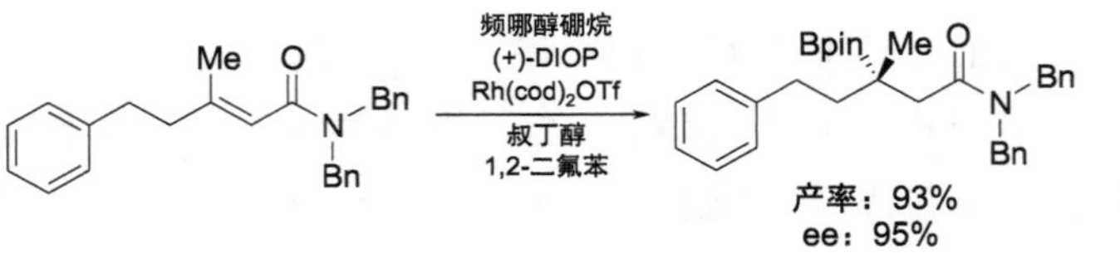

**具体反应示例**[3]：

在充满氩气的手套箱中称量 α,β-不饱和酰胺(0.25mmol，1.0eq.)、Rh(cod)$_2$OTf [2.3mg，2.0%(摩尔分数)] 和(+)-DIOP [3.0mg，2.4%(摩尔分数)]，随后，通过注射器加入 1,2-二氟苯(2.5mL)、叔丁醇(4.0mol%)和频哪醇硼烷(80.0mg，2.5eq.)，用含有聚四氟乙烯的螺帽封住小瓶，然后，将反应瓶从手套箱中取出，在-24℃下搅拌 24h，待反应完成后，减压浓缩，粗产物经 $^1$H NMR 分析，最后以乙酸乙酯和正己烷混合物为洗脱剂，经色谱柱分离纯化得到硼氢化产物。

**参考文献**

[1] Kagan H B, Dang T P. *J. Am. Chem. Soc.*, **1972**, *94*: 6429-6433.
[2] Koschker P, Lumbroso A, Breit B. *J. Am. Chem. Soc.*, **2011**, *133*: 20746-20749.
[3] Gao T T, Lu H X, Gao P C, et al. *Nat. Commun.*, **2021**, *12*: 3776.

## 6.2.10　(*S,S*)-f-Binaphane 配体的合成及应用实例

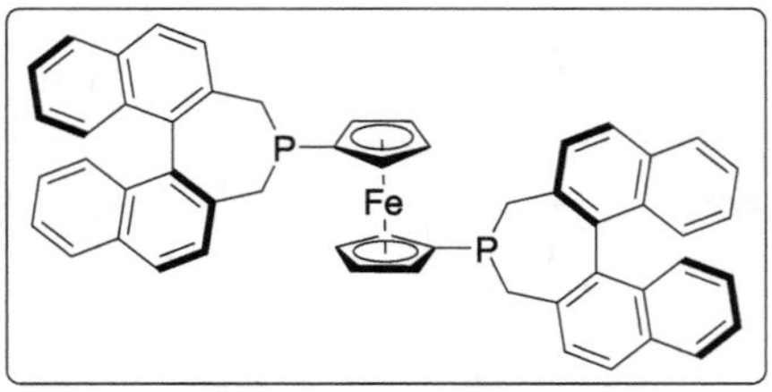

## (a) 配体的合成反应过程

**步骤1：** 化合物1的合成

氮气保护下，在室温下0.5h内向正丁基锂（25mL，2.5mol/L，62.5mmol）和四甲基二胺（9.5mL，62.5mmol）的10mL己烷溶液中滴加二茂铁（4.65g，25mmol）己烷溶液200mL，将反应体系在室温下搅拌16h，$N_2$保护下过滤，并用20mL×3己烷洗涤。得到二茂铁的二锂四甲基二胺配合物，产物为橙色粉末（4.40g，收率70%）。

**步骤2：** 化合物2[1,1′-双（二氯膦基）二茂铁]的合成

氮气保护下，在0℃下将二乙氨基双氯磷（0.46mL，2.2mmol）加入二茂铁的二锂四甲基二胺络合物（314mg，1mmol）的5mL乙醚溶液中，而

后将反应体系加热至室温，继续搅拌 2h。于 0℃下，加入 10mL 盐酸乙醚溶液(1mol/L, 10mmol)搅拌 2h。在 $N_2$ 下过滤出沉淀物，将固体用 10mL×2 乙醚洗涤，合并有机相，减压浓缩，并用 3mL 戊烷重结晶，得到的 1,1′-双(二氯膦基)二茂铁产物为浅棕色固体，收率 75%(290mg)。

### 步骤 3：(S,S)-f-Binaphane 的合成

氮气保护下，-78 ℃下，将 (S)-2,2-二甲基-1,1-联萘基(615mg, 1.5mmol)的二锂四甲基二胺配合物加入四氢呋喃(1mL)溶液中，滴加 1,1′-双(二氯膦基)二茂铁(290mg, 0.75mmol)的四氢呋喃(5mL)溶液之后，将反应体系移至室温搅拌过夜。反应结束后，用 10mL 水淬灭反应，减压除去溶剂，加入 20mL 二氯甲烷萃取，分离有机相，并用 20mL×3 水洗涤，无水硫酸钠干燥，过滤，减压除去溶剂，通过二氯甲烷/正己烷=1:10 重结晶，得到橙色粉末状的固体 489mg，收率 81%。

### (b) (S,S)-f-Binaphane 配体的结构表征信息[1]

$^1H$ NMR($CD_2Cl_2$, 360 MHz)δ：8.07~7.99(6H, m; Ar-H)，7.87~7.85 (2H, d, $J$ = 8.34Hz)，7.80~7.78 (2H, d, $J$ = 8.31Hz)，7.53~7.50 (4H, m; Ar-H)，7.31~7.28(8H, m; Ar-H)，7.07~7.04(2H, d, $J$ = 8.37Hz)，4.49(2H, s; Cp-H)，4.45(2H, s; Cp-H)，4.26(2H, s; Cp-H)，3.56(2H, s; Cp-H)，3.10~3.06(2H, m; $ArCH_2$)，2.82~2.72(4H, m; $ArCH_2$)，2.63~2.59(2H, m; $ArCH_2$)。

$^{13}C$ NMR($CD_2Cl_2$, 90MHz)δ：135.79, 135.18, 134.28, 133.57, 133.21, 133.05, 132.96, 132.67, 129.28, 129.00, 128.94, 128.10, 128.04, 127.29, 127.19, 126.58, 126.45, 125.73, 125.46, 76.22(d, $J_{C,P}$ = 21.92Hz)，75.62(d, $J_{C,P}$ = 32.00Hz)，72.43, 71.63, 71.14, 34.60(d, $J_{C,P}$=20.90Hz)，31.36(d, $J_{C,P}$= 11.74Hz)。

$^{31}$P NMR $(CD_2Cl_2)$ $\delta$: -1.26。

**MS**(ESI): $m/z$(%) = 807 [M+H]$^+$。

#### (c)((S,S))-f-Binaphane 的应用实例

2010 年，张绪穆等人[2]将(S,S)-f-Binaphane 配体成功应用于不对称铱催化氢化反应，可实现无保护基团的 β-烯胺酯类化合物的高立体选择性氢化转化，得到 β-氨基酸(ee值高达 95%)。

2020 年 Schaub 课题组[3]将(S,S)-f-Binaphane 配体应用于钌催化的不对称还原胺化，用碘化铵作为胺源，对简单脂肪族酮进行直接伯胺化转化，可得到 84% 的产率和 68% 的 ee 值。

**具体反应示例**[3]：

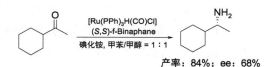

在手套箱中，将催化剂 [Ru(PPh)$_3$H(CO)Cl] (0.01mmol，9.5mg)、(S,S)-f-Binaphane(0.015mmol，12.1mg)、碘化铵(2.5mmol，363mg)、甲苯/甲醇(1∶1，10mL)和酮(1.0mmol)先后加入一个配备有磁性搅拌棒的约 40mL Premex 高压反应器中，然后在手套箱外于室温下用 3MPa 的 H$_2$ 对高压釜加压。将反应置于 80℃下搅拌 36h，反应结束后，冷却至室温，小心释放压力。反应后的粗混合物通过硅藻土过滤，并用气相色谱法分析监测转化率。

#### 参考文献

[1] Xiao D M, Zhang X M. *Angew. Chem., Int. Ed.*, **2001**, 40: 3425-3428.
[2] Hou G H, Li W, Ma M F, et al. *J. Am. Chem. Soc.*, **2010**, 132: 12844-12846.
[3] Ghosh T, Schaub T. *Eur. J. Org. Chem.*, **2020**, 30: 4796-4800.

## 6.2.11 Josiphos 配体的合成及应用实例

## (a) 配体的合成反应过程

### 步骤1：化合物1的合成

氩气保护下，将 N,N-二甲基-1-二茂铁基乙胺(1.028g, 4.0mmol, 1.0eq.)加入四氢呋喃(20mL)溶液中，随后将反应体系置于-78℃下缓慢滴加正丁基锂(4.7mL, 8.0mmol, 1.7mol/L)。反应30min后，将反应体系移至-25℃再搅拌1h后缓慢滴加二苯基氯化膦(5.0mmol, 896μL, 1.2eq.)。将该溶液缓慢加热至室温。3h后用水(10mL)淬灭反应。二氯甲烷萃取，分离有机层。合并的有机层用无水硫酸钠干燥，过滤，减压除去溶剂，剩余粗品通过柱色谱法(硅胶，石油醚：乙酸乙酯：三乙胺 = 1：1：0.5%)纯化，得到橙色粉末固体，收率90%。

### 步骤2：Josiphos的合成

氩气保护下，向产物 **1**(1.55g, 3.512mmol)的蒸馏和脱气的乙酸(20mL)溶液中加入二环己基膦(785mg, 4.215mmol)，将所得混合物脱气3次。80℃油浴中加热反应过夜，减压除去溶剂，其通过快速柱色谱法[使用石油醚/乙酸乙酯/三乙胺(1：1：0.5%)作为洗脱剂]进行纯化，得到黄色泡沫状的化合物1.81g，收率88%。

## (b) Josiphos 配体的结构表征信息 [1]

$^1$**H NMR**(259 MHz, CDCl$_3$)δ：0.92～1.22(br m, 11 Cy H)，1.34～1.72

(br m, 11 Cy H), 1.54(dd, $J$ = 7Hz, 3Hz, 3H), 3.13(qd, $J$ = 2Hz,2Hz, 1H), 3.77(s, 5 Cp H), 3.93～3.97(m, 1 Cp H), 4.20～4.25(m, 1 Cp H), 4.28～4.33(m, 1 Cp H), 7.07～7.18(m, 5 Ph H), 7.26～7.34(m, 3 Ph H), 7.54～7.62(m, 2 Ph H)。

**$^{31}$P NMR**(101 MHz, CDCl$_3$)$\delta$: +15.7(d, $J_{PP}$ = 30, PCy$_2$), −25.8(d, PPh$_2$)。

**MS**(ESI): $m/z$(%)=594[M+H]$^+$。

## (c) Josiphos 的应用实例

自 1951 年二茂铁被发现以来，因其热稳定性、构象刚性和面手性等优点一直是最重要的配体骨架之一。二茂铁骨架的膦配体在过渡金属催化的偶联反应中非常重要，非手性的二茂铁膦，如 1,10-二(二苯膦基)二茂铁，已成功地应用于各种过渡金属催化的偶联反应(Suzuki-Miyaura 反应、Buchwald-Hartwig 反应、Stille 偶联反应等)。手性二茂铁膦配体已成为不对称合成不可或缺的有力工具。Josiphos 作为一种手性二茂铁膦配体已经广泛应用于药物合成中。

2018 年，Ilan Marek 课题组[2] 报道了一个三元碳环的环丙烯类直接官能化不对称转化反应，可方便地合成手性环丙烷类化合物。该方法具有高效、经济、非对映选择性和对映选择性高的特点。如下示例所示，在 5%(摩尔分数)的 ($R$,$S$)-Josiphos 存在下，不对称铑催化的芳基环丙烯与各种芳基硼酸的芳基化反应，可高效合成具有季碳立体中心的手性二芳基环丙烷。

2021 年，Dong 课题组[3] 报道了 Josiphos 控制的不对称铑催化环丙烯的巯基化反应，可获得产率90%、dr＞20∶1 和 ee值90% 的优异结果。

## 具体反应示例[3]：

在充满 $N_2$ 的手套箱中，将 [Rh(cod)Cl]$_2$(1.2mg，0.0025mmol)、Josiphos (2.7mg，0.0050mmol)和乙腈(0.50mL)加入装有搅拌棒的反应瓶中，得到的混合反应液搅拌 10min，而后依次加入硫醇(11.0mg，0.10mmol)和环丙烯(0.10mL，1.2mol/L 乙腈溶液，0.12mmol)。将混合物保持在 30℃反应 6h，直到 TLC 监测反应结束。通过对未提纯的反应混合物进行 $^1$H NMR 分析，确定了区域选择性比，浓缩后的粗品经快速柱色谱法分离得到纯品。

**参考文献**

[1] Togni A, Breutel C, Schnyder A, et al. *J. Am. Chem. Soc.*, **1994**, *116*: 4062-4066.
[2] Dian L, Marek I. *Angew. Chem. Int. Ed.*, **2018**, *57*: 3682-3686.
[3] Nie S Z, Lu A, Kuker E L, et al. *J. Am. Chem. Soc.*, **2021**, *143*: 6176-6184.

## 6.2.12 DIPAMP 配体的合成及应用实例

### (a) 配体的合成反应过程

**步骤 1:** 化合物 2 的合成

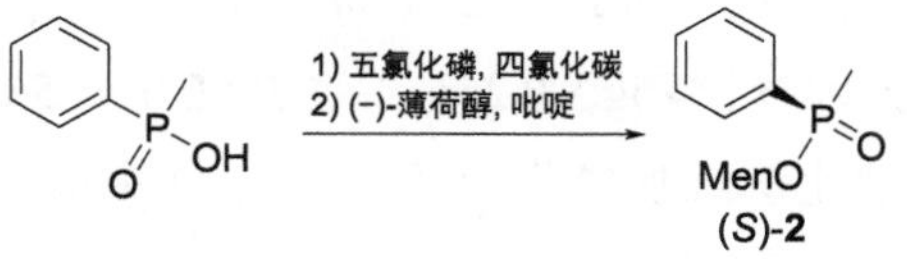

在氮气氛下,将甲基苯基膦酸(340g)溶于四氯化碳(1500mL)中,在 40℃下滴加五氯化磷(418g),将混合物搅拌过夜。除去溶剂,蒸馏残余物,得到 318g 甲基苯基膦酰氯。将甲基苯基膦酰氯(296g,1.7mol)溶于无水乙醚(400mL)中,加入吡啶(121g,1.53mol)和(−)-薄荷醇(239g,1.53mol),搅拌过夜,过滤混合物以除去吡啶盐酸盐,除去溶剂,然后将残留物无色油状液体溶于己烷中并在 5℃下保存几天,将粗产物用正己烷多次重结晶得到 40g 产物(S)-2,通过冷却原始母液,可以得到少量纯的非对映体(R)-2。

**步骤 2:** 化合物 3 的合成

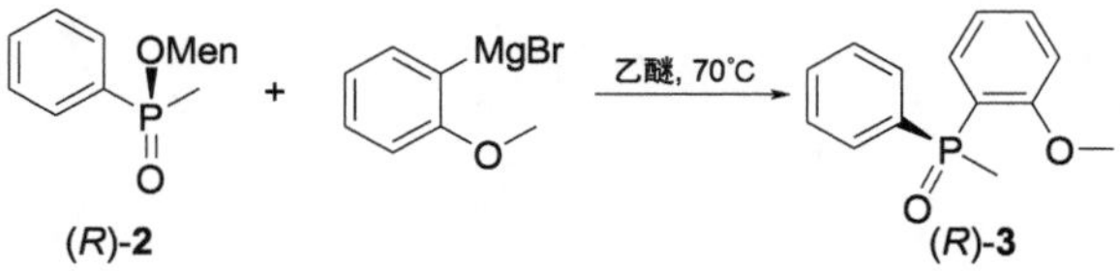

将(R)-2(7g,0.024mol)溶于苯(90mL)中,滴加到 β-甲氧基苯基溴化镁的无水乙醚溶液(50mL)中。蒸馏除去乙醚,升温至 70℃,并将溶液加热回流 18h。再将该混合物冷却至 0℃,加入饱和氯化铵溶液(100mL),用氯仿萃取六次,合并的有机物用无水硫酸钠干燥。蒸发溶剂得到 8.7g 残留物,通过在 110℃下蒸馏将薄荷醇除去,残余物通过己烷重结晶得到化合物(R)-3(2.30g,收率 32.8%)。

**步骤 3:** 化合物 4 的合成

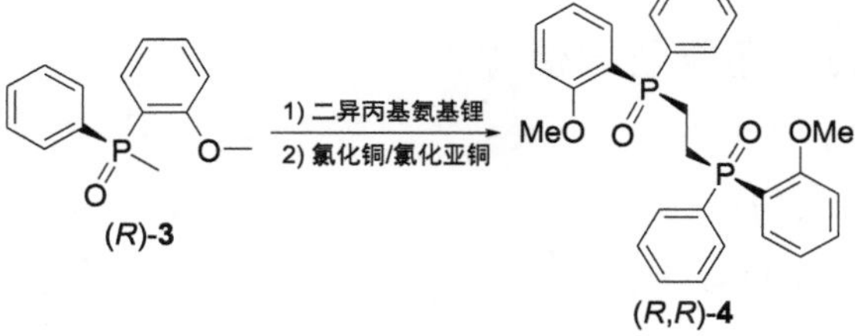

将化合物(R)-3(49.2g, 0.2mol)加入100mL四氢呋喃中,加入二异丙基氨基锂溶液[制备方法是将91.6mL的2.4mol/L丁基锂己烷溶液加入溶于100mL四氢呋喃的二异丙胺(24.5g, 0.24mol),并在0~5℃下搅拌半小时]。然后在0~5℃下按比例添加氯化亚铜(20.0g, 0.2eq.)。搅拌半小时后添加氯化铜(26.9g, 0.2eq.),然后加热1h至20~25℃,并在该温度下保持半小时。产物在15~20℃下用浓盐酸淬灭。沉降后,弃去己烷层。然后将有机物用氯仿萃取,并用氨水将氯仿层中的硫酸盐洗净,最后用水冲洗。将有机层的溶剂蒸干然后加入200mL乙酸叔丁酯,加热至116℃以除去氯仿,然后缓慢冷却至0~5℃。过滤产物,用50mL冷乙酸丁酯洗涤,并在100℃下真空干燥。得到化合物(R,R)-4(33.5g,收率68.1%)。

**步骤4:** DIPAMP 的合成

在氮气保护下将(R,R)-4(60.0g, 0.122mol)溶解在450mL的干燥乙腈中,加入150g干燥的三丁胺,将反应液加热至65~70℃。反应1h后加入97g三氯硅烷,将温度保持在70℃搅拌2h。然后将其冷却至30~40℃,加入360mL的25%氢氧化钠溶液。在骤冷结束时,将温度升至45~50℃以促进分离。分离有机相,并在氮气气氛下再用150mL 25%质量分数的氢氧化钠溶液洗涤。将由两层组成的有机相在45~55℃下浓缩,直到仅保留高沸点三丁胺为止。加入甲醇(50mL)辅助结晶,冷却至0~5℃并过滤,滤饼用两份40mL的冷甲醇洗涤并于60℃下干燥,得到DIPAMP(50.8g,收率84.6%)。

## (b) DIPAMP 的结构表征信息[1]

**mp**:102~104℃;

$[\alpha]_D^{20}$: -85 (c=1.0,氯仿);

**MS**(ESI):m/z(%)=458[M+H]$^+$。

### (c) DIPAMP 的应用实例

Barbara Mohar 等人[2]将铑（Ⅰ）-DIPAMP 型配体应用于不对称催化氢化 α-乙酰氨基苯乙烯，可制备各种手性 α-芳基胺，具有较好的对映选择性和反应活性。

同时，手性膦配体 DIPAMP 也可以作为有机小分子催化剂应用于不对称催化反应，如 2017 年，张俊良课题组[3]将手性双膦配体(R,R)-DIPAMP 成功应用于 γ-芳基取代联烯酸酯与三氟甲基取代不饱和烯酮的 [3+2] 环加成反应中，高效构建了一系列含有三个连续手性中心的三氟甲基取代环戊烯类化合物。

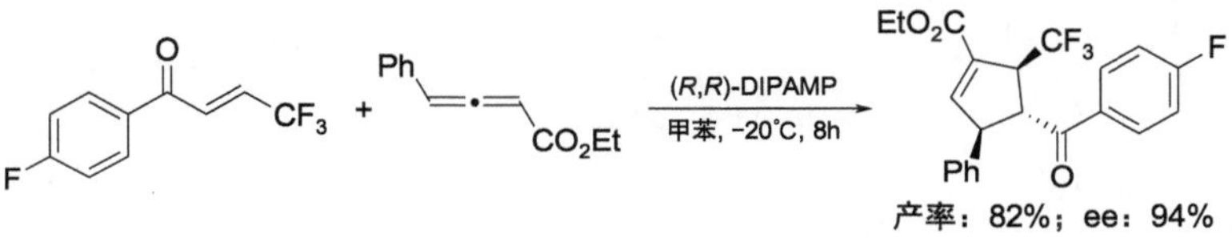

**具体反应示例**[3]：

在氮气氛围下，依次向干燥的史莱克管中加入三氟甲基取代的不饱和烯酮(0.2mmol)、γ-芳基取代联烯酸酯(0.3mmol)和干燥的甲苯(2mL)，然后将上述混合溶液冷却至-20℃。随后，加入(R,R)-DIPAMP (0.02mmol)，将反应混合物在-20℃下搅拌，TLC 监测反应。反应结束后减压除去溶剂，以石油醚 / 乙酸乙酯为洗脱剂直接进行硅胶色谱纯化，得到预期的环加成产物。

### 参考文献

[1] Vineyard B D, Knowles W S, Sabacky M J, et al. *J. Am. Chem. Soc.*, **1977**, 99: 5946-5952.
[2] Zupancic B, Mohar B, Stephan M. *Tetrahedron Letters*, **2009**, 50:7382-7384.
[3] Wang H M, Tao M N, Zhu C Z, et al. *Chem. Sci.*, **2017**, 8: 4660-4665.

## 6.2.13　(R, S)-Binaphos 配体的合成及应用实例

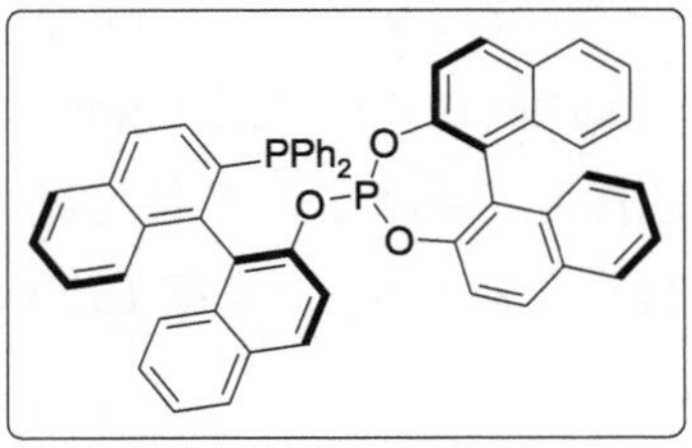

## (a) 配体的合成反应过程

**步骤1：** 化合物 2 的合成

氮气保护下，将(R)-1,1′-联二萘酚(5g，17.4mmol)、2,6-二甲基吡啶(5.58g，52.2mmol)和4-二甲基氨基吡啶(0.955g，7.83mmol)溶于二氯甲烷(26mL)中。0℃下加入三氟甲磺酸酐(14.7g，52.2mmol)，将反应体系移至室温下搅拌反应23h。反应完成后，减压除去溶剂，剩余物质通过柱色谱法(硅胶，二氯甲烷)纯化，得到产物9.56g，收率100%。

## 步骤 2：化合物 3 的合成

氩气保护下，将化合物 2(2.74g, 4.98mmol)、二苯基氧膦(1.99g, 9.82mmol)溶解于二甲基亚砜(26mL)中，加入醋酸钯(110mg, 0.491mmol)、1,3-双(二苯基膦基)丙烷(203mg, 0.491mmol)、乙基二丙胺(5.1mL)和甲酸钠(33mg, 0.491mmol)，室温下搅拌反应 20min。将反应体系在 90℃下搅拌 19h，移至室温。向反应液中加入 250mL 乙醚和 150mL 水，搅拌混合物以分离有机层和水层。分离后，用 125mL 水清洗有机相 4 次，用 125mL 5% 稀盐酸清洗 2 次，用 50mL 水清洗 2 次，用 125mL 饱和碳酸氢钠水溶液清洗 1 次，最后用 125mL 盐水清洗 1 次。合并有机相，无水硫酸镁干燥，过滤，减压除去溶剂。剩余物质通过柱色谱法(硅胶，甲苯：乙腈 = 3：1)纯化，得到产物 2.51g，收率 83%。

## 步骤 3：化合物 4 的合成

氮气保护下，将化合物 3(400mg, 0.664mmol)、三甲胺(1.21g, 12mmol)和三氯氢硅(1.62g, 12mmol)溶解在二甲苯(22mL)中，在 120℃下搅拌反应 17h。反应完成后，向反应体系中加入 4.4mL 35% 氢氧化钠水溶液，继续搅拌反应 2h，萃取，分液。用 30mL 盐水洗涤有机相 2 次，无水硫酸镁干燥，过滤，减压除去溶剂，从而获得粗产品。将粗产物溶解于 7mL 四氢呋喃中，向其中添加氢氧化锂一水合物(335mg,

7.98mmol) 置于 2.4mL 水中的溶液。将所得混合物在室温下搅拌 15h，向其中加入 50mL 乙醚和 15mL 5% 稀盐酸以分离有机层和水层。有机层用水洗涤两次后，合并有机相，无水硫酸镁干燥，过滤，减压除去溶剂。剩余物质通过柱色谱法(硅胶，正己烷:乙酸乙酯 =5 : 1)纯化，得到产物 153mg，收率 51%。

## 步骤 4：化合物 5 的合成

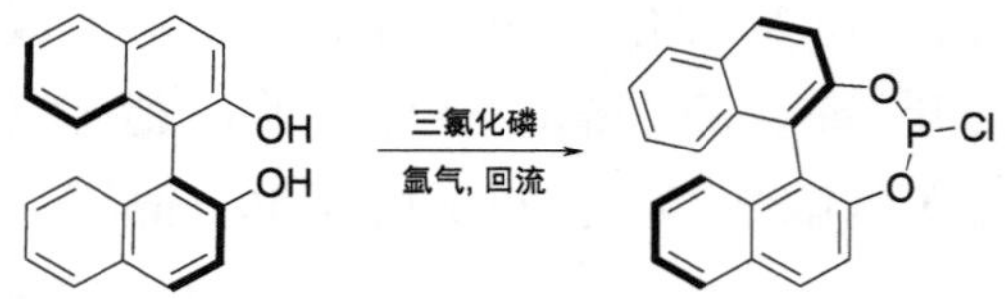

1b
(S)-1,1′-联二萘酚
5

向装有回流冷凝器的 50mL 烧瓶中加入(*S*)-1,1′- 联二萘酚(4.94g，17.3mmol)和三氯化磷(113g，0.83mol)，在氩气流中回流混合物 4h。将反应混合物冷却至室温，减压蒸馏除去未反应完的三氯化磷，得到白色晶体 5.56g，收率 92%。

## 步骤 5：(*R*, *S*)-Binaphos 的合成

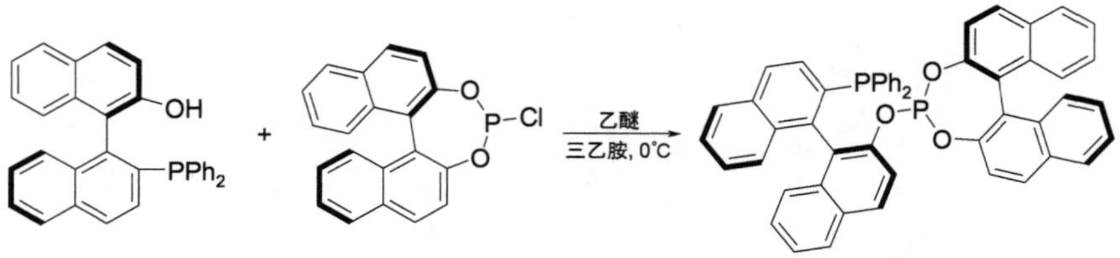

4　　5　　(*R*,*S*)-Binaphos

氮气保护下，将化合物 **4**(430mg，0.946mmol) 和化合物 **5**(662mg，1.89mmol) 溶解在乙醚(30mL) 中，0 ℃下，向反应体系加入三乙胺(191mg，1.89mmol)。在室温下将混合物搅拌 15h 后，加入 20mL 水以停止反应。分离有机相和水相，无水硫酸镁干燥，过滤，减压除去溶剂。剩余物质通过柱色谱法(硅胶，正己烷:二氯甲烷 = 1 : 1)纯化，得到产物 712mg，收率 98%。

## (b) (R,S)-Binaphos 配体的结构表征信息 [1,2]

$^{31}$P NMR(CDCl$_3$)$\delta$：-13.3(d, $J_{p-p}$ = 29.0Hz),146.2(d)。

## (c) (R,S)-Binaphos 的应用实例

2007 年，Nozaki 课题组[3]以 Rh(Ⅰ)-Binaphos 为催化剂，研究了乙烯基噻吩的不对称氢甲酰化反应(如下图所示)。研究结果表明，在 Rh(Ⅰ)-Binaphos 作用下，乙烯基噻吩的不对称氢甲酰化反应以较高的收率和对映选择性合成了相应的醛。

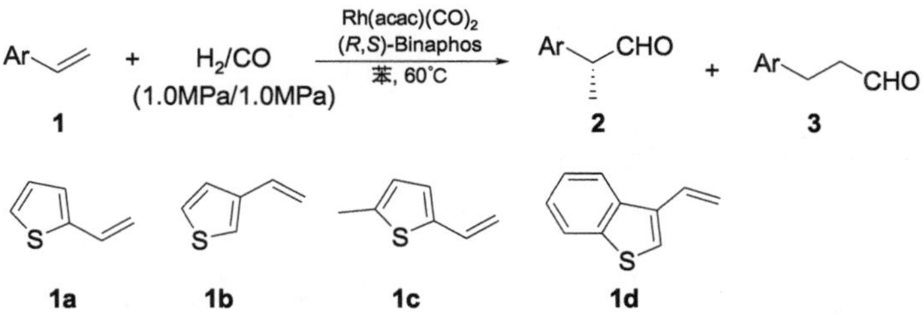

### 具体反应示例[3]：

以 5-甲基-2-乙烯基噻吩为原料，使用如下典型的反应流程：Rh(acac)(CO)$_2$(2.8mg, 0.010mmol)、(R,S)-MeO-Binaphos(33mg, 0.040mmol)和 5-甲基-2-乙烯基噻吩(2.0mmol)溶于苯(1.0mL)中形成反应溶液。将该混合溶液通过真空泵循环脱气，并在氩气氛围下转移到 50mL 高压釜中。在反应釜内加入氢气(1.0MPa)和 CO(1.0MPa)，并将所得混合物在 60℃下搅拌适当的时间。反应结束后，释放 H$_2$/CO 压力后，所得混合物的样品通过 $^1$H NMR 分析以确定支链醛和线型醛的比率。通过硅胶柱色谱法(正己烷：乙酸乙酯=10：1, $R_f$ = 0.42)纯化粗产物：产量为 281mg(产率 92%)，为无色油状物。2-(5-甲基噻吩-2-基)丙醛的 ee 值为 95%。

**参考文献**

[1] Sakai N, Mano S, Nozaki K, et al. *J. Am. Chem. Soc.*, **1993**, 115: 7033-7034.
[2] Yan Y J, Zhang X M. *J. Am. Chem. Soc.*, **2006**, 128: 7198-7202.
[3] Tanaka R, Nakano K, Nozaki K. *J. Org. Chem.*, **2007**, 72: 8671-8676.

# 6.3
## 手性膦-杂原子双齿配体的代表性合成及应用实例

### 6.3.1 *i*Pr-BiphPHOX 配体的合成及应用实例

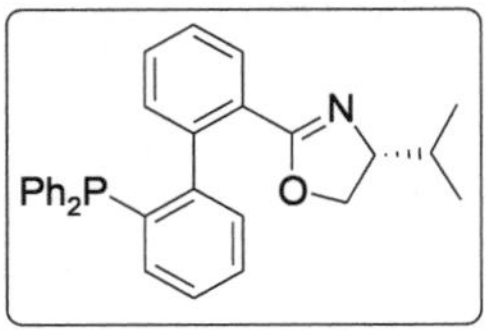

(a) 配体的合成反应过程

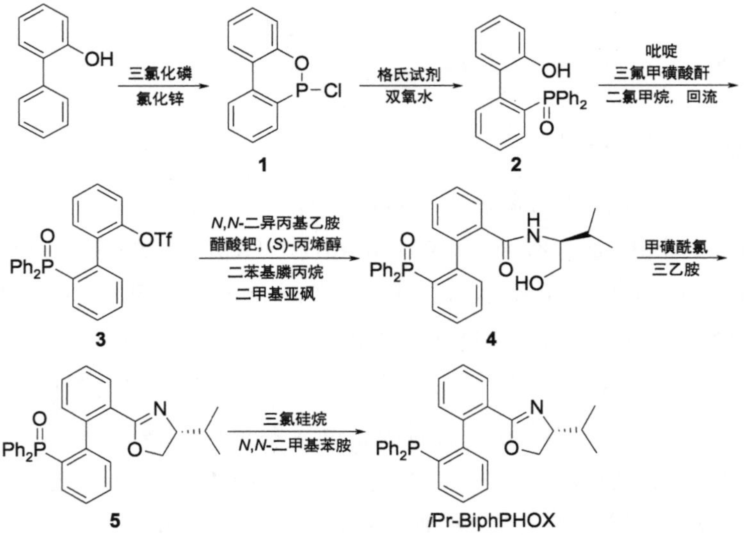

**步骤 1：** 化合物 1 的合成

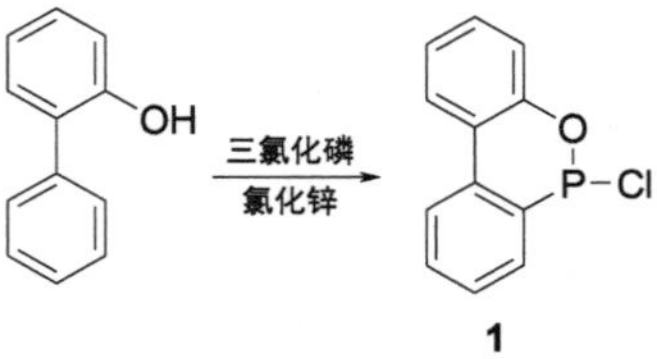

氮气保护下，将邻苯基苯酚(10.2g，60mmol)和三氯化磷(6.6mL，75mmol)加入反应瓶中，然后将反应液在 5h 内缓慢加热到 140℃，反应过程中有氯气产生，最后再加入氯化锌(0.06g，0.44mmol)，在 210℃下反应 3h。反应结束后，减压除去溶剂，减压蒸馏的方法纯化得到产物 12.6g，收率 90%。

**步骤 2：化合物 2 的合成**

氮气保护下，-50℃条件下，在 50mL 三口烧瓶中，依次加入芳基溴的溶液(21.3mmol)、镁悬浮液(0.62g，25.5mmol)，加入 5mL 超干四氢呋喃(或乙醚)，反应体系加热回流，在搅拌状态下通过滴液漏斗加入 2mL 的芳基溴的四氢呋喃/乙醚溶液，加料完成后，在回流条件下搅拌得到的深褐色溶液，反应 2h，制得格氏试剂。在含 **1**(2.0g，8.5mmol)的四氢呋喃溶液中滴加格氏试剂，温度控制在 25～40℃。反应过夜，用冰水冷却到 0℃，用饱和氯化铵溶液淬灭反应。分离的有机相，用稀盐酸水溶液洗涤，水相用二氯甲烷(10mL×3)萃取。合并有机相，并用无水硫酸镁干燥，过滤，减压除去溶剂。通过硅胶柱色谱法纯化得到产物，收率 83%。

**步骤 3：化合物 3 的合成**

氮气保护下，将 **2**(2mmol)、超干吡啶(0.64mL，8mmol)加入二氯甲烷(20mL)溶液中。混合均匀后，0℃下，缓慢加入三氟甲磺酸酐(0.66mL，4.0mmol)，加热回流反应 12～72h，TLC 监测反应。反应结束后，反应体系变为深棕色溶液，冰水浴冷却。稀盐酸洗涤，用二氯甲烷(20mL×3)萃取。有机相用无水硫酸镁干燥，过滤，减压除去溶剂，残余物通过硅胶柱纯化得到产物，收率 58%。

### 步骤 4：化合物 4 的合成

CO 保护下，将 **3**(2.0mmol)、*N,N*-二异丙基乙胺(0.93mL，5.6mmol)、(*S*)-丙烯醇(0.41g，4.0mmol)、醋酸钯(Ⅱ)(44.9 mmg,0.2mmol)和二苯基膦丙烷(82.1mg，0.2mmol)加入二甲基亚砜(12mL)溶液中，110℃油浴中加热搅拌 18h。反应结束后，冷却到室温，用二氯甲烷(15mL×3)萃取，无水硫酸镁干燥，过滤，减压除去溶剂。用硅胶柱进行纯化，收率 60%。

### 步骤 5：化合物 5 的合成

氮气保护下，将 **4**(0.9mmol)、三乙胺(0.44mL，3.1mmol)加入二氯甲烷(6.0mL)溶液中，0℃下加入甲磺酰氯(90μL)。搅拌反应 30min，然后将反应体系移至室温搅拌反应 1h，TLC 监测反应，反应结束后，用二氯甲烷稀释，用稀盐酸水溶液洗涤，无水硫酸镁干燥，过滤，减压除去溶剂，硅胶柱纯化得到产物，收率 97%。

### 步骤 6：*i*Pr-BiphPHOX 的合成

氮气保护下，将三氯硅烷(0.2mL，2.0mmol)、**5**(0.2mmol)和 *N,N*-二甲基

苯胺(0.25mL，2.0mmol)加入甲苯(2mL)中，反应体系加热回流，反应过夜。冷却至 0℃后，用氢氧化钠水溶液淬灭反应，乙酸乙酯稀释，加水分层，无水硫酸镁干燥，过滤，减压除去溶剂。得到黏稠油状产物，收率 80%。

## (b) *i*Pr-BiphPHOX 的结构表征信息 [1]

$^1$H NMR(400 MHz，CDCl$_3$)(主要的：次要的 = 55：45)δ：7.89(dd，1H，*J* = 7.7Hz、1.5Hz)，7.88(dd，1H，*J* = 7.7Hz、1.5Hz)，7.07 ~ 7.39(m，32H)，6.88(d，1H，*J*=8.0Hz)，6.83(d，1H，*J* = 8.0Hz)，3.93 ~ 4.06(m，32H)，3.75 ~ 3.89(m，3H)，0.81(s，9H)，0.73(s，9H)。
$^{31}$P NMR(161 MHz，CDCl$_3$)δ：-14.95，-15.02。

## (c) *i*Pr-BiphPHOX 的应用实例

张万斌小组开发的新型联苯膦唑啉配体 *i*Pr-BiphPHOX 已成功地应用于富电子和缺电子的外环烯烃的不对称氢化反应，如 3-苯基-1-甲苯磺酰基-2,5-二氢-1*H*-吡咯的不对称催化氢化反应为例，*i*Pr-BiphPHOX 作为铱催化剂的手性配体，在常温常压下进行，可以得到 99% 的产率和 92% 的对映选择性。当然如果将手性配体 *i*Pr-BiphPHOX 改为 In-BiphPHOX，对映选择性可以进一步提升到 98%[2]。

2021 年，张万斌课题组[3] 使用 *i*Pr-BiphPHOX 作为配体，完成了高效的铱催化 2-亚烷基-4-苄基苯并[*b*][1,4]噁嗪-3-酮的对映选择性氢化反应，该反应过程显示出良好的官能团兼容性，以非常高的产率(高达 99%)和优异的对映选择性(ee值高达 99%)得到相应产物，同时反应能够在低催化剂负载量 [0.1%(摩尔分数)] 下得以实现，而且能够保持高的对映选择性。

**具体反应示例**[3]：

将铱和配体的络合物(3.2mg，0.002mmol，0.01eq.)和反应物(0.2mmol，1.0eq.)加入装有磁力搅拌棒的5mL试管中。然后将该管放入充氮容器中的高压反应器。在氮气气氛下向混合物中添加溶剂(2mL)。用氢气吹扫三次(低于所需压力)，最后加压至3MPa。反应混合物在室温下搅拌24h。反应结束后将氢气缓慢释放，减压除去溶剂，用 $^1$H NMR 检测转化率，粗品用硅胶柱色谱分离，采用手性高效液相色谱法测定 ee 值。

**参考文献**

[1] Liu Y, Y, Yang G Q, Yao D M, et al. *Sci China Chem.*, **2011**, *54*: 87-94.
[2] Meng K, Xia J Z, Wang Y Z, et al. *Org. Chem. Front*, **2017**, *4*: 1601-1605.
[3] Nie Y, Li J, Yan J, et al. *Org. Lett.*, **2021**, *23*: 5373-5377.

## 6.3.2　MeO-MOP 配体的合成及应用实例

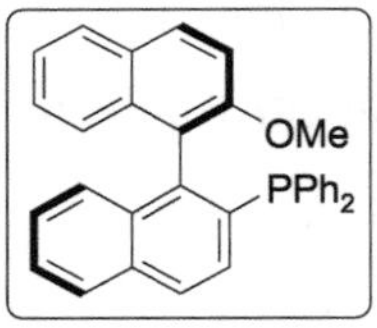

### (a) 配体的合成反应过程

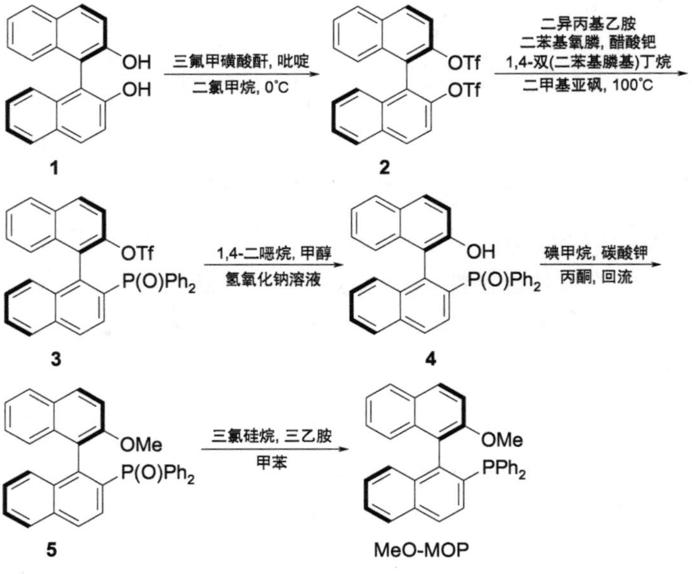

**步骤 1：化合物 2 的合成**

在无水无氧的条件下，向一个装有磁力搅拌棒和 30mL 恒压滴液漏斗的 200mL 两颈圆底烧瓶中，加入 14.3g(50.0mmol) 的 **1**、12.0mL(148mmol) 吡啶，和 100mL 二氯甲烷。并置于 0℃的冰水浴中，在 10min 内，逐滴加入 20.0mL(33.5g, 119mmol) 三氟甲磺酸酐，在 0℃下反应 6h。反应完成后，将反应液用旋转蒸发仪浓缩后用 200mL 乙酸乙酯稀释，然后转移到 500mL 的分液漏斗中。有机相用 5% 盐酸(70mL)、饱和碳酸氢钠(70mL)和饱和氯化钠(70mL)洗涤。再用无水硫酸钠干燥，并用旋转蒸发仪减压浓缩。将残余物在含有硅胶(700g)的色谱柱上进行柱色谱分离(10cm×20cm 色谱柱)。用二氯甲烷洗脱该柱，并用 30% 二氯甲烷-己烷作为洗脱剂，得到白色粉末 26.3g，收率 96%。

**步骤 2：化合物 3 的合成**

在无水无氧的条件下，向一个装有磁力搅拌棒的 500mL 史莱克管中加入 25.0g(45.4mmol) 的 **2**、18.4g(91.0mmol) 二苯基氧膦、1.02g(4.54mmol) 醋酸钯、1.94g(4.55mmol)1,4-双(二苯基膦基)丁烷、23.4g(181mmol) 二异丙基乙胺和 200mL 二甲基亚砜，在 100℃下反应 12h。反应结束后，冷却至室温，用旋转蒸发仪浓缩后用 400mL 乙酸乙酯稀释，然后转移到 1L 的分液漏斗中，依次用水(100mL×2)、5% 盐酸(100mL)、饱和碳酸氢钠(100mL)和饱和氯化钠(100mL)洗涤。有机相用无水硫酸钠干燥，过滤后用旋转蒸发仪减压浓缩。将残余物在含有硅胶(1.5kg)的色谱柱上进行柱色谱分离(14cm×30cm 色谱柱)。用 50% 乙酸乙酯-己烷作为洗脱剂,得到白色粉末化合物 3(23.8g,收率 87%)。

**步骤 3:** 化合物 4 的合成

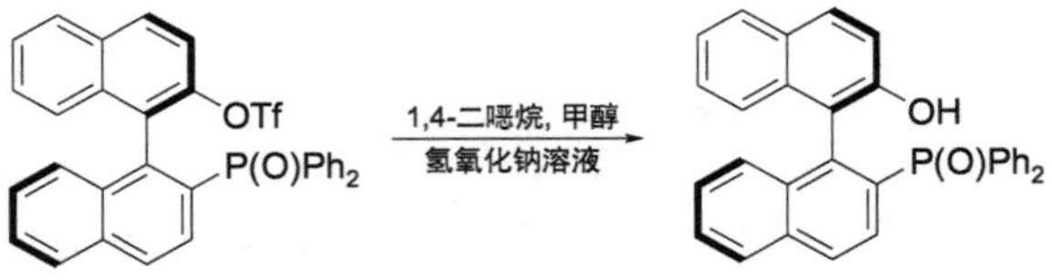

向一个装有磁力搅拌棒的 100mL 圆底烧瓶中加入 6.07g(10.1mmol) 的 3、30mL 1,4-二噁烷和 14mL 甲醇,并在室温下,向溶液中加入 14.1mL 的 3mol/L 氢氧化钠水溶液。该反应液搅拌 12h 后,加入几滴浓盐酸将其酸化到 pH=1。将混合物转移至分液漏斗中,用乙酸乙酯萃取两次,有机相用无水硫酸镁干燥,过滤后用旋转蒸发仪浓缩,得到 6.19g 的固体 4。该粗产品无需纯化,即可直接用于下一步。

**步骤 4:** 化合物 5 的合成

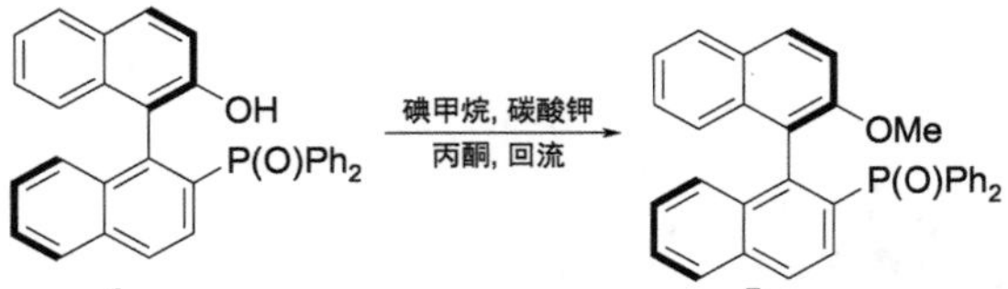

向一个 250mL 的圆底烧瓶中加入粗品 4(5.55g,40.2mmol)、碳酸钾、66mL 丙酮,最后加入(2.5mL,40.2mmol)碘甲烷。将反应混合物回流 3h 后,将其冷却至室温,用硅藻土过滤,用乙醚洗涤滤渣,收集滤液,用旋转蒸发仪浓缩滤液,得到 6.88g 棕色固体 5。该粗物质无需纯化,即可直接用于下一步。

**步骤 5:** MeO-MOP 的合成

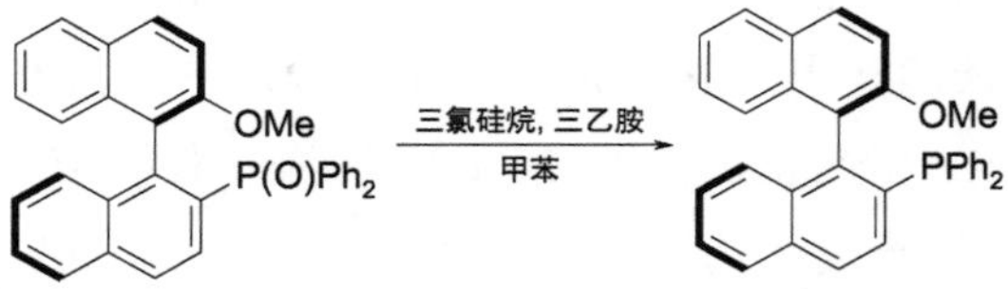

向一个装有磁力搅拌棒的 250mL 圆底烧瓶中加入粗品 **5**(7mL,50mmol)、三乙胺和 84mL 甲苯。将混合物冷却至 0℃后，用注射器加入三氯硅烷（4mL，40mmol）。将反应加热至 120℃搅拌 5h 后，将其冷却至室温，用乙醚稀释，用饱和碳酸氢钠水溶液淬灭。将得到的悬浊液用硅藻土过滤，用乙醚洗涤滤渣。有机相用无水硫酸镁干燥，用旋转蒸发仪浓缩得到 4.94g 黄色固体。该固体用少量二氯甲烷溶解后，在含有硅胶(615g)的色谱柱(10cm×30cm)上进行柱色谱分离。该柱用乙醚洗脱，得到灰白色粉末 4.08g，最后三步的总产率为 86%。

## (b) MeO-MOP 的结构表征信息

$^1$H NMR(300MHz，CDCl$_3$)$\delta$：3.67(s，3H)，6.95～8.08(m，22H)。
**IR**(KBr) $\nu$(cm$^{-1}$)：2968，2875，1656，1383，1230，1060，886。
**mp**：174～177℃。

## (c) MeO-MOP 的应用实例

Hayashi 课题组[1,2] 利用 MeO-MOP 配体和 [Pd(C$_3$H$_5$)Cl]$_2$ 通过链状末端烯烃的不对称硅氢化反应和 Fleming-Tamao 氧化反应成功以高收率、高对映选择性和高区域选择性合成手性仲醇。

2010 年，张俊良课题组[3] 开发了一种金(Ⅰ)/银(Ⅰ)双金属催化 4-苯基-3-苯乙炔基-3-丁烯-2-酮与 1,3-二苯基苯并[C]呋喃的串联双环化反应，在温和的反应条件下为构建手性多环化合物提供了一种实用的反应策略。

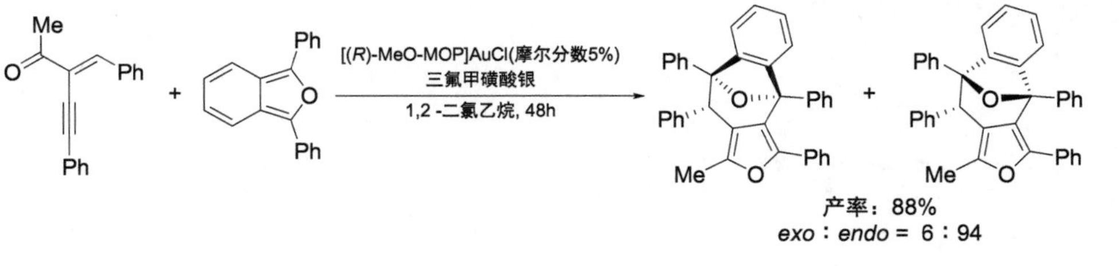

产率：88%
exo : endo = 6 : 94

**具体反应示例**[3]：

将催化剂 (R)-MeO-MOP/AuCl(10.5mg，0.015mmol) 溶于二氯乙烷(1mL)，在氩气下加入三氟甲磺酸银(3.8mg，0.015mmol)，室温搅拌10min。然后将4-苯基-3-苯乙炔基-3-丁烯-2-酮（73.8mg，0.3mmol）和1,3-二苯

基苯并[C]呋喃(105.3mg，0.39mmol)溶于二氯乙烷2mL中，再将此混合溶液加入反应瓶中。在30℃下搅拌48h后，经TLC监测反应完全。反应结束后，将反应液减压浓缩，再经柱色谱法分离纯化，得纯品，合并两种构型的产物总收率为88%，exo:endo = 6:94。

**参考文献**

[1] Uozumi Y, Hayashi T. *J. Am. Chem. Soc.*, **1991**, *113*: 9887-9888.
[2] Uozumi Y, Lee S Y, Hayashi T. *Tetrahedron Lett.*, **1992**, *33*: 7185-7188.
[3] Gao H Y, Wu X X, Zhang J L. *Chem.Commun.*, **2010**, *46*: 8764-8766.

## 6.3.3　PHOX 配体的合成及应用实例

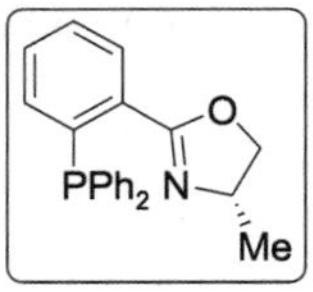

### (a) 配体的合成反应过程

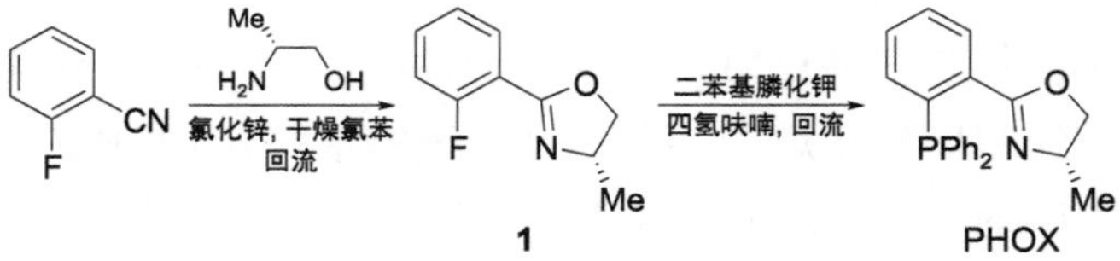

**步骤 1**：化合物 1 的合成

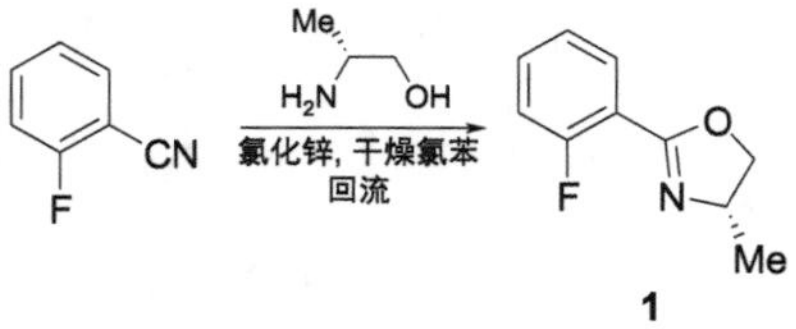

取一带有回流装置的三颈圆底烧瓶，加入催化量的氯化锌（摩尔分数 5%）和干燥氯苯（20mL）、邻氟苯甲腈（5mmol）与手性氨基醇（1.2eq.，6mmol），回流 8h。用乙酸乙酯和水萃取，无水硫酸钠干燥，旋干后过硅胶柱，得到化合物 **1**，产率 47%。

## 步骤 2: PHOX 的合成

N$_2$ 氛围下，在火焰干燥的 50mL 双颈烧瓶中，通过注射器加入二苯基膦化钾 (1mmol) 的四氢呋喃 (2mL) 溶液中。然后将该反应体系加热至回流，并加入化合物 **1**，将混合物在回流下搅拌 2h，红色溶液褪色为淡黄色。TLC 监测反应，反应完成后，然后反应体系通过二氯甲烷 (20mL) 和水 (20mL) 进行萃取得到有机层，有机层用硫酸镁干燥，然后在真空中去除溶剂。剩余物质通过柱色谱法 (硅胶，石油醚：乙酸乙酯 =3：1) 分离纯化，得到无色固体，产率 80%。

## (b) PHOX 的结构表征信息 [1]

**$^1$H NMR** (400 MHz, CDCl$_3$) $\delta$ : 0.95 (d, $J$ = 6.5Hz, 3H), 3.54 (t, $J$ = 7.6Hz, 1H), 4.08 ~ 4.21 (m, 2H), 6.84 (m, 1H), 7.20 ~ 7.70 (m, 12H), 7.90 (m, 1H)。

**$^{13}$C NMR** (100 MHz, CDCl$_3$) $\delta$ : 20.6, 61.7, 73.4, 127.7, 128.1, 128.2, 128.3, 128.5, 128.6, 128.8, 130.2, 130.4, 130.6, 132.3, 133.2, 133.6, 133.8, 133.9, 134.0, 134.3, 163.3。

**HRMS** (APCI) : $m/z$ (%) = 345.1303 [M+H]$^+$;

**mp** : 93 ~ 95℃；

**$[\alpha]_D^{25}$** : −7.5 ($c$ = 2.0, 氯仿);

**IR $v_{max}$** /cm$^{-1}$ : 1650。

## (c) PHOX 的应用实例

2018 年，Steven J. Malcolmson 课题组报道了一例高对映选择性的 Pd-PHOX 催化的 1,3-二烯的氢烷基化反应，例如 $\beta$-二酮和丙二腈，在温和的反应条件下就可以高达 96% 的收率和 95% 的 ee 值获得烷基化产物 [2]。

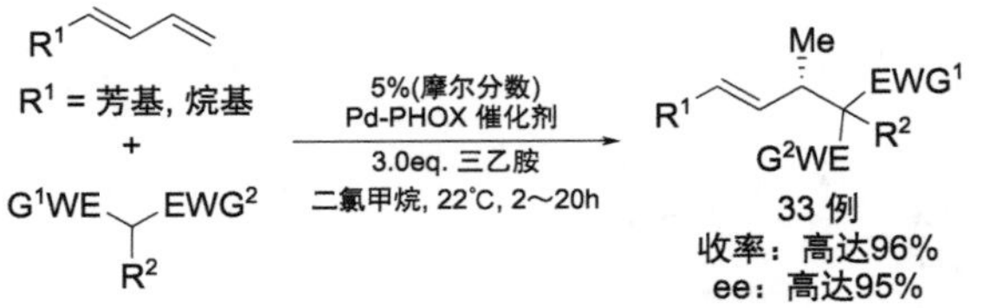

2018 年，Cheng 课题组[3]报道了手性配体 PHOX 应用于钴催化合成 1-氨基茚的反应体系，PHOX 表现出很高的区域选择性和优异的对映选择性。

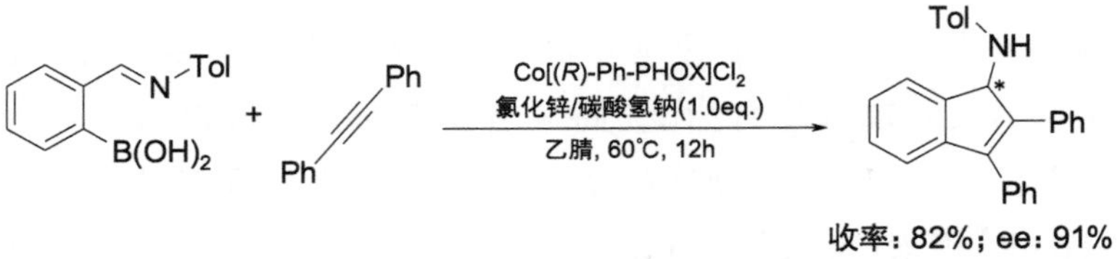

### 具体反应示例[3]：

将 Co[(R)-Ph-PHOX]Cl$_2$（摩尔分数 10%）、碳酸氢钠（1.0eq.）、氯化锌（摩尔分数 20%）和芳基硼酸（0.2mmol，1.0eq.）置于史莱克管中，将其脱气并用氮气吹扫 3 次，然后，在氮气下通过注射器将二苯乙炔（0.3mmol，1.5eq.）和超干的乙腈（1.0mL）加入密封管中。将反应混合物在 60℃下搅拌 12h，反应完全后进行冷却，用二氯甲烷稀释，混合物通过硅藻土和硅胶垫过滤。随后浓缩滤液，残余物通过柱色谱纯化，使用己烷和乙酸乙酯作为洗脱剂，得到预期 *S*-构型产物。

### 参考文献

[1] Dawson G J, Frost C G, Williams J M J. *Tetrahedron Lett.*, 1993, 34: 3149-3150.
[2] Adamson N J, Wilbur K C E, Malcolmson S J. *J. Am Chem. Soc.*, 2018, 140: 2761-2764.
[3] Chen M H, Hsieh J C, Lee Y H, et al. *ACS Catal.*, 2018, 8: 9364-9369.

## 6.3.4　SIPHOX 配体的合成及应用实例

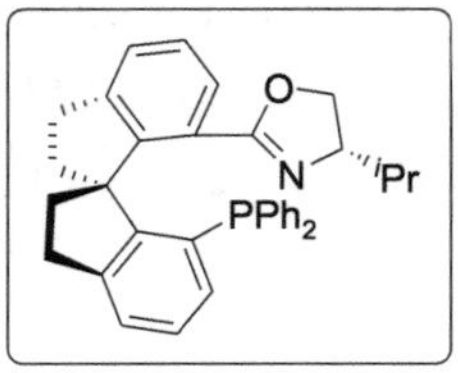

## (a) 配体的合成反应过程

## 步骤1：化合物2的合成

氮气保护下，将SPINOL(5.0g，19.8mmol)、吡啶(7.0mL，86.7mmol)加入二氯甲烷(100mL)溶液中，0℃下，恒压滴液漏斗滴加三氟甲磺酸(8.2mL，43.7mmol)。添加完毕，将反应体系移至室温搅拌过夜。减压除去溶剂，剩余物质用80mL乙酸乙酯溶解，转移至分液漏斗，用5%盐酸溶液、饱和碳酸氢钠溶液、饱和食盐水依次洗涤，无水硫酸钠干燥。过滤，减压除去溶剂，用适量二氯甲烷溶解，二氯甲烷作洗脱剂，过硅胶短柱。得到白色固体9.9g，收率97%。

**步骤 2：化合物 3 的合成**

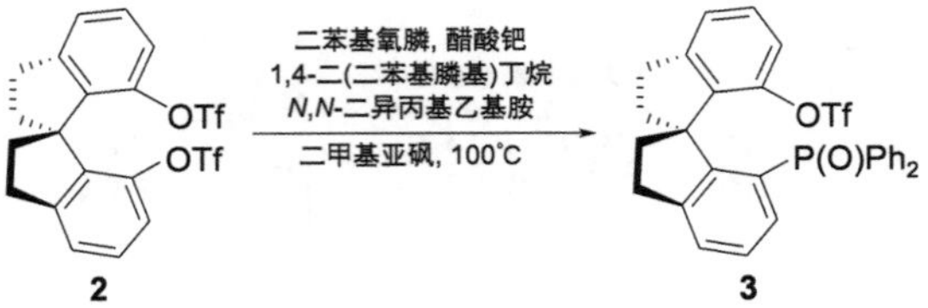

氮气保护下，100mL 反应瓶中加入 **2**(4.0g，7.75mmol)、二苯基氧膦(3.13g，15.5mmol)、醋酸钯(87mg，0.39mmol)、1,4-二(二苯基膦基)丁烷(dppb，166mg，0.39mmol)，用 25mL 脱气 DMSO 将其溶解。磁力搅拌使其充分混匀，加入 N,N-二异丙基乙基胺(4.1g，32mmol)后，100℃油浴加热反应 6h。冷却至室温，乙酸乙酯萃取，有机层用 5% 盐酸溶液、饱和食盐水、饱和碳酸氢钠溶液、饱和食盐水依次洗涤，无水硫酸钠干燥。过滤，减压除去溶剂，剩余物质通过柱色谱法(硅胶，石油醚:乙酸乙酯 = 3:1)分离纯化，得到白色固体 4.0g，收率 90%。

**步骤 3：化合物 4 的合成**

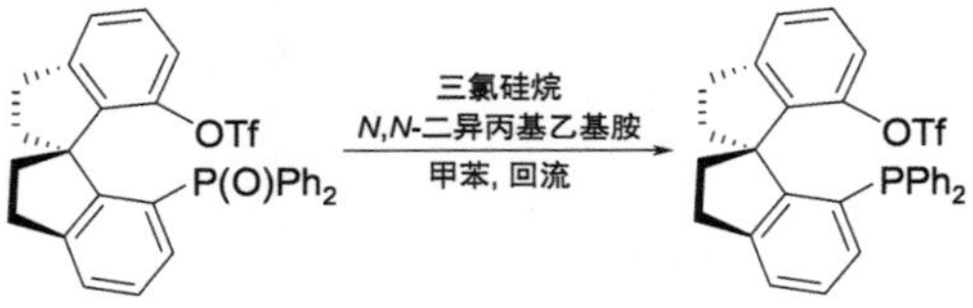

氮气保护下，将 **3**(1.4g，2.5mmol)、N,N 二异丙基乙基胺(13.2g，102mmol) 加入甲苯(50mL) 溶液中，0℃搅拌下加入三氯硅烷(4.0mL，39mmol)，移至 110℃油浴中搅拌反应 5d。冷却至室温，乙醚稀释，饱和氢氧化钠溶液淬灭反应，过滤，乙醚洗涤，无水硫酸钠干燥。过滤，减压除去溶剂，剩余物质通过柱色谱法(硅胶，石油醚:乙酸乙酯 = 15:1)分离纯化，得到白色固体 1.26g，收率 91%。

**步骤 4：化合物 5 的合成**

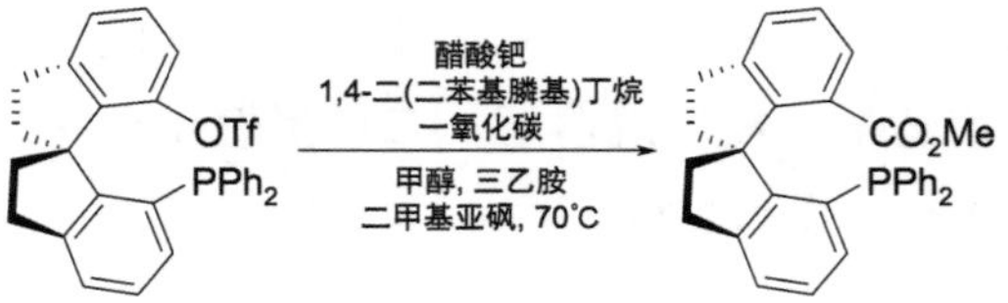

向装有抽气头和反口塞的 250mL 两口瓶中加入甲醇(60mL)、二甲基亚砜(90mL)和三乙胺(24mL)，混合均匀后，在一氧化碳氛围下脱气三次。在另外一个装有磁力搅拌、反口塞、回流冷凝管、抽气头和温度计的四颈烧瓶中加入 4(6.27g，11.3mmol)、醋酸钯(381.7mg，1.70mmol)和 1,4-二(二苯基膦基)丁烷(701.1mg，1.70mmol)，反应体系用一氧化碳保护，即刻用双针头将脱气的溶液小心移入。70℃油浴加热，搅拌反应 4~6h，体系由橙色变为灰褐色。TLC 监测直至反应完全。减压除去溶剂，剩余物质通过柱色谱法(硅胶，石油醚:乙酸乙酯 = 16:1)分离纯化，得到无色黏稠液体 4.8g，收率 91%。

**步骤 5：** 化合物 6 的合成

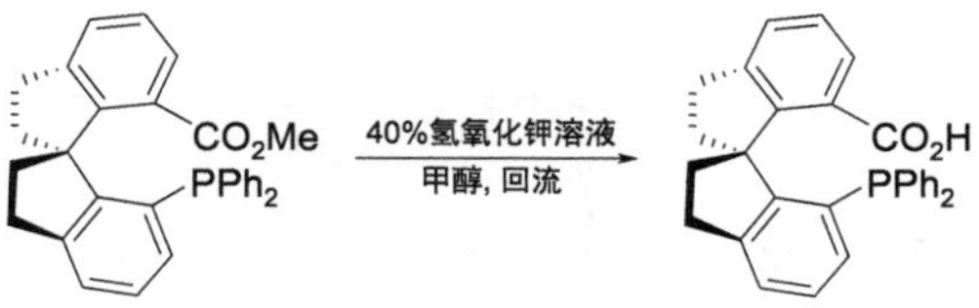

氮气保护下，将 5(3.0g，6.49mmol)加入甲醇(75mL)搅拌使其溶解。随后缓慢滴加 40% 氢氧化钾水溶液(15mL)。开始有白色絮状物生成，然后缓慢溶解。置于 100℃油浴中使反应体系微微回流，此时溶液变澄清。保持该温度反应 24h，TLC 跟踪监测，直至反应完全。将其置于冰水浴，缓慢向反应体系中滴加浓盐酸至 pH = 2，有大量白色沉淀生成。用 100mL 水稀释，乙酸乙酯(100mL×3)萃取、分液，合并有机相，用饱和食盐水洗涤，无水硫酸钠干燥。过滤，减压除去溶剂，剩余物质通过柱色谱法(硅胶，石油醚:乙酸乙酯 = 5:1)分离纯化，得到白色固体 2.88g，收率 99%。

**步骤 6：** 化合物 7 的合成

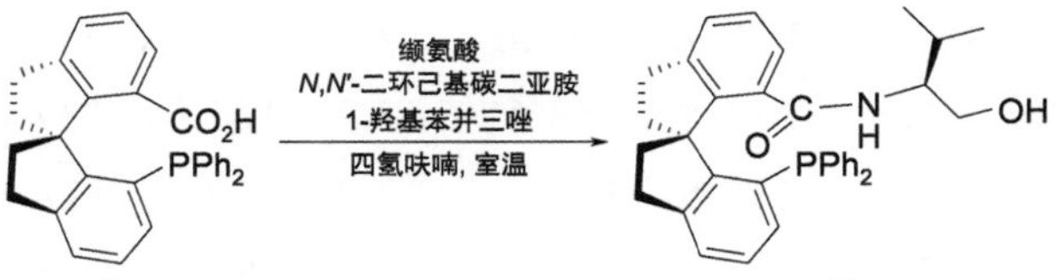

氮气保护下，0℃下，将 6(500mg，1.12mmol)、缬氨酸(360mg，3.50mmol)、1-羟基苯并三唑(380mg，2.48mmol)和 N,N′-二环己基碳二亚胺(664mg，3.22mmol)加入四氢呋喃(60mL)，随后将反应体系移至室温搅拌反应，体系中有大量白色沉淀生成。TLC 跟踪反应，直至转化完全。向体系中加入 10g 硅胶，减压除去溶剂，剩余物质通过柱色谱法(硅胶，石油醚:乙酸乙酯 = 2∶1)分离纯化，得到黏稠液体 0.6g，收率 100%。

**步骤 7:** SIPHOX 的合成

氮气保护下，将 7(600mg，1.13mmol)和 4-二甲氨基吡啶(5mg，0.041mmol)加入二氯甲烷(70mL)溶液中，0℃下，依次加入二氯甲烷(0.34mL)和甲基磺酰氯(170μL)，并搅拌反应 30min，加入三乙胺(1.45mL)，将反应体系移至室温搅拌反应过夜。TLC 跟踪反应，直至转化完全。向体系中加入 10g 硅胶淬灭反应，减压除去溶剂，剩余物质通过柱色谱法(硅胶，石油醚:乙酸乙酯 = 10∶1)分离纯化，得到白色泡沫状固体 0.4g，收率 69%。

### (b) SIPHOX 配体的结构表征信息 [1]

$^1$H NMR(300 MHz，CDCl$_3$)δ：7.77(d，J = 7.5Hz，1H，Ar-H)，7.40～6.85(m，20H，Ar-H)，4.78(dd，J = 10.2Hz、6.9Hz，1H，CH)，3.67(dd，J = 8.1Hz、7.2Hz，1H，CH$_2$)，3.37(t，J = 8.4Hz，1H，CH$_2$)，3.12～2.94(m，3H，CH$_2$)，2.86～2.78(m，1H，CH$_2$)，2.65(dd，J = 22.2Hz、10.5Hz，1H，CH$_2$)，2.29～1.95(m，3H，CH$_2$)。

$^{13}$C NMR(75 MHz，CDCl$_3$)δ：165.6，154.9，154.6，149.6，145.1，144.7，144.6，142.7，138.8，138.6，137.9，137.7，134.2，134.1，133.9，133.8，133.6，133.5，132.4，129.2，128.5，128.2，128.1，128.0，127.9，127.2，126.7，126.6，126.5，73.8，69.1，63.6，40.7，

40.6,38.4,31.0,30.7。
**³¹P NMR**(121 MHz,CDCl₃)δ:−20.6(s)。
**MS**(EI):m/z(%)=549[M+H]⁺。

### (c) SIPHOX 的应用实例

在不对称催化反应中，手性膦-噁唑啉类型配体具有广泛的用途。这类配体与铱金属的配合物常被用于催化非官能化烯烃和亚胺等的不对称氢化反应中，但这类催化剂催化效率普遍较低，原因在于在氢化条件下，这类手性催化剂会因为自身聚合生成无催化活性的三聚体。周其林课题组[1,2]利用螺环骨架本身的刚性来抑制铱类催化剂的自聚情况，从而设计合成了具有螺二氢茚骨架的手性螺环膦-噁唑啉配体。研究结果表明，这是一类具有高效催化活性和高选择性的手性螺环铱催化剂，该铱催化剂稳定性非常好，在反应条件的氢气氛围下没有发生自聚而失活。其中，($S_a,S$)-SIPHOX 的铱催化剂在 N-芳基亚胺的不对称氢化反应中呈现出色的催化活性和选择性，此类反应可以在常温、常压下进行，可以得到 ee 值为 90%~97% 的对映选择性。

2020 年，周其林和朱守非课题组[3]使用 SIPHOX 的铱催化剂实现了脂肪族 γ-酮酸的一步氢化环化反应，合成得到的手性内酯具有高达 99% 的产率与 99% 的 ee 值。

**具体反应示例**[3]：

a) 催化剂的制备：

在手套箱中，将 SIPHOX(628.5mg，0.75mmol)、[Ir(cod)Cl]$_2$(296mg，0.44mmol) 和 NaBArF·3H$_2$O(882mg，0.94mmol) 加入充满氩气的 50mL 史莱克管中，用 13mL 的 CH$_2$Cl$_2$ 溶解，并将混合物加热回流 30min。冷却至室温后，减压浓缩，通过硅胶色谱柱提纯，淋洗剂为二氯甲烷/石油醚(体积比为 2∶1)得到橙黄色固体的 Ir-SIPHOX 配体。

b) 反应过程：

随后向充入 H$_2$ 的史莱克管中加入脂肪族 $\gamma$-酮酸(0.25mmol)、Ir-SIPHOX 催化剂(10.0mg，0.005mmol)、三乙胺(12.5mg，0.125mmol) 和甲醇(2mL)。将高压釜置换 H$_2$ 5 次后将史莱克管放入，随后向高压釜中充入 H$_2$ 至 0.6MPa，65℃搅拌 24h。反应结束后放出氢气，溶液用 5% 氢氧化钠(0.5mL) 溶液处理，水层用乙醚(8mL)洗涤，1mol/L 盐酸(pH 5~6)酸化 30min(不同脂肪族 $\gamma$-酮酸反应时间不同)，然后用乙醚(8mL×3)萃取。合并有机层，饱和氯化钠洗涤，硫酸钠干燥，真空蒸发得到产物，GC 或 HPLC 分析以确定 ee 值。

**参考文献**

[1] Zhu S F, Xie J B, Zhang Y Z, et al. *J. Am. Chem. Soc.*, **2006**, *128*: 12886-12891.
[2] 谢建华, 周其林. 神奇的手性螺环配体 [J]. 化学学报, **2014**, *72*: 778-797.
[3] Li M L, Li Y, Pan J B, et al. *ACS Catal.*, **2020**, *10*: 10032-10039.

## 6.3.5　TF-BiphamPhos 配体的合成及应用实例

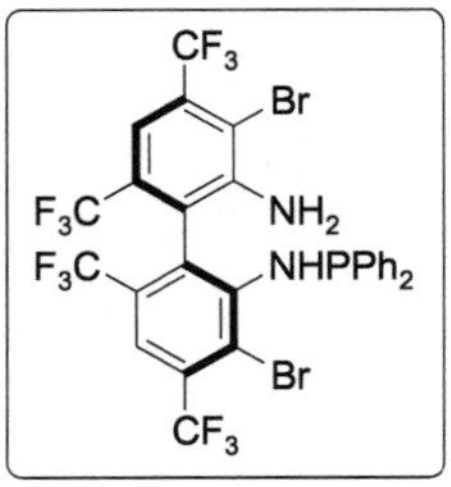

## (a) 配体的合成反应过程

**步骤1：** 化合物 2 的合成

氮气保护下，将 **1**(10g，21.92mmol)、三乙胺(15mL，109.6mmol)加入二氯甲烷(300mL)，随后将反应体系置于 0℃下缓慢滴加(+)-10-樟脑磺酰氯(13.5g，54.8mmol)，将反应体系移至室温下搅拌 4～6h。反应结束后，反应体系用 1mol/L 盐酸洗涤，二氯甲烷萃取。合并有机层，无水硫酸钠干燥，过滤，减压除去溶剂，剩余物质通过硅胶柱色谱法纯化，得到 **2**(9.21g，收率 > 99%)。

**步骤 2**：化合物 3 的合成

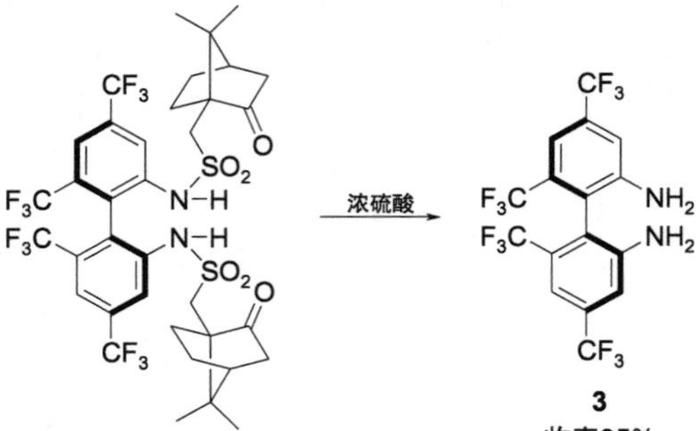

氮气保护下，将化合物 **2** 加入 5mL $H_2O$ 和 60mL 95% 浓硫酸混合体系中，反应体系在 60℃下反应 4～5h。冷却至室温后，将反应液缓慢地倒入冰水中，用 6mol/L 氢氧化钠溶液中和，二氯甲烷萃取，合并有机层，无水硫酸钠干燥，过滤，减压除去溶剂，剩余物质通过硅胶柱色谱法纯化，收率 95%。

**步骤 3**：化合物 4 的合成

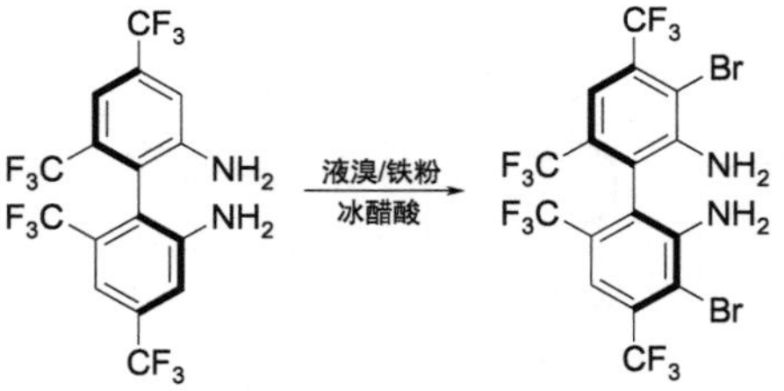

氮气保护下，向含有 **3**(2.28g，5.0mmol) 的圆底烧瓶中加入 40mL 冰醋酸、铁粉(20mg) 和液溴(1.03mL，20.0mmol)。反应体系在室温下搅拌反应 4h，反应结束后，加入 2mol/L 氢氧化钠水溶液调至 pH > 8。二氯甲烷萃取，合并有机相，依次用饱和碳酸氢钠、饱和氯化钠水溶液洗涤，无水硫酸钠干燥。过滤，减压除去溶剂，剩余物质通过柱色谱法(硅胶，乙酸乙酯：石油醚 = 1：5) 纯化，得到棕色固体 2.95g，收率 96%。

**步骤 4**: TF-BiphamPhos 的合成

氮气保护下，在 0℃下，将化合物 **4**(1.0mmol)加入二氯甲烷(10mL)溶液中，加入 4-二甲氨基吡啶(12.2mg，0.1mmol)、三乙胺(1.0mL，7.2mmol)和二苯基氯化膦(1.5mmol)，将反应体系移至室温搅拌过夜，减压除去溶剂，剩余物质通过快速氧化铝色谱法纯化得到白色固体 0.57g，收率 72%。

### (b) TF-BiphamPhos 配体的结构表征信息 [1-3]

**¹H NMR**(CDCl₃，TMS，300 MHz)$\delta$：7.90(d，$J$=3.6Hz，1H)，7.50(s，1H)，7.23～7.34(m，11H)，7.10(s，1H)，4.21(d，$J$ = 6.3Hz，1H)，3.75(br，2H)。

**¹³C NMR**(CDCl₃，TMS，75 MHz)$\delta$：147.2，147.0，146.5，139.2，139.0，138.4，138.3，131.8～133.2(m)，131.4，130.6～131.4(m)，129.7，129.6，128.8，128.7，123.3(q，$J_{C-F}$ = 271.7Hz)，123.2(q，$J_{C-F}$ = 271.6Hz)，121.9，121.2，118.1，115.8，115.4，115.2，114.2，113.0。

**³¹P NMR**(CDCl₃，85% H₃PO₄，242.86 MHz)$\delta$：32.17。

**HRMS**(ESI)：$m/z$(%) = 641.1023[M+H]⁺。

### (c) TF-BiphamPhos 的应用实例

2008 年，王春江课题组 [1,2] 发现铜 / TF-BiphamPhos 络合物可作为新型且高效的不对称 1,3-偶极环加成反应的催化剂，底物普适性好，具有较为理想的立体选择性和收率。最近，该课题组进一步将铜(Ⅰ)/(S)-TF-BiphamPhos 催化体系应用于三氟甲基丙烯酸叔丁酯的不对称 1,3-偶极环加成反应 [3]，以高达 98% 的产率、> 98 : 2 的 dr 值以及 97%～99% 的 ee 值合成了含季碳手性中心的手性吡咯烷衍生物。

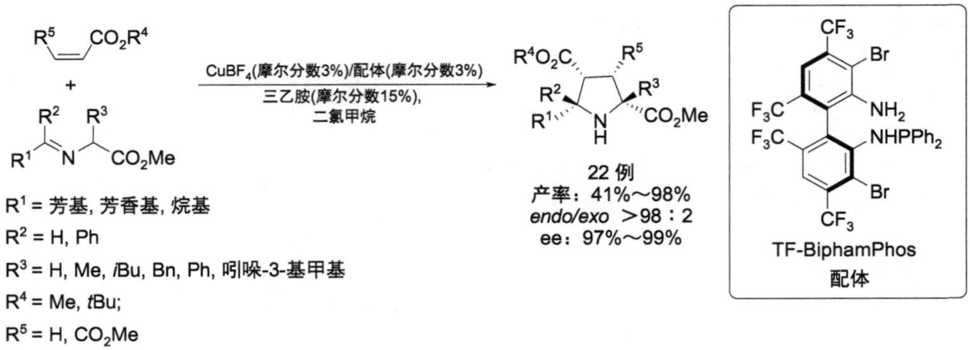
CuBF$_4$(摩尔分数3%)/配体(摩尔分数3%)
三乙胺(摩尔分数15%),
二氯甲烷
R$^1$ = 芳基, 芳香基, 烷基
R$^2$ = H, Ph
R$^3$ = H, Me, iBu, Bn, Ph, 吲哚-3-基甲基
R$^4$ = Me, tBu;
R$^5$ = H, CO$_2$Me
22 例
产率：41%～98%
endo/exo ＞98：2
ee：97%～99%
TF-BiphamPhos
配体

## 具体反应示例[3]：

在氮气保护下，将(*S*)-TF-BiphamPhos(摩尔分数3%)和$CuBF_4$(摩尔分数3%)溶解在二氯甲烷(2mL)中，并在室温下搅拌0.5h。然后，依次加入甘氨酸席夫碱底物(1.5eq.)、三乙胺(摩尔分数15%)和丙烯酸叔丁酯-2-(三氟甲基)酯(0.23mmol)。反应通过TLC监测，反应结束后，混合物通过硅藻土过滤，滤液浓缩至干。通过柱色谱法纯化产物，得到相应的环加成产物。


**参考文献**

[1] Wang C J, Liang G, Xue Z Y, et al. *J. Am. Chem. Soc.*, **2008**, *130*: 17250-17251.
[2] Wang C J, Gao F, Liang G. *Org. Lett.*, **2008**, *10*: 4711-4714.
[3] Wang C J, Dong X Q, Cheng X. *Asian. J. Org. Chem.*, **2020**, *9*: 1567-1570.


## 6.3.6 Spiro Phosphino-oxazine 配体的合成及应用实例

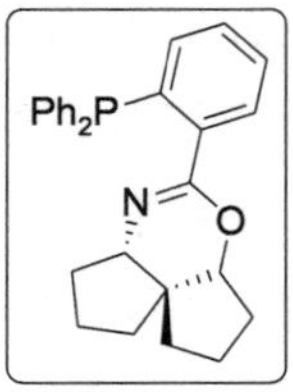
Ph$_2$P
N
O

## (a) 配体的合成反应过程

## 步骤1：化合物3的合成

氮气保护下，向装有回流冷凝器的烧瓶中加入镁(24.2g，9996mmol)和碘粒，并溶于四氢呋喃(750mL)。将1,4-二溴丁烷(41.8g，194mmol)的四氢呋喃(100mL)溶液在30min内缓慢滴加到反应体系，回流，直至红褐色消散。再加热回流24h，接着在1h内将δ-戊内酯(17.8g，178mmol)的四氢呋喃(100mL)溶液加入反应体系中，将所得的浅灰色悬浮液加热回流5h，冷却至0℃，加入水200mL淬灭反应，乙酸乙酯(200mL)洗涤，加入10%的盐酸水溶液直至浑浊消失，乙酸乙酯(300mL×3)萃取，先后用饱和碳酸氢钠水溶液(200mL)和盐水(200mL)洗涤，硫酸镁干燥并真空浓缩得到棕褐色液体的二醇，无需纯化即可使用。将其溶于二甲基亚砜(400mL)和水(80mL)中。加入草酸二水合物(22.6g，179mmol)，然后将反应液在120℃下搅拌2h。冷却至室温后，溶液用乙醚萃取(150mL×4)，用盐水(200mL×3)洗涤，硫酸镁干燥并浓缩得到棕褐色液体。Kugelrohr 蒸馏(0.25 Torr，70℃)得到无色液体17.8g，收率71.4%。

## 步骤 2：化合物 4 的合成

氮气保护下，向戴斯-马丁氧化剂(44.2g, 104mmol)的二氯甲烷(300mL)的淡黄色悬浮液中，加入 **3**(11.1g, 78.9mmol)的二氯甲烷(100mL)悬浮液。搅拌 1h 后，混合物变浓稠，产生白色沉淀。将其倒入乙醚(350mL)、饱和碳酸钠水溶液(350mL)和饱和硫代硫酸钠水溶液(350mL)中并搅拌 1h。水相用乙醚(300mL×2)萃取，合并有机相用盐水(200mL)洗涤，无水硫酸镁干燥，过滤，减压除去溶剂，得到浅黄色液体状的醛，其无需纯化即可使用。将其溶解在乙醚(250ml)中，并加入碳酸钠(25.2g, 238mmol)和盐酸羟胺(14.0g, 201mmol)的水溶液中，剧烈搅拌 15h 后，乙醚(200mL×2)萃取，有机相用盐水(200mL)洗涤，经硫酸镁干燥并真空浓缩。用 Kugelrohr 器蒸馏(26.6Pa, 85℃)得到蜡状白色固体 11.4g，收率 94.1%。

## 步骤 3：化合物 5 的合成

氮气保护下，将 5.25% 的次氯酸钠水溶液(20.0mL)添加到 **4**(850mg, 5.55mmol)的二氯甲烷(25mL)溶液中，并剧烈搅拌 15h。二氯甲烷(25mL×3)萃取，盐水(20mL)洗涤，无水硫酸镁干燥，过滤，减压除去溶剂，剩余物质通过柱色谱法(硅胶，正己烷:乙酸乙酯＝9:1)纯化，得到浅色黄色液 430mg，收率 51.2%。

## 步骤 4：化合物 6 的合成

氮气保护下，0℃下 **5**(1.13g，7.45mmol)的四氢呋喃(15mL)溶液加入 LiAlH₄(380mg，10.0mmol)的四氢呋喃(15mL)悬浮液中，加热至22℃并搅拌15h。将灰色悬浮液冷却至0℃，并用10%的氢氧化钠水溶液(5mL)淬灭。移至室温后，加入10%的氢氧化钠水溶液(100mL)以溶解铝盐。用乙醚(75mL×3)萃取，用盐水(75mL)洗涤，无水硫酸镁干燥，过滤，减压除去溶剂，得到白色蜡状固体1.00g，收率86.3%。

**步骤5：** 化合物7的合成

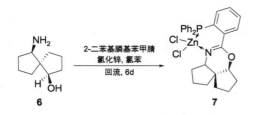

将2-二苯基膦基苯甲腈(535mg，1.86mmol)、**6**(402mg，2.59mmol)和氯化锌(504mg，3.70mmol)在氯苯(8mL)中回流6d。冷却至室温后，所得溶液用硅胶(5cm)过滤，以乙酸乙酯为洗脱剂洗脱。真空浓缩得到米色固体，其无需进一步纯化即可使用。

**步骤6：** Spiro Phosphino-oxazine 的合成

氮气保护下，将2,2'-联吡啶(299mg，1.91mmol)和 **7** 溶于干燥的氯仿(15mL)中，并在室温搅拌1h。将所得溶液通过二氧化硅(5cm)过滤，并且用 CHCl₃(100mL)冲洗，真空浓缩得到蓬松黏稠白色固体746mg，收率89.5%。

**(b) Spiro Phosphino-oxazine 配体的结构表征信息**

¹**H NMR**(400 MHz，CDCl₃)$\delta$：7.77(dd，$J$ = 7.1Hz、3.6Hz，1H)，7.40～7.20(m，12H)，6.85(dd，$J$ =7.5Hz、4.3Hz，1H)，3.79(d，

$J$ = 4.4Hz, 1H), 3.46(t, $J$ = 8.2Hz, 1H), 2.23～2.12(m, 1H), 1.89～1.76(m, 1H), 1.73～1.18(m, 9H), 0.91～0.80(m, 1H)。
**$^{13}$C NMR**(100 MHz, CDCl$_3$)$\delta$: 155.5(C), 139.3(C, d, $J$ = 20.7Hz), 138.9(C, d, $J$ = 21.7Hz), 138.8(C, d, $J$ = 20.9Hz), 137.2(C, d, $J$ = 20.9Hz), 134.0(CH, d, $J$ = 19.7Hz), 133.8(CH, d, $J$ = 20.4Hz), 130～128(CH, m), 80.2(CH), 60.3(CH), 47.8(C), 36.2(CH$_2$), 35.4(CH$_2$), 35.1(CH$_2$), 30.5(CH$_2$), 21.7(CH$_2$), 20.7(CH$_2$)。
**$^{31}$P NMR**(160 MHz, CDCl$_3$)$\delta$: 6.3。
**MS**(VG7070)$m/z$: 425(2, M$^+$), 261(26), 225(28), 208(34), 183(73), 153(23), 152(22), 127(21), 125(22), 113(26), 111(31), 107(24), 99(32), 97(38), 95(35), 79(47), 71(68), 69(40), 67(34), 58(100), 56(45), 45(52), 43(78);
**HRMS** (ESI): $m/z$(%)= 425.18906[M+H]$^+$;
**IR**(薄膜) $\nu_{max}$(cm$^{-1}$): 3066, 3051, 2952, 2930, 2865, 2209, 1663, 1650, 1582, 1555, 1461, 1432, 1345, 1314, 1273, 1254, 1198, 1177, 1142, 1096, 1070, 1026, 908, 777, 742, 692, 667, 545, 501;
$[\alpha]_D^{21}$: +68.6($c$ = 1.25, 氯仿)。

## (c) Spiro Phosphino-oxazine 的应用实例

　　1,2-氨基醇合成的多种手性膦-噁唑啉配体，在过渡金属催化反应中应用广泛，包括钯催化的烯丙基烷基化、烯丙基胺化、Heck 反应和 Diels-Alder 反应，铂催化的烯丙基烷基化，铜催化的共轭加成反应和 Diels-Alder 反应，铑催化的硅氢加成反应和镍催化的 Grignard 交叉偶联反应。Brian A. Keay 课题组在 2003 年报道了螺-氨基醇的合成[1]，可以很容易地拆分成两个对映体，其 ee 值＞99%。2004 年该课题组在此基础上报道了螺-膦-噁嗪(Spiro Phosphino-oxazine)的合成[2]，这是第一个包含稠合螺系统的膦-噁嗪，可应用于钯催化的丙二酸甲酯与 1,3-二苯基烯丙基乙酸酯的烷基化反应。研究结果表明该配体 Spiro Phosphino-oxazine 在钯催化的烯丙基烷基化反应中具有较好的手性诱导能力，可获得高达 91% ee 的立体选择性。

## 参考文献

[1] Lait S M, Parvez M, Keay B A. Tetrahedron: Asymmetry, **2003**, *14*: 749-756.
[2] Lait S M, Parvez M, Keay B A. Tetrahedron: Asymmetry, **2004**, *15*: 155-158.

# 6.4 手性多官能化多齿型膦配体的代表性合成及应用实例

## 6.4.1 Xing-Phos 配体的合成及应用实例

### (a) 配体的合成反应过程

**步骤 1：** 化合物 2 的合成

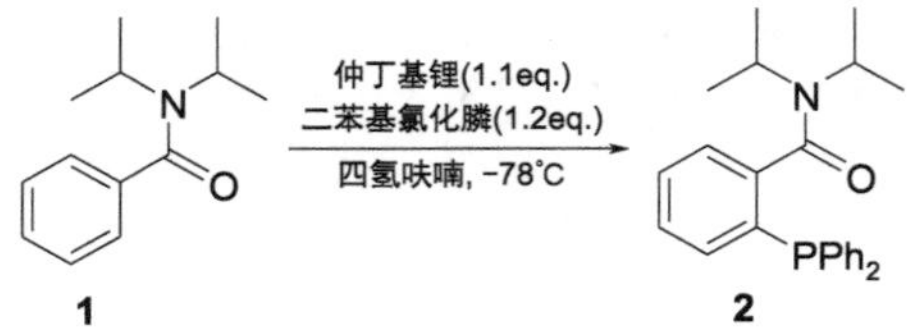

氮气保护下，将 N,N-二异丙基苯甲酰胺(化合物 **1**，2.09g，10.2mmol，1.0eq.)加入到四氢呋喃(60.0mL)中，随后将反应体系置于-78℃下缓慢滴加仲丁基锂(1.3mol/L，8.5mL，11.1mmol，1.1eq.)，反应 30min 后，逐滴加入二苯基氯化膦(2.2mL，12.3mmol，1.2eq.)，继续反应 1h，将反应体系移至室温下继续搅拌，TLC 监测至反应结束，反应体系用饱和氯化铵溶液淬灭，乙酸乙酯萃取，有机层用盐水洗涤，无水硫酸镁干燥，过滤，减压除去溶剂，剩余物质通过硅胶柱色谱法纯化，得到白色固体 1.96g，收率 94%。

**步骤 2：** 化合物 3 的合成

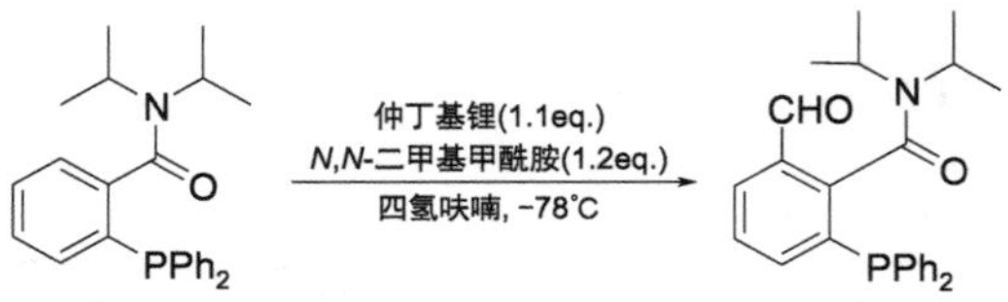

氮气保护下，将化合物 **2**(1.91g，4.9mmol，1.0eq.)加入四氢呋喃(20mL)中，随后将反应体系置于-78℃下缓慢滴加仲丁基锂(1.3mol/L，4.2mL，1.1eq.)，反应 1h 后，逐滴加入 N,N-二甲基甲酰胺(0.46mL，6.0mmol，1.2eq.)，反应 15min 后，将反应体系移至室温下继续搅拌 2~3h，将反应液缓慢倒入水中进行淬灭，乙酸乙酯萃取，有机层用盐水洗涤，无水硫酸镁干燥，过滤，减压除去溶剂，剩余物质通过柱色谱(硅胶，石油醚:乙酸乙酯=10:1)纯化，得到白色固体 1.39g，收率 73%。

**步骤 3：** 化合物 3 的合成

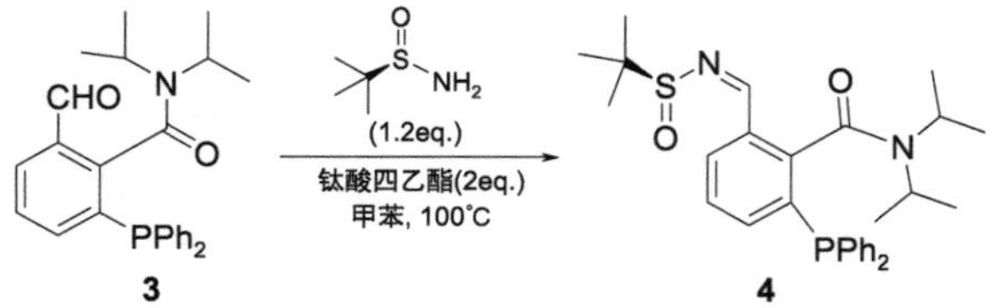

氮气保护下，将化合物 **3**(1.21g，2.9mmol，1.0eq.)、*R*-(+)- 叔丁基亚磺酰胺(0.42g，3.5mmol，1.2eq.)溶于 10mL 甲苯中，然后加入钛酸四乙酯(1.2mL，5.8mmol，2.0eq.)，在 100℃下反应 2～3h，TLC 监测至反应结束，将反应体系冷却至室温后，加入等量的饱和食盐水，过滤，用乙酸乙酯萃取滤液，萃取所得有机层用盐水洗涤，无水硫酸镁干燥，过滤，减压除去溶剂，剩余物质通过柱色谱法(硅胶，石油醚：乙酸乙酯 = 10∶1)纯化，得到淡黄色固体 0.99g，收率 82%。

**步骤 4:** Xing-Phos 的合成

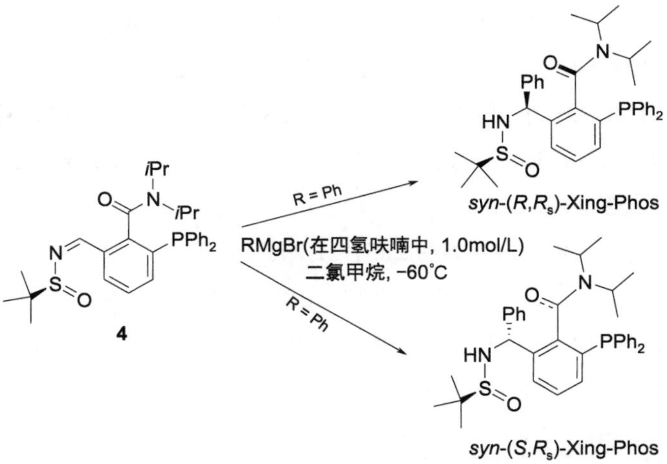

Xing-Phos 的一般合成过程：

氮气保护下，将化合物 **4**(1.0eq.)溶于二氯甲烷中，随后将反应体系置于-60℃下逐滴加入苯基格氏试剂(2.0eq.)，反应 1h 后，将反应体系移至室温下继续搅拌 1～2h。反应完成后，反应体系用饱和氯化铵溶液淬灭，乙酸乙酯萃取，萃取所得有机层用盐水洗涤，无水硫酸镁干燥，过滤，减压除去溶剂，剩余物质通过柱色谱法纯化。

*syn*-(*R*,*R*s)-Xing-Phos 和 *syn*-(*S*,*R*s)-Xing-Phos 的合成：

将化合物 **4**(1.04g，2.0mmol，1.0eq.)溶于 5mL 的二氯甲烷溶液中后，在-60℃下逐滴加入苯基格氏试剂(1.0mol/L 在四氢呋喃中，4mL，4.0mmol，2.0eq.)，余下过程与上述一般合成过程条件与操作相同。TLC

监测反应(石油醚:乙酸乙酯=3:1)有两个新的化合物点出现,其中极性较弱的点为 syn-$(R,R_s)$-Xing-Phos(0.52g,收率43%,$R_f \approx 0.3$),极性较强的点为 syn-$(S,R_s)$-Xing-Phos(0.42g,收率35%,$R_f \approx 0.2$)。

### (b) Xing-Phos 的结构表征信息 [1]

$^1$H NMR(400 MHz,CDCl$_3$)$\delta$:7.69(d,$J$ = 7.8Hz,1H),7.56(d,$J$ = 7.5Hz,2H),7.26(dd,$J$ = 19.9Hz、13.3Hz,15H),7.08(dd,$J$ =7.4Hz、2.9Hz,1H),5.73(s,1H),3.89~3.70(m,1H),3.64~3.59(m,2H),1.71(dd,$J$ = 30.1Hz、6.7Hz,6H),1.39~1.23(m,15H)。
$^{13}$C NMR(101 MHz,CDCl$_3$)$\delta$:167.79,143.67,143.30,140.62,138.87,138.78,137.95,137.83,136.51,136.40,134.93,134.76,134.52,133.54,133.34,133.12,132.94,128.63,128.52,128.45,128.43,128.39,128.37,128.30,128.17,127.90,127.52,57.77,55.70,51.17,46.39,22.67,21.61,21.00,20.93,20.75,19.93。
$^{31}$P NMR(202 MHz,CDCl$_3$)$\delta$:-13.94。
HRMS(APCI):$m/z$(%)= 599.2857[M + H]$^+$;
IR(KBr)$v_{max}$(cm$^{-1}$):3447,3297,3058,2963,2927,2869,1629,1475,1446,1433,1368,1339,1319,1063,1037,763,743,695。
$[\alpha]_D^{30}$:-68.2($c$ = 0.61,氯仿)。

### (c) Xing-Phos 的应用实例

2015年徐利文课题组[1]设计合成了命名为Xing-Phos的新型膦配体,该膦配体在银催化的不对称[3+2]环加成反应中表现出高的非对映选择性和对映选择性,可得到非常高的产率和ee值,为各种取代基的手性吡咯烷酮提供了高效的催化合成策略。如在苯基亚甲胺酯与活性烯烃的[3+2]环加成反应中,可以很高的非对映选择性以及对映选择性(ee值高达99%)得到相应的吡咯烷衍生物。

2018 年法国 Jubault 和 Bouillon 等人[2]利用徐利文提供的 Xing-Phos 解决了银催化的超吸电子五氟化硫取代基丙烯酸酯的不对称 [3+2] 环加成反应，Xing-Phos 在该反应中表现出非常好的立体选择性，为合成带有四个连续碳立体中心的五氟化硫取代的手性吡咯烷提供了一种有效的策略，底物普适性好，对映选择性高。

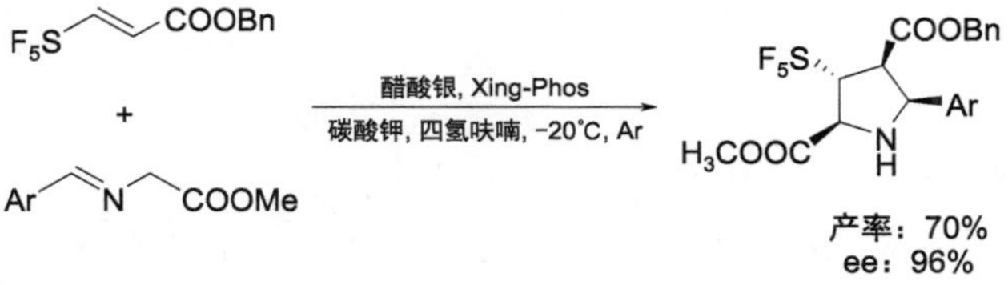

产率：70%
ee：96%

**具体反应示例**[2]：

在氩气下将 AgOAc(0.05eq.，摩尔分数 5%)和手性配体 Xing-Phos(0.11eq.，摩尔分数 11%)溶于干燥的四氢呋喃(2.0mL)，室温搅拌 10min，然后依次加入两种底物和碳酸钾(1.5eq.，0.15mmol)。在-20℃搅拌直到反应完成后，在减压下除去溶剂。粗产物经硅胶柱色谱法(洗脱液：从石油醚/乙酸乙酯 = 9∶1 到石油醚/乙酸乙酯 = 1∶1)分离，得到对应的手性产物。

**参考文献**

[1] Bai X F, Song T, Xu Z, et al. *Angew. Chem. Int. Ed.*, 2015, 54: 5255-5259.
[2] Zhao Q, Vuong T M H, Bai X F, et al. *Chem. Eur. J.*, 2018, 24: 5644-5651.

## 6.4.2　Tao-Phos 配体的合成及应用实例

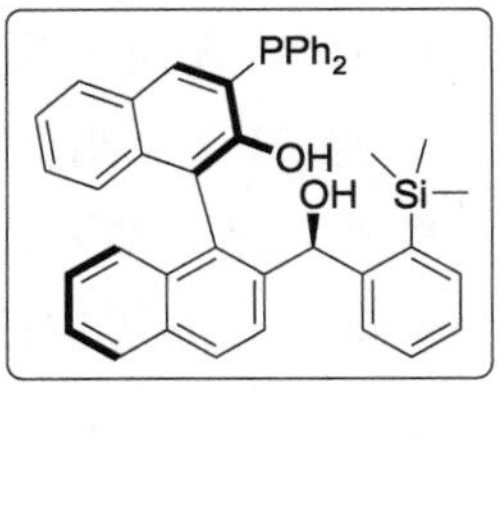

## (a) 配体的合成反应过程

## 步骤1：化合物2的合成

称取(R)-BINOL(1.0eq.)、碳酸钾(1.2eq.)置于烧瓶中，然后加入丙酮，将烧瓶放置于70℃油浴锅中搅拌回流1h后，将邻溴苄溴(1.1eq.)缓慢滴加入烧瓶中，继续在70℃下搅拌回流2h，经TLC监测反应完全后，将反应瓶取下冷却至室温，将反应体系进行抽滤，用乙酸乙酯洗涤。取下层滤液，减压旋干得到浓缩的有机相，经色谱柱分离法(石油醚：乙酸乙酯＝5∶1)得到化合物2，收率65%。

**步骤 2：化合物 3 的合成**

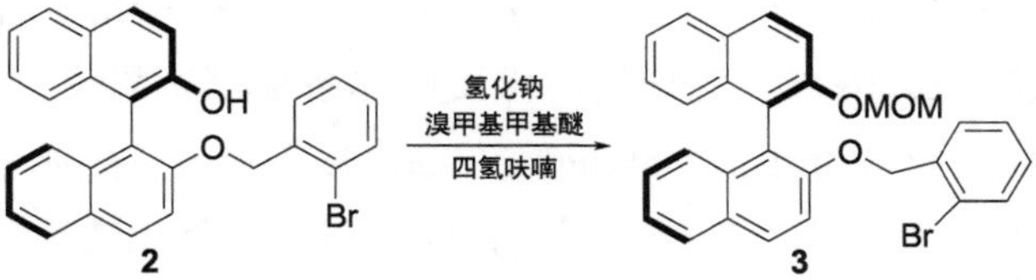

氮气保护下，称取氢化钠（1.2eq.）溶于入 50.0mL 四氢呋喃（THF）。将反应体系置于 0℃，将化合物 **2**（4.91g，10mmol，1.0eq.）溶于四氢呋喃中，缓慢滴加入烧瓶中，继续在 0℃下搅拌 1h 后，再抽取溴甲基甲基醚（1.5eq.）缓慢滴加到反应体系中，随后将反应体系转移到室温搅拌过夜，经 TLC 监测反应完全后，在低温下用饱和的氯化铵水溶液进行淬灭，乙酸乙酯进行萃取，合并有机相，减压旋干得到浓缩的有机相，经色谱柱分离（石油醚：乙酸乙酯 = 1：5）得到化合物 **3**，收率 67%。

**步骤 3：化合物 4 的合成**

氮气保护下，将化合物 **3**（4.91g，10mmol，1.0eq.）溶于四氢呋喃（50.0mL）中，随后将反应体系置于 -78℃下并缓慢滴加正丁基锂（2.5mol/L，4.8mL，12mmol，1.2eq.）。于 -78℃下反应 1h 后，逐滴加入氯硅烷（1.04mL，12mmol，1.2eq.），搅拌一段时间后，将反应体系移至室温下继续搅拌。TLC 监测，反应结束后反应体系用饱和氯化铵溶液淬灭，乙酸乙酯萃取，有机层用盐水洗，无水硫酸钠干燥，过滤，减压除去溶剂，粗产物通过色谱柱（硅胶，石油醚：乙酸乙酯 = 20：1）纯化，得到白色固体 3.198g，收率 65%。

**步骤 4：化合物 5 的合成**

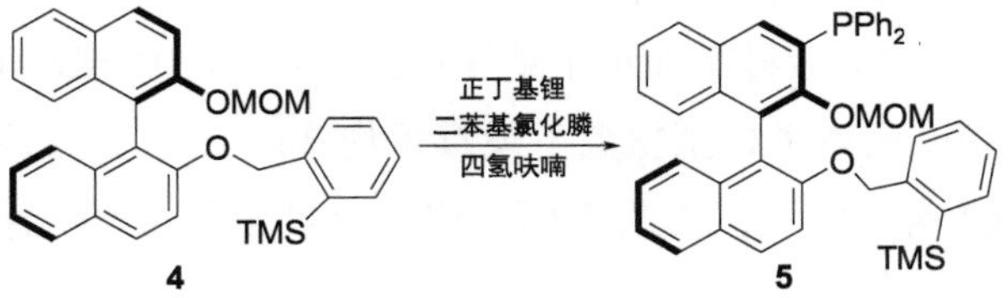

氮气保护下，化合物 4(3.198g，6.5mmol，1.0eq.)溶于 THF(30.0mL)中，随后将反应体系置于-78℃下缓慢滴加正丁基锂(2.5mol/L，3.12mL，7.8mmol，1.2eq.)，待滴加完毕后，将反应体系移至室温反应 1h 后，然后取二苯基氯化膦(1.4mL，7.8mmol，1.2eq.)逐滴加入反应体系中，继续搅拌，通过 TLC 监测至反应结束，反应体系用饱和氯化铵溶液淬灭，乙酸乙酯萃取，有机层用盐水洗涤，无水硫酸钠干燥，过滤，减压除去溶剂，剩余物质通过色谱柱(硅胶，石油醚：乙酸乙酯 =10：1)纯化，得到黄色油状液体 3.29g，收率 75%。

**步骤 5：** 化合物 6 的合成

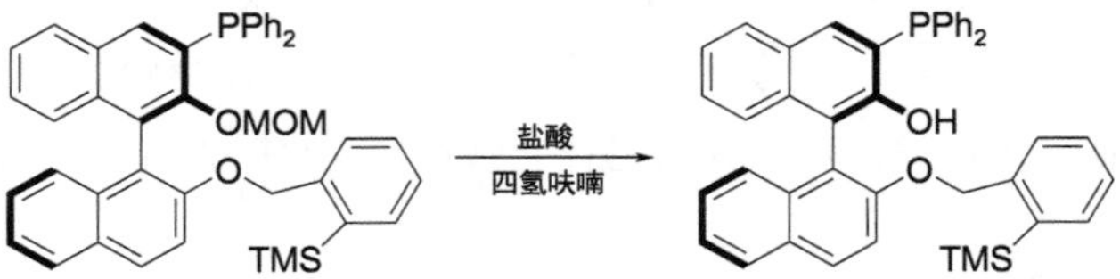

氮气保护下，化合物 5(3.29g，4.87mmol，1.0eq.)溶于四氢呋喃(20.0mL)中，加入盐酸(3mL，12mol/L)并于 40℃下反应 2h。反应结束后，用饱和碳酸氢钠溶液中和反应体系，用乙酸乙酯提取有机相，无水硫酸钠干燥，过滤，减压除去溶剂，粗产物通过色谱柱(硅胶，石油醚：乙酸乙酯 = 5：1)纯化，得到黄色固体 2.844g，收率 92%。

**步骤 6：** Tao-Phos 的合成

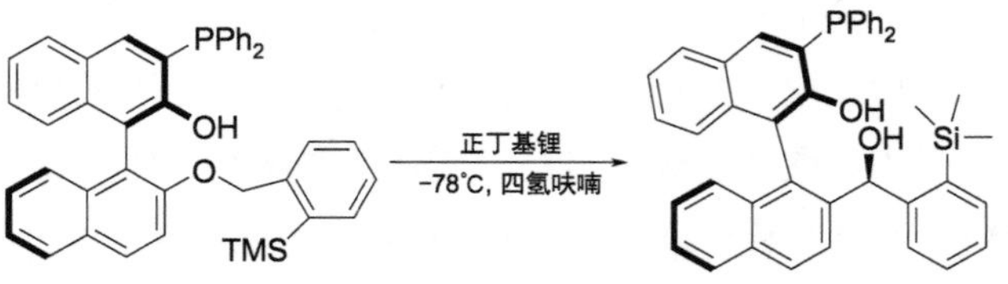

化合物 6(2.844g，4.5mmol，1.0eq.)溶于四氢呋喃(20.0mL)中，随后将反应体系置于 -78℃缓慢滴加正丁基锂(1.23mol/L，10.9mL，13.4mmol，1.2eq.)，然后将反应体系移至室温下继续反应 2h。反应完成后，反应体系用饱和 $NH_4Cl$ 溶液淬灭，乙酸乙酯萃取，有机层用盐水洗涤，无水硫

酸钠干燥，过滤，减压除去溶剂，剩余物质通过色谱柱(硅胶，石油醚：乙酸乙酯 = 5 : 1 ～ 10 : 1)纯化，得到产物 2.56g，收率 90%。

## (b) Tao-Phos 的结构表征信息 [1]

$^1$**H NMR**(400 MHz，CDCl$_3$)$\delta$：7.89(d，$J$ = 4.8Hz，1H)，7.85(d，$J$ = 4.8Hz，1H)，7.63～7.58(m，2H)，7.49～7.12(m，19H)，6.96(t，$J$ = 7.6Hz，1H)，6.82(d，$J$ = 8.8Hz，1H)，6.45(s，1H)，6.10(s，1H)，3.11(s，1H)，0.05(s，9H)。

$^{13}$**C NMR**(100 MHz，CDCl$_3$)$\delta$：151.9，151.8，146.5，140.5，140.2，138.1. 134.3，133.9，133.5，133.5，133.3，132.9，132.7，132.6，132.3，130.1，128.3，128.2，128.0，127.9，127.9，127.8，127.7，127.7，127.3，127.2，127.1，125.9，125.8，125.7，125.5，125.3，124.5，122.8，117.5，72.8，0.3。

$^{31}$**P NMR**(203 MHz，CDCl$_3$)$\delta$：-17.15。

$^{29}$**Si NMR**(99MHz，CDCl$_3$)：4.29。

**IR**(KBr)$\nu$(cm$^{-1}$)：3446.4，3055.0，2952.9，1618.0，1507.0，1434.2，1384.6，1026.3，743.3。

**HRMS**(ESI-TOF)：$m/z$(%) = 633.2375[M+H]$^+$；

$[\alpha]_D^{30}$：-32.07($c$ = 1.0，氯仿)。

## (c) Tao-Phos 的应用实例

徐利文课题组[1,2]设计合成了一种新型膦配体 Tao-Phos，并利用 Tao-Phos 和氟化铜的催化剂体系成功实现了不对称 Huisgen 类型的 [3+2] 环加成反应，此类不对称 Click 化学可便捷地得到含有琥珀酰亚胺和连三唑取代基团的含季碳手性中心的官能化分子，所得产物具有良好的产率(60%～80%)和 ee 值高达 99% 的对映选择性。

2018 年，徐利文课题组[3] 首次报道了 Tao-Phos 能够在纳米氧化铜催化的叠氮化物-炔烃不对称 Huisgen[3+2] 环加成反应中发挥手性控制能力，生成高对映选择性(ee 值高达 90%)和良好产率(61%)的产物，表现出与常规单齿或双齿膦配体不同的手性诱导能力。

er高达95：5
s/b高达20：1

**具体反应示例**[3]：

在 $N_2$ 气氛下，向干燥的史莱克管中加入 Tao-Phos (25.3mg，0.04mmol) 和氧化亚铜 (20nm) (1.4mg，0.01mmol)，然后添加 1.0mL 乙腈。溶液在 25℃下搅拌 1h 后，添加溶于 1mL 乙腈中的二炔底物 (0.2mmol)，并将混合物搅拌 0.5h。添加 DIPEA (5μL，0.05mmol，0.25eq.) 和叠氮化物 (0.22mmol) 后，将溶液冷却至 0℃。将反应液继续搅拌 72h，直到通过 TLC 监测反应结束。反应完成后，用饱和氯化铵水溶液 (1mL) 将其淬灭并剧烈搅拌 5min。用乙酸乙酯 (5mL×3) 萃取水相。有机层用无水硫酸钠干燥并在减压下浓缩。通过硅胶柱色谱法 (石油醚：乙酸乙酯 =3：1 ～ 5：1) 分离纯化获得手性产物。

**参考文献**

[1] Zheng Z J , Xu Z, Xu L W, et al. Chem. Eur. J., **2015**, 21: 554-558.
[2] Chen M Y, Song T, Zheng Z J, et al. RSC Adv., **2016**, 6: 58698-58708.
[3] Chen M Y, Xu Z, Xu L W, et al. ChemCatChem., **2018**, 10: 280-286.

## 6.4.3 HZNU-Phos 配体的合成及应用实例

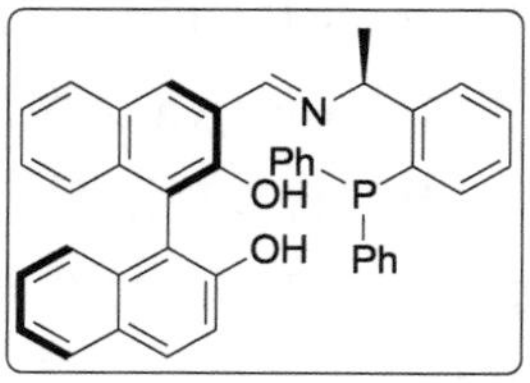

## (a) 配体的合成反应过程

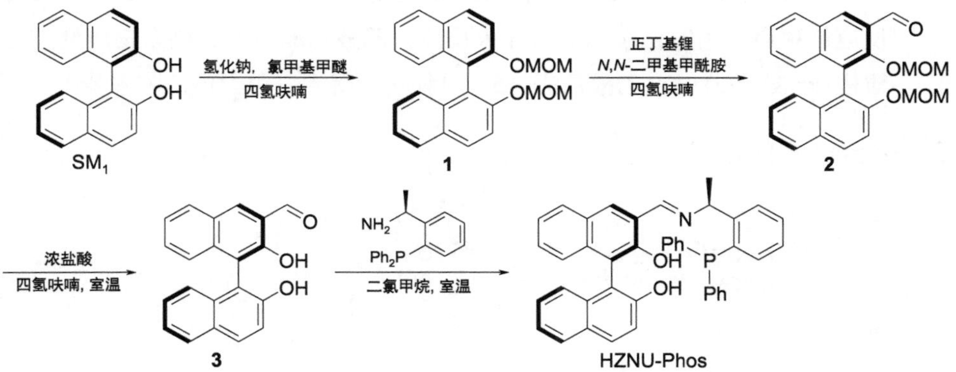

**步骤 1：化合物 1 的合成**

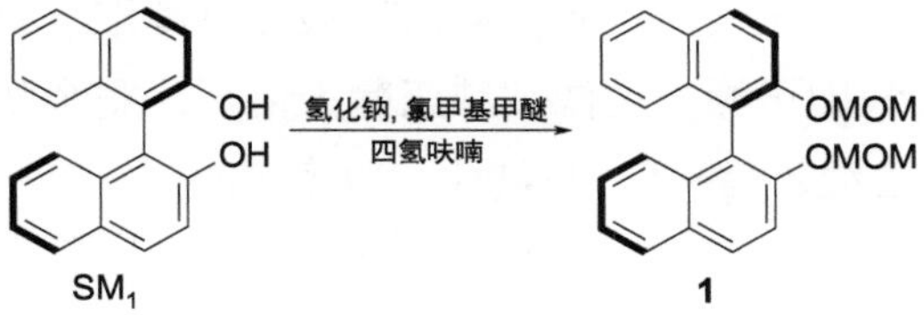

氮气保护下，将氢化钠(60%，46mmol，1.84g)、$SM_1$(20mmol，5.72g)分别加入单口瓶。在盛有 $SM_1$ 的瓶中加入 20mL THF，室温搅拌溶解。氢化钠瓶中加入 60mL 四氢呋喃，0℃下搅拌。将溶解好的 $SM_1$ 缓慢滴加于氢化钠中，0℃搅拌反应 1h。然后加入氯甲基甲醚(50mmol，3.79g)，继续搅拌 4～5h。反应结束后，加入氯化铵水溶液淬灭反应，乙酸乙酯萃取，剩余物质通过色谱柱(硅胶，石油醚：乙酸乙酯 = 10：1)纯化。得到白色固体，收率 75%。

**步骤 2：化合物 2 的合成**

氮气保护下，将化合物 **1**(10mmol，3.74g)加入四氢呋喃(40mL)中，

在-78℃条件下缓慢加入正丁基锂(12mmol, 4.8mL)反应 1h, 加入 N,N-二甲基甲酰胺(15mmol, 1.15mL)搅拌 15min 之后, 移至室温下搅拌反应 2~3h, 反应结束后, 用氯化铵水溶液淬灭反应, 乙酸乙酯萃取, 剩余物质通过色谱柱(硅胶, 石油醚:乙酸乙酯 = 5:1)纯化。得到白色固体, 收率 86%。

**步骤 3：** 化合物 3 的合成

氮气保护下, 将化合物 **2**(5.236mmol, 2.11g)加入四氢呋喃(20mL)中, 加入 6mL 浓盐酸(1mol/L), 在室温下搅拌反应 3~4h, 反应结束后, 用碳酸氢钠淬灭反应, 至无气泡产生, 加入乙酸乙酯萃取, 氯化钠饱和水溶液洗涤, 无水硫酸钠干燥, 过滤, 减压除去溶剂, 剩余物质通过色谱柱(硅胶, 石油醚:乙酸乙酯 = 5:1)纯化。收率 83%。

**步骤 4：** HZNU-Phos 的合成

氮气保护下, 取化合物 **3**(1mmol, 0.3141g)溶于 20mL 二氯甲烷, 将其升温至 50℃, 在 30min 内将(S)-1-[2-(二苯基膦)苯基]乙胺(2.5mmol, 0.6109g)溶于 10mL 乙醇中, 加入上述溶液中, 在 50℃下搅拌反应 4h, 反应结束后, 冷却至室温, 过滤除去沉淀, 加入乙醇(2mL)洗涤黄色固体, 收集滤液, 减压浓缩, 将残余物加入很少量的二氯甲烷中, 加入大量乙醇, 最终形成黄色沉淀物。收率 50%。

### (b) HZNU-Phos 配体的结构表征信息 [1-3]

$^1$H NMR(400MHz, CDCl$_3$)δ: 13.48(s, 1H), 8.13(s, 1H), 7.94~7.78(m, 3H), 7.68~7.56(m, 2H), 7.40~7.17(m, 17H), 7.16~7.05(m, 3H),

6.88~6.80(m, 1H)，5.54~5.40(m, 1H)，1.49(d, J=6.4Hz, 3H)。

**13C NMR**(100MHz，CDCl₃)δ：155.6，151.5，147.6，147.4，136.5，136.4，136.2，136.1，135.2，134.6，134.4，134.3，134.3，134.1，133.6，133.5，130.1，129.8，129.3，129.1，128.9，128.8，128.7，128.3，127.6，127.2，126.6，124.9，124.7，123.9，123.4，121.0，117.8，114.6，113.4，65.8，65.6，24.6。

**31P NMR**(162MHz，CDCl₃)δ：-15.4(s)。

**IR**，ν(cm⁻¹)：3538，3067，1631，1434，1345，1191，1131，971，815，743，696。

**HRMS**(FAB)：m/z(%)=606.2243 [M+H]⁺。

### (c) HZNU-Phos 的应用实例

　　带有多个立体中心的多功能配体 HZNU-Phos，是一类具有代表性的多官能化手性膦配体。徐利文课题组[2]将该类配体成功应用于铜催化的二乙基锌与α,β-不饱和酰基硅烷的共轭加成反应，可以得到良好对映选择性的酰基硅烷，具备中等至良好的产率，对映选择性可以达到 85%，该类多官能化 HZNU-Phos 配体，具有膦中心和 BINOL 基团的醇酚官能团，研究表明每一类官能化基团都在实现该反应的选择性控制方面发挥了至关重要的作用。

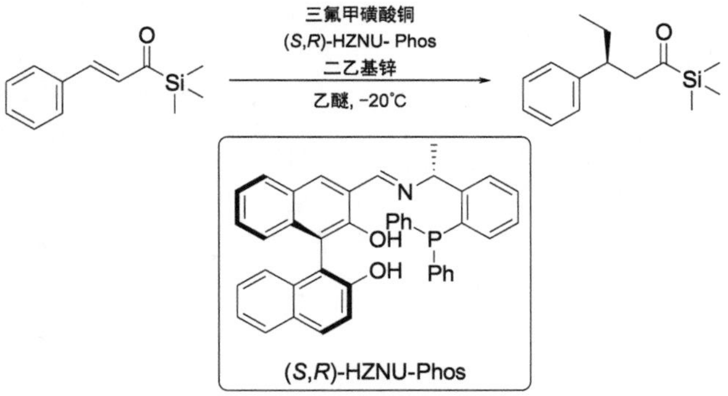

### 具体反应示例[3]：

　　氮气保护下，在干燥的史莱克管中加入三氟甲磺酸铜(3.6mg，0.01mmol，摩尔分数 2.0%)和(S,R)-HZNU-Phos(10.0mg，0.016mmol，摩尔分数 3.2%)，将混合物溶解在干燥的乙醚(3.0mL)中。在室温下搅拌溶液 30min，然后冷却至-20℃。向上述溶液中滴加二乙基锌(3.0mmol，3.0mL 1mol/L 甲苯溶液，6eq.)。然后，立即将不饱和酰基硅烷(0.5mmol)添加到透明黄色溶液中。混合物在-20℃下搅拌 9～12h，然后用饱和氯化铵水溶液淬火。分离各层，用乙酸乙酯(5mL×2)萃取水层。合并有机层，无水硫酸钠干燥，减压浓缩。残余物通过硅胶柱纯化，得到产物。通过手性 HPLC 测定产物的对映体过量。

**参考文献**

[1] Ye F, Zheng Z J, Deng W H, et al. *Chem. Asian J.*, **2013**, *8*: 2242-2253.
[2] Lv J Y, Zheng Z J, Li L, et al. *RSC Adv.*, **2017**, 7: 54934-54938.
[3] Gao P S, Zhang J L, Li N, et al. *Appl Organometal Chem.*, **2018**, *32*: e4166.

## 6.4.4　Ming-Phos 配体的合成及应用实例

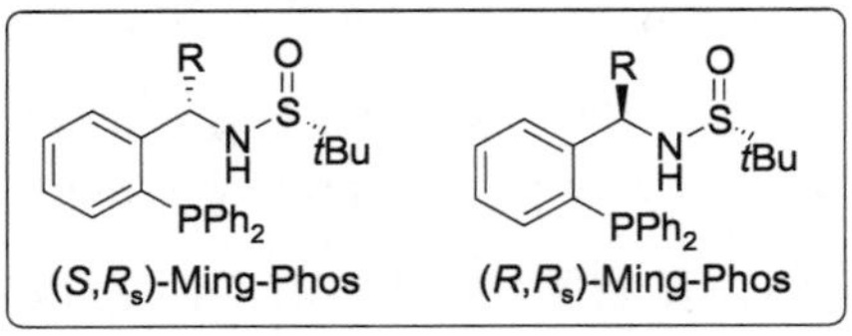

### (a) 配体的合成反应过程

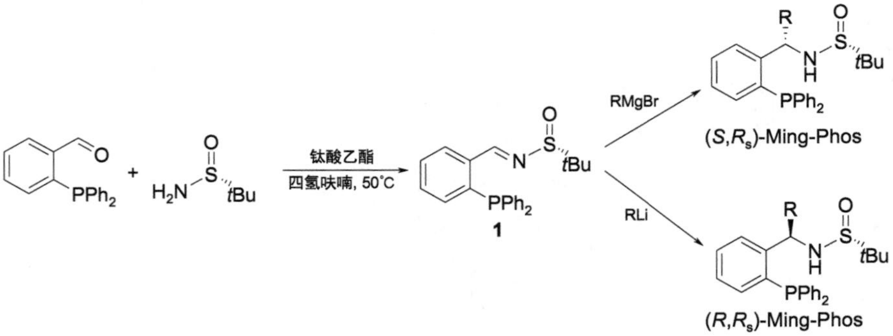

**步骤 1：**化合物 1 的合成

氮气保护下，将叔丁基亚磺酰胺(97.0mg，0.800mmol，1.0eq.)，2-二苯基膦苯甲醛(232mg，0.800mmol，1.0eq.)加入四氢呋喃(3.2mL)中将其溶解，缓慢滴加钛酸乙酯(0.37mL，1.6mmol，2.0eq.)，置于50℃油浴中反应，TLC监测至反应结束，将反应体系冷却至室温，用乙酸乙酯稀释，快速搅拌之后倒入盐水中，将所得悬浮液用硅藻土过滤，用乙酸乙酯冲洗，滤液转移到分液漏斗中，并用等量的盐水冲洗有机层，有机相用无水硫酸镁干燥，过滤，减压除去溶剂，剩余物质通过色谱柱(硅胶，正己烷:乙酸乙酯=85:15)纯化，得到产物240mg，收率77%。

**步骤 2:** ($S,R_s$)-Ming-Phos 的合成

氮气保护下，将化合物 1(1.0eq.)溶于二氯甲烷中，随后将反应体系置于-48℃下逐滴加入格氏试剂(2.0eq.)，反应 4～6h 后，将反应体系移至室温下搅拌过夜。反应完成后，反应体系用饱和氯化铵溶液淬灭，乙酸乙酯萃取，分离有机相，水相用乙酸乙酯萃取 2 次，合并有机相，无水硫酸钠干燥，过滤，减压除去溶剂，剩余物质通过色谱柱纯化。

**步骤 3:** ($R,R_s$)-Ming-Phos 的合成

氮气保护下，将化合物 1(1.0eq.)溶于甲苯中，随后将反应体系置于-78℃下逐滴加入有机锂试剂(2.0eq.)，反应 2～4h。反应完成后，0℃下向反应液中滴加氯化铵溶液淬灭，乙酸乙酯萃取，分离有机相，水相用乙酸乙酯萃取两次，合并有机相，无水硫酸钠干燥，过滤，减压除去溶剂，剩余物质通过色谱柱纯化。

## (b) Ming-Phos 配体的结构表征信息 [1]

### ($S,R_s$)-Ming-Phos

$^1$H NMR(400 MHz，CDCl$_3$)$\delta$：7.66(dd，$J$ = 7.5Hz、4.3Hz，1H)，7.40(t，$J$ = 7.2Hz，1 H)，7.34～7.29(m，3 H)，7.28～7.15(m，8 H)，7.13～6.97(m，6 H)，6.66(dd，$J$ = 8.6Hz、3.6Hz，1 H)，3.89(d，$J$ = 3.2Hz，1 H)，1.21(s，9 H)。

$^{13}$C NMR(100 MHz，CDCl$_3$)$\delta$：146.60($J_{C,P}$ = 25Hz)，141.90，136.94($J_{C,P}$ = 11Hz)，135.97，135.81，135.72，134.87，133.78($J_{C,P}$ = 14Hz)，133.59($J_{C,P}$ = 13Hz)，129.23，128.45($J_{C,P}$ = 5Hz)，128.33($J_{C,P}$ = 6Hz)，128.21($J_{C,P}$ = 4Hz)，128.16，127.86，127.85，127.59，127.24，59.84($J_{C,P}$ = 28Hz)，55.90，22.61。

$^{31}$P NMR(162 MHz，CDCl$_3$)$\delta$：-18.40。

**MS**(ESI)：$m/z$(%) = 472.1874[M+H]$^+$。

$[\alpha]_D^{20}$：-46.3($c$ = 0.50，氯仿)。

### ($R,R_s$)-Ming-Phos

$^1$H NMR(400 MHz，CDCl$_3$)$\delta$：7.63(dd，$J$ = 7.7Hz、4.3Hz，1 H)，7.35(t，$J$ = 7.5Hz，1 H)，7.29～7.11(m，11 H)，7.09～7.01(m，5 H)，6.99(dd，$J$ = 7.7Hz、4.1Hz，1 H)，6.54(dd，$J$ = 7.8Hz、4.1Hz，1 H)，4.03(d，$J$ = 3.9Hz，1 H)，1.18(s，9 H)。

$^{13}$C NMR(100 MHz，CDCl$_3$)$\delta$：147.09($J_{C,P}$ = 24Hz)，140.54，136.40($J_{C,P}$ = 11Hz)，135.81($J_{C,P}$ = 9Hz)，135.30($J_{C,P}$ = 15Hz)，134.83，133.65($J_{C,P}$ = 19Hz)，133.45($J_{C,P}$ = 20Hz)，129.44，128.45，128.30($J_{C,P}$ = 7Hz)，128.22，128.16，128.09($J_{C,P}$ = 7Hz)，127.89，127.65，127.60，126.98，59.50($J_{C,P}$ = 26Hz)，55.78，22.49。

$^{31}$P NMR(162 MHz，CDCl$_3$)$\delta$：-18.49。

**MS**(ESI)：$m/z$(%) = 472.1872[M+H]$^+$。

$[\alpha]_D^{20}$：-12.9($c$ = 0.50，氯仿)。

## (c) Ming-Phos 的应用实例

新型 Ming-Phos 膦配体在金催化 α-炔基共轭烯酮的环加成反应中表现出非常好的非对映选择性和对映选择性，以较高的产率和 ee 值得到多取代手性呋喃衍生物[2]。

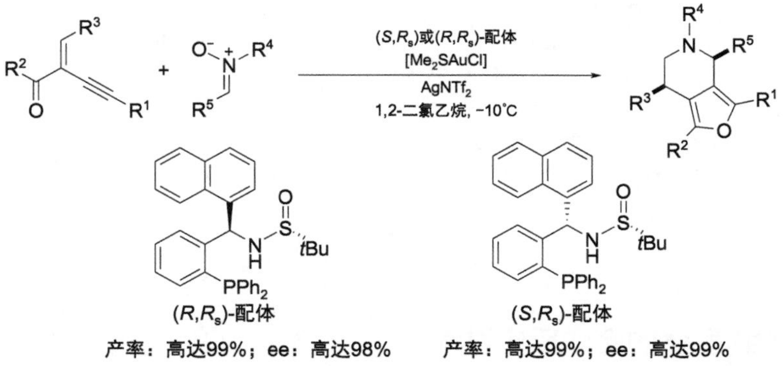

2020 年，麻生明和张俊良课题组[3]将 Ming-Phos 应用于手性烯的合成，实现了 74% 的产率和 93% 的 ee 值。

## 具体反应示例[3]：

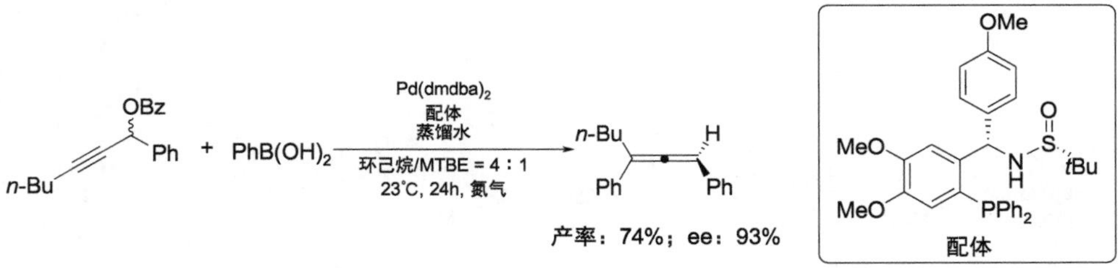

在干燥的 20mL 小瓶中加入 PhB(OH)$_2$（61.3mg，0.5mmol）和 (S, R$_s$)-Ming-Phos（34.4mg，0.06mmol），然后转移到手套箱中。加入 Pd(dmdba)$_2$（39.7mg，0.05mmol）、蒸馏水（18μL，1.0mmol）、炔烃（365.4mg，1.25mmol）、MTBE（3mL）和环己烷（12mL）后，移出含氮气的手套箱。在 23℃下继续搅拌 24h 后，反应混合物经乙酸乙酯（10mL×3）洗脱，硅胶短柱滤掉催化剂等物质。真空去除溶剂后，直接用硅胶快速柱色谱法进行分离纯化，得到 92.1mg 产物（产率 74%，ee 值 93%）。

**参考文献**

[1] Schenkel L B, Ellman J A. *Org. Lett.*, **2003**, 5: 545-548.
[2] Zhang Z M, Chen P, Li W B, et al. *Angew. Chem. Int. Ed.*, **2014**, 53: 4350-4354.
[3] Wang H N, Luo H B, Zhang Z M, et al. *J. Am. Chem. Soc.*, **2020**, 142: 9763-9771.

## 6.4.5 由金鸡纳碱衍生的手性膦酰胺配体的合成及应用实例

### (a) 配体的合成反应过程

**步骤1:** 化合物1的合成

氮气保护下,将奎尼丁(1622.3mg, 50mmol)、4-二甲氨基吡啶(31.0mg, 0.25mmol)和三乙胺(2.5mL, 17.5mmol)、对甲苯磺酰氯(1926.2mg, 10.0mmol)溶解在干燥的二氯甲烷(20mL)中,回流过夜后将反应物冷却至室温,用饱和碳酸氢钠水溶液淬灭。二氯甲烷萃取有机相,盐水洗涤,无水硫酸钠干燥,过滤,减压除去溶剂,剩余物质通过色谱柱(硅胶,乙酸乙酯:正己烷:三乙胺 = 50:50:4)纯化,得到白色固体 2216.1mg,收率 95%。

**步骤 2:** 化合物 2 的合成

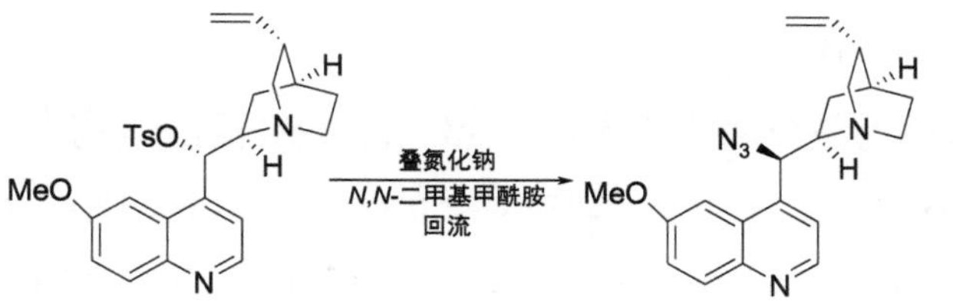

氮气保护下,将 **1**(955.3mg,2.0mmol)加入 N,N-二甲基甲酰胺(10mL)中,再将叠氮化钠(260.1mg,4.0mmol)加入,将反应混合物在 90℃下回流 5h。反应结束后,将反应混合物冷却至室温,用水淬灭。乙酸乙酯萃取,有机相用盐水洗涤后用无水硫酸钠干燥,过滤,减压除去溶剂。剩余物质通过色谱柱(硅胶,乙酸乙酯:正己烷:三乙胺 =20:80:4)纯化,得到白色固体 575.4mg,收率 82%。

**步骤 3:** 膦酰胺配体的合成

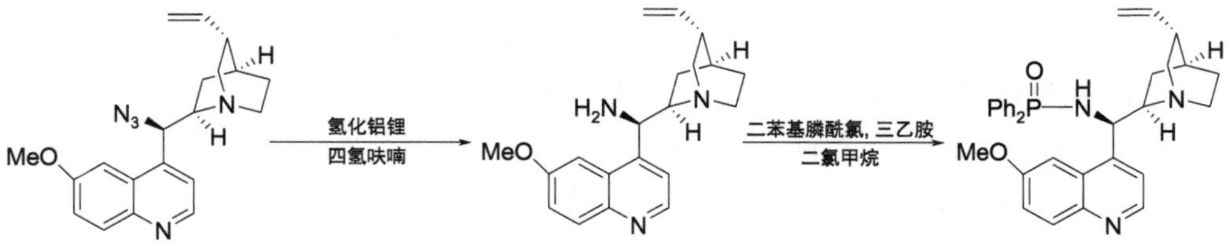

氮气保护下,将氢化铝锂(200.2mg,5.0mmol)加入四氢呋喃(4mL)中,将 **2**(699.1mg,2.0mmol)溶于四氢呋喃(8mL)中,0℃下,逐滴滴加氢化铝锂溶液,将反应体系移至室温。在室温下搅拌 2h 后,用饱和硫酸钠溶液淬灭,减压过滤,旋蒸得到残余物,将其溶于二氯甲烷/10% 盐酸(15mL/15mL)中。用二氯甲烷分离和洗涤水层,然后加入氨水调节其 pH 值至碱性。用二氯甲烷萃取得到的有机相,盐水洗涤,无水硫酸钠干燥,过滤,减压除去溶剂,剩余物质通过色谱柱(硅胶,乙酸乙酯:甲醇 =

9∶1)纯化得到黄色油状物 **3**。将所得的 **3** 和三乙胺(850μL，6.0mmol)溶于二氯甲烷(8mL)中，在 0℃下滴加二苯基膦酰氯(560μL，3.0mmol)(滴加 15min)。将混合物升温至室温，在室温下搅拌 5h 后，溶液用饱和碳酸钠水溶液淬灭。用二氯甲烷萃取得到的有机相，用盐水洗涤后旋蒸浓缩，剩余物质通过色谱柱(硅胶，乙酸乙酯∶正己烷∶三乙胺 = 60∶40∶40)纯化，得到白色固体 568.3mg，收率 53%。

(b) 金鸡纳碱衍生的手性膦酰胺配体的结构表征信息 [1]

$^1$**H NMR**(400MHz，CDCl$_3$)(外消旋体)$\delta$: 0.82～0.93(m, 1H)，1.15～1.29(m, 3H)，1.51～1.62(m, 2H)，2.23～2.41(m, 1H)，2.91～3.12(m, 3H)，3.18～3.30(m, 1H)，3.88(s, 2.2H)，4.00(s, 0.8H)，4.53(t, 0.3H)，5.02～5.19(m, 2H)，5.18～5.25(m, 0.7H)，5.40～5.51(m, 1H)，5.73～5.99(m, 1H)，6.71～6.78(m, 0.7H)，6.79～6.85(m, 1.3H)，6.93～7.02(m, 1H)，7.03～7.08(m, 1H)，7.18～7.23(m, 1H)，7.30～7.45(m, 5H)，7.64(d, 0.6H)，7.76～7.91(m, 3.1H)，7.99～8.02(m, 0.3H)，8.26(d, 0.3H)，8.63(d, 0.7H)。

$^{13}$**C NMR**(100MHz，CDCl$_3$)(外消旋体)$\delta$: 24.7, 26.3, 27.3, 38.4, 46.0, 49.1, 55.3, 59.7, 61.2, 100.2, 103.9, 114.9, 120.5, 120.7, 121.8, 123.2, 126.9, 127.2, 128.2, 128.4, 130.6, 131.1, 131.5, 131.7, 131.9, 139.8, 140.1, 144.1, 145.7, 146.8, 147.2, 157.4。

**HRMS**(ESI)：$m/z$(%)= 524.2461[M+H]$^+$。

$[\alpha]_D^{22.6}$: 84.4($c$ = 1.00，二氯甲烷)。

**mp**：55～56℃。

**参考文献**

[1] Shen B, Huang H Y, Bian G L, et al. *Chirality*., **2013**, *25*: 561-566.

## 6.4.6　Fei-Phos 配体的合成及应用实例

### (a) 配体的合成反应过程

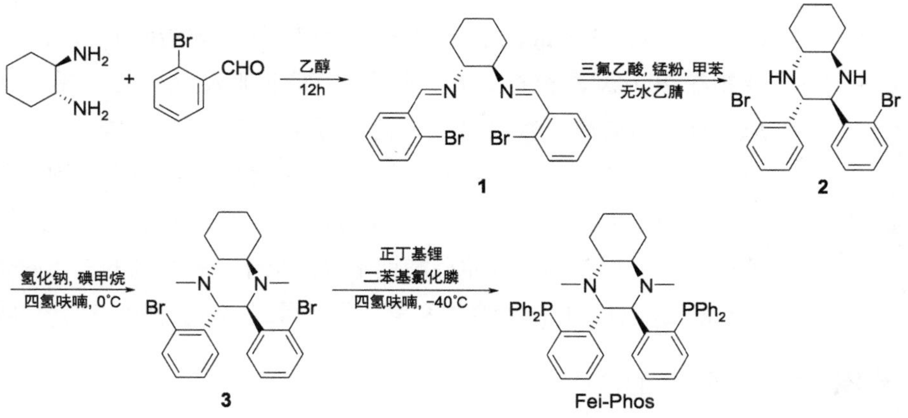

**步骤 1：化合物 1 的合成**

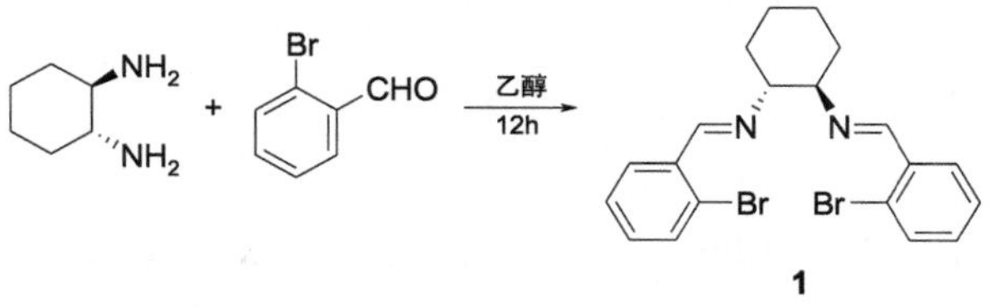

将邻溴苯甲醛(9.2g，50mmol，2.0eq.)溶于乙醇(25mL)，在室温下滴加到反式二氨基环己烷(2.85g，25mmol，1.0eq.)的乙醇/蒸馏水(1.5/1)溶液(50mL)中，并于 1h 内滴加完。随后在回流温度下反应 12h，将蒸馏水(30mL)加入混合物中并将反应体系冷却至 0℃。在 0℃下保持 5h 后，过滤沉淀物，用冷却的乙醇和蒸馏水洗涤滤渣，滤液蒸发溶剂后得到油状黏稠化合物 8.92g，收率 78%。

**步骤 2：化合物 2 的合成**

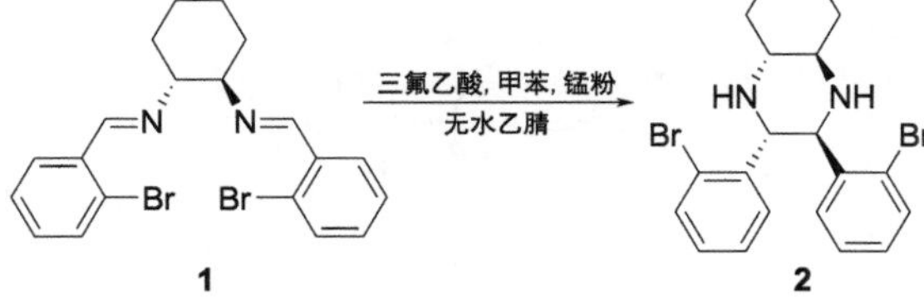

氮气保护下，将化合物 **1** 溶于无水乙腈(200mL)和甲苯(20mL)的混合物中，并向该混合体系中加入锰粉(300目，2.2g，40mmol)，冷却至 0℃，然后在 30min 内滴加三氟乙酸(7.6mL，80mmol，2eq.)。将反应混合物在 20℃剧烈搅拌 24h，随后在 0℃下加入 2eq. 的三氟乙酸，继续反应。反应结束后，静置 2h，过滤所得混合物，用石油醚(10mL×2)洗涤残余物，得到白色固体。将固体溶解在蒸馏水(10mL)中并用饱和碳酸氢钠溶液调节 pH 值至 8。用二氯甲烷(20mL×3)萃取水溶液。将合并的有机层用蒸馏水(10mL)洗涤，无水硫酸钠干燥，浓缩，将得到的白色固体用石油醚和乙酸乙酯(1∶1)的混合体系重结晶，得到无色针状物 5.37g，收率 60%。

### 步骤 3：化合物 3 的合成

将化合物 **2**(5.37g，12mmol，1eq.)溶于四氢呋喃(20mL)中，在 0℃下滴加到含氢化钠(80%，0.86g，28.8mmol)的四氢呋喃溶液中，搅拌 3h。随后将碘甲烷(1.86mL，30mmol，2.5eq.)加入上述混合物中，滴加结束后反应移至室温下搅拌 12h。反应完成后，加入 1mol/L 亚硫酸氢钠水溶液淬灭反应。用二氯甲烷(20mL×3)萃取水相，合并的有机层用硫酸钠干燥并减压浓缩。通过硅胶柱色谱法(石油∶乙酸乙酯＝5∶1 ～ 20∶1)纯化残余物，得到白色固体 **3**(4.85g，收率 85%)。

### 步骤 4：Fei-Phos 的合成

氮气保护下，将化合物 **3**(4.85g，10mmol，1.0eq.)溶于 20mL 四氢呋

喃，在-40℃下将2.5mol/L的正丁基锂(10mL，25mmol，2.5eq.)缓慢加入上述混合体系中。将反应体系在-40℃下搅拌1h，然后滴加二苯基氯化膦(4.5mL，25mmol)的四氢呋喃(20mL)溶液。将反应混合物在-40℃搅拌3h，然后升至室温搅拌4h。反应结束后，缓慢加入1mol/L氯化铵水溶液，直至反应混合物在两相中变澄清。用二氯甲烷(20mL×3)萃取水相，合并有机层用硫酸钠干燥并减压浓缩。将残渣用硅胶柱色谱法(石油：乙酸乙酯＝5∶1～20∶1)纯化，得到白色固体5.51g，收率80％。

### (b) Fei-Phos 的结构表征信息 [1]

$^1$H NMR(400 MHz，CDCl$_3$)$\delta$：7.38～6.99(m，21H)，6.95～6.70(m，4H)，6.61～6.51(m，3H)，4.88(d，$J$ = 6.0Hz，2H)，2.25～2.00(m，4H)，1.75～1.67(m，2H)，1.61(s，6H)，1.28～1.23(m，2H)，1.18～1.10(m，2H)。
$^{13}$C NMR(100 MHz，CDCl$_3$)$\delta$：147.2，146.9，138.8，138.6，137.2，136.4，134.8，134.3，134.1，132.7，132.5，130.3，129.7，128.5，128.3，128.2，127.9，127.3，126.8，67.5，39.3，29.6，26.9，24.9。
$^{31}$P NMR(202 MHz，CDCl$_3$)$\delta$：-20.5(s)。
IR $\nu$(cm$^{-1}$)：2923.3，2851.2，2782.8，1464.9，1432.8，1307.0，1077.8，1006.9，761.8，740.1，694.2。
HRMS(FAB)：$m/z$(%) = 688.3213 [M + H]$^+$。
mp：89～93℃。
$[\alpha]_D^{25}$：+106.3($c$ = 1.0，氯仿)。

### (c) Fei-Phos 的应用实例

2013年，徐利文课题组设计合成一种新型多官能化的多齿膦配体CycloN$_2$P$_2$-Phos，也称之为Fei-Phos[1]。该类配体在钯催化的烯丙基化反应中表现优异，如在钯催化的脂肪醇或硅烷醇的不对称烯丙基醚化(AAE)中，可得到优异的收率和最高水平的对映选择性(最高99％)[2,3]。

2018 年，徐利文课题组在构筑硅手性中心的硅氢加成/醇解反应中，发现已有的商品化手性膦配体均无法解决其硅手性中心的选择性控制难题，仅有 Fei-Phos 具备一定的对映选择性（高达 32%）[4]。

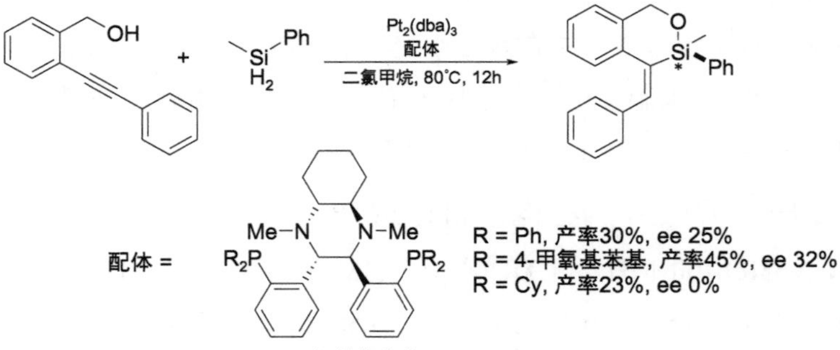

配体 = Fei-Phos 与其衍生物

R = Ph，产率30%，ee 25%
R = 4-甲氧基苯基，产率45%，ee 32%
R = Cy，产率23%，ee 0%

**具体反应示例**[4]：

将无水二氯甲烷（4mL）加入 $Pt_2(dba)_3$（4.4mg，0.004mmol）和手性配体 Fei-Phos（摩尔分数 2%）的混合物中，在室温下搅拌 30min，而后加入 2-苯乙炔基苄醇（0.4mmol）和甲基苯基硅烷（0.48mmol）。在 80℃下搅拌 12h 后冷却至室温，用乙醚稀释并用硅藻土过滤，有机相用无水硫酸钠干燥，浓缩后用硅胶柱色谱法（石油醚：乙酸乙酯 = 50:1）分离纯化，得到的产物用高效液相色谱分析其对映选择性。

**参考文献**

[1] Ye F, Zheng Z J, Li L, et al. *Chem. Eur. J.*, **2013**, *19*: 15452-15457.
[2] Xu J X, Ye F, Bai X F, et al. *RSC Adv.*, **2016**, *6*: 45495-45502.
[3] Xu J X, Ye F, Bai X F, et al. *RSC Adv.*, **2016**, *6*: 70624-70631.
[4] Long P W, Bai X F, Ye F, et al. *Adv. Synth. Catal.*, **2018**, *360*: 2825-2830.

## 6.4.7 SIOCPhox 配体的合成及应用实例

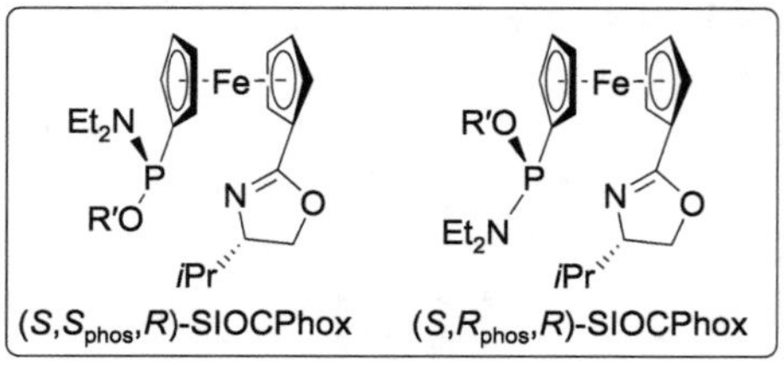

($S,S_{phos},R$)-SIOCPhox    ($S,R_{phos},R$)-SIOCPhox

## (a) 配体的合成反应过程

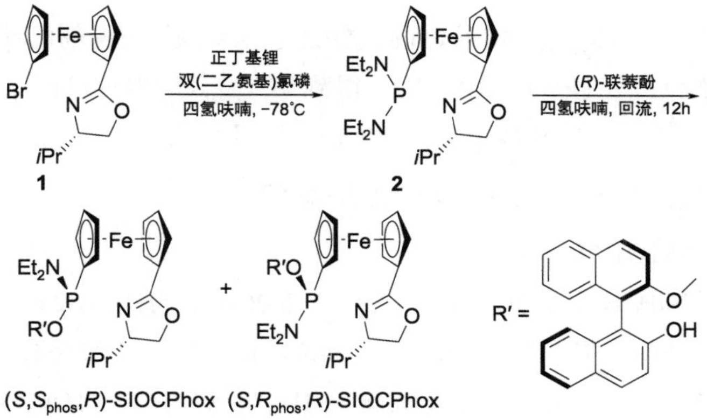

**步骤 1：化合物 2 的合成**

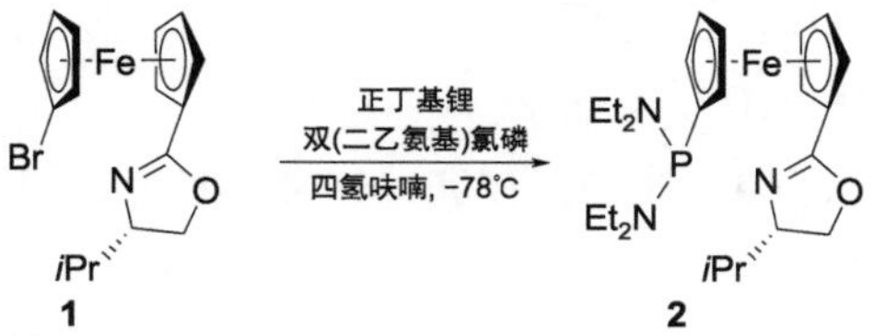

氩气保护下，将化合物 **1**(2.26g，6mmol) 加入四氢呋喃(40mL) 中，随后将反应体系置于-78℃下缓慢滴加正丁基锂(4.2mL，6.6mmol，1.6mol/L)并将所得深红色溶液搅拌 20min。然后加入双(二乙氨基)氯磷(1.7mL，8mmol)，持续搅拌所得混合物并在 30min 内升温至室温。反应混合物用乙醚(20mL)稀释，用蒸馏水、盐水洗涤并经无水硫酸钠干燥，过滤，减压除去溶剂，剩余物质通过柱色谱法(硅胶，石油醚：乙酸乙酯：三乙胺 = 10:1:1)纯化，得到橙色油状液体 1.75g，收率 62%。

**步骤 2：SIOCPhox 的合成**

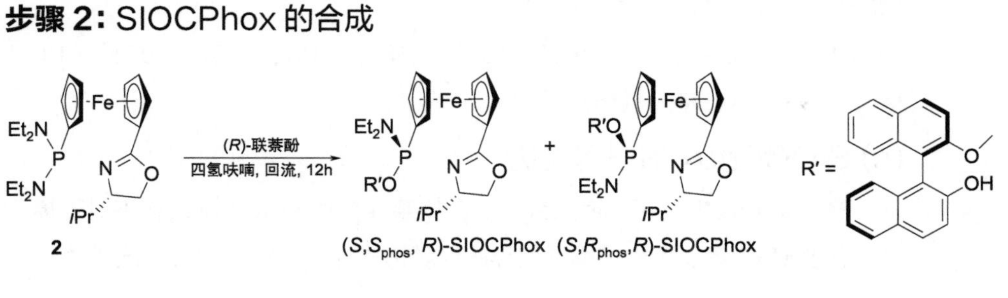

6　手性膦配体的合成及应用实例

氮气保护下，将化合物 **2**(471mg，1mmol)和(*R*)-联萘酚(286mg，1mmol)加入四氢呋喃(40mL)中，加热回流反应 12h。反应结束后，减压除去溶剂，得到的粗产物通过硅胶快速色谱纯化(洗脱剂配比为乙酸乙酯：石油醚：三乙胺=1：10：1)，得到(*S*,*R*$_{phos}$,*R*)-SIOCPhox(274mg，收率 40%)和(*S*,*S*$_{phos}$,*R*)-SIOCphox(280mg，收率 41%)。

### (b) SIOCPhox 配体的结构表征信息[1]

#### (*S*,*S*$_{phos}$,*R*)-SIOCPhox

$^1$**H NMR**(400MHz，CDCl$_3$)$\delta$：7.83～8.06(m，5H)，7.21～7.37(m，7H)，5.21(br，1H)，4.55(m，1H)，4.40(m，1H)，4.22(dd，*J*=8.5Hz、9.2Hz，1H)，4.13(s，1H)，4.08(s，1H)，3.87～4.04(m，4H)，3.71(m，1H)，3.38(s，1H)，2.85(t，*J*=7.5Hz，4H)，1.82(m，1H)，0.98(d，*J*=6.8Hz，3H)，0.91(d，*J*=6.7Hz，3H)，0.75(t，*J*=7.0Hz，6H)。

$^{31}$**P NMR**(161.92 MHz，CDCl$_3$)$\delta$：127.87。

**MS**(ESI)：*m/z*(%)= 684 [M+H]$^+$。

**IR**(KBr)$\nu$(cm$^{-1}$)：3056，2960，1645，1506，1465，1237，1127。

$[\alpha]_D^{25}$：-410(*c* = 0.37，氯仿)。

#### (*S*,*R*$_{phos}$,*R*)-SIOCPhox

$^1$**H NMR**(400 MHz，CDCl$_3$)$\delta$：9.90(br，1H)，7.79～7.99(m，5H)，7.12～7.41(m，7H)，5.17(t，*J* = 1.2Hz，1H)，4.57(t，*J* = 1.1Hz，1H)，4.31～4.43(m，4H)，4.20～4.26(m，1H)，4.11(t，*J* = 8.0Hz，1H)，4.06(m，1H)，4.00(m，1H)，3.77(m，1H)，2.38～2.61(m，4H)，1.84(m，1H)，1.00(d，*J* = 6.8Hz，3H)，0.92(d，*J* = 6.7*Hz*，3H)，0.50(t，*J* = 6.9Hz，6H)。

$^{31}$**P NMR**(161.92 MHz，CDCl$_3$)$\delta$：117.94。

**MS**(ESI)：*m/z*(%)= 684 [M+H]$^+$。

**IR**(KBr)$\nu$(cm$^{-1}$)：3051，2962，1640，1589，1461，1232，1024，810。

$[\alpha]_D^{25}$：493(*c* = 0.54，氯仿)。

### (c) SIOCPhox 的应用实例

戴立信和侯雪龙等报道的系列二茂铁手性配体 SIOCPhox，在不对称烯丙基取代及 Heck 等反应中取得了优异的区域选择性、非对映和对映选择性[2,3]。

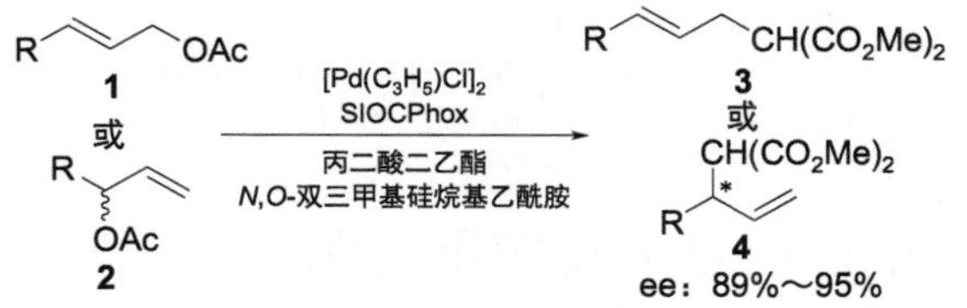

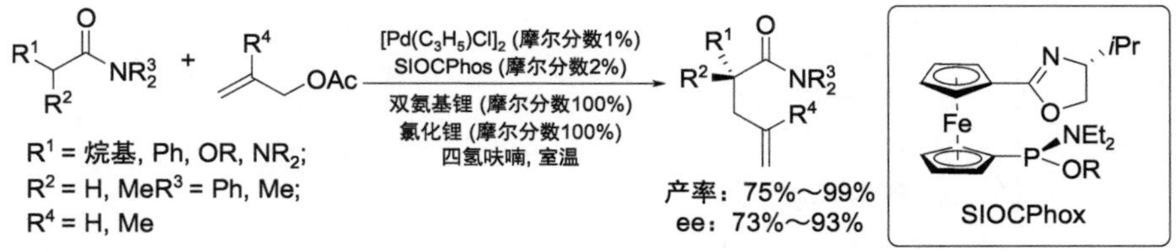

2018年侯雪龙课题组[4]以SIOCPhox为手性配体，研究了钯催化非环酰胺与单取代的烯丙基碳酸酯或多烯基碳酸酯的反应，可得到具有3个手性中心的环丙烷衍生物，其dr值为60%～92%，ee值为83%～97%。

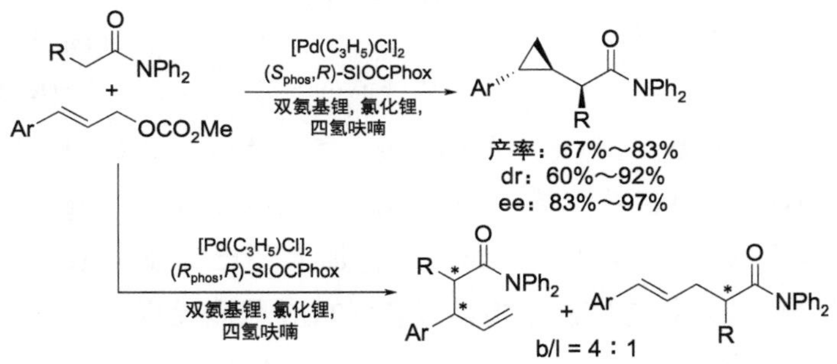

**具体反应示例**[4]：

将含有氯化锂（4.25mg，0.10mmol）的干燥史莱克管进行火焰干燥并用氩气置换后，依次加入酰胺（0.10mmol）、[Pd($\eta^3$-C$_3$H$_5$)Cl]$_2$（0.92mg，0.0025mmol）、手性配体($S_{phos}$,R)-SIOCPhox（3.3mg，0.005mmol）和甲基碳酸烯丙酯（0.115mmol）。然后，在0℃下依次加入四氢呋喃（3mL）和双氨基锂（1.0mol/L 四氢呋喃溶液，0.12mL，0.12mmol）。在0℃下搅拌10min后在20℃下继续搅拌。反应完成后（TLC监测，约4h），真空浓缩反应混合物，得到油状粗品（利用GC测定区域选择性和非对映选择性），用硅胶柱色谱法分离得到混合产物，HPLC测定其对映选择性。可通过重结晶或制备型HPLC得到单一的环丙烷产物。

**参考文献**

[1] You S L, Zhu X Z, Luo Y M, et al. *J. Am. Chem. Soc.*, **2001**, 123: 7471-7472.

[2] Dai L X, Tu T, You S L, et al. *Acc. Chem. Res.*, **2003**, *36*: 659-667.
[3] Zhang K, Peng Q, Hou X L, et al. *Angew. Chem. Int. Ed.*, **2008**, *47*: 1741-1744.
[4] Huang J Q, Liu W Y, Zheng B H, et al. *ACS Catal.*, **2018**, *8*: 1964-1972.

### 6.4.8 Trost 配体的合成及应用实例

#### (a) 配体的合成反应过程

将 2-(二苯基膦基)苯甲酸(4.83g, 15.78mmol, 2.2eq.)、4-二甲氨基吡啶和 1-(3-二甲氨基丙基)-3-乙基碳二亚胺盐酸盐(3.30g, 17.19mmol, 2.4eq.)溶于无水二氯甲烷(50mL)中，加入环己二胺(0.82g, 7.17mmol, 1.0eq.)，将混合物在室温下搅拌过夜。TLC 监测反应，反应完成后，向反应体系中加入乙醚(100mL)，有机层依次用 10% 的盐酸(100mL×3)、水(100mL×1)、饱和碳酸氢钠水溶液(100mL×3)、水(100mL×1)洗涤，无水硫酸镁干燥并真空浓缩得到浅棕色固体(4.5g，产率为 92%)。再用乙腈重结晶得到白色固体(3.66g，产率 74%)。

#### (b) Trost 配体的结构表征信息[1]

$^1$**H NMR**(200 MHz, CDCl$_3$)$\delta$: 7.58～7.48(m, 2H), 7.26～7.16(m, 24H) 6.94～6.88(m, 2H), 6.45(bd, $J$ = 7.3Hz, 2H), 3.87～3.70(m, 2H), 1.90～1.78(m, 2H), 1.70～1.58(m, 2H), 1.30～0.90(m, 4H);

$^{13}$**C NMR**(50 MHz, CDCl$_3$)$\delta$: 169.2, 140.7, 137.7, 136.6, 134.2, 133.8, 130.1 128.7, 128.5, 128.4, 127.5, 53.8, 31.9, 24.6。

$^{31}$**P NMR**(162 MHz, CDCl$_3$)$\delta$: -8.3(s)。

$[\alpha]_D^{24}$: +55.6($c$ = 2.3, 二氯甲烷)。

**mp**: 136～139℃。

## (c) Trost 配体的应用实例

Trost 配体是不对称催化烯丙基化反应的优势配体之一，有大量的报道显示其具有非常好的手性诱导能力。最近 Trost 教授课题组[2]报道了钯催化的苯并或非苯并稠合的 δ-戊内酰胺化合物的脱羧不对称烯丙基烷基化(Pd-DAAA)，实现了一步构建含有 C3-立体中心的手性内酰胺，具有很好的对映选择性(高达 99%)。

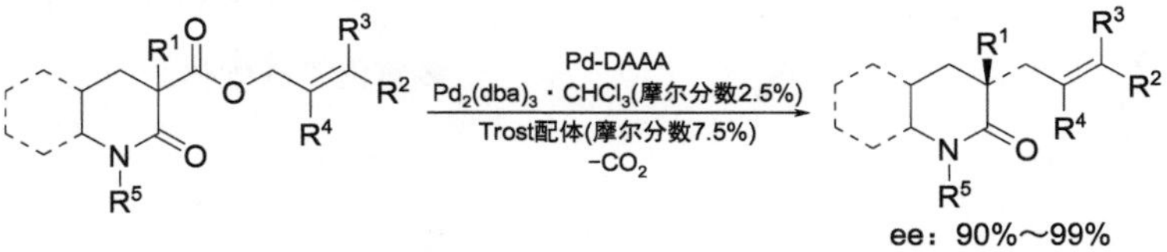

ee: 90%~99%

2020 年，钟国富课题组[3]将 Trost 配体应用于不对称烯丙基脱芳烷基化反应，得到 81% 的产率和 84% 的 ee 值。

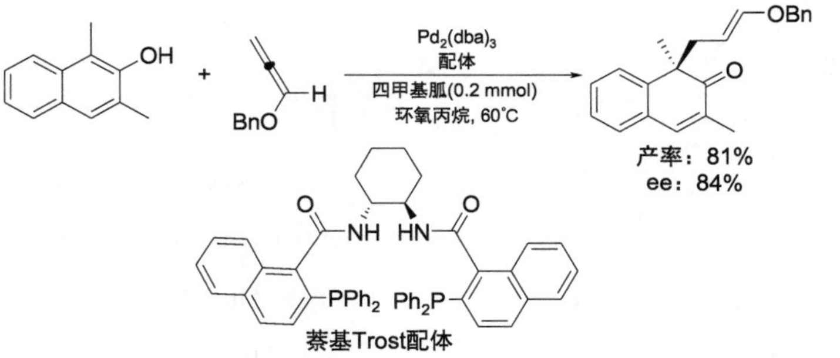

产率：81%
ee：84%

萘基Trost配体

**具体反应示例[3]：**

在惰性气体保护下，向 5mL 密闭管中加入溶于 2mL 环氧丙烷的 0.2mmol 1,3-二甲基-β-萘酚和 0.3mmol 苄氧基联烯，再分别加入 Pd$_2$(dba)$_3$(摩尔分数 2.5%)、四甲基胍(TMG，0.2mmol)和 Trost 配体(摩尔分数 5%)，在 60℃油浴中反应至反应完全(TLC 监测)。所得到的反应混合物在减压下浓缩，再用快速柱色谱法分离纯化(洗脱剂为乙酸乙酯/正己烷)，得到含有一个季碳手性中心的产物。

**参考文献**

[1] Fuchs S, Berl V, Lepoittevin J P. *Eur. J. Org. Chem.*, **2007**, 1145-1152.
[2] Trost B M, Nagaraju A, Wang F J, et al. *Org. Lett.*, **2019**, *21*: 1784-1788.
[3] Hu J X, Pan S L, Zhu S, et al. *J. Org. Chem.*, **2020**, *85*: 7896-7904.

# 索引

## A
氨基酸 .................................................. 233

## B
钯催化 .................................................. 085
吡咯烷类化合物 ...................................... 240
丙二烯化反应 .......................................... 133
不对称催化 ............................................ 014
不对称催化反应 ...................................... 006
不对称合成 ............................................ 129
不对称环氧化反应 .................................. 008
不对称氢化反应 ...................................... 002
不对称双羟化 ........................................ 175

## C
稠环化合物 ............................................ 124
串联加成反应 ........................................ 034

## D
单膦配体 ................................................ 002
对映体过量 ............................................ 007
对映选择性 ............................................ 125

多官能化 ................................................ 232
多官能 P,N 配体 ...................................... 233
多官能 P,O 配体 ...................................... 238

## E
二茂铁 .................................................. 144
1,6-二烯环化 .......................................... 037

## F
Aldol 反应 .............................................. 111
Aza-Heck 反应 ........................................ 037
Diels-Alder 反应 .................................... 117
ene 反应 ................................................ 126
Mannich 反应 ........................................ 107
Pauson-Khand 反应 ................................ 114
$\alpha$-芳基化 ...................................... 020
$N$-芳基亚磷酰胺配体 .......................... 307
芳香杂环化合物 .................................... 004
非等价配位 ............................................ 232
非共价相互作用 .................................... 232
分子间氢键 ............................................ 232

## G

格氏试剂 ... 089

共轭加成反应 ... 030

共轭作用 ... 232

钴催化 ... 114

硅氢还原 ... 073

硅氢还原反应 ... 073

硅手性 ... 252

过渡金属 ... 006

## H

含 S 多齿单膦配体 ... 239

还原偶联反应 ... 081

环丙烷化反应 ... 034

环化二聚反应 ... 129

[2+2] 环加成反应 ... 241

[4+2] 环加成反应 ... 028

Huisgen 环加成反应 ... 252

环异构化反应 ... 128

## J

DFT 计算 ... 235

季碳手性中心 ... 085

C—H 键活化反应 ... 024

C—N 键偶联反应 ... 021

Si—C 键偶联反应 ... 045

金催化 ... 125

金鸡纳生物碱 ... 233

金属催化 ... 096

酒石酸衍生的二醇 ... 030

## K

开环反应 ... 021

## L

铑催化 ... 003

Taddol 类亚磷酰胺配体 ... 301

联苯二酚 ... 030

联萘二酚 ... 030

钌 ... 003

膦 - 杂原子双齿配体 ... 168

膦/氮配体 ... 168

膦/硫配体 ... 199

膦/氧配体 ... 203

磷手性中心配体 ... 026

卤代反应 ... 134

螺环二酚 ... 030

## N

纳米氧化亚铜 ... 252

镍催化 ... 142

# O

1,3-偶极环加成反应 ...................... 125

A3 偶联反应 ................................. 218

Kumada 偶联反应 ....................... 016

Suzuki-Miyaura 偶联反应 ............ 026

# P

BINAP 配体 ................................. 313

Binaphane 配体 ........................... 344

Binaphos 配体 ............................. 353

C$_3$-TunePhos 配体 ...................... 327

DIOP 配体 ................................... 341

DIPAMP 配体 ............................. 350

DuanPhos 配体 ........................... 331

DuPhos 配体 ............................... 338

Fei-Phos 配体 .............................. 402

Feringaphos 配体 ........................ 309

HZNU-Phos 配体 ........................ 392

Josiphos 配体 .............................. 347

Me-BI-DIME 配体 ...................... 304

Ming-Phos 配体 .......................... 396

Monophosphine-olefin 配体 ......... 292

MOP 配体 ................................... 362

Ph-BINEPINE 配体 ..................... 288

PHOX 配体 ................................. 366

P,N 配体 ..................................... 184

P,N,N 配体 .................................. 249

QuinoxP 配体 .............................. 335

SDP 配体 .................................... 319

SegPhos 配体 .............................. 316

SIOCPhox 配体 ........................... 406

SIPHOX 配体 .............................. 368

SKP 配体 .................................... 323

Spiro Phosphino-oxazine 配体 ..... 378

SPO 配体 .................................... 295

Tao-Phos 配体 ............................. 387

TF-BiphamPhos 配体 .................. 374

Trost 配体 ................................... 410

Xing-Phos 配体 ........................... 383

配位反应 ..................................... 006

平面手性 ..................................... 172

# Q

亲核加成反应 .............................. 027

亲核试剂 ..................................... 092

氢胺化反应 ................................. 077

氢硅化反应 ................................. 032

氢化联萘二酚 ............................. 030

氢磺酰化 ..................................... 038

氢甲酰化 ..................................... 017

氢甲酰化反应 ............................. 149

氢硫醇化反应 ............................. 079

氢硼化 ......................................... 079

氢硼化反应 ................................. 003

氢烷氧化反应 ................................. 078

氢酰化反应 ..................................... 127

氢酰基化反应 ................................. 036

去对称化 ......................................... 130

炔丙基化反应 ................................. 081

## S

三齿酰胺膦配体 ............................. 235

手性口袋 ......................................... 263

手性膦配体 ..................................... 002

手性螺环 ......................................... 140

手性螺环化合物 ............................. 125

手性配体 ......................................... 002

手性诱导作用 ................................. 171

手性杂环化合物 ............................. 243

双核铜 ............................................. 252

双膦配体 ......................................... 002

## T

碳 - 氢键活化 ................................. 019

碳 - 碳键活化反应 ......................... 043

羰基化反应 ..................................... 028

羰基化合物 ..................................... 004

铜催化 ............................................. 103

脱羧基化反应 ................................. 115

## W

烷基化反应 ..................................... 101

Friedel-Crafts 烷基化反应 ............. 106

## X

烯丙基化 ......................................... 003

烯丙基化反应 ................................. 020

烯醇硅醚 ......................................... 101

烯烃 ................................................. 004

新骨架 ............................................. 003

P,P,N,N 型多齿配体 ...................... 264

N,P,N 型手性配体 .......................... 250

N,P,O 型手性配体 .......................... 257

N,P,P,N 型多齿配体 ...................... 270

P,N,N 型手性配体 .......................... 246

P,N,N,P 型配体 .............................. 266

P,N,O 型手性配体 .......................... 253

P,N,P 型手性配体 .......................... 260

P,N,S 型手性配体 .......................... 257

P,O,O 型手性配体 .......................... 250

P,P,P 型三齿配体 ........................... 272

## Y

亚胺化合物 ..................................... 004

亚次膦酰胺 ..................................... 179

亚砜 ................................................. 194

亚磷酸酯配体 .................................. 038

亚磷酰胺酯 ...................................... 038

亚膦酰胺酯 ...................................... 180

铱催化 ............................................... 072

异构化反应 ...................................... 107

异构体 ............................................... 124

阴离子交换 ...................................... 232

银催化 ............................................... 244

银配合物 .......................................... 112

有机硅 ............................................... 089

有机硼 ............................................... 089

有机钛 ............................................... 089

有机锌 ............................................... 089

## Z

轴手性 ...................................... 062, 171

转移氢化反应 .................................. 019

唑啉 ................................................... 169

## 其他

BINAP ............................................... 054

TADDOL ........................................... 038

Xing-Phos ........................................ 244

# 元素周期表

IUPAC 2013

氧化态(单质的氧化态为0，未列入；常见的为红色) — +2 +3 +4 +5 +6
95 — 原子序数
Am — 元素符号(红色的为放射性元素)
镅▲ — 元素名称(注▲的为人造元素)
$5f^{7}7s^{2}$ — 价层电子构型
以 $^{12}C=12$ 为基准的原子量(注✦的是半衰期最长同位素的原子量) — 243.06138(2)✦

s区元素 | p区元素 | d区元素 | ds区元素 | f区元素 | 稀有气体

| 族 / 周期 | 1<br>ⅠA | 2<br>ⅡA | 3<br>ⅢB | 4<br>ⅣB | 5<br>ⅤB | 6<br>ⅥB | 7<br>ⅦB | 8<br>ⅧB(Ⅷ) | 9<br>ⅧB(Ⅷ) | 10<br>ⅧB(Ⅷ) | 11<br>ⅠB | 12<br>ⅡB | 13<br>ⅢA | 14<br>ⅣA | 15<br>ⅤA | 16<br>ⅥA | 17<br>ⅦA | 18<br>ⅧA(0) | 电子层 |
|---|---|---|---|---|---|---|---|---|---|---|---|---|---|---|---|---|---|---|---|
| 1 | −1 +1<br>1 **H** 氢<br>$1s^{1}$<br>1.008 | | | | | | | | | | | | | | | | | 2 **He** 氦<br>$1s^{2}$<br>4.002602(2) | K |
| 2 | +1<br>3 **Li** 锂<br>$2s^{1}$<br>6.94 | +2<br>4 **Be** 铍<br>$2s^{2}$<br>9.0121831(5) | | | | | | | | | | | +3<br>5 **B** 硼<br>$2s^{2}2p^{1}$<br>10.81 | −4 +2 +4<br>6 **C** 碳<br>$2s^{2}2p^{2}$<br>12.011 | −3 −2 −1 +1 +2 +3 +4 +5<br>7 **N** 氮<br>$2s^{2}2p^{3}$<br>14.007 | −2 −1<br>8 **O** 氧<br>$2s^{2}2p^{4}$<br>15.999 | −1<br>9 **F** 氟<br>$2s^{2}2p^{5}$<br>18.998403163(6) | 10 **Ne** 氖<br>$2s^{2}2p^{6}$<br>20.1797(6) | L K |
| 3 | −1 +1<br>11 **Na** 钠<br>$3s^{1}$<br>22.98976928(2) | +2<br>12 **Mg** 镁<br>$3s^{2}$<br>24.305 | | | | | | | | | | | +3<br>13 **Al** 铝<br>$3s^{2}3p^{1}$<br>26.9815385(7) | −4 +2 +4<br>14 **Si** 硅<br>$3s^{2}3p^{2}$<br>28.085 | −3 −2 +1 +3 +5<br>15 **P** 磷<br>$3s^{2}3p^{3}$<br>30.973761998(5) | −2 +2 +4 +6<br>16 **S** 硫<br>$3s^{2}3p^{4}$<br>32.06 | −1 +1 +3 +5 +7<br>17 **Cl** 氯<br>$3s^{2}3p^{5}$<br>35.45 | 18 **Ar** 氩<br>$3s^{2}3p^{6}$<br>39.948(1) | M L K |
| 4 | −1 +1<br>19 **K** 钾<br>$4s^{1}$<br>39.0983(1) | +2<br>20 **Ca** 钙<br>$4s^{2}$<br>40.078(4) | +3<br>21 **Sc** 钪<br>$3d^{1}4s^{2}$<br>44.955908(5) | −1 0 +2 +3 +4<br>22 **Ti** 钛<br>$3d^{2}4s^{2}$<br>47.867(1) | 0 ±1 +2 +3 +4 +5<br>23 **V** 钒<br>$3d^{3}4s^{2}$<br>50.9415(1) | −3 0 ±1 ±2 +3 +4 +5 +6<br>24 **Cr** 铬<br>$3d^{5}4s^{1}$<br>51.9961(6) | −2 0 ±1 +2 ±3 +4 +5 +6 +7<br>25 **Mn** 锰<br>$3d^{5}4s^{2}$<br>54.938044(3) | −2 0 ±1 +2 +3 +4 +5 +6<br>26 **Fe** 铁<br>$3d^{6}4s^{2}$<br>55.845(2) | 0 ±1 +2 +3 +4 +5<br>27 **Co** 钴<br>$3d^{7}4s^{2}$<br>58.933194(4) | 0 ±1 +2 +3 +4<br>28 **Ni** 镍<br>$3d^{8}4s^{2}$<br>58.6934(4) | +1 +2 +3 +4<br>29 **Cu** 铜<br>$3d^{10}4s^{1}$<br>63.546(3) | +1 +2<br>30 **Zn** 锌<br>$3d^{10}4s^{2}$<br>65.38(2) | +1 +3<br>31 **Ga** 镓<br>$4s^{2}4p^{1}$<br>69.723(1) | +2 +4<br>32 **Ge** 锗<br>$4s^{2}4p^{2}$<br>72.630(8) | −3 +3 +5<br>33 **As** 砷<br>$4s^{2}4p^{3}$<br>74.921595(6) | −2 +2 +4 +6<br>34 **Se** 硒<br>$4s^{2}4p^{4}$<br>78.971(8) | −1 +1 +3 +5 +7<br>35 **Br** 溴<br>$4s^{2}4p^{5}$<br>79.904 | +2 +4<br>36 **Kr** 氪<br>$4s^{2}4p^{6}$<br>83.798(2) | N M L K |
| 5 | −1 +1<br>37 **Rb** 铷<br>$5s^{1}$<br>85.4678(3) | +2<br>38 **Sr** 锶<br>$5s^{2}$<br>87.62(1) | +3<br>39 **Y** 钇<br>$4d^{1}5s^{2}$<br>88.90584(2) | +1 +2 +3 +4<br>40 **Zr** 锆<br>$4d^{2}5s^{2}$<br>91.224(2) | 0 ±1 +2 +3 +4 +5<br>41 **Nb** 铌<br>$4d^{4}5s^{1}$<br>92.90637(2) | 0 ±1 ±2 +3 +4 +5 +6<br>42 **Mo** 钼<br>$4d^{5}5s^{1}$<br>95.95(1) | 0 ±1 +2 +3 +4 +5 +6 +7<br>43 **Tc** 锝▲<br>$4d^{5}5s^{2}$<br>97.90721(3)✦ | 0 +1 ±2 +3 +4 +5 +6 +7 +8<br>44 **Ru** 钌<br>$4d^{7}5s^{1}$<br>101.07(2) | 0 ±1 +2 +3 +4 +5 +6<br>45 **Rh** 铑<br>$4d^{8}5s^{1}$<br>102.90550(2) | 0 +1 +2 +3 +4<br>46 **Pd** 钯<br>$4d^{10}$<br>106.42(1) | +1 +2 +3<br>47 **Ag** 银<br>$4d^{10}5s^{1}$<br>107.8682(2) | +1 +2<br>48 **Cd** 镉<br>$4d^{10}5s^{2}$<br>112.414(4) | +1 +3<br>49 **In** 铟<br>$5s^{2}5p^{1}$<br>114.818(1) | +2 +4<br>50 **Sn** 锡<br>$5s^{2}5p^{2}$<br>118.710(7) | −3 +3 +5<br>51 **Sb** 锑<br>$5s^{2}5p^{3}$<br>121.760(1) | −2 +2 +4 +6<br>52 **Te** 碲<br>$5s^{2}5p^{4}$<br>127.60(3) | −1 +1 +3 +5 +7<br>53 **I** 碘<br>$5s^{2}5p^{5}$<br>126.90447(3) | +2 +4 +6 +8<br>54 **Xe** 氙<br>$5s^{2}5p^{6}$<br>131.293(6) | O N M L K |
| 6 | −1 +1<br>55 **Cs** 铯<br>$6s^{1}$<br>132.90545196(6) | +2<br>56 **Ba** 钡<br>$6s^{2}$<br>137.327(7) | 57~71<br>**La~Lu**<br>镧系 | +1 +2 +3 +4<br>72 **Hf** 铪<br>$5d^{2}6s^{2}$<br>178.49(2) | 0 ±1 +2 +3 +4 +5<br>73 **Ta** 钽<br>$5d^{3}6s^{2}$<br>180.94788(2) | 0 ±1 ±2 +3 +4 +5 +6<br>74 **W** 钨<br>$5d^{4}6s^{2}$<br>183.84(1) | 0 ±1 +2 +3 +4 +5 +6 +7<br>75 **Re** 铼<br>$5d^{5}6s^{2}$<br>186.207(1) | 0 +1 +2 +3 +4 +5 +6 +7 +8<br>76 **Os** 锇<br>$5d^{6}6s^{2}$<br>190.23(3) | 0 ±1 +2 +3 +4 +5 +6<br>77 **Ir** 铱<br>$5d^{7}6s^{2}$<br>192.217(3) | 0 +2 +3 +4 +5 +6<br>78 **Pt** 铂<br>$5d^{9}6s^{1}$<br>195.084(9) | +1 +2 +3 +5<br>79 **Au** 金<br>$5d^{10}6s^{1}$<br>196.966569(5) | +1 +2 +3<br>80 **Hg** 汞<br>$5d^{10}6s^{2}$<br>200.592(3) | +1 +3<br>81 **Tl** 铊<br>$6s^{2}6p^{1}$<br>204.38 | +2 +4<br>82 **Pb** 铅<br>$6s^{2}6p^{2}$<br>207.2(1) | −3 +3 +5<br>83 **Bi** 铋<br>$6s^{2}6p^{3}$<br>208.98040(1) | −2 +2 +3 +4 +6<br>84 **Po** 钋<br>$6s^{2}6p^{4}$<br>208.98243(2)✦ | ±1 +5 +7<br>85 **At** 砹<br>$6s^{2}6p^{5}$<br>209.98715(5)✦ | +2<br>86 **Rn** 氡<br>$6s^{2}6p^{6}$<br>222.01758(2)✦ | P O N M L K |
| 7 | +1<br>87 **Fr** 钫▲<br>$7s^{1}$<br>223.01974(2)✦ | +2<br>88 **Ra** 镭<br>$7s^{2}$<br>226.02541(2)✦ | 89~103<br>**Ac~Lr**<br>锕系 | 104 **Rf** 𬬻▲<br>$6d^{2}7s^{2}$<br>267.122(4)✦ | 105 **Db** 𬭊▲<br>$6d^{3}7s^{2}$<br>270.131(4)✦ | 106 **Sg** 𬭳▲<br>$6d^{4}7s^{2}$<br>269.129(3)✦ | 107 **Bh** 𬭛▲<br>$6d^{5}7s^{2}$<br>270.133(2)✦ | 108 **Hs** 𬭶▲<br>$6d^{6}7s^{2}$<br>270.134(2)✦ | 109 **Mt** 鿏▲<br>$6d^{7}7s^{2}$<br>278.156(5)✦ | 110 **Ds** 𫟼▲<br>281.165(4)✦ | 111 **Rg** 𬬭▲<br>281.166(6)✦ | 112 **Cn** 鿔▲<br>285.177(4)✦ | 113 **Nh** 鿭▲<br>286.182(5)✦ | 114 **Fl** 𫓧▲<br>289.190(4)✦ | 115 **Mc** 镆▲<br>289.194(6)✦ | 116 **Lv** 𫟷▲<br>293.204(4)✦ | 117 **Ts** 鿬▲<br>293.208(6)✦ | 118 **Og** 鿫▲<br>294.214(5)✦ | Q P O N M L K |

| | | | | | | | | | | | | | | | |
|---|---|---|---|---|---|---|---|---|---|---|---|---|---|---|---|
| ★ 镧系 | +3<br>57 **La** ★ 镧<br>$5d^{1}6s^{2}$<br>138.90547(7) | +2 +3 +4<br>58 **Ce** 铈<br>$4f^{1}5d^{1}6s^{2}$<br>140.116(1) | +3 +4<br>59 **Pr** 镨<br>$4f^{3}6s^{2}$<br>140.90766(2) | +2 +3 +4<br>60 **Nd** 钕<br>$4f^{4}6s^{2}$<br>144.242(3) | +3<br>61 **Pm** 钷▲<br>$4f^{5}6s^{2}$<br>144.91276(2)✦ | +2 +3<br>62 **Sm** 钐<br>$4f^{6}6s^{2}$<br>150.36(2) | +2 +3<br>63 **Eu** 铕<br>$4f^{7}6s^{2}$<br>151.964(1) | +3<br>64 **Gd** 钆<br>$4f^{7}5d^{1}6s^{2}$<br>157.25(3) | +3 +4<br>65 **Tb** 铽<br>$4f^{9}6s^{2}$<br>158.92535(2) | +3 +4<br>66 **Dy** 镝<br>$4f^{10}6s^{2}$<br>162.500(1) | +3<br>67 **Ho** 钬<br>$4f^{11}6s^{2}$<br>164.93033(2) | +3<br>68 **Er** 铒<br>$4f^{12}6s^{2}$<br>167.259(3) | +2 +3<br>69 **Tm** 铥<br>$4f^{13}6s^{2}$<br>168.93422(2) | +2 +3<br>70 **Yb** 镱<br>$4f^{14}6s^{2}$<br>173.045(10) | +3<br>71 **Lu** 镥<br>$4f^{14}5d^{1}6s^{2}$<br>174.9668(1) |
| ★ 锕系 | +3<br>89 **Ac** ★ 锕<br>$6d^{1}7s^{2}$<br>227.02775(2)✦ | +3 +4<br>90 **Th** 钍<br>$6d^{2}7s^{2}$<br>232.0377(4) | +3 +4 +5<br>91 **Pa** 镤<br>$5f^{2}6d^{1}7s^{2}$<br>231.03588(2) | +2 +3 +4 +5 +6<br>92 **U** 铀<br>$5f^{3}6d^{1}7s^{2}$<br>238.02891(3) | +3 +4 +5 +6 +7<br>93 **Np** 镎<br>$5f^{4}6d^{1}7s^{2}$<br>237.04817(2)✦ | +3 +4 +5 +6 +7<br>94 **Pu** 钚<br>$5f^{6}7s^{2}$<br>244.06421(4)✦ | +2 +3 +4 +5 +6<br>95 **Am** 镅▲<br>$5f^{7}7s^{2}$<br>243.06138(2)✦ | +3 +4<br>96 **Cm** 锔▲<br>$5f^{7}6d^{1}7s^{2}$<br>247.07035(3)✦ | +3 +4<br>97 **Bk** 锫▲<br>$5f^{9}7s^{2}$<br>247.07031(4)✦ | +2 +3 +4<br>98 **Cf** 锎▲<br>$5f^{10}7s^{2}$<br>251.07959(3)✦ | +2 +3<br>99 **Es** 锿▲<br>$5f^{11}7s^{2}$<br>252.0830(3)✦ | +2 +3<br>100 **Fm** 镄▲<br>$5f^{12}7s^{2}$<br>257.09511(5)✦ | +2 +3<br>101 **Md** 钔▲<br>$5f^{13}7s^{2}$<br>258.09843(3)✦ | +2 +3<br>102 **No** 锘▲<br>$5f^{14}7s^{2}$<br>259.1010(7)✦ | +3<br>103 **Lr** 铹▲<br>$5f^{14}6d^{1}7s^{2}$<br>262.110(2)✦ |